International Conference on History, Theory and Methodology of Learning (ICHTML 2020)

SHS Web of Conferences Volume 75 (2020)

Kryvyi Rih, Ukraine
13 – 15 May 2020

Editors:

Vita Hamaniuk
Serhiy Semerikov
Yaroslav Shramko

ISBN: 978-1-7138-0861-9

Printed from e-media with permission by:

Curran Associates, Inc.
57 Morehouse Lane
Red Hook, NY 12571

Some format issues inherent in the e-media version may also appear in this print version.

For additional information, please contact EDP Sciences – Web of Conferences at the address below.

EDP Sciences – Web of Conferences
17, Avenue du Hoggar
Parc d'Activité de Courtabœuf
BP 112
F-91944 Les Ulis Cedex A
France

Phone: +33 (0) 1 69 18 75 75
Fax: +33 (0) 1 69 28 84 91

contact-edps@webofconferences.org

Additional copies of this publication are available from:

Curran Associates, Inc.
57 Morehouse Lane
Red Hook, NY 12571 USA
Phone: 845-758-0400
Fax: 845-758-2633
Email: curran@proceedings.com
Web: www.proceedings.com

TABLE OF CONTENTS

ICHTML 2020 – How learning technology wins coronavirus

Vita Hamaniuk, *Serhiy* Semerikov*, and *Yaroslav* Shramko

Kryvyi Rih State Pedagogical University, 54 Gagarina Ave., Kryvyi Rih, 50086, Ukraine

Abstract. This is an introductory text to a collection of papers from the ICHTML 2020: The International Conference on History, Theory and Methodology of Learning, which held in Kryvyi Rih State Pedagogical University, Kryvyi Rih, Ukraine, on the May 13-15, 2020. It consists of short introduction, conference review and some observations about the event and its future.

Introduction: from social distancing to distant learning

The rapid spread of the coronavirus that causes COVID-19 has sparked worldwide alarm and media hype [1]. In the Ukraine, the Ministry of Healthcare is advising people to be prepared for disruptions to daily life that will be necessary if the coronavirus spreads within communities. On March 11, 2020 the Cabinet of Ministers of Ukraine introduced a three-week nationwide quarantine in connection with a pandemic, and all public events in the country have been canceled: "In accordance with the article 29 of the Law of Ukraine On Protection of the population from the infectious diseases to prevent the spread in the territory of Ukraine of coronavirus Covid-19 and considering the decision of the State Commission on technogeneous and ecological security and emergency dated March 10, 2020, the Cabinet of Ministers ordered to initiate the quarantine from March 12 until April 3, 2020" [2].

In 1665, following an outbreak of the bubonic plague in England, Cambridge University closed its doors, forcing Isaak Newton to return home to Woolsthorpe Manor. It was a version of "social distancing" of 17th century, a public health tool making a comeback this year as school and universities, including Kryvyi Rih State Pedagogical University, send people home to try to slow the spread of the novel coronavirus.

Without his professors to guide him, Newton apparently thrived: the papers he wrote on mathematical problems he had begun at Cambridge became early calculus; the experiments with a prisms in his bedroom give us the theories on optics; and even apple tree inspired the gravity laws: "In the year he retired again from Cambridge on account of the plague to his mother in Lincolnshire & whilst he was musing in a garden it came into his thought that the same power of gravity (which made an apple fall from the tree to the ground) was not limited to a certain distance from the earth but must extend much farther than was usually thought – *Why not as high as the Moon* said he to himself" [3].

"Newton returned to Cambridge in 1667, theories in hand. Within six months, he was made a fellow; two years later, a professor. So if you're working or studying from home over the next few weeks, perhaps remember the example Newton set", wrote Gillian Brockell for The Washington Post's history blog [4].

These risk-control decisions [2] have temporary led millions of students in Ukraine and other countries into massive open distance learning using contemporary educational technologies (Fig. 1):

1. To help minimize the impact of the coronavirus outbreak on students, the Coursera community is launching Coursera Coronavirus Response Initiative, a global effort to assist universities and colleges to deliver courseware online [5]. Educational institutions may enroll students in 3800 courses and 400 specializations through July 31, 2020 free of charge.

2. University of Massachusetts Amherst plans to shift their instruction to an online modality and help students to continue making progress in their studies with disruption-resilient instruction [6]. Instructors are asked to follow next steps:

– use the Learning Management Systems (LMS) of Moodle and Blackboard for communicating with students about your course;
– use the LMS to share course content, such as lecture notes, PowerPoint slides, assignments, and other course materials that can be uploaded to LMS course site;
– use Zoom to stream or record class lectures: faculty members may choose to schedule a Zoom class meeting with their students during their normally scheduled course day/time (these Zoom class lectures can also be recorded to be viewed at a later time if desired);
– adapt the course content, pedagogy, and assessment, if necessary, to fit this new modality of instruction.

The base of learning technologies is learning sciences, an interdisciplinary field that works to critical theoretical understanding of cognition and learning as well as design and implementation of learning innovations, and the improvement of learning/instructional methodologies.

* Corresponding author: semerikov@gmail.com

https://doi.org/10.1051/shsconf/20207500001

Note @75

chris 3/9/20

Online Discussions and Office Hours

Hi all,

as you probably already anticipated, I will move my discussion sections and office hours online, starting tomorrow. I'm sure your biggest concern right now is how we can continue to have class. Don't worry – I am prepared:

- The default method we'll use is Zoom – you'll have to download the client at https://zoom.us. I will post the data you need to join the meeting on bcourses.
- In case Zoom doesn't work for some reason (e.g. because the service is overloaded), we will use Google Meet. If that happens, I'll announce it beforehand. If you don't use an @berkeley.edu email address, I will have to invite you individually, so keep that in mind.
- If Google Meet also doesn't work anymore, I'll use a site called https://explaineverything.com, which is hosted AWS. You'll be able to see a whiteboard and talk to me directly in your web browser, but there's no video.
- In case that also ceases to work, we'll have audio-only discussion sections on my mumble server. I will send the details when it comes to that.
- In the case of bigger infrastructure disruptions (like a large scale power outage), we're going to have sections using amateur radio: I will use the Mt Diablo repeater station at 147.060 MHz with a PL tone of 100 Hz. My call sign is KN6CDY.
- In case the Bay area becomes uninhabitable, I will move to the wilderness. I plan to still be reachable via APRS, but we'll probably have to move sections to shortwave. In that case, we'll use CW transmission, so practice your Morse code.
  Note that a nuclear attack in the upper atmosphere can create very strong electromagnetic fields, so if you want to prepare for that, either use vacuum tube based radios, or store them in a Faraday cage.

I hope this assures you that no matter what happens this semester, I will not let it prevent you from learning quantum mechanics.

Best wishes,
Christian

logistics
Good note | 0

Fig. 1. The end of the world is apparently no excuse to miss class.

ICHTML 2020 at a glance

The International Conference on History, Theory and Methodology of Learning (ICHTML) is a peer-reviewed international conference, which covers interdisciplinary research on education, learning and training, and applications of theories and philosophies used in the sciences of learning and adjacent sciences (Fig. 2).

Fig. 2. ICHTML 2020 logos.

The ICHTML occupies contributions in all aspects of psychology of learning, learning theories, learning technologies and tools, paradigms and models and related fields of interest with an emphasis on human and machine learning. The main problematic field of the conference is the current and future issues of modern pedagogical science: psychological and pedagogical, philosophical, socio-cultural aspects of education, learning and training, modern theories, technologies and teaching aids, the emergence of which is determined by globalization, integration processes, social transformations, humanitarian and scientific and technological development. There is urgent general need for principled changes in postclassic education elicited by current theories, models, tools, services, networks and communications.

ICHTML has two presentation levels, Main Conference and Doctoral Consortium. The purpose of the Doctoral Consortium is to orient the PhD students to the world standards of quality of scientific work and to give an opportunity to present their achievements standalone.

ICHTML topics of interest are grouped into 4 tracks:

Track 1: History of Education, Learning and Training

– Evolution of Education, Learning and Training

– Socio-Cultural Determinants of Modern Education Development

– Comparative pedagogy

– Personalia

– World Trends in Learning, Training and Education Development

Track 2: Theories of Learning, Education and Training

– Analytical Psychology and Learning
– Anthropology of Learning and Cognition
– Comparative Psychology
– Development and Learning
– Developmental Cognitive Neuroscience and Learning
– Discourse and the Production of Knowledge
– Neuropsychology of Learning
– Philosophy of Learning
– Psychology of Learning
– Problems of Preschool, Primary, Secondary, Higher, Professional and Postgraduate Education
– Lifelong Learning
– Theory and Practice of Reforming the Education System
– Standardization and Monitoring in Education
– Gender Pedagogy
– Inclusive Education
– Childhood: Problems, Realities and Perspectives
– Formation and Development of Personality
– Theory and practice of profile education

Track 3: Discourses of Learning, Education and Training

– Philosophy of Education in the System of Socio-Humanitarian Knowledge
– Education and Civil Society: Trends in Interrelation
– Socio-Cultural Aspects of Teaching Humanitarian Disciplines in Terms of the Globalization
– Sociological and Economic Factors of Education Development
– Peculiarities of the Child Development
– Psychological Support of Personality Development
– Psychological Aspects of Educational Content in Terms of Inclusion
– Educational Management

Track 4: Methodology of Learning, Education and Training

– Learning Technology
– Artificial Intelligence, Knowledge Engineering, and Intelligent Tutoring Systems
– Machine Learning, Robot Learning and Artificial Learning
– Theory of Curricula
– Didactic Systems and Technologies
– Development of the Core Educational Competences
– Assessment in Learning
– Interdisciplinary Dimensions of Learning, Education and Training
– Methodology of Informatization in Education
– Theory and Methodology of Art Education
– STEAM Education
– Methods of Teaching

This volume contains the papers presented at ICHTML 2020: The International Conference on History, Theory and Methodology of Learning held on the May 13-15, 2020 in Kryvyi Rih, Ukraine.

There were 133 submissions. Each submission was reviewed by at least 3, and on the average 3.9, program committee members. The committee decided to accept 44 papers.

ICHTML 2020 venue

Kryvyi Rih State Pedagogical University has 90-year experience in training generations of specialists for the state and public needs (Fig. 3). Nowadays Kryvyi Rih State Pedagogical University is a center of innovative educational and methodological research, scientific and cultural center of the Kryvyi Rih and Dnipropetrovsk region, entrusted with the mission of training highly competitive professional teachers for all parts of the education sector. University graduates hold the administrative, state, cultural, environmental protection positions. They work effectively in different spheres such as education, chemistry, sociology, psychology, technology, design, tourism, translation, journalism, publishing and sports [7].

Fig. 3. Kryvyi Rih State Pedagogical University main building

ICHTML 2020 program committee

Dr. **Andrii Abdula**, Kryvyi Rih State Pedagogical University, Ukraine

WWW: https://kdpu.edu.ua/personal/aiabdula.html

Dr. **George Abuselidze**, Professor of Economics and Business, Department of Finance, Banking and Insurance, Batumi Shota Rustaveli State University, Batumi, Georgia

George Abuselidze, from the Batumi Shota Rustaveli State University (in 2001) and Ivane Javakhishvili Tbilisi State University, Georgia, in 2005, and a Doctor of Economics Sciences degree (Dr. habil.) from the National Academy of Sciences of George, in 2005. Since 2002, he has been working in the Finance and Banking at the Ivane Javakhishvili Tbilisi State University, since 2006 – Batumi Shota Rustaveli State University where he is currently Head department of Finance, Banking and Insurance. His research interests include Economics, Econometrics, Finance and Social Sciences (miscellaneous). He has published a number of papers in international journals and volumes in book series, is a member of editorial or/and review boards of Oeconomia Copernicana, Journal of Financial Economic Policy, International Journal of Economics and Finance, journal of Science and studies of accounting and finance: problems and perspectives, Management Studies and etc. He also played instrumental role in different prestigious internal collaborative research project with USA, Canada, Lithuania, Poland, Ukraine, Turkey and etc.

WWW: https://orcid.org/0000-0002-5834-1233
E-mail: george.abuselidze@bsu.edu.ge

Dr. **Svitlana Amelina**, Doctor of Education, Professor, Head of the Department of Foreign Philology and Translation, National University of Life and Environmental Sciences of Ukraine, Kyiv, Ukraine

Svitlana Amelina, born in 1961, received a Candidate of Philological Sciences degree (Dr. phil.) in 1994 and a Doctor of Pedagogical Sciences degree (Dr. habil.) in 2009. She studied at the Humboldt-Universität zu Berlin, trained at the largest universities in Germany and France. Since 2012, she has been working at the National University of Life and Environmental Sciences of Ukraine. Her research interests include theory and methodology of professional education; pedagogy of higher school; methods of teaching foreign languages and translation; syntax, semantics and pragmatics of the Germanic languages. She participated in numerous scientific-theoretical and scientific-practical conferences. She was a member of numerous scientific-theoretical and scientific-practical conferences. She is the author of many scientific publications, including monographs, articles, and reports.

E-mail: svetlanaamelina@ukr.net

Dr. **Vira Andriievska**, Associate Professor, Department of Informatics, H. S. Skovoroda Kharkiv National Pedagogical University, Kharkiv, Ukraine

Vira Andriievska, born in 1980, received a Candidate of Pedagogical Sciences degree (2009) and a Doctor of Pedagogical Sciences degree (2019) from the H.S. Skovoroda Kharkiv National Pedagogical University. Since 2004, she has been working in the field of primary education at the H.S. Skovoroda Kharkiv National Pedagogical University. Her research interests include training future primary school teachers to use ICT in their professional activities. Present research interests include the project-based learning and STEM education. She has published a number of papers in international journals.

WWW: http://kafinfo.org.ua
E-mail: andvera80@gmail.com

Prof. **Ruslana Balabai**, Kryvyi Rih State Pedagogical University, Ukraine

WWW: https://kdpu.edu.ua/personal/rmbalabai.html

Dr. **Liudmyla Bilousova**, Full Professor of Computer Science Department at H. S. Skovoroda Kharkiv National Pedagogical University, Ukraine

Liudmyla Bilousova got a M.A. in Radio-physics from Kharkiv State University, USSR (1959), and PhD in physical-mathematical science from Kharkiv State University, USSR (1970). Liudmyla Bilousova headed Computer Science Department at G.S. Skovoroda Kharkiv National Pedagogical University for 27 years (1991-2018). She is a founder and a head of her scientific school on the problems of development of Informatics education and the implementation of innovative technologies in educational process. In the framework of the school a series of PhD and doctoral research has been completed. Dr. Liudmyla Bilousova is a scientific supervisor of 21 PhD theses. She is an author of about 400 scientific and methodical works including 1 collective monograph and 64 tutorials. Liudmyla Bilousova is a program committee member of International Conference ICTERI-2019 and International Workshop CTE.

WWW: http://hnpu.edu.ua/uk/bilousova-lyudmyla-ivanivna
E-mail: lib215@gmail.com

Dr. **Olga Bondarenko**, Candidate of Pedagogical Sciences, Associate Professor, Department of Economic and Social Geography and Methods of Teaching, Kryvyi Rih State Pedagogical University, Kryvyi Rih, Ukraine

Olga Bondarenko, born in 1979, in 2001 graduated with honors from the geographical faculty of Krivoy Rog State Pedagogical University, majoring in "Pedagogy and Methodology of Secondary Education. Geography and Biology", acquired Bachelor Degree. In 2002 she received a Master Degree with honors in the major "Pedagogy and Methodology of Secondary Education. Geography", qualified as a teacher of geography. In 2009, she successfully defended her PhD thesis at the Republican Higher Educational Institution "Crimean Humanities University" (Yalta) and received a PhD in Pedagogical Sciences. Since 2005 he has been working at Kryvyi Rih State Pedagogical University. Author of a number of scientific publications on vocational education, training of future teachers for pedagogical activity. Her research interests include teacher training, the use of ICTs and GIS technologies in the educational process.

WWW: https://kdpu.edu.ua/personal/ovbondarenko.html
E-mail: bondarenko.olga@kdpu.edu.ua

Prof. **Allen V. Del Carmen**, University of St. La Salle, Philippines

WWW: https://www.linkedin.com/in/allen-del-carmen-a3b36519/

Dr. **Yannis Dimitriadis** is Full Professor of Telematics Engineering and ex Dean of the Doctoral School, University of Valladolid, Spain. He is also the coordinator of the GSIC/EMIC research group, an inter-disciplinary group, integrating over 20 researchers and practitioners from the field of Information and Communications Technologies (ICT) and Pedagogy.

His recent research work has focused on learning analytics and smart learning environments, alignment of learning design and learning analytics, design patterns, conceptual and technological support to the orchestration of computer-supported collaborative learning processes, active pedagogies at scale, and across-spaces (Web, 3D worlds and augmented reality) learning. He has participated in more than 50 competitive research projects on technology-enhanced learning, co-authored more than 80 journal papers and 200 conference papers, and organized several workshops and symposia, such as the ECTEL 2019 workshop on Hybrid Learning Spaces, CSCL 2015 invited symposium on computer science challenges in CSCL, or the ICLS 2012 workshop on classroom orchestration. Dr. Dimitriadis is a senior member of IEEE, and member of ISLS, and spent his most recent sabbatical year (2017-2018) at Berkeley, University of Edinburgh, and EPFL.

WWW: https://www.gsic.uva.es/members/yannis
E-mail: yannis@tel.uva.es

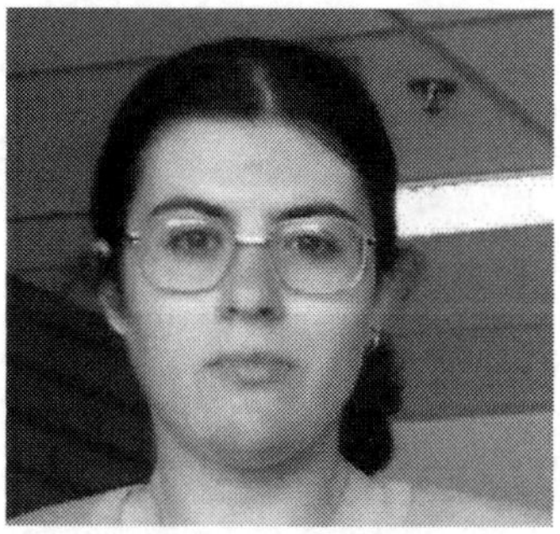

Prof. **Irina Georgescu**, Academy of Economic Studies, Romania

Irina Georgescu holds a PhD in Economics from Turku Centre for Computer Science, Åbo Akademi, Turku, Finland. Currently she is with the Department of Economic Informatics and Cybernetics at Faculty of Cybernetics, Statistics and Economic Informatics from the Bucharest Academy of Economic Studies. Her research interests lie in the area of fuzzy sets and systems, social choice and welfare and risk theory. She has published 2 books in Springer Verlag and about 30 journal papers on fuzzy choice functions and possibilistic risk theory.

WWW: https://www.researchgate.net/profile/Irina_Georgescu

Dr. **Olena Glazunova**, Professor of ICT in Education, Department of Information Technologies, National University of Life and Environmental Sciences of Ukraine

Olena Glazunova, received a Candidate of Pedagogical Sciences degree (Dr. phil.) in 2003, and a Doctor of Pedagogical Sciences degree (Dr. habil.) from the Institute of Informational Technologies and Tools in Education NAPS of Ukraine in 2015. Since 2003, she has been working in the field of design and development of cloud-based scientific-educational environment of the university, use of technologies of inquiry-based and project-based activities and implementation of teamwork approaches in IT-students learning, transfer and modification of educational methods to specialized teaching of research masters' programs and courses of e-learning management.

WWW: https://nubip.edu.ua/IT.NUBIP
E-mail: o-glazunova@nubip.edu.ua

https://doi.org/10.1051/shsconf/20207500001

Dr. **Liudmyla Gryzun**, Full Professor of Computer Science Department at H. S. Skovoroda Kharkiv National Pedagogical University, Ukraine.

Liudmyla Gryzun earned a M.A. in Applied Mathematics from the Kharkiv State University, USSR (1986); PhD and Second Doctoral Degree in Pedagogical science from G.S. Skovoroda Kharkiv National Pedagogical University (Ukraine). The sphere of her research is focused on the curriculum and educational content design in higher education, the process of curriculum disciplines structuring, based on scientific knowledge integration; AI application to pedagogical problems solution; IT tools for inquiry-based and holistic learning etc. She is an author of more than 110 scientific and methodical works including 1 monograph and 5 tutorials. Liudmyla Gryzun has delivered a number of Keynote presentations at the International conferences: 2018 ICTEL (Rome, Italy), 2018 ICRTEL (Barcelona, Spain), 2019 ICSTR (Rome, Italy) and others. She is also a reviewer of the foreign journals: Athens Journal of Education, IJIRES (International Journal of Innovation and Research in Educational Sciences), and PEOPLE: International Journal of Social Sciences.

Profile Weblinks:
http://hnpu.edu.ua/uk/gryzun-lyudmyla-eduardivna
https://www.linkedin.com/in/liudmyla-gryzun-68769280/
https://www.researchgate.net/profile/Liudmyla_Gryzun
E-mail: Lgr2007@ukr.net

Dr. **Yasemin Gulbahar**, Professor of Computer Education, Department of Computer Education and Instructional Technology, Ankara University, Ankara, Turkey

Dr. Gulbahar has got her BS degree from Department of Mathematics of Science Faculty at Middle East Technical University (METU) in 1992. Same year she started working as a programmer at METU Computer Center. Then, in 1998, she became a research assistant to the Department of Computer Education and Instructional Technologies (CEIT) in the Faculty of Education, METU while studying her MS degree at the same department. She earned his MS degree in the field of Science Education at METU Graduate School of Science in 1999 and she received her PhD in Department of Computer Education and Instructional Technologies (CEIT) from Graduate School of Sciences in 2002. After, she worked for Başkent University Faculty of Education Department of Computer Education and Instructional Technologies (CEIT) for about 9 years. Between years 2011-2017 she worked in Ankara University as the head of Department of Informatics and meanwhile served as the Chairman of Distance Education Center. Since 2018 she is a faculty member of Faculty of Educational Sciences in the Department of Computer Education and Instructional Technologies (CEIT). Dr. Gulbahar has got her Associate Professor degree in 2009 and full Professor Degree in 2014.

Yasemin Gulbahar has lectured on many topics such as programming languages, problem solving and algorithms, instructional technologies, instructional design, material design and development, distance learning, web design, measurement and evaluation, research methods, teaching methods, software development, technology integration and planning both in undergraduate and graduate level. She has also many national and international publications as books, book chapters, journal articles and proceeding papers.

WWW: http://cv.ankara.edu.tr/gulbahar@ankara.edu.tr
E-mail: gulbahar@ankara.edu.tr

Prof. **Vita Hamaniuk**, Doctor of science (pedagogical sciences), professor of the Department of German, Literature and Didactics, Kryvyi Rih State Pedagogical University, Kryvyi Rih, Ukraine

Vita Hamaniuk, born in 1965, Education – Dnipropetrovsk State University (1988), received a Candidate of Pedagogical Sciences degree (PhD) from the Charkiv National Pedagogical Skovoroda University, Kryvyi Rih, Ukraine, in 1995, a Doctor of Sciences degree from East-Ukrainian National Wolodymyr Dahl University, Lugansk, Ukraine, in 2013. Since 2002 has been working at the Department of German, Literature and didactics at the Kryvyi Rih State Pedagogical University where she is currently professor. Her research interests include foreign languages teaching and learning, didactics of multilingualism, e-learning, blended learning, comparative researches in Education. She has published a number of papers in Ukrainian and international journals, actively participates in international conferences and projects.

WWW: https://kdpu.edu.ua/personal/vagamanuk.html
E-mail: vitana65@gmail.com

Dr. **Anna Iatsyshyn**, Leading Researcher, Department of open education and scientific information systems, State Institute of Information Technologies and Learning Tools of the National Academy of Educational Sciences of Ukraine, Kyiv, Ukraine.

Anna Iatsyshyn, born in 1984, received a Candidate of pedagogical sciences degree from the Institute of Teacher Education and Adult Education of the National Academy of Educational Sciences of Ukraine, in 2009. A. Iatsyshyn is actively engaged in scientific activities in investigates the use of digital systems for: postgraduate training, research, scientometrics, image formation of scientists and organization. Features of ICT implementation in inclusive education and teacher training. She has published a number of papers in journals and monographs, is a member of associate editors of Information Technologies and Learning Tools is a bimonthly peer-reviewed e-journal in educational sphere, published full-text articles online with immediate open-access.

WWW: http://iitlt.gov.ua/structure/departments/science/detail.php?ID=53
E-mail: anna13.00.10@gmail.com

Dr. **Hale Ilgaz**, Ankara University, Turkey

WWW: https://www.researchgate.net/profile/Hale_Ilgaz

Prof. **Michael Kerres**, University Duisburg-Essen, Germany

Since 2001, Michael Kerres is full professor of educational science at University Duisburg-Essen, Germany. He holds the chair of educational media and knowledge management and is director of the university's Learning Lab. From 1998-2001, he has been a professor of educational psychology at Bochum University, and from 1990 to 1998 professor of educational technology at Furtwangen University of Applied Sciences. In 1983, he has earned a doctoral degree in psychology at Bochum University and in 1988, he finished his habilitation (post-doctoral thesis) at University of Education in Freiburg. His present research interests include digital transformation of education, instructional designs for digital learning and learning infrastructures.

WWW: https://learninglab.uni-due.de/

 https://doi.org/10.1051/shsconf/20207500001

Prof. **Arnold Kiv**, Ben-Gurion University of the Negev, Israel

Arnold Kiv received the D. Sc. (Dr. Hab.) degree in solid state physics from Tartu Institute of Physics, in 1978. From 1964 to 1982, he was a Senior Researcher and a Head of the Laboratory of Radiation Effects, Institute of Nuclear Physics, Academy of Sciences, Uzbekistan. From 1983 to 1998, he was a Head of the Department of Theoretical Physics, South-Ukrainian National Pedagogical University. In 1997 he was an Invited Professor, Western Ontario University, Canada. From 1999 to the present, he is a Professor-Researcher in the Department of Materials Engineering, Ben-Gurion University of the Negev, Israel. In 1996 and 2011 he was co-Director of NATO Advanced research Workshops and an Editor of two NATO Series books. He has about 300 publications, three monographs and three Invention Certificates in the field of radiation effects in solid state electronics. His research interests include mechanisms of formation of radiation defects in solids, radiation methods in microelectronics, computer simulation in natural and humanitarian sciences.

WWW: https://www.researchgate.net/profile/Arik_Kiv

Dr. **Oleksandr Kolgatin**, Professor of Information System Chair, Department of Economic Informatics, Simon Kuznets Kharkiv National University of Economics, Kharkiv, Ukraine

Oleksandr Kolgatin received a Candidate of Technical Sciences degree (PhD) from the Low Temperature Physics and Engineering of the National Academy of Sciences of Ukraine in 1995 and a Doctor of Pedagogical Sciences degree from the Institute of Information Technologies and Learning Tools of the National Academy of Educational Sciences of Ukraine in 2011. His research interests include computational modelling, information and communication technologies in education, and pedagogical diagnostics. He has published a number of papers in Ukrainian and international journals, is a member of editorial boards of Journal of Information Technologies in Education and an associate editor of Information Technologies and Learning Tools.

WWW: https://orcid.org/0000-0001-8423-2359

Prof. **Zhanna Koloiz**, Kryvyi Rih State Pedagogical University, Ukraine

WWW: https://kdpu.edu.ua/personal/zhvkoloiz.html

Prof. **Oleksandr Konoval**, Kryvyi Rih State Pedagogical University, Ukraine

WWW: https://kdpu.edu.ua/personal/oakonoval.html

Dr. **Svitlana Kovpik**, Doctor of Philological Sciences, Professor of the Department of the Ukrainian and World Literatures, Kryvyi Rih State Pedagogical University, Kryvyi Rih, Ukraine

Svitlana Kovpik, born in 1977, received the Candidate of Philological Sciences degree (Dr.phil.) from Kherson State University in 2005, and Doctor of Philological Sciences degree (Dr. habil.) from Taras Shevchenko National University of Kyiv, in 2011. Since 2001, she has been working in the field of Philology at Kryvyi Rih State Pedadogical University, where she is currently a Professor. Her research interests include poetics of fiction. She has published a number of articles in international journals and she is the editor-in-chief of the scientific journal "Literatures of the World: Poetics, Mentality and Spirituality".

E-mail: kovpiks@ukr.net

Dr. **Nadiia Kozachenko**, Kryvyi Rih State Pedagogical University, Ukraine

Professor Nadiia Kozachenko is chair of Department of Philosophy at Kryvyi Rih State Pedagogical University. She got PhD in logic from the Institute of Philosophy of the NAS of Ukraine in 2010. The main directions of Dr. Kozachenko' research is logic and philosophy of information society.

WWW: https://kdpu.edu.ua/personal/n_p_kozachenko.html

Dr. **Tetiana Kramarenko**, Associate Professor of the Department of Mathematics and Methods of its Teaching, Kryvyi Rih State Pedagogical University, Kryvyi Rih, Ukraine

Tetiana Kramarenko, born in 1962, received a Candidate of Pedagogical Sciences in speciality 13.00.02 "Theory and methods of studies (Mathematics)" degree (Dr. phil.) from the National Pedagogical Dragomanov University, Kyiv, Ukraine, in 2008. Associate Professor (2011). Scientific interests: Education, Math education, STEM education, ICT in education, Methodology for teaching Mathematics, Probability theory, Mathematical statistics.

WWW: https://kdpu.edu.ua/personal/tkramarenko.html
E-mail: kramarenko.tetyana@kdpu.edu.ua

Dr. **Volodymyr Kukharenko**, Professor of Technical Cryophisics Department, National Technical University "Kharkiv Polytechnic Institute", Kharkiv, Ukraine

Volodymyr Kukharenko, born in 1947, received a Candidate of Technical Sciences degree from Physical Technic Institute of Low Temperature National Academy of Sciences of Ukraine. Since 1976, he has been working in the field of low temperature at the National Technical University "Kharkiv Polytechnic Institute", where he is professor of Technical Cryophysics Department and academician of International Academy of Refrigeration UD. His research interests include distance learning. He has published a number of papers in international journals and six books from creating distance courses, about tutor, blended learning.

WWW: https://dl.khpi.edu.ua
E-mail: kukharenkovn@gmail.com

Dr. **Volkan Kukul**, Amasya University, Turkey

WWW: https://egitim.amasya.edu.tr/personel/akademik-personel?id=1659

https://doi.org/10.1051/shsconf/20207500001

Dr. **Olena Kuzminska**, Associate Professor of Information Technology and Distance Learning Department, National University of Life and Environmental Sciences of Ukraine, Kyiv, Ukraine

Olena Kuzminska, born in 1970, received a Candidate of Pedagogic Sciences degree (PhD) from the National Pedagogical Dragomanov University (Kyiv) in 2008. Since 2008, she has been working at the National University of Life and Environmental Sciences of Ukraine, where she is currently head of Information Technology and Distance Learning Department. She has experience in research and teaching in: design and development of innovative blended and distant e-learning technologies, instructional design; development of programs and organization of trainings for teachers and students on efficient use of information & communicational technologies (ICT) in education and research; use of technologies of inquiry-based and project-based activities and implementation of teamwork approaches as a tool for development of digital competences; transfer and modification of education methods to specialized teaching of research masters' programs and courses of e-learning management and scholarly communication. She research interests also include design and integration of scholarly communication tools to digital educational environment of university; integration of Microsoft and Google cloud services with e-learning process and postgraduate training. She has published a number of papers in international journals and volumes in book series, is a member of program committee of ICTERI (International Conference on ICT in Research, Education and Industrial Applications).

WWW: https://nubip.edu.ua/node/3900, https://cutt.ly/9thm8Un
E-mail: o.kuzminska@nubip.edu.ua

Dr. **Olena Lavrentieva**, Full Professor, Head of General Sciences and Vocational Training Department, Kryvyi Rih State Pedagogical University, Kryvyi Rih, Ukraine

Olena Lavrentieva, born in 1968, received a Candidate of Pedagogical Sciences degree (Dr. phil.) from the Lesya Ukrainka East European National University, Ukraine, in 2005 by specialty 13.00.09 "Theory of Education", and a Doctor of Pedagogical Sciences degree (Dr. habil.) from the Institute of Teacher Education and Adult Education of the National Academy of Sciences of Ukraine by specialty 13.00.04 "Theory and Methods of Professional Education", in 2015. Since 2003, she has been working in the Kryvyi Rih State Pedadogical University, where she is currently head of department. Her research interests include didactic and methodology issues of vocational training process. She has published a number of papers in international journals and volumes in book series, is a member of editorial boards of Physical and Mathematical education.

E-mail: helav68@gmail.com

Prof. **Olena Lokshyna**, Dr. Sc., Head of the Department of Comparative Education, Institute of Pedagogy of the National Academy of Educational Sciences of Ukraine, Kyiv, Ukraine

Olena Lokshyna, born in 1959, received a Candidate of Educational Sciences degree (Dr. phil. in Education) in 1992, and a Doctor of Sciences degree (Dr. habil. in Education) from the Institute of Pedagogy of the National Academy of Educational Sciences of Ukraine, in 2011. She has been working in the area of comparative and international education at the Institute of Pedagogy of the National Academy of Educational Sciences of Ukraine since 1990. She received her Professor rank in 2015 and was elected as a Corresponding Member of the National Academy of Educational Sciences of Ukraine in 2019.
Her research interests focus on developmental trends and innovations in education in Ukraine and abroad through the lens of comparison, with special interest in reforms, curricula, competences, students' assessment, education quality

monitoring, VET, methodology of comparative education research, education policy analyses. Olena Lokshyna is an author/co-author of about 300 works – monographs and analytical studies, textbooks and course outlines, journal papers and conference abstracts. She teaches courses on comparative and international education for PhD students at the Institute of Pedagogy and at Borys Grinchenko Kyiv University. She is a member of the editorial boards of the Ukrainian Pedagogical Journal, Education Modern Discourses Journal and Studies in Comparative Education Journal.

WWW: http://undip.org.ua/
E-mail: luve2001@hotmail.com

Dr. **Iryna Lovianova**, Doctor of Pedagogical Sciences, Professor of the Department of Mathematics and Methods of its Teaching, Kryvyi Rih State Pedagogical University, Kryvyi Rih, Ukraine

Irina Lovianova received a Candidate of Pedagogical Sciences degree (Dr. ph.) from the Institute of Pedagogical of the National Academy of Sciences of Ukraine, in 2006, and a Doctor of Philosophical Sciences degree (Dr. habil.) from the The Bohdan Khmelnytsky National University of Cherkasy in 2015. Since 2001, she has been working in the field of didactics of mathematics at the Kryvyi Rih State Pedadogical University. The range of interests is the issues of Education, math education, methodology for teaching mathematics. She has published a number of papers in international journals and collective monographs.

E-mail: lovyanova.iryna@kdpu.edu.ua

Dr. **Svitlana Lytvynova**, Head of the Open Learning Environment Technology Department, Institute of Information Technologies and Learning Tools of the National Academy of Education Sciences of Ukraine, Kiev, Ukraine

Svitlana Lytvynova, born in 1964, in 2011 received a degree of Candidate of Pedagogical Sciences (PhD) at the Institute of Information Technologies and Learning Tools of the National Academy of Education Sciences of Ukraine. In 2017, received a degree of Doctorate of Pedagogical Sciences at the Institute of Information Technologies and Learning Tools of the National Academy of Education Sciences of Ukraine. Since 2004, she has been working on the development, implementation and use of information and communication technologies in educational practice. Her research interests include improving the quality of the educational environment of educational institutions and the continuous development of educators. She is a MIE Expert Microsoft Education. She has published a number of articles in international scientific journals and book series volumes, is a member of the editorial boards of the journal "Information Technologies and Teaching Tools" (Web of Science) and the professional publication "Computer at School and Family".

WWW: https://www.facebook.com/slytvynova
E-mail: s.h.lytvynova@gmail.com

Dr. **Iryna Makarenko**, Kryvyi Rih State Pedagogical University, Ukraine

WWW: https://kdpu.edu.ua/personal/i_e_makarenko.html

Dr. **Andriy Matviychuk**, Professor of Economic and Mathematical Modeling, Kyiv National Economic University named after Vadym Hetman, Kyiv, Ukraine

Andriy Matviychuk, born in 1978, became a Candidate of Science in Economics (PhD) by specialty Economic and Mathematical Modelling at the Technological University of Podillia (Khmelnitskiy) in 2003, and a Doctor of Economics (DSc) with a degree in Mathematical Methods, Models and Information Technologies in Economics at the Kyiv National Economic University named after Vadym Hetman in 2008.
Since 1990 he has been working in the field of AI & ML at the Kyiv National Economic University named after Vadym Hetman, where he is currently professor of Economic and Mathematical Modeling Department, director of Institute of Modeling and Informational Technologies in Economics and CEO at KNEU Science Park.
His research interests cover the field of mathematical modeling of complex systems, primarily using tools of neural networks and fuzzy logic. He is a Chief Editor of scientific and analytical journal Neuro-Fuzzy Modeling Techniques in Economics and a member of editorial boards of a number of scientific journals.

WWW: https://kneu.edu.ua/ua/depts9/k_ekon_matematychn_modeljuvannja/vykladachi_kmm/Matvijchuk.A.V/
E-mail: editor@nfmte.com

Dr. **Pavlo Nechypurenko**, Assistant Professor of Department of Chemistry and Methods of its Teaching, Kryvyi Rih State Pedagogical University, Kryvyi Rih, Ukraine

Pavlo Nechypurenko, born in 1981, received a Magister of Teaching of Chemistry from Kryvyi Rih State Pedagogical University, Ukraine, in 2004, and a Candidate of Pedagogical Sciences degree (Dr. phil.) from the Luhansk Taras Shevchenko National University, Ukraine, in 2017. Since 2004, he has been working in the field of analytical chemistry and method of solving chemical problems at the Kryvyi Rih State Pedagogical University. His research interests include using of ICT on Chemistry education, Analytical Chemistry, Technique of chemical experiment. He has published a number of papers in Ukrainian and international journals and developed a series of virtual laboratory work to teaching chemistry.

E-mail: acinonyxleo@gmail.com, acinonyxleo@kdpu.edu.ua

Dr. **Yulia Nosenko**, Leading Researcher, Department of Cloud-Oriented Systems of Education Informatization, Institute of Information Technologies and Learning Tools of NAES of Ukraine, Kyiv, Ukraine

Yulia Nosenko, born in 1984, received a Candidate of Pedagogical Sciences degree (Dr. phil.) in 2011. In 2010-2015 worked at Taras Shevchenko National University of Kiev (part time). Since 2010 has been working at the Institute of Information Technologies and Learning Tools of National Academy of Educational Sciences of Ukraine, where she is currently leading researcher. Her research interests relates to implementation and use of cloud services in education, formation and development of educators' digital competence, use of ICT as a tool for supporting inclusive learning.
She has published over 70 scientific papers, including articles in international journals, is a member of editorial board of peer-reviewed e-journal "Information Technologies and Learning Tools".

WWW: http://iitlt.gov.ua/structure/departments/cloud/detail.php?ID=48
E-mail: nosenko@iitlt.gov.ua

Dr. **Vasyl Oleksiuk**, PhD (pedagogical sciences), associate professor of the Department of Computer Science and Teaching Techniques, Ternopil Volodymyr Hnatiuk National Pedagogical University, Ternopil, Ukraine

Vasyl Oleksiuk, born in 1980, received a Candidate of Pedagogical Sciences degree (Dr. phil.) from the National Pedagogical University, Kyiv, Ukraine, in 2007. Since 2003, he has been working Department of Computer Science and Teaching Techniques at Ternopil Volodymyr Hnatiuk National Pedagogical University, where he is currently associate professor. His research interests include computer networks, cloud computing, e-learning, electronic libraries. He has published a number of papers in Ukrainian and international journals, is a member of editorial boards of the journals Information Technologies and Learning Tools (Institute of Information Technologies and Learning Tools of NAES of Ukraine) and The Scientific Issues of Ternopil Volodymyr Hnatiuk National Pedagogical University (Series: pedagogy).

WWW: http://kafinf.tnpu.edu.ua/en/teachers/oleksuk_v-p/#
E-mail: oleksyuk@fizmat.tnpu.edu.ua

Olena Olifer, assistant lecturer, Department of Phylosophy, Kryvyi Rih State Pedadogical University, Kryvyi Rih, Ukraine

Olena Olifer, born in 1993, received a master's degree from Kryvyi Rih National University in 2015. At the moment she is working on her PhD thesis "Language bases of socio-cultural personal identity". Her research interests include analytic philosophy, British philosophy, and philosophy of language.

E-mail: olifer.olena@kdpu.edu.ua

Dr. **Kateryna Osadcha**, Full professor, Department of Computer Science and Cybernetics, Bogdan Khmelnitsky Melitopol State Pedagogical University, Melitopol, Ukraine

Kateryna Osadcha, born in 1977, received a Candidate of Pedagogical Sciences (PhD in Education) from the Vinnytsia State Pedagogical University named after Mykhailo Kotsiubynsky, Ukraine, in 2010. Since 2011, she has been working as a professor of the Department of Computer Science and Cybernetics at the Bogdan Khmelnitsky Melitopol State Pedagogical University.
Her research interests include: computer science, network technology, programming, ICT, e-learning, engineering education, educational technology, tutoring. She is author of about a hundred scientific works, including textbooks, monographs, author's certificates. She is a member of editorial boards of "Ukrainian Journal of Educational Studies and Information Technology" (Ukraine), "International Conference on Higher Education Advances" (Spain), "Transactions of Kremenchuk Mykhailo Ostrohradskyi National University" (Ukraine), "Professional Education: Methodology, Theory and Technologies" (Ukraine).

WWW: https://www.facebook.com/kosadca
E-mail: okp@mdpu.org.ua

Dr. **Viacheslav Osadchyi**, Professor of Department of Computer Science and Cybernetics, Bogdan Khmelnitsky Melitopol state pedagogical university, Melitopol, Ukraine

Viacheslav Osadchyi, born in 1975, received a Candidate of Pedagogical Sciences (PhD in Education) from the Vinnytsia State Pedagogical University named after Mykhailo Kotsiubynsky, Ukraine, in 2006, and a Doctor of Pedagogical Sciences from the Vinnytsia State Pedagogical University named after Mykhailo Kotsiubynsky, Ukraine, in 2013.
Since 1999, he has been working in the field of information technology and vocational education at the Bogdan Khmelnitsky Melitopol State Pedagogical University. Now – Head of the Department of Computer Science.
His research interests include: computer science, information networks, ICT, programming, software development, information systems, data science. He has published a number of papers in international journals.
He is a member of editorial boards of "Ukrainian Journal of Educational Studies and Information Technology" (Ukraine), "International Conference on Higher Education Advances" (Spain), "Transactions of Kremenchuk Mykhailo Ostrohradskyi National University" (Ukraine), "Information Technologies and Learning Tools" (Ukraine), "Scientific papers of Berdyansk State Pedagogical University Series: Pedagogical sciences" (Ukraine).

WWW: http://osadchyi.mdpu.org.ua/
E-mail: poliform55@gmail.com

Dr. **Nataliia Ovcharenko**, Professor of Music Art and Pedagogy, Department of Music Education, Vocal and Choir Conducting, Kryvyi Rih State Pedadogical University, Kryvyi Rih, Ukraine

Nataliia Ovcharenko, born in 1965, received a Candidate of Pedagogical Sciences degree (Dr. Phil. (Music Art)) from H. S. Skovoroda Kharkiv National Pedagogical University, Ukraine, in 1994, and a Doctor of Pedagogical Sciences degree (Dr. habil.) from Borys Grinchenko Kyiv University, Ukraine, in 2016. Since 1990, she has been working in the field of music art and pedagogics at Kryvyi Rih State Pedagogical University. Her research interests include vocal performance, pedagogics and inclusive music education.
She has published the papers in international journals and monographs in Ukraine, is a member of National All-Ukrainian Musical Association.

WWW: https://kdpu.edu.ua/personal/naovcharenko.html
E-mail: shvager77@gmail.com

Dr. **Liubov Panchenko**, Professor at the Department of Sociology, National Technical University of Ukraine "Igor Sikorsky Kyiv Polytechnic Institute", Kyiv, Ukraine

Liubov Panchenko was awarded a Candidate of Pedagogical Sciences degree (Dr. phil.) from H. S. Skovoroda Kharkiv National Pedagogical University, Kharkiv, Ukraine, in 1995, and a Doctor of Pedagogical Sciences degree (Dr. habil.) from the Luhansk Taras Shevchenko National University, in 2012. Since 1993, she has been working in the field of information and communication technology in education. Since 2016 she has been a Professor at the Department of Sociology, National Technical University of Ukraine "Igor Sikorsky Kyiv Polytechnic Institute". Her research interests include information and communication technology in education, university's educational environment, MOOCs, data analysis and multivariate methods in scientific research, digital storytelling, adult education. She has published a number of papers and text books ("Computer data analysis", "Data analysis practicum") and is an editorial board member of the Ukrainian journals "Information Technologies and Learning Tools" (associated editor), "e-Environment of Modern University", and "Humanization of the educational process".

WWW: http://www.sociology.kpi.ua/en/faculty-2
E-mail: lubov.felixovna@gmail.com

Dr. **Oksana Pershukova**, Professor of Aviation English Department, National Aviation University, Kyiv, Ukraine

Oksana Pershukova received a Candidate of Pedagogical Sciences degree (Dr. phil.) from the Institute of Pedagogy of the NAPS of Ukraine (Kyiv) in 2002, and a Doctor of Pedagogical Sciences degree (Dr. habil.) from the Institute of Pedagogical Education and Education of Adults of the NAPS of Ukraine (Kyiv), in 2016. Since 1995, she has been working in the field of comparative education in the Institute of Pedagogy, Kyiv. Since 2017 she works at the National Aviation University, Kyiv, at the position of a Professor of Aviation English Department. Her research interests are related to improving the quality of bilingual and multilingual education, developing and preserving multilingualism in the context of the educational space of European countries and the US. She has been analyzing the opportunities and finding ways to apply positive experiences in these fields in Ukraine. The range of her scientific interests also includes features of mastering a foreign language at the university level (especially ESP), as well as peculiarities of forming students' autonomy and finding the ways to increase students' motivation for foreign languages learning.

E-mail: pershoks@gmail.com

Dr. **Larysa Petrenko**, Senior Researcher, Professor of the Department of Vocational and Higher Education, University of Educational Management, Kyiv, Ukraine.

Larysa Petrenko, born in 1951, received a Candidate of Pedagogical Sciences degree (Ph.D.) in 2006 from the Institute of Pedagogy of the National Academy of Educational Sciences of Ukraine, and a Doctor of Pedagogical Sciences from the Institute of Vocational Education of NAESU in 2014. Since 2019, she has been a professor in the field of vocational and higher education at the University of Educational Management. Her research interests include the informational culture, information and analytical competence of pedagogical, scientific-pedagogical staff and heads of educational institutions of various types. She has published a number of articles in domestic and foreign collections of scientific papers and monographs were published. She is also a member of the editorial board of researcher's digest "Scientific Herald of the Institute of Vocational Education and Training of NAESU. Professional Pedagogy".

WWW: http://umo.edu.ua/institutes/cippo/struktura/kafedra-upop/sklad
E-mail: laravipmail@gmail.com

Dr. **Lukáš Pichl**, Senior Associate Professor of Information Science, Department of Natural Sciences, International Christian University, Mitaka, Tokyo, Japan

Lukáš Pichl, born in 1974, received his PhD from the Graduate University of Advanced Studies in 2000. Since 2001 he taught computer science in the University of Aizu, before joining the International Christian University in 2005. His current field of specialization is time series analysis and machine learning algorithms for the analysis of economic data including cryptocurrencies. Dr. Pichl has published a number of papers in international journals; he is a member of ACM and IEEE Computer Society, and currently serves as an associate editor of the Journal of Economic Interaction and Coordination (Springer).

E-mail: lukas@icu.ac.jp

Dr. **Olga P. Pinchuk**, Deputy Director for Scientific Experimental Work, Leading Researcher, PhD (in Pedagogics), Senior Researcher in the field of information and communication technologies in education. Education: M.P. Drahomanov Kyiv State Pedagogical Institute in specialties of Mathematics, Computer Science and Computer Engineering teacher.

Currently the experience in teaching is 25 years. Since 2005 I work in the Institute of Information Technologies and Learning Tools of the NAES of Ukraine. I worked on the implementation of the tasks of the scientific research works "Scientific and methodological foundations use of computer oriented tools in teaching natural and mathematical subjects in profile School", "Scientific and methodological principles of organization of distance learning environment in secondary schools" , "Methodology of design network resource centers of distance education of secondary schools", "Formation of information and educational environment for learning high school students through technology electronic social networks" (Head of Scientific Research). I have more than 80 published scientific works, the author of collective monographs, manuals. I also obtain the post of co-editors-in-chief of "Information Technologies and Learning Tools", a bimonthly peer-reviewed e-journal in educational sphere, publishing full-text articles online with immediate open-access.

WWW: http://iitlt.gov.ua/
E-mail: opinchuk@iitlt.gov.ua

Dr. **Nataliia O. Ponomarova**, Doctor of Pedagogical Sciences, Professor of Department of Informatics, H. S. Skovoroda Kharkiv National Pedagogical University, Kharkiv, Ukraine

Nataliia O. Ponomarova, born in 1972, received a Candidate of Pedagogical Sciences degree (Dr. phil.) from the H. S. Skovoroda Kharkiv National Pedagogical University, in 1998, and a Doctor of Pedagogical Sciences degree (Dr. habil.) from the H. S. Skovoroda Kharkiv National Pedagogical University, in 2018. Since 2005, she has been working in the field of preparation of teachers of informatics at H. S. Skovoroda Kharkiv National Pedagogical University, where she is currently Dean of Department of Physics and Mathematics. Her research interests include innovative pedagogical technologies; use of information and communication technologies in education; theoretical and methodological foundations of professional training of future teachers of informatics; career guidance of students. She has published a number of papers in international journals and volumes in book series.

WWW: http://hnpu.edu.ua/uk/ponomarova-nataliya-oleksandrivna
E-mail: ponomna@gmail.com

Dr. **Volodymyr Proshkin**, Professor of Department of Computer Science and Math, Borys Grinchenko Kyiv University, Kyiv, Ukraine

Expert of the National Agency for Higher Education Quality Assurance. Deputy Chairman of the Specialized Academic Council (Borys Grinchenko Kyiv University), member of the specialized Academic Council (Donbass State Pedagogical University). Executor of the international project "High school teacher competence in change" with the assistance of the Visegrad Fund and the Ministry of Foreign Affairs of the Netherlands.
Author of 10 articles in journals included in the databases Scopus, WoS. Member of editorial boards of journals: "Open educational e-environment of modern University", "Cybersecurity: Education, Science, Technique" (Borys Grinchenko Kyiv University), "Transactions of Kremenchuk Mykhailo Ostrohradskyi National University".

WWW: http://fitu.kubg.edu.ua/pro-fakultet/kafedry
E-mail: v.proshkin@kubg.edu.ua

Dr. **Oleg Pursky**, Professor of Computer Science and Information Systems, Head of Department of Computer Science and Information Systems, Kyiv National University of Trade and Economics, Kyiv, Ukraine

Oleg Pursky, born in 1967, received a Candidate of Sciences in Physics and Mathematics degree (Dr. phil.) from the Institute for Low Temperature Physics and Engineering of the National Academy of Sciences of Ukraine, in 2001, and a Doctor of Sciences in Physics and Mathematics degree (Dr. habil.) from the Taras Shevchenko National University of Kyiv, Ukraine, in 2010. His research interests include informational systems development, computer simulation and modeling of socio-economic systems. He has published a number of papers in international journals, monographs and volumes in book series, is a member of editorial board of International Journal of Economic Theory and Application, reviewer of scientific journals International Journal of Modern Physics (B) and Heat Transfer and certified Data Science & Machine Learning specialist. He is a member of Scientific Council section of Ukrainian Ministry of Education and Science on the specialty "Informatics and Cybernetics". Currently, he is working as a Head of Department of Computer Science and Information Systems, Kyiv National University of Trade and Economics.

WWW: https://knute.edu.ua/blog/read/?pid=12695&uk
E-mail: Pursky_O@ukr.net

Dr. **Serhiy Semerikov**, Professor of Computer Science and Educational technology, Kryvyi Rih State Pedagogical University, Ukraine

Serhiy Semerikov is professor of Department of Computer Science and Applied Mathematics at Kryvyi Rih State Pedagogical University. He got both PhD and DSc in education (informatics) from the National Pedagogical Dragomanov University in 2001 and 2009, respectively. The main directions of Dr. Semerikov' research is methods of learning and educational technology.

WWW: https://kdpu.edu.ua/semerikov/
E-mail: semerikov@gmail.com

Dr. **Yevhenii Shapovalov**, a researcher at Junior Academy of Science of Ukraine, Chief Specialist at Ministry of Digital Transformation of Ukraine

Yevhenii Shapovalov was born in 1992. He defended Ph.D. in 2019 in the field of Biotechnology at the National University of life and environmental sciences of Ukraine. Since 2014 worked in the National Academy of Science of Ukraine where he provided chemical laboratory, work in science in the field of computer sciences. Science 2020 starts working at the Ministry of Digital Transformation of Ukraine. He was a participant of 3 international Jean Monnet projects related to implementing European standards in Educational programs on Environmental Sciences and Sustainable development.

WWW: http://www.nas.gov.ua/UA/PersonalSite/Pages/default.aspx?PersonID=0000026333
E-mail: sjb@man.gov.ua

Dr. **Yaroslav Shramko**, Professor of Logic and Philosophy, Department of Philosophy, Kryvyi Rih State Pedagogical University, Kryvyi Rih, Ukraine

Yaroslav Shramko, born in 1963, received a Candidate of Philosophical Sciences degree (Dr. phil.) from the Lomonosov Moscow State University, USSR, in 1990, and a Doctor of Philosophical Sciences degree (Dr. habil.) from the Institute of Philosophy of the National Academy of Sciences of Ukraine, in 1998. Since 1990, he has been working in the field of logic and analytic philosophy at the Kryvyi Rih State Pedagogical University, where he is currently rector. His research interests include non-classical logic and analytic philosophy. He has published a number of papers in international journals and volumes in book series, is a member of editorial boards of Studia Logica, European Journal of Mathematics, Logic and Logical Philosophy.

WWW: https://kdpu.edu.ua/shramko/yse.htm
E-mail: shramko@rocketmail.com

Dr. **Mariya Shyshkina**, Head of Department of Cloud-based Systems for Education Informatization of the Institute of Information Technologies and Learning Tools of the National Academy of Educational Sciences of Ukraine, Kyiv, Ukraine

Mariya Shyshkina, born in 1969, received a Candidate of Philosophical Sciences degree (PhD, Philosophy) from the Institute of Philosophy of the National Academy of Sciences of Ukraine, Kyiv, Ukraine, 1999, and a Doctor of Educational Sciences degree (Dr. habil.) from the Institute of Information Technologies and Learning Tools of the National Academy of Educational Sciences of Ukraine, 2016, Kyiv, Ukraine. Since 1999, she has been working in the field of information and communication technologies for education at the Institute of Information Technologies and Learning Tools of NAES of Ukraine where she is currently a chief of a department. Her research interests include philosophy of education, ICT for learning and research. She has published a number of papers in international journals and volumes in book series, is a member of editorial board of Information Technologies and Learning Tools Journal.

WWW: http://iitlt.gov.ua/eng/structure/departments/cloud/detail.php?ID=269
E-mail: shyshkina@iitlt.gov.ua

Dr. **Oleksandra Sokolyuk**, Institute of Information Technologies and Learning Tools of the NAES of Ukraine, Ukraine

WWW: http://iitlt.gov.ua/eng/structure/departments/technology/detail.php?ID=274

Prof. **Vladimir Soloviev**, Kryvyi Rih State Pedagogical University, Ukraine

WWW: https://kdpu.edu.ua/personal/vmsoloviov.html

Dr. **Oleg Spirin**, Doctor of Pedagogy, Full Professor, Corresponding Member of the National Academy of Pedagogical Science of Ukraine, Vice Rector for Research and Digitalization of the University of Educational Management, Kyiv, Ukraine

Oleg Spirin, born in 1965, 1989 – graduated from Zhytomyr Ivan Franko State University, Ukraine, majoring in Mathematics and Physics. Scientific degree: Ph.D (2002), Thesis “Differentiated approach to the Study of the Foundations of Artificial Intelligence in Computer Science Course Physics and Mathematics in Higher Educational Institutions”; Doctor of Pedagogical Sciences (2009), Thesis “Theoretical and Methodological Basis of Credit-modular System of Future Teachers of Informatics Training”. Academic status: Associate Professor of Computer Science (2004). Professor of the specialty 13.00.10 – Information

Technologies in Education (2013). Prof. Spirin is an expert in the informatization of education and science and information training of students. He has published a number of papers in international journals and volumes in book series, is Deputy Editor-in-Chief of Electronic scientific edition "Information Technologies and Learning Tools", member of the Editorial Boards of the specialized journals "Information Technologies in Education", "Computer in School and Family".

WWW: http://umo.edu.ua/en/university/leadership/spirin-olegh-mikhajlovich
E-mail: oleg.spirin@gmail.com

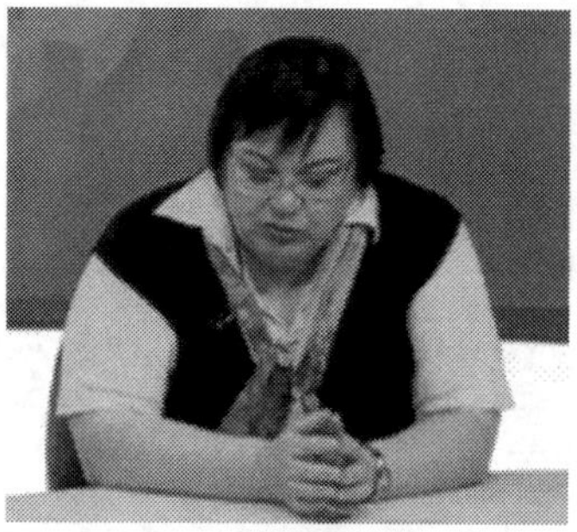

Dr. **Iryna Trubavina** – Doctor of Pedagogical sciences, professor of Department of General pedagogics and pedagogic of High education, H. S. Skovoroda Kharkiv national pedagogical university, consultant of the La Strada – Ukraina (2000-2015), scientific consultant of the State Social services for families, children and youth (2000-2006).

Field of scientific interest: social pedagogics, social work, education, family, pedagogics skills. She is Editor of the journal "Social Work in Ukraine: Theory and Practice".

WWW: http://hnpu.edu.ua/uk/trubavina-iryna-mykolayivna
E-mail: trubavina@gmail.com

Dr. **Tetiana Vakaliuk**, professor of the department of Software Engineering, Zhytomyr Polytechnic State University, Zhytomyr, Ukraine.

Tetiana Vakaliuk, born in 1983, received a Candidate of Pedagogical Sciences degree from the National Pedagogical Dragomanov University, Ukraine, in 2013, and a Doctor of Pedagogical Sciences degree from the Institute of Information Technologies and Learning Tools of the National Academy of Sciences of Ukraine, in 2019. Since 2019, she has been working in the field of information technologies at the Zhytomyr Polytechnic State University. Her research interests include information technologies, ICT in Education, Cloud technologies. She has published a number of papers in international journals, is a member of editorial boards of Information Technologies and Learning Tools, Zhytomyr Ivan Franko State University Journal: Pedagogical Sciences, Collection of Scientific Papers of Uman State Pedagogical University.

WWW: https://sites.google.com/view/neota
E-mail: tetianavakaliuk@gmail.com

Dr. **Maryna Vardanian**, Full Professor of Children's Literature and Comparative Literature, Faculty of Foreign Languages, Kryvyi Rih State Pedagogical University, Kryvyi Rih, Ukraine

Maryna Vardanian, born in 1979, received a Candidate of Philological Sciences degree (PhD in Philology) from The Volodymyr Vynnychenko State Pedagogical University of Kirovograd (Kropyvnytskyi, Ukraine), in 2010, and a Doctor of Philological Sciences degree (Dr. habil.) from The Bohdan Khmelnytsky National University of Cherkasy, Ukraine, in 2019. Since 2012, she has been working in the field of children's literature and comparative literature at the Kryvyi Rih State Pedagogical University, where she is currently Dean of Faculty of Foreign Languages. Her research interests include literary imagology, translating studies and literature written by Ukrainian Diaspora. She has published a number of papers in journals, is a member of International Research Society for Children's Literature, and editorial board of Literatures of the World: Poetics, Mentality and Spirituality.

WWW: https://kdpu.edu.ua/personal/pm_rectora.html
E-mail: maryna.vardanian@gmail.com

https://doi.org/10.1051/shsconf/20207500001

Dr. **Vladyslav Velychko**, Full Professor of the Chair of Teaching Methods of Mathematics and Teaching Methods of Computer Sciences, State Higher Educational Institution "Donbas State Pedagogical University", Sloviansk, Ukraine

Vladyslav Velychko, born in 1973, received a Candidate of Physical and Mathematical Sciences (PhD.) from the Taras Shevchenko National University of Kyiv, Ukraine, in 2007. Since 1994 he has been working in the field of algebra and number theory and in the field of information and communication technologies in education at Donbas State Pedagogical University. His research interests include quasi-ideal of semigroups, free software in education, open electronic educational resources. He has published a number of papers in international journals, and is a member of the editorial boards of the Collection of Scientific Papers of the Faculty of Physics and Mathematics of Donbas State Pedagogical University and the electronic journal of E-Learning TeXnology.

E-mail: vladislav.velichko@gmail.com

Dr. **Kateryna Vlasenko**, Professor of Mathematics, Head of the Mathematics and Modeling Department, Donbas State Engineering Academy, Kramatorsk, Ukraine

Kateryna Vlasenko, born in 1966, received a Candidate of Pedagogical Sciences degree (PhD) from the National Pedagogical Dragomanov University, Ukraine, in 2004, and a Doctor of Pedagogical Sciences degree (D.Sc. in Educational Science) from the Bohdan Khmelnytsky National University of Cherkasy, in 2011. Since 2008, she has been working in the field of mathematical and pedagogical modeling at Donbas State Engineering Academy, where she is currently head of the Mathematics and Modeling Department. Her research interests include the issues of mathematics education. She has published a number of papers in international journals and volumes in book series, is a member of editorial boards of Innovative Solutions in Modern Science, Topical Issues of Natural and Mathematical Education Sumy State Pedagogical University named after A. Makarenko.

WWW: http://formathematics.com/
E-mail: vlasenkokv@ukr.net

Dr. **Nataliia Volkova**, Doctor of Pedagogy, Professor, Head of the Department of Pedagogy and Psychology, Alfred Nobel University, Dnipro, Ukraine

Nataliia Volkova born in 1959. Year 2006 – defended Doctoral Dissertation on "Theoretical and Methodological Principles of Training Future Teachers for Vocational and Pedagogic Communication", specialization 13.00.04 – "Theory and Methodology of Professional Education". Year 2008 - awarded the title of "Professor of the Department of Pedagogy". Leading expert in pedagogy of higher education in Ukraine. Author of more than 160 printed works, three monographs, two textbooks with a stamp of the Ministry of Education and Science of Ukraine, three study guides with a stamp of the Ministry of Education and Science of Ukraine, 12 educational and practical study guides, articles in leading Ukrainian and foreign journals.

WWW: https://duan.edu.ua/university-ukr/kafedry/15-pages/332-kafedra-pedahohiky-ta-psykholohii.html
E-mail: npvolkova@yahoo.com, npvolkova@duan.edu.ua

Dr. **Yuliia Yechkalo**, Associate professor, Department of Physics, Kryvyi Rih National University

Yuliia Yechkalo, born in 1981, received a Candidate of Pedagogical Sciences degree from the Kirovograd State Vladimira Vinnichenka Pedagogical University, Ukraine, in 2013. Since 2005, she has been working at the National Metallurgical Academy of Ukraine. She has been working at the Kryvyi Rih National University since 2012. Her research interests include theory and methods of education (physics) and information and communication technologies in education.

E-mail: uliaechk@gmail.com

Conclusion

The vision of the ICHTML 2020 is to create a leading interdisciplinary platform for researchers, practitioners and educators, to present and discuss the most recent innovations, trends, and concerns as well as practical challenges encountered and solutions adopted in the fields of learning.

The conference is a successfully performing forum to transferring and discussing research result among the academics, students, teachers, government, private sector or industries. Participants and presenters from several countries such as Israel, Italy, Lithuania, Moldova, Netherlands, Poland, Romania, Russia, Sweden and Ukraine have attended the conference to share their significant contribution in research related to the History, Theory and Methodology of Learning.

We are thankful to all the authors who submitted papers and the delegates for their participation and their interest in ICHTML as a platform to share their ideas and innovation. Also, we are also thankful to all the program committee members for providing continuous guidance and efforts taken by peer reviewers contributed to improve the quality of papers provided constructive critical comments, improvements and corrections to the authors are gratefully appreciated for their contribution to the success of the conference. Moreover, we would like to thank the developers and other professional staff of EasyChair, who made it possible for us to use the resources of this excellent and comprehensive conference management system, from the call of papers and inviting reviewers, to handling paper submissions, communicating with the authors, and creating the conference proceedings.

We are looking forward to excellent presentations and fruitful discussions, which will broaden our professional horizons. We hope all participants enjoy this conference and meet again in more friendly, hilarious, and happiness of further ICHTML 2021.

References

1. V. Richardson, 'Inducing panic': Media under fire for driving coronavirus hype to epidemic levels (The Washington Times, New York, 2020), https://www.washingtontimes.com/news/2020/mar/11/media-fueling-coronavirus-panic-under-fire/. Accessed 16 Mar 2020.
2. Cabinet of Ministers of Ukraine, Regulation No. 211 dated March 11, 2020, published on the governmental portal (2020), https://www.kmu.gov.ua/npas/pro-zapobigannya-poshim110320rennyu-na-teritoriyi-ukrayini-koronavirusu-covid-19. Accessed 16 Mar 2020
3. J. Conduitt, Draft account of Newton's life at Cambridge (2004), http://www.newtonproject.ox.ac.uk/view/texts/normalized/THEM00167. Accessed 16 Mar 2020
4. G. Brockell, During a pandemic, Isaac Newton had to work from home, too. He used the time wisely (The Washington Post, 2020), https://www.washingtonpost.com/history/2020/03/12/during-pandemic-isaac-newton-had-work-home-too-he-used-time-wisely. Accessed 16 Mar 2020
5. Coursera Coronavirus Response Initiative | Coursera for Campus (2020), https://www.coursera.org/coronavirus. Accessed 16 Mar 2020
6. Contingency Planning For Moving Courses Online | UMass Amherst Response to the Coronavirus (COVID-19) (2020), https://www.umass.edu/coronavirus/news/contingency-planning-moving-courses-online. Accessed 16 Mar 2020
7. Kryvyi Rih State Pedagogical University (2020), https://kdpu.edu.ua/en/. Accessed 16 Mar 2020

Statement of Peer review

In submitting conference proceedings to *Web of Conferences*, the editors of the proceedings certify to the Publisher that

1. They adhere to its **Policy on Publishing Integrity** in order to safeguard good scientific practice in publishing.
2. All articles have been subjected to peer review administered by the proceedings editors.
3. Reviews have been conducted by expert referees, who have been requested to provide unbiased and constructive comments aimed, whenever possible, at improving the work.
4. Proceedings editors have taken all reasonable steps to ensure the quality of the materials they publish and their decision to accept or reject a paper for publication has been based only on the merits of the work and the relevance to the journal.

Title, date and place of the conference

The International Conference on History, Theory and Methodology of Learning (ICHTML 2020)

May 13-15, 2020 | Kryvyi Rih, Ukraine

Proceedings editor(s):

Vita Hamaniuk

Serhiy Semerikov

Yaroslav Shramko

Date and editor's signature

March 18, 2020

Development of media education in Canada: a brief history

Hanna Tsvietkova[1], *Olena* Beskorsa[2,*], and *Liudmyla* Pryimenko[3]

[1]National Pedagogical Dragomanov University, Department for Pedagogy and Psychology of Preschool Education and Children's Creativity, Kyiv, 01601, Ukraine
[2]Donbas State Pedagogical University, Department of Primary Education Theory and Practice, Sloviansk, 84100, Ukraine
[3]Donbas State Pedagogical University, Department of Pedagogy of Higher School, Sloviansk, 84100, Ukraine

Abstract. The article focuses on a holistic retrospective analysis of the history of media education in Canada, which has been done for the first time. Based on the theoretical findings of Canadian media educators, the authors substantiate the periodization, identify the trends, the periodization criteria and three main periods of establishment and development of Canadian media education in the context of socio-political and socio-pedagogical determinants. The historical preconditions of the development of media education in Canada are revealed. The essence of media education is regarded as a theory study and development of practical skills for mastering modern mass media, which is considered as part of a specific, autonomous field of knowledge in pedagogical theory and practice. The authors determine that media education is associated with all types of media, which include the set of information and communication tools that each person interacts with in everyday life: printed (newspapers, magazines), auditive (radio, audio) and screen or audiovisual media (cinema, TV, video, multimedia, Internet, etc.); they identify the essential characteristics of media education, determine that media education is the form of media literacy and media culture of an individual; as well as they find out, characterize and systematize the gaps in media education in Canada. The article presents the evolution of media education programs, techniques as well as media education associations. In the context of studying the experience of Canadian media theorists and practitioners, the necessity of applying positive Canadian experience to solve the problems of implementing media education in Ukraine in terms of reforming and humanizing its educational space has been substantiated.

1 Introduction

Media culture is an integral part of modern informational educational environment. The modern world is full of different media, which greatly affects the lives of people. Speaking of the media, we mean the whole set of information and communication tools which each person uses in everyday life. The field of modern media functioning includes printed means (newspapers, magazines), audio means (radio, sound recordings) and screen or audio-visual means (cinema, TV, video, multimedia, Internet, etc.). The world of a modern person, which is less and less in contact with reality, being surrounded by its representative images as well as images produced by human thinking with the help of the media, is forming under the influence of these information and communication means. So today we can say that the environment that surrounds a person, and the processes in it become largely virtual. Due to the fact that rather large experience has accumulated in the field of information and communication technologies, it allows to implement journalistic, promotional, advertising and artistic functions of the media, the main task of the development of personality is the exploitation of the language sphere, the language culture of new media, with the help of which it becomes possible to provide a competent interaction with society.

Having signed the Bologna Declaration, Ukraine has joined the process of harmonizing the higher education system of many European countries, aimed at ensuring its high quality. The implementation of the Bologna Process and new educational standards is the subject of lively discussion in the domestic scientific community. In particular, to realize such a large-scale integration project, the native higher educational institutions should modernize the educational process. In terms of the world and European educational community, the introduction of media education is an important requirement for the training of specialists in higher educational institutions. This is due to the fact that in order to protect the audience from manipulating the media, modern people need to have special knowledge, communicative and informative skills and abilities, and the capacity of critique (i.e. media literacy) that can be formed through organized and purposeful media education.

Moreover, the methodological recommendations of the Canadian Degree Qualifications Framework are based on the need to develop new competences of different levels, new personal qualities whose absence is detrimental to the comfortable and productive existence

* Corresponding author: beskorsyhelen@gmail.com

of a person in the information environment. The development of a well-cultured person, his information competence and literacy became relevant to society.

In this regard, the development of university students' professional competence involves both the use of media education technologies in traditional education and the fundamental restructuring of the entire system, the restructuring of the structure and organization of the educational process, the implementation of appropriate methodological and didactic principles, and the development of pedagogically innovative, media educational and information training technologies.

Consequently, due to the discrepancy between the content and forms of education and the present stage of society development, researchers and educators around the world are searching for a new model of today's higher education institution, effective methods of training aimed at overcoming monotonous learning, which would allow to increase the students' cognitive activity on the basis of their own initiatives when using theoretical knowledge and practical skills.

The implementation of media education as a serious and important area is associated with the majority of the social structures in a democratic society and has been used around the world for over sixty years, particularly in countries such as Canada, Australia, the United States and the United Kingdom. According to the Human Development Index (2013) conducted by UN experts, Canada ranks the first in the world in terms of living standards, the second in terms of the area. Canada is one of the leading countries in the world in terms of the number of computers connected to the Internet, with the help of which the media education process in the country is conducted. The investment in the Canadian education is at a high level (7.1-7.9% of GDP in 2010–2013).

Today, Canada is one of the most advanced countries in the field of media pedagogy. The media culture in this country is an integral part of education. Media education courses are available at almost all Canadian universities. Almost every Canadian province has its own Media Education Association, which holds scientific and methodological conferences, publishes media products (newspapers, magazines), and prints teaching materials. So, studying the experience of Canadian media theorists and practitioners is necessary for the Ukrainian education system to adequately solve the problems of implementing media education and developing the ways of improvement of the Ukrainian educational system in the context of reforming, humanizing and globalizing the educational space.

In this perspective, it is useful to trace the history of the evolution of media education in Canada. Knowledge of historical development can serve as an advisor in the development of modern concepts for the implementation of media education in the in educational institutions in Ukraine.

2 Analysis of the Relevant Research and Publications

Media education aims at using the most of the educational, technical and communicational potential of modern mass media and, as UNESCO defines, it is a separate area of education, since it helps people to understand ways of using mass communication in society; to analyze media texts and critically evaluate the values, political, social, commercial and cultural interests, as well as to create and disseminate their own media texts through mass media [1].

In the definitions of UNESCO [2], media education and media competence are consistently associated with the development of democratic thinking and the development of civic responsibility of an individual. According to the outcome documents of many international forums, such as the Alexandrian Proclamation on Information Literacy and Lifelong Learning [3], which discuss the problems of media education, the media process in the world is based on civil responsibility, humanism and democracy.

Canada is the home country of the famous researchers who contributed to developing media theory – H. Innis, M. McLuhan, R. Morgan, D. Buckingham, and D. Smythe. M. McLuhan was the first who developed in Canada the special course on media culture in the 1950s. Robert Morgan, a cultural studies professor, made a survey among the teachers and characterized problems that emerged while teaching media. The subject of his scientific research was studying how to teach culture and media education theory [4]. Educators Sh. Turkle and D. Buckingham emphasized on the importance of studying the audience who are taught the media theory. They also highlighted the issue of the necessity of implementing media education in the process of teacher training as teachers should deal with the ways students process media. Teachers should decide what the starting points for studying media are [5, 6, 7].

The interconnection and mutual influence of education, the mass media and culture are presented in the research works of foreign scientists: French scientists J. Gonnet, E. Bevort and L. Porsche, American media theorists J. Lull and J. Peters; the psychological aspects of mass media are dealt with in the papers of the English media expert R. Harris.

American media education is also based on the work of the famous Canadian media theorist M. McLuhan, the British media educators L. Masterman and K. Bazalgette and other English-speaking researchers, as well as on such basic concepts of media education as "category", "agency", "language", "technology", "audience" and "representation" [8]. At the present stage, media education is integrated into the curricula of all 50 states. There are also a number of media education associations in the United States. Among them there are the Media Education Center in Washington DC, the National Council for Television and Media in Virginia, the Media Literacy Center in Los Angeles, and others. Some professional associations of teachers are trying to achieve the inclusion of media education in the state standards [9].

In today's media education practice, American educators widely use the potential of the Internet and television, and focus on the fact that the PBS (Public Broadcasting System) model is one of the examples of effective media and education interaction between universities. Note that in the American media education, unlike Canadian one, the most important aspect is successfully realized, i.e. the media education of teachers and future teachers. So, media education in the United States is also at a rather high level.

Summing up, we may state that the ideas of M. McLuhan significantly influenced the development of media education all over the world and as N. Andersen claims that media education theory in Canada is special and unique and it should be studied for further researches [8].

3 The purpose of the article

The purpose of this article is to overview key media education movements and organizations within Canada; to identify the main stages of the development of media education in Canada; to present the evolution of media education programs, methodologies, and media education associations; to determine the perspective for developing positive Canadian experience of implementing media education in teacher education in the conditions of globalizing and humanizing the educational space.

The investigated period (since 1940 to present) is full of promising unique programs, significant achievements in the development of educational innovative technologies in distance education, media pedagogy, which are based on respect for human rights, critical thinking, the formation of a well-cultured personality, his information competence and literacy.

4 Methods

In the article a set of research methods was used: general scientific methods (analysis, synthesis, comparison, analogy, comparison and contrasting), which allowed to synthesize and systematize the views of Ukrainian and Canadian scientists on the research issue, as well as to justify the original theoretical and methodological points; scientific-historical methods (chronological one, the method of periodization, logical-historical and historical-genetic method) that provided the study of the preconditions for the formation of media education, the elucidation of the stages and leading trends in the development of media education in the practice of the studied period; formalized methods: scientometric analysis, methods of interpretation, systematization, generalization for the disclosure of the features of the development of media education in Canada. Terminological analysis of scientific literature made it possible to define the basic concepts and categories according to the subject of the research. Theoretical generalization, methods of functional-structural analysis were used to formulate the main points and conclusions of the study on the role of media education and the generalization of the experience of overcoming the problems of media education in Canada.

As the criteria for the periodization of the historical development of media education in Canada were chosen: socio-economic, scientific and technical, cultural and educational development of society. In accordance with this, three main periods of development of media education were identified and presented below.

5 Results

The logic of our study requires understanding and clarifying the concept of "periodization" and defining it in relation to the historical and pedagogical issues.

As the scientists claimed, periodization should be understood as the logical delineation of the investigated period in accordance with its qualitative characteristic in relation to certain stages. Such a characteristic should reflect both the general laws of development of public life and specific in its content, significant moments of the basic methods and forms of the genesis of the problem and the determining stages of its development.

"Periodization is a basic construction that we use in our imagination, which is based on a strict scientific analysis of the research objects that is a diverse set of sources that forms the preliminary analytical work and finds its expression in the creation of a hypothetical image of the cognitive object. Periodization is a definite scheme (core, skeleton), on which we "collect" a set of events – the results of our cognitive activity in the form of "facts-knowledge", which in their totality represent the image of a certain phenomenon that acts for us as an object periodization – the object of knowledge".

The periodization is considered as the distribution of the whole process of scientific development into segments of time, which differ from each other with specific features established on the basis of objective criteria and principles. This distribution may be general in nature and relate to the whole science or its part, which has certain distinctive characteristics that determine the more detailed allocation of time limits. Therefore, periodization can be structurally represented as follows: era, period, stage. The historical era is not a logical abstraction. In the case of periodization of education history, it should be noted that education history synthesizes a variety of socio-economic, didactic, psychological and pedagogical phenomena and processes. Understanding the historical epoch includes the definition of the leading trends in the development of certain educational phenomena. It should be added that the establishment of certain limits of the historical era cannot be "absolutized", because they are conditional and dynamic. Consequently, the epoch is a more general concept compared with the period. In turn, the period is a more general concept than a stage. The period is the length of time during which a process occurs. The stage is a period of time in the development of any movement or process. The stage acts as part of the period and may differ because it includes the main and the most important events of the period, that is why it carries its essence. However, there may be the stages of a preparatory nature, the completion of what has been done during the main stage, the stages of comfort, and others. We accept such an interpretation of the concepts of "period" and "stage", especially since it coincides with the position of most historians, in particular, historians of education.

Based on the theoretical developments of Canadian media theorists (M. McLuhan, M. Mitchell,

 https://doi.org/10.1051/shsconf/20207501001

A. Hodgkinson, G. Moore, C. Worsnop, N. Andersen, B. Duncan, J. Pungente, et al.) and emphasized the following criteria for periodization – socio-economic, scientific and technical, cultural and educational development of society, we consider it appropriate to identify the three main periods of the development of media education in Canada. The first period covers the 1940s – the beginning of the 1970s (this is the period of the birth and establishment of media education in Canada). The second period is since 1972 to the beginning of the 1990s (the period of development of media education and its integration into the system of secondary and higher education in Canada) and the third period is the 2000s to the present (this is the modern period of media education development) (Table 1).

The first period of the "Birth and development of media education in the Canadian education system" is divided into three stages. The first stage of this period is "Emergence of the foundations of media education in Canada". This is the time when the media education movement in the North American continent was still in its infancy. Its borders cover the 1940s – the beginning of the 1950s.

According to the respectable Canadian scholars such as N. Andersen, B. Duncan, J. Pungente, the emergence of media education in Canada at this stage was caused, first of all, by two main factors [10, 11]:

1) intensive expansion of Canadian society by the American "pop culture";
2) flexible and progressive education system that created favourable conditions for the emergence of new educational paradigms.

Table 1 Historical periodization of the development of media education in Canada

Period	Stage	Stage Title	Trends and representatives
Birth and establishment of media education in the Canadian education system	1940s – 1950s	Emergence of the foundations of media education in Canada	Marshall McLuhan's theory was based on the idea that "every media has its own" technological grammar "and creates a unique message.
	1950s – 1960s	Development of the media education system	Development of the "Art Concep" in Canadian secondary school and visual literacy by M. McLuhan.
	1960s – the beginning of the 1970s	Media education Integration in Canadian secondary and higher education	Domination of the aesthetic theory of media education. Influence of the ideas of M. Mitchell and A. Hodgkinson.
Development of media education and its integration into the system of higher and secondary education in Canada	1972 – 1977	Media education regression in Canada	Principle of 'back to the foundations', media education was considered as "unnecessary luxury".
	1978 – 1989	Renewal of the positive trends in the development of media education in the Canadian education system	Development of semiotic and cultural theory of media education. The brightest representatives and theorists of which are C. Wornstop, J. Pungente and B. Duncan. Foundation of the Jesuit Communication Project.
	1990 – 1999	Sequential integration of compulsory media education in the Canadian education system	Publication of the textbook on media literacy (B. Duncan). The outstanding representatives are N. Andersen, J. Pungente. The establishment of media associations in all provinces of Canada.
Modern period of media education development	2000s – 2009	Education reform, reflection of the best practices in media education in secondary and higher schools in Canada	International forums and world conferences on media education were held in Toronto and Montreal.
	2010s – present times	Modernization, further improvement of media education programs and technologies	Development of modern media literacy and media education programs. The brightest representatives are C. Wing and C. Wilson.

The birth of media education in Canada, as in most other countries of the world (the USA, the UK, France, Australia), took place on the basis of cinematographic materials in the 1940s – 1950s. According to the survey, a common Canadian student had spent about 15,000 hours in front of the TV screen before he left school and had watched at least 500 films [12]. Therefore, the rapid development of television and the high intensity of students' contacts with the new audiovisual media forced the teachers to reflect on the consequences of this impact.

A well-known Canadian scientist and educator Herbert Marshall McLuhan made a significant contribution to the development of media education at this stage. He emphasized the educators' attention to the growing popularity of the new visual culture in the lives of young people and encouraged for the active use of the rich potential of audiovisual culture for teaching students through non-verbal communication methods. Marshall McLuhan was one of the first educators who used the concept "media" to determine mass media. The scientist claimed that the media have a certain impact on society (in particular, the children's audience). The famous quote by Marshall McLuhan "Medium is a message" quite clearly reflects the scholar's opinion that media texts can be perceived differently. In accordance with this, there was a need for the development of media culture [13].

Thus, the first official mention of the birth of the media education movement in Canada can be attributed to the 1950s, when the concept of screen education was formed by British activists who established Society for Education in Film and Television in 1950. In general, due to the fact that television began to develop intensively in the 1940s and 1950s, there was a problem of regulation of interaction between students and the media. Canadian educators understood the futility of attempts to shield students from the ever-increasing influence of television and cinema, and they understood the need to develop and implement special courses and programs to help students adapt better in the media world as well as optimize the learning process itself.

The second stage "Development of the media education system" is characterized by the emergence of basic concepts of media education in Canada, which has now grown into an international educational movement that began in the second half of the 1950s. The media education theorist and teacher M. McLuhan proposed the concept of "Global village", which laid the ground for the dialogue of cultures in the information society of the era of globalization. "Globalization" and "Information Society" are two complementary phenomena that determine the nature of the modern era [8].

Canadian educators are increasingly concerned about the rapidly spreading and ever-increasing impact of the media on the lives of the young generation. Leading scholars recognized the existence of competition between school and the media, which was identified as a "parallel school". However, at that time, no constructive decisions about the need to include the "media education" course in schools or higher education institutions were not adopted. In 1959, M. McLuhan developed a curriculum for the 11th form of secondary schools for media education (Toronto), which undoubtedly became the basis for the further development of media education programs and courses for both the secondary and higher education in Canada [8].

The third stage entitled "Media education integration in Canadian secondary and higher education" is known as the stage of "Screen Education". The rapid development of communication and information technology in the second half of the twentieth century led to the fact that the original term "Film Teaching" as the study of cinema that reflected the narrower orientation of the course transformed into the term "Screen Education".

In 1965, the Toronto Ministry of Education initiated a one-day conference on cinema education, which in those years became very popular among educators and civil activists. In 1968, the Ontario Ministry of Education provided the position of assistant coordinating work in the field of "Screen Education", which led to three similar one-day conferences, each of them numbered about 200 participants. The main achievement of these conferences at this stage was the creation of an information committee, which eventually became a harbinger of the future media education association. Approximately at the same time, student film festivals and conferences devoted to media education were actively held around the world. So in 1968, the first international conference "International Conference on Communication Media Education" took place under the patronage of British professor A. Hodkinson. The conference addressed the issues of media education in different countries and analyzed the national experience of various countries, such as the United States, Britain, Australia and Canada. Also in the same year in Halifax (Nova Scotia province), the first student film festival took place, where young participants from Canada, the USA, Italy, Great Britain, Australia and other countries of the world demonstrated their achievements. Immediately after the festival in Ontario, the School Council even initiated a space for films created by Canadian students.

In 1969, thanks to the enthusiasm of a creative group and an information committee that disseminated the idea of the need to create an official media education structure, it was founded the Canadian Association for Screen Education (CASE) [10]. Already in June 1969, at the York University in Toronto, thanks to the efforts of this association, the first conference was held throughout the country. The result of the event was a course with the introduction of the aesthetic theory of media education. After all, "the purpose of aesthetic education is to teach the younger generation to think critically... to learn to see the difference between good and bad films and TV shows" [14].

A significant support from the National Film Board (NFB) for providing a resource base for "Screen Education" had also positively influenced the introduction of media education in the Canadian education system. In the period of 1960s – 1970s, about 40 representatives of the National Film Board functioned on the territory of Canada.

The "Screen Education" course was available at a number of Canadian universities, for example, at Loyola College, McGill University and Sir George Williams University.

However, despite all the positive changes in implementing media education in the educational process, the second period "Development of media education and its integration into the system of higher and secondary education in Canada" began with the stage of media education regression. Analysing the process of development of "Screen Education" in the 1960s, we identified a number of problems that complicated the work of teachers on the introduction of "Screen Education" in the educational process of the educational institutions in Canada. The overload of curricula that did not allow for the inclusion of a course in the schedule, orientation of the majority of teachers and official representatives of the educational institutions in the print media, the lack of proper funding of media education programs for a centralized information resource, and the lack of common standards for the development of media culture hampered the process of developing and implementing media education in the Canadian education system.

Mostly only in Ontario, media education projects were implemented. In the 1970s, the course for implementing media literacy was called "unnecessary luxury" [15].

But even in rather difficult socio-economic conditions, at the regression stage, media education was carried out at the expense of individual educators-enthusiasts who tried to integrate the "media education" course in a school curriculum based on the cinematic material, television and press.

However, in 1978, the second largest association – the Association for Media Literacy (AML) – was founded in the country in order to support media educators that created favorable conditions for the further development of media education at the next stage, entitled "Restoration of the positive trends in the development of media education in the Canadian education system".

At the end of the 1980s, media education improved significantly. An example of such improvement was the emergence of a significant number of regional associations in the field of media education. In January 1988, a guiding conference made a decision to create the Media Literacy Association in Saskatchewan (MLS). Mike Ellis (the first Canadian who received the degree of media education at London University, England), and other teachers in Saskatoon created the Media Literacy Saskatchewan. The aim was to identify and communicate between teachers, support the development and integration of media literacy in the educational curriculum, to influence education policy, to provide professional support, to maintain contact with Canadian and international organizations. The members of the Media Literacy Saskatchewan developed three programs – "TV media", "News media" and "Children's Media" for use in schools, the written part of the program for studying film and literature for the senior English course and published the "Handbook of media literacy from primary to high school". This course included the subjects that were integrated with all aspects of the curriculum for primary and secondary levels. The Saskatchewan media literacy organization published a quarterly bulletin "Media View", which included practical information, bibliography, review of books and audiovisual resources, and the development of lessons. Saskatchewan had six categories of collaborative research that were integrated into all courses of study. Two of them – "Communication Skills" and "Critical and Creative Considerations". Using these two studies, the MLS developed the proposals how to integrate media education throughout the primary and secondary curriculum. Media literacy proposals consisted of such key concepts as text, audience, representation, and products. Saskatchewan arts policy required media literacy, which in the future would integrated into the curriculum not only as part of basic education, but also as a support for the foundations of the language art. Even so, several schools in Saskatchewan offered such courses because there were no trainings for teachers in this field. After all, the MLS worked for the interest of schools and provided them with programs for improving their teachers' qualifications.

The Jesuit Communication Project (JCP) that was formed in 1985 in Toronto had a great impact on the development of media education in the country. Since 1985, the TV channel CHUMTelevision also actively supported the development of media literacy. It was a channel (the first in the world) that introduced a special department on media education. In 1999, the leaders of CHUMTelevision were invited by the UNESCO World Organization to participate in a conference on the exchange of experiences.

In 1989, the first Media Literacy Resource Guide was published with the support of the Ontario Ministry of Education and the Media Literacy Association (AML). The main purpose of the textbook was to help teachers use their media potential [16]. The publication of the first textbook on media education in some way became the basis for the introduction of this course into the curriculum of Canadian educational institutions.

Generally, the development of Canadian media education since the 1980s was very productive. Provincial organizations throughout the country organized conferences, developed curricula and programs, held workshops and trainings, lectures on various aspects of media education.

Only in 1973, UNESCO proposed that the media should be studied through higher and secondary education curricula. And in 1977, UNESCO announced media education as a priority area for the coming decade.

It should be noted that the Federal Ministry of Education did not exist at the national level in the country. Unlike other developed countries, Canada does not have a nationwide system of education: according to the constitution and administrative structure, each of ten provinces and three territories has its own educational structures. This provision makes it difficult to meet the uniform requirements for provincial educational institutions. At the same time, this country has been the world's leading one in the field of media education for more than twenty-five years. In Canada, the study of media culture became an integral part of the school curriculum for all secondary school students in September 1999.

The third stage of this period is "Sequential integration of compulsory media education in the Canadian education system". The constantly growing role of the media in the life of modern society led to the rapid development of information technology and computer technology in the 1990s. It inspired a new impetus for the development of media education in Canada. There was a need for purposeful integration of media education into the educational process.

The intensive development of modern information and telecommunication technologies contributed to the formation of new approaches to education. In the 1990s, the computerization of schools was intensively taking place in all provinces of Canada. In 1999, the first part of the Schoolnet program was started. It connected 16,000 Canadian schools and public libraries to the Internet. We draw attention to the fact that the last school that joined the Schoolnet program was a school on the island of Pictou (Nova Scotia province), where only three students studied. So today, the Schoolnet is a unique educational resource (more than 7,000 teaching materials) for Canadian educators. Schoolnet also acts as a virtual educational center for both educators and students who are trying to master the skills of using the computer and the Internet.

It the 1990s, the new educational standards appeared that included the mandatory introduction of media education in the curriculum of school programs. Since 1993, departments of education in the British Columbia, Alberta, Manitoba, Saskatchewan, Yukon, Nunavut and Northern Territories have been actively collaborating within the framework of the Western Canadian Protocol for Collaboration in Basic Education (WCP). Due to this cooperation, at this stage, the specialized methodological centers were created to develop the unified theoretical foundations of curricula that included the media education course. Since 1997, the representatives of provincial departments of education have started developing their own programs and implementing relevant documents that provide the teachers with the necessary pedagogical strategies in the field of media education.

The official documents of the Western Canadian Protocol on Cooperation in Basic Education highlighted the relevance of the media education course: "Understanding the meaning of verbal and other forms of media texts is important for living in democracy. Thus, by studying various media texts, students have the opportunity to gain a wealth of experience, to learn other cultures. In the process of working with media texts, students analyze, synthesize, evaluate and create their own media products" [9].

In the east of the country, a similar organization operated, it was the Atlantic Provincial Education Foundation (APEF), which included Nova Scotia, Labrador, Newfoundland, New Brunswick, and Prince Edward Island. In 1995, the eastern provinces declared their desire to integrate the media education elements into the program "English Language and Arts". In 1996, the curriculum was successfully tested, and since 1997, its implementation has started. The successful integration of this program into the "English language and Art" course is due to the fact that the concept of literacy in the 1990s went beyond the scope of written language and has included the ability to use and understand the visual and technological means of communication. Consequently, according to the above-mentioned documents, at the end of the twentieth century, media education became an important part of the English language and art training course, as well as a real media education reform through the Atlantic Provincial Education Fund and the Western Canada Canadian Cooperation in Basic Education in Canadian Provinces.

The Ministries of Education in Quebec and Ontario were not part of these organizations, but had similar functions for the implementation of the media education initiatives in the educational process of their provinces.

Ontario was the first province where media education became a compulsory part of the curriculum. The Ontario Media Education course consisted of verbal and visual communication, with emphasis on the need for further knowledge, skills and competences: the classification of media genres and the selection of media elements related to it. In April 1995, the Ontario Ministry of Education approved two formal documents: "Curriculum: Concept and Learning Outcomes from Grades 1 to 9" and "Ontario Provincial Standards: Language and Literature, Grades 1-9". The first document clearly identified the amount of knowledge and skills that students should receive, the other document provided objective and logical indicators to determine the success of student learning. A compulsory course in English and Literature from grades 1 to 9 should include listening and speaking, reading, writing, reviewing and presenting.

The regulations of Ontario province set the new standards for education and emphasized the need to introduce media education as part of a traditional curriculum of native language and literature. According to these standards, one third of the English language in middle and upper grades of the school should be devoted to the study of the media. In the seventh and eighth grades, at least 10 percent of the time of education should be allocated to the study of media education. In addition, students were given the opportunity to independently choose an autonomous media course – one of five special courses – within the framework of the English language curriculum [10].

Since 1993, the New Technologies of Montreal University have successfully conducted the large-scale experiments in the field of media education. The results of these fundamental experiments contributed to the adoption of strategic decisions in the field of the media industry. While working on the project "Children and Media" under the supervision of A. Caron, the problems of contemporary media education were studied: they analyzed the television preferences of Canadian children, the family perception of television broadcasts, stereotypes and violence on the screen, the educational potential of television and interactive technology for children, regulation of advertising in children's programs, social advertising and promotion of a healthy lifestyle [17]. The main purpose of these studies was to obtain reliable data for the objective assessment of the perception of the media space for the children's audience.

 https://doi.org/10.1051/shsconf/20207501001

The rather complex nature of the Canadian education system certainly influenced the peculiarities of the development of the media education process in the country, but in turn, it allowed each province to carry out media education reforms independently. A characteristic feature of Canadian media education, which cannot be ignored, is the presence of many associations that operate in every province of the country. First of all, at this stage, the following media educational associations were formed:

1. Manitoba Association of Media Literacy (MAML) was founded in October 1990, thanks to the results of the conference "Special Areas Group" (SAG), funded by the Association of Teachers of the Humanities. At that time, several representatives interested in media education met with members of the Ontario Media Literacy Association Neil Andersen and John Pungente. The result of this meeting was the formation of MAML. The purpose of MAML is to promote the goals of media education and to help investigate the role of media in society. MAML enables to provide the following opportunities: to develop the skills, knowledge, attitudes needed to explain the media, develop social, cultural, political and economic consciousness, develop an evaluation and aesthetic understanding of the media. To accomplish these goals, MAML sponsors conduct workshops for teachers, parents, participate in the development of media literacy programs for schools, provide opportunities for teachers to improve their qualifications, and publish manuals and quarterly bulletins. The Canadian province of Manitoba has an official regional media education policy. Teachers of the Humanities research the media as a part of their English-language course. In 1992, Manitoba University offered the summer media education courses for teachers to improve their knowledge.
2. The Association for Media Education in Quebec (AMEQ) appeared in September 1990 with an initiative group of teachers of English and French-language secondary school teachers, university professors got interested in media literacy, so they met at the Protestant School in Montreal to create the Media Education Association. French researchers J. Gonnet and R. La Borderi had a significant impact on the development of media education in Quebec [18]. At the end of the 1990s, the Quebec Ministry of Education reformed the curriculum of junior and secondary education and implemented a multidisciplinary approach to the integration of media education. In connection with the new school reform aimed at modernizing the entire provincial education system, teaching media education is incorporated into the curriculum of most major disciplines and is one of the leading areas of education. AMEQ also sponsors various student media festivals, media literacy conferences, workshops for teachers and parents. The association representatives regularly report to the Quebec Ministry of Education according to the curriculum changes, as well as to the Canada Telecommunication Commission for Radio and Television.
3. The Canadian Association for Media Education in Vancouver (CAME) and the British Columbia Association for Media Education were founded in 1991 (BCAME). In 1994, the associations organized summer courses for teachers and began to publish training manuals and recommendations for media educators. In the same year, the associations attracted the Ministry of Education by their projects. This led to the fact that in autumn 1996, British Columbia became the first province to introduce the compulsory study of the media into the educational process (the courses in Art and English);
4. The Association for Media Literacy in Nova Scotia (AML-NS) was founded in 1992. The first original objective of the association was to provide information, lesson plans, knowledge and experience, professional development of teachers`, pupils` and parents` media literacy.
5. The Canadian Association for Media Education Organizations (CAMEO) was established in 1992. The purpose of the association wass to coordinate the efforts of media educators from different provinces.
6. The Alberta Association for Media Awareness (AAMA) appeared in the spring of 1993. A group of teachers and media professionals organized this association. The purpose of the work was to promote the media consciousness of education and understanding, to provide information support, to organize forums for information, discussions, publishing activities, preparing conferences and trainings for teachers, and to support the resource centre.

The main strategic goal of all the above-mentioned associations was to promote the development and integration of media education among children, adolescents, and adult audiences.

Thus, since 1999, the study of the media culture course throughout the country has become an integral part of the school curriculum of secondary education institutions. Between 1989 and 2004, the most provincial departments and ministries of education gradually introduced media education to the courses in the arts and the English language at higher educational institutions.

The third period "Modern period of media education development" consists of two stages "Education reform, reflection of the best practices in media education in secondary schools and higher educational institutions in Canada" and "Modernization, further improvement of media education programs and technologies".

During this period, UNESCO has been actively involved in developing the foundations for media and information literacy and has supported UNESCO Member States, including Canada, for achieving the goals set out in the Grunwald Declaration on Media Education (1982), the Alexandrian Declaration on Information Literacy and Education Throughout Life (2005) and the Paris Summit of UNESCO (2007).

Let's consider the first stage of the third period "Education reform, reflection of the best practices in media education in secondary schools and higher educational institutions in Canada". With the growing demands for education reform (new types of literacy, updating of educational technology and communication, creation of larger educational institutions), the renewal of the teacher's role in the educational process becomes important. The teachers of the new educational structure should not only fulfil their basic responsibilities: mentoring and management of the educational process, the transfer of values and social guidance, monitoring the conditions due to which the learning process and educational communication take place, assessment of progress, and the process of managing the strategy. But the future teacher must master additional competencies:

- ability to model contexts, situations and circumstances for the educational process by using the potential of information and communication technologies;
- ability to upgrade, adapt, obtain and develop educational materials based on a multimedia basis;
- ability to organize effective communication of the school community with the use of information and communication tools and the involvement of new media;
- ability to work with information and communication technologies;
- ability to create virtual co-operative networks;
- ability to realize the psychological impact of new media on students;
- ability to adapt to new relationships related to the sharing of duties, abilities and energy that are created on the basis of educational activities as a result of the transformation and the advanced education system.

The key event in Canadian media space during this stage was the worldwide largest scientific and methodological conference "Summit 2000: Children, Youth and the Media – Beyond the Millennium", held in Toronto in 2000. The conference was attended by more than 1500 representatives from 54 countries of the world. The sponsors of the event were the leading television organizations and companies related to the media and education. The main goals of the conference were to support the effective media education of the children, youth and adult audiences, stimulate the dissemination of the best educational programs and practical methods, converge the views of media workers (producers, film and television script writers) and media education (teachers and university lecturers). Such events enable the development of cooperation between media professionals from different countries, scientific schools and technological approaches.

At the beginning of the XXIst century, two media educational associations were created. The Media Literacy Association in New Brunswick (AMLNB) was founded in 2001, and the Association for Media Literacy in Newfoundland and Labrador Media (AML-NL) in 2002. The purpose was to develop the media education system in their own provinces.

The major event for Canadian media education became the Canadian National Week of Media Literacy in 2006, with the support of the Media Awareness Network and Media Smarts and the Canadian Teachers' Federation. The main goal of the current event is to encourage teachers and students in the field of media education and support the integration of media education into curricula and programs. Now, for the twelfth year, the Week of Media Literacy is being held, which has become not only a Canadian, but also an international event in the life of media educators.

"The Understanding of Media Literacy: Inside Plato's Cave" course has been integrated since 2009 into the curriculum of Atbauche Unversity (Alberta, Canada). This course includes introducing media literacy, examples of media education programs from leading Canadian universities and core modules.

The stage "Modernization, further improvement of media education programs and technologies" had a great event for the development of media education in Canada. It was the organization "Generation of Young Canadians in the World of Internet. Phase III: Online Life" in 2014. The program focuses on the attitude of the younger generation to the global Internet network, and aims at exploring what young people are "doing online", which sites are actively using, their attitude to "online security", home rules for using the Internet, and the disconnection of electronic technologies. On the site of the presented program Media Information Network has published a large number of sources on media literacy on a variety of topics: from a gender image in the media space to "cyber bullying", marketing and consumer protection for parents, teachers and students in both the official languages (English and French).

6 Conclusions

The experience of famous Canadian media educators provides grounds for arguing that the media plays a crucial role in the Canadian education system (especially in the educational process in secondary schools and higher educational institutions).

The research that we have conducted with regard to the historical periodization of the development of media education in Canada, based on the following criteria for periodization, such as socio-economic, scientific and technical, cultural and educational development of society, allow us to distinguish three main periods. The first period covers three stages (the 1940s – the beginning of the 1970s (this is the period of the birth and establishment of media education in the Canadian education system). The second period, which also encompasses three stages, is 1972 – the beginning of the 1990s. This is the period of the media education development and its integration into the system of secondary and higher education in Canada and the third one is the 2000s – up to date. This is the modern period of the media education development, which includes two stages.

In 1959, a famous Canadian researcher created the first media education program for 11th grade students and included it into a school curriculum. It was Marshall McLuhan. In the early 1960s, the system of high quality specialists for the film and cinema education was still in the stage of formation. Nevertheless, the part-time evening courses that trained instructors for screen culture were organized on the basis of colleges and universities. The peak of media education in Canada was in the end of the 1980s, which led to the establishment of media education associations in all provinces, but media education was not compulsory in the school curriculum due to the lack of a unified state centralized education system. However, today the Canadian media literacy network has become well-known around the world thanks to the active work of the media education associations that operate throughout the country and produce educational films, publish professional journals, and offer seminars, training programs, manuals, etc. As a result of the theoretical analysis of the determined periods, we have characterized the process of establishment and development of media education in the Canadian education system. Consequently, the Canadian media education received its modern status in the process of evolution, and the established methodological and theoretical basis contributed to the consistent transition from optional to compulsory study of the media education foundations in Canadian educational institutions. And now Canada is a recognized media education leader in the world.

However, our study does not cover the problems of media education as a scientific, historical and educational phenomenon, as the study of theory and practical skills for mastering modern mass media, that are considered as part of a specific, autonomous field of knowledge in education theory and practice. To the directions of further explorations, we include a review of foreign experiments in the field of media education; an analysis of the works by Western educators on the ways of implementing media education and the achievement of media literacy; the search for common ways with other countries in the training of specialists in the field of media education; studying and implementing the experience of the media education organizations for improving media culture.

References

1. W. J. Potter, *Media Literacy. Thousand Oaks* (Sage Publication, London, 2001)
2. Literacy: A UNESCO Perspective (2003), http://unesdoc.unesco.org/images/0013/001318/131817eo.pdf. Accessed 14 Jul 2019
3. S.D. Garner (ed.), *High-Level Colloquium on Information Literacy and Lifelong Learning* (2006), https://www.ifla.org/files/assets/wsis/Documents/high-level-colloquium.pdf. Accessed 21 Mar 2019
4. R. Morgan, in *Teaching the Media: International Perspective,* ed. by A. Hart. (Lawrence Erlbaum Associates, New Jersey, 1998), pp. 145–167
5. Sh. Turkle, *Life Behind the Screen: Identity in the Age of the Internet* (Simon & Schuster, London, 1995)
6. D. Buckingham, *Watching Media Learning* (Falmer Press – Gage, Canada, 1990)
7. D. Buckingham, Child. Geogr. **5**(1–2), 43–54 (2007)
8. M. McLuhan, *Understanding Media: Intentions of Man* (McGraw-Hill, New York, 1964)
9. K. Tyner, J. of Med. Lit. **1** (146), 4 (2000)
10. N. Andersen, in: *Children and Media: Image. Education. Participation,* ed. by C. von Feilitzen, U. Carlsson (1999), pp. 139–162
11. B. Duncan, *Media Literacy Resource Guide. Ontario Ministry of Education and the Association for Media Literacy* (Queen's Printer for Ontario, Toronto, 1989)
12. G.J. Moore, *The Case for Screen Education in Canadian Schools* (Canadian Education Association, Toronto, 1969)
13. M. McLuhan, *The Mechanical Bride: Folklore of Industrial Man* (Vanguard Press, New York, 1951)
14. F.K. Stewart, J. Nuttall, *Screen Education in Canadian Schools* (Canadian Education Association, Toronto, 1969)
15. K. Uing, *What is the Internet literacy? The handbook on the freedom of mass information on the Internet* (Wien, 2004)
16. *Media literacy resource guide, intermediate and senior divisions* (Ministry of Education, Ontario, 1989)
17. A. Caron, *Centre for Youth and Media Studies* (Montreal, 1998)
18. J. Gonnet, *Education aux medias: Les Controversies fecondes* (CNDP, Hachette, Paris, 2001)

Translating children's literature of Ukrainian Diaspora as an implementation of the educational ideal of Ukrainian abroad

Maryna Vardanian*

Kryvyi Rih State Pedagogical University, Department of Foreign Languages, 54 Gagarina Ave., Kryvyi Rih, 50086, Ukraine

Abstract. This essay examines the implementation of the educational ideal of Ukrainian Abroad in the Ukrainian Diasporic translation of children's literature by such authors as Dhan Gopal Mukerji, Louis Henri Boussenard, Oscar Wilde, Daniel Defoe, Rudyard Kipling, etc. I claim that this ideal was formed in the 20th century in different cultural centers, united by Ukrainian World Congress. The Second Ukrainian World Congress approved main principles of the educational ideal in 1973, summarizing previous achievements. It is based on such human values as love, respect, and tolerance. All of these values got a Ukrainian sense. My focus in this article is on how these principles of the educational ideal are fulfilled in the translation of children's literature of Ukrainian Diaspora.

Introduction

The subject of the Ukrainian educational ideal is relevant for Ukrainian modern society. This issue is enunciated in *The New Ukrainian School*, the policy document of the Ministry of Education and Science of Ukraine, the conceptual principles of which are based on the viewpoint that Ukrainian school leaver is a personality, patriot, and innovator. The first element includes the notion of a fully developed all-round personality, capable of critical thinking. The second element relates to the development of a patriot taking an active role, acting according to ethical principles and capable of making responsible decisions, respecting the dignity and the rights of people. The third element reveals a school leaver's ability to change the world around him / her. Thus, these components need a meaningful filling. In this regard, teachers are given a creative freedom to choose approaches for that purpose [1].

For the realization of the second component, namely the development of a patriot, it is considered expedient to examine the experience of Ukrainians abroad. Living in exile, outside of Ukraine, these Ukrainians remained the patriots of their country, and actively implemented the concept of the educational ideal at different levels of material and spiritual culture. These are Ukrainian cultural centers, schools, universities, archives, museums, churches, and a creative heritage (such as pedagogical articles, literature, including translated literature for children, etc.).

The goal of the article is to trace the implementation of the educational ideal of Ukrainian in the translated children's literature.

1 Bibliography analysis on the issue

The issue was not a special subject of research of Ukrainian scientists. Some of its aspects were considered in the monograph *The Self – the Other in the Children's Literature of the Ukrainian Diaspora: National Sphere of Concepts, Imagological Models* [2]. Moreover, O. Dzhus studied the national and patriotic education of children and youth in Ukrainian Diaspora in her *National Patriotic Education of Children and Youth in Ukrainian Diaspora: Experience, Attempts of Rethinking* [3]. The researcher examined the issue of the educational ideal of Ukrainian abroad in the works *Nationalism and Internationalism in an Educational* by S. Rusova, *The Educational Ideal* by H. Vashchenko, *Native Language and Intellectual Development of Children* by V. Simovych, *The Psycho-didactic Aspects of Ukrainian Education in Diaspora* by O. Kulchytskyi, *Ukrainian Nature and our Educational Ideal* by V. Yaniv, etc. They were mainly Ukrainian teachers, and scientists, who emigrated during the second wave (between the two world wars in the 20th century), or the third wave after World War II (these emigrants were named the generation of the Second World War). Hence, issues of their heritage include the study of the place and role of the native language; national and cultural heritage in the education of the young generation of Ukrainians; building national school; comprehension of foreign pedagogical ideas, etc.

Similar issue was the subject of the dissertation *Ukrainian Pedagogical Press of Canada as the Factor of Development of Native Language Education in Diaspora (second half of the 20th century)* by H. Byhar. There, she examines the development of Ukrainian school and pedagogical idea in Ukrainian Diaspora of Canada [4]. In particular, the researcher analyzed the Ukrainian Diasporic teachers' works, which highlighted the role of native language for human, society, and nation, its purpose for ethnic community in Diaspora.

Another aspect of my research is a collaboration of children's literature and translation. Actually, it is a general scientific issue that was not detailed in Ukrainian scientific works, but some of its components were explored by foreign researchers such as Emer O'Sullivan in *Comparative Children's Literature,* Ritta Oittinen in *Translating for Children,* G. Klingberg in *Children's Fiction in the Hands of the Translator,* etc. The Routledge Companion to translation studies published the collection entitled *Handbook of Translation Studies*.

Gillian Lathey, and Cecilia Alvstad studied the issue in the articles of the handbooks. "Translation of children's literature is characterized by a series of traits", emphasized C. Alvstad. Among these, she defined the

* Corresponding author: maryna.vardanian@gmail.com

most common subjects treated by scholars in the field: cultural context adaptation, and ideological manipulation [5]. The approach named 'cultural context adaptation' by G. Klingberg includes domestication of foreign names, coinage and foodstuffs, literary references, foreign languages, historical background, flora and fauna, and other culture-specific phenomena.

Translation strategy of ideology in the adoptation attracted the attention of a number of scholars. "Fernández López (2000) discusses intercultural ideological factors in the translation into Spanish of the work of Roald Dahl and Enid Blyton. The eradication of racist and sexist language from the work of these authors during the 1970s and 1980s was ignored in Spanish translations that returned to earlier, 'unpurified' versions of source texts, thereby creating a mismatch between English and Spanish versions published in the same period," said G. Lathey in her *Children's Literature* [6]. Such adaptations called "purification" by G. Klingberg, is translation which is adapted to adhere to the adults' (parents', teachers', etc.) supposed sets of values [5].

Here, I will not discuss all of these works in detail. It is a subject of another research. On the one hand, national and patriotic education was a cross-cutting and important issue for Ukrainian emigrants, who sought to preserve their national identity outside of Ukraine. On the other hand, Ukrainian translators used both approaches, which I mentioned before.

Therefore, in this article I wonder about the principles of the educational ideal that were formed by the Ukrainian Diaspora and how they were implemented in the translated children's literature.

In this connection, it should be emphasized that I chose the translation of children's literature of Ukrainian Diaspora for several reasons. *Firstly*, the formation of personality begins from childhood. In this process, children's literature played the major role for Ukrainian Diaspora. *Secondly*, Ukrainians abroad were learning the world's best experience, using its practices through the translated literature. *Thirdly*, Ukrainian Diaspora interpreted human values of the best world writings for children, using Ukrainian senses.

Thus, the translators realized in translated children's literature the national and spiritual basics of Ukrainian identity or mentality, their vision of the world picture. To this effect, my essay consists of two parts, following as:

(1) the concept of the educational ideal of Ukrainian, its formation in Diaspora;

(2) the implementation of the educational ideal of Ukrainian in translated literature for children and youth.

2 The formation of the concept of educational ideal of Ukrainian Abroad

In different periods of the 20th century, the Ukrainian Diaspora established different centers in the countries of its new settlement. (Fig. 1). Cultural community consolidated in these centers, which included such organizations as:

- *educational institutions* (in particular, kindergartens, Ukrainian Saturday Schools, universities such as Ukrainian Free University, and Canadian Institute of Ukrainian Studies),
- *scientific institutions* (in particular, Ukrainian Free Academy of Sciences, Shevchenko Scientific Society, etc.);
- *youth unions and movements*, including Plast (Ukrainian Scouting Organisation), CYM (Ukrainian Youth Association), etc.;
- *women organizations* (for example, Union of Ukrainian Americans, Canada, etc.);
- *literary associations* (in particular, Ukrainian Writers' Association in exile "Slovo", Leonid Hlibov Association of Children's Literature).

Fig. 1. The cultural centers in Munich (top – the building of Ukrainian Free University; middle – the building of "Ridna Schkola" ("Native School"); bottom – the building of Exarchy's Apostle)

The primary task of these centers was to preserve national, cultural and religious identity, crafts and traditions of Ukrainian people. These organizations unite into the Ukrainian World Congress. The Ukrainian World Congress has the World Coordination Educational Council, which aim is to coordinate the educational process in the Diaspora.

Here, I primarily turn to Leonid Hlibov Association of Children's Literature, because it is related to both notions of the article, including the educational ideal and translating literature. First of all, the Association was founded at the conference of writers and teachers, held on March 26-27, 1946 near Munich (Germany). As provided in the association's Statute, its members should create a magazine for children and periodicals on issues about children's literature and pedagogy; publish original (native) and translated children's books; call for teachers, parents and community to unite and work for Ukrainian children (taken from the collection of

TsDAZU in Ukraine – The Central State Archive of Foreign Ukrainica).

The association's members were both writers and teachers, and all of them worked for the same purpose, namely the formation of the educational ideal of Ukrainian abroad.

The pedagogical line was implemented, in particular, in the magazine *Ukrainian School*. In addition to the resolutions of Leonid Hlibov Association of Children's Literature, it covered information on methodological teacher's conferences and congresses, as well as statistics of Ukrainian schools in Germany at that time.

However, its main aim was to discuss various issues in theory, history and methodology of learning. The magazine was published in 1947-1948 in Augsburg, Germany. Now it is a rarity edition and a valuable find for researchers of the history of Ukrainian pedagogy (Fig. 2).

The national and patriotic issues were widely discussed in the magazine, in particular, in such articles and reports as *A teacher-national and the principles of his / her work* by prof. Leonid Biletskyi (*Ukrainian School*, 1947), *Educational issue of our nowadays* by prof. Borys Lysianskyi (*Ukrainian School*, 1948), *The Concept of Ukrainian Human as the Educational Ukrainian Ideal* by prof. Oleksandr Kulchytskyi (*Ukrainian School*, 1948). Thus, according to the emigrants of that time, the Ukrainian educational ideal is "a universal human ideal taken through Ukrainian perception," O. Kulchytskyi says [7].

Another magazine by Leonid Hlibov Association of Children's Literature is *We and Our Children*. The newsletter was first published in the second half of 20th century in Toronto, Canada and New York, USA. Here, the issues of literary text were considered through the moral, ethical, national and patriotic upbringing (Fig. 3).

The children's literature had a special role in this magazine. In particular, it was considered as a spiritual and patriotic inspirer for Ukrainian children and young adult of the Diaspora. Actually, the literature was called the "literature with mission", which aim was to educate the true patriot of Ukraine's future.

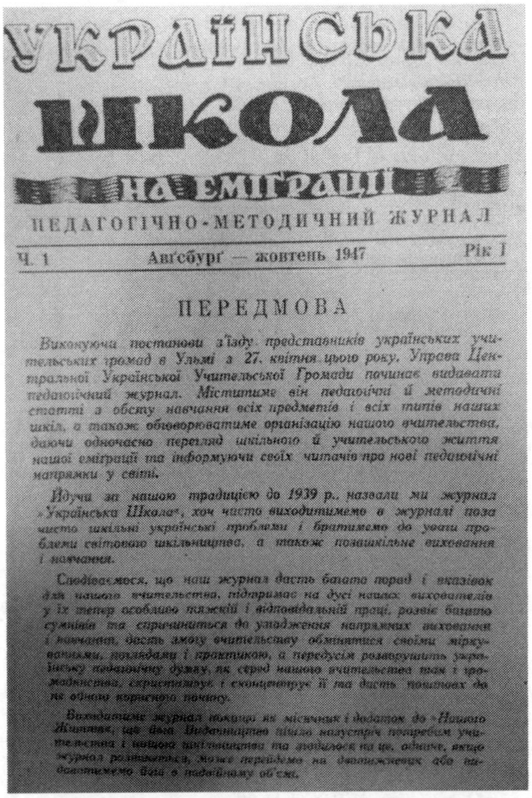

УКРАЇНСЬКА ШКОЛА НА ЕМІГРАЦІЇ

ПЕДАГОГІЧНО-МЕТОДИЧНИЙ ЖУРНАЛ

Ч. 1 Авґсбурґ — жовтень 1947 Рік I

ПЕРЕДМОВА

Виконуючи постанови з'їзду представників українських учительських громад в Ульмі з 27. квітня цього року, Управа Центральної Української Учительської Громади починає видавати педагогічний журнал. Міститиме він педагогічні й методичні статті з обсягу навчання всіх предметів і всіх типів наших шкіл, а також обговорюватиме організацію нашого вчительства, даючи одночасно перегляд шкільного й учительського життя нашої еміграції та інформуючи своїх читачів про нові педагогічні напрямки у світі.

Йдучи за нашою традицією до 1939 р., назвали ми журнал »Українська Школа«, хоч часто виходитимемо в журналі поза чисто шкільні українські проблеми і братимемо до уваги проблеми світового шкільництва, а також позашкільне виховання і навчання.

Сподіваємося, що наш журнал дасть багато порад і вказівок для нашого вчительства, підтримає на дусі наших вихователів у їх тепер особливо тяжкій і відповідальній праці, розвіє багато сумнівів та спричиниться до упрядження напрямних виховання і навчання, дасть змогу вчительству обмінятися своїми міркуваннями, поглядами і практикою, а передусім розворушить українську педагогічну думку, як серед нашого вчительства так і громадянства, скристалізує і сконцентрує її та дасть поштовх до не одного корисного почину.

Виходитиме журнал покищо як місячник і додаток до »Нашого Життя«, [illegible] видаватимемо його в подвійному об'ємі.

Fig. 2. The title of the first number of the magazine *Ukrainian School* (Germany, 1947)

Fig. 3. The title of the newsletter by Leonid Hlibov Association of Children's Literature (Toronto, 1969)

One of the ideologists of Leonid Hlibov Association of Children's Literature B. Hoshovskyi in his articles proclaimed that "the soul of a Ukrainian child" is a fundamental value of Ukrainian nation in children's literature of Ukrainian Diaspora, and "a writer is the creator of the soul" [8].

This point was reflected in the choice of heroes and motifs. The images had to implement the aesthetic concept of human in the artistic world of Ukrainian Diasporic writers. This aesthetic concept was based on the idea of the educational ideal of the Ukrainian. Similarly, motifs had to express themselves in the preservation and continuation of the national idea. Therefore, the educational idea and poetic form were combined in the writings of Ukrainian Diasporic children's literature. Such combination means, on the one hand, the preservation of Ukrainian literary traditions, keeping in touch with the historical Motherland, on the other hand, it is an ensuring continuity and sustainable development of the Ukrainian idea.

Actually, the idea of the educational ideal of the Ukrainian in the Diaspora was approved at the Second World Congress of Free Ukrainians (Toronto, Canada, 1973). However, the work on the concept had started in 1967 by the just-established World Coordination Educational Council. According to the materials of the Second World Congress of Free Ukrainians (Toronto – New York, 1986) the idea of the educational ideal is based on the following principles:

1) love for the native Ukrainian language and culture heritage;
2) respect to the national and personal dignity of Ukrainians;
3) Christian tolerance as the antithesis of class struggle;
4) ethical and moral values as the antithesis of the current trends of materialism, atheism, nihilism;
5) respect and tolerance to the cultures of the new settlement.

This idea was detailed in the monograph *Parents and Children* by Antonina Horokhovych, the leader of the teachers' movement of the Ukrainians of Canada, and writer. As the researcher concludes, "the purpose of Ukrainian education is a human, who is nationally conscious, has a strong character, is not assimilated, but spiritually rich, all of which is drawn from ancestral culture. As the spiritually rich personality, he / she is the bearer of the idea, and lives with understanding that he / she needs to share his / her spiritual goodness with others" [9]. Together with Ukrainian Diasporic writers, A. Horokhovych emphasized that the educational ideal of the Ukrainian was reflected in literature.

3 The implementation of the educational ideal of Ukrainian in translated literature for children and youth

Translating children's literature had an important role for Ukrainian Diaspora of the 20th century, who actively translated the world literature. When Diasporic Ukrainians were outside of their Motherland, they discovered the culture and language of the countries of their new settlement. The new reality influenced the writers' works. On the one hand, these writings reproduced the understanding of the world of the other culture, and on the other hand, they were related to the idea of the abandoned Motherland, and discovered the Ukrainian mental, national and spiritual world.

In its connections with the culture of various nations, the literature played several roles. B. Hrinchenko talked about two of them in his time. The writer wrote the article *Press for Children*, in which he reflected about the role of translation in children's magazines. The researcher emphasized that translated works, on the one hand, "discovered the life of other nations," and on the other hand, they directed Ukrainian writers "in learning from world classics" [10].

Moreover, translating literature in exile could discover for young reader the world of another culture as well as the national identity of Ukrainian. It was its third role. Because the internal factors of the translated literature of Ukrainians abroad were to discover another culture and to find the national identity. Evidently, the choice of texts for translation was guided by the following conditions: first, the text had to be close to the moral, ethical, national and spiritual values of the Ukrainians, and to their idea of educational ideal; second, the artistic work needed to be popular in the "cultural world" as well as to be translated into different languages.

With the first condition is connected the issue of understanding, i.e. how the translator understands the author of the original text. It is met, when intentions and meanings of original and translation are the same. With regard to the second thought, it appears that Ukrainians wished to promote the Ukrainian language as distinctive and equal with others, and to attract the attention of children to reading the best books in their native language.

Primarily, Ukrainian writers in exile translated literature to promote their native language, which they called their **mother tongue**. In Ukrainian context, A. Horokhovych related three elements with language: national prestige, the soul of people, and their belonging to the family as well as the mother (that is why the Ukrainian language is called the mother's tongue) [9].

In particular, the publisher Yurii Teodorovych suggests in the preface of the translated book *Gay Neck, the Story of a Pigeon* by Dhan Gopal Mukerji that Ukrainian translation will help young adults to "learn more about the native language of Ukrainians, their nature and spirit" [11].

The book was translated by Stefaniia Nahorna and published in 1961 in Chicago, USA. It tells about interesting adventures and implements values such as humanity and brotherhood.

The two-volume edition of *The Fairy Tales* (1956 – 1958) by Wilhelm Hauff, published by the community "Tryzub" in Winnipeg, Canada, had the same aim. Reading these books, children of Ukrainian Diaspora could enlarge their Ukrainian vocabulary. In the fairy tale *Little Muck*, translators reproduced a teasing rhyme in Ukrainian styling. They used Ukrainian vocabulary

https://doi.org/10.1051/shsconf/20207501002

and phraseology, reproducing it in the tradition of Ukrainian children's folklore (Fig. 4).

Як тільки він виходив на вулицю, то ми підіймали
такий галас, аж в ухах лящало.

— Малий Мук! Малий Мук! — Ми навіть зложили про нього веселий вірш, який ми, бувало гуртом виспівували:

„Малий карлик, Малий Мук!
Гарний з тебе парубчук
Голова завбільшки в гору,
А штани, мов Чорне море…
Сам живеш в великій хаті
І виходиш погуляти
Раз на місяць. Озирнись бо,
Глянь на нас та побіжи бо —
Чи не вловиш. Карлик Мук
Любий, гарний парубчук!

Fig. 4. The illustration of the fairy tale *Little Muck* by Wilhelm Hauff with a teasing rhyme

In their adaptations and translations, translators often used such Ukrainian culture-specific elements as common national lexis. Each fiction piece for children had a national styling, which intended to implement the educational idea with the regard to the native language and Ukrainian culture.

Factually, these artistic translations for children and youth were based on Ukrainian cultural tradition. Revealing before children previously unknown horizons and new cultures, the translators presented the experience through **the Ukrainian national vision**. Therefore, most translations for children are pervaded with the following motifs: (1) unity of the Ukrainians in the struggle for the historical Motherland; (2) the exile and search of a better fate. In general, Ukrainian Diasporic children's literature has these motifs.

The realization of the educational idea of respect to **the national dignity of Ukrainians** became the dominant for most translations of children's literature. It was achieved by addressing to the important topic of the struggle of Ukrainians for the independence of their own country, colonized during centuries.

Foreign fairy tales received reinterpretation in Ukrainian context. In particular, the novel *Robinson Crusoe* by Daniel Defoe is an adaptation, published by "Hoverla" in 1956, New York. According to aesthetic taste of the translator, the adaptation is "a didactic novel for young adults". The translator of this novel is M. Sydir-Chartoryiskyi, the author of works on educational issues and the founder of the publishing house. His translated novel has Ukrainian styling. In particular, Daniel Defoe is named Danyla. It is evident that Defoe's views on the reforming of society were close to the translator's, because the latter called the writer "a poet of truth and justice."

Ukrainian translation of animal stories by the English author Rudyard Kipling also has national particularities. There were translated two short stories from Kipling's collection *The Jungle Book*, namely the story about the boy or "man-cub" Mouwgli, published in Ukrainian in 1947 by the title *Mouwgli's Fellows*, and the story about the adventures of a valiant young mongoose, titled *Rikki-Tikki-Tavi* (1948). Here, the Ukrainian national features are highlighted with language codes such as "free", "kin", "struggle", etc. The translator actualized the motifs of unity, harmony, and conflict resolution between Ukrainians.

Ukrainian character is expressed in the short story *Captain Daredevil* (in French *Le Capitaine Casse-Cou*) by Louis Henri Boussenard translated in Ukrainian as *The Fighters for Freedom* ("Bortsi za Voliu", New York, 1965). Here, the translator, whose name is unknown, freely unfolds the plot of the original text. It is demonstrated even by the translated title of the writing.

If we compare the original and the translated texts, we can see a disparity. The original text portrayed a daredevil captain, who took part in the Boer War against the British rule. However, Ukrainian translation emphasized the struggle for freedom. Similar national struggle for Ukrainian independence occurred in 1917-1921. Therefore, this idea was the most important for Ukrainians in exile.

There are some analogies with Ukraine in the adaptation, particularly in the struggle of the Ukrainian people for their independence against various forms of colonization (economic, politic, cultural and language assimilation), which was described through the Boer-English war.

Such analogy demonstrates the map at the beginning of the adaptation (Fig. 5). The translator wrote: "Today the whole South Africa is colonized by England, though it has its own parliament and self-government. Now look up! There you will see our Black Sea and a strip of our motherland. It is Ukraine! Reading this book guess in it your Ukraine!" [12].

Similar ideas can be found in the collection *The World of Big-Hearted Animals* translated from English by Svitlana Kuzmenko and published by Ukrainian Women's Organization of Canada (Toronto, 1995). The fairy tales *The Big-Hearted Moose Tidwick* and *The Clumsy Elephant* by the American writers are translated close to the ideas of the desire of Ukrainians for their statehood and respect for national dignity.

V. Barahura highlighted the specifics of Kuzmenko's translations in his review of the mentioned book. The researcher emphasized that both short stories had implemented Ukrainian idea. In particular, the first story about the elephant reflected the ideas of "national value", and "unity of people in the nation, their strength of spirit." In the second fairy tale about the moose, Barahura reflected on the restoring of historical justice, and summarized: "The story reminds us about the historical fate of our nation, particularly the occupation of our mother land by the neighbors, lack of national unity, and the exploitation of our historical land. Four

https://doi.org/10.1051/shsconf/20207501002

years ago, Ukraine finally got rid of the occupiers and proclaimed its independence. Since then Ukraine has been leading an independent national-state life among free nations" (taken from the collection of TsDAZU in Ukraine – The Central State Archive of Foreign Ukrainica).

Fig. 5. The illustration of a map from the translated book *The Fighters for Freedom* ("Bortsi za Voliu") by Louis Henri Boussenard

Another national issue is the returning to the Motherland. This sacred dream of the Ukrainians in exile is also implemented in their adaptations and translated works for children. The elder generation believed that Ukrainian Diasporic young adults would eventually return to the Motherland to make Ukraine independent. The novel for youth *Little Lord Fauntleroy* by the English-American writer Frances Hodgson Burnett (published in 1958, Winnipeg and translated by V. Litynska) also has similar motifs.

In the original, the author describes a character, young Cedric Errol who lives with his mother in New York City in genteel poverty. They are here in exile after the death of Cedric's father. Soon the boy returns to the "father's homeland" in England. There he must be a good master for his native land. Although the events happen in America and England, the Ukrainian reader perceives "his / her own" text: to follow the trade of the father and return to the motherland.

For Ukrainian Diaspora national and religious issues are interlinked. The works on religious issues highlighted **the idea of Christian tolerance**. Significantly, Ukrainian Diaspora has given a special place to the issue of faith, adding it to the basic elements of Ukrainian spirituality. In particular, H. Vashchenko in *The Traits of Our Human Ideal* includes the following: "strong faith in God, the devotion to the will of God, faith in providence, humility, love for one's neighbor, honesty, truthfulness, loyalty to the oath, love for the Motherland, bravery, love for work and knowledge" (from archives of Ukrainian Free University, the collection by H. Vashchenko).

Such senses fill Oscar Wilde's fairy tales. The fairy tales *The Star-child* and *The Selfish Giant* were translated by M. Lototskyi and published in 1947 by "Ukrainske Slovo" (Regensburg, Germany). The works promote such Christian virtues as kindness, mercy, and empathy. With the images of children, giants, beggars, and wizards are revealed the human vices of cruelty, indifference, selfishness, and pride.

The same as in the original text of *The Selfish Giant*, the translator shows the image of Jesus Christ and recreates the idea of atonement and paradise. However, the comprehensibility of the translation for Ukrainian reader is achieved through Ukrainian lexis and reproduction of the culture-specific elements. Even illustrations to the book have Ukrainian coloring and stylistics. Thus, children are portrayed in Ukrainian national clothing (Fig. 6).

Fig. 6. The title of the translated book *The Selfish Giant* by Oscar Wilde

Both translated books were published in 1947 after the Second World War. At that time Ukrainian emigrants and other nations experienced the terrible impact of war. Therefore, the issues of charity, peace and tranquility in a human soul were relevant for Ukrainians and for the world in general, thus creating a new system of **moral and ethical values**.

The issues of post-war period such as the traumas of war, fear, self-destruction and mental illness are discussed in the book *Gay Neck, the Story of a Pigeon* by Dhan Gopal Mukerji. The first edition in Ukrainian was published in Regensburg, Germany in 1947, and the second one, called in the preface by Y. Teodorovych as "the Nobel pedagogical and artistic work" was translated by S. Nahirna in 1961.

Besides the purpose of learning the Ukrainian mother tongue, the translation discovers the Ukrainian attitude towards evil, hatred, war, courage, peace and truth in tune with the ideas of the Indian writer. The translation embodies the idea of solidarity, which is implemented through understanding of the traumatic experience of the other people as the result of the war. At the same time, the translator expressed the idea of creating "bridges" to overcome alienation and misunderstanding.

Already mentioned S. Kuzmenko in his other translation of the collection *The World of Big-Hearted Animals* creatively implemented **the idea of tolerance**. Each story is a story of another, distinguished by its otherness. Thus, the texts interpret the meaning of the uniqueness of each being. Relations between different cultures are realized in the idea of respect for different cultures.

The issue of respect for the kin and the family took the leading place in the system of moral and ethical values of Ukrainian Diaspora. The theme is interpreted by L. Bryzghun-Shanta in her free translations from English and French. The titles of the works are self-revealing: *The Beavers' Family*, *The Ducks' Family*, and *The Bears' Family*. The main characters in her stories are families of animals or birds, and ducks even get the nickname Kachurovski (from Ukrainian "kachur", which means "duck").

There the author shows the habits of animals and birds, which distinguishes animal stories among other genres of children's literature. Through her small characters L. Bryzghun-Shanta highlights the issues of respect and reverence for elders, obedience of children, mastering the experience of adults. These issues reproduce the traditions of Ukrainian family upbringing.

Conclusions

To conclude, I emphasize that the principles of national and patriotic education were formed and implemented through the "concept of the educational ideal of the Ukrainian abroad".

1. The idea is based on the following principles: love for the native Ukrainian language and culture heritage; respect to the national and personal dignity of Ukrainians; respect and tolerance to the various cultures; Christian tolerance; ethical and moral values. They were at the center of the pedagogical reflections by H. Vashchenko, S. Rusova, O. Kulchytskyi, L. Biletskyi, A. Horokhovych, etc.

2. Particularly, the educational idea was implemented in translation of children's literature by such writers as Dhan Gopal Mukerji, Louis Henri Boussenard, Oscar Wilde, Daniel Defoe, Rudyard Kipling, etc. Such translations and adaptations are filled with the Ukrainian national colouring, including the names of heroes, common national lexis, passages from Ukrainian children's folklore, illustrations, etc.

3. Ukrainian translators have implemented the idea through the formula "universal values in the national sense." There was a specificity in translated children's literature. The Ukrainian native language or mother tongue, national dignity, Christian tolerance and cultural solidarity are those components of the educational ideal that are worth adoption by modern youth in Ukraine.

References

1. M. Gryshchenko (ed.), *The New Ukrainian School* (Ministry of Education and Science of Ukraine, Kyiv, 2016), https://mon.gov.ua/storage/app/media/zagalna%20serednya/Book-ENG.pdf. Accessed 21 Mar 2020
2. M. Vardanian, *Svii – Chuzhyi v Ukrainskii Diaspornii Literaturi dlia Ditei ta Yunatstva: Natsionalna Kontseptosfera, Imaholohichni modeli* (The Self – the Other in the Children's Literature of the Ukrainian Diaspora: National Sphere of Concepts, Imagological Models). (Dionat, Kryvyi Rih, 2018)
3. O. Dzhus. Science and Education. **10**, 76 (2016)
4. H. P. Byhar, Dissertation, Yuriy Fedkovych Chernivtsi National University, 2006
5. C. Alvstad, in *Handbook of Translation Studies*, vol. 1, ed. by Y. Gambier, L. van Deorslaer (John Benjamins Publishing Company, Amsterdam, 2010), pp. 22–23
6. G. Lathey, in *Routledge Encyclopedia of Translation Studies*, ed. by M. Baker, G. Saldanha (Routledge, London New York, 2009), p. 33
7. O. Kulchytskyi. Ukrainian School. **5–6**, 23 (1948)
8. B. Hoshovskyi, *Ukrainska Dytiacha Literatura Sproba Ohliadu i Problematyka* (An Outline of Ukrainian Children's Literature). (Ass'n of Ukrainian Writers for Children, Toronto, 1966)
9. A. Horochowych, *Batky ta Dity* (Parents and Children) (World Coordination Educational Council, Winnipeg Toronto, 1990)
10. B. Hrinchenko. We and Our Children **1**, 261 (1965)
11. D.G. Mukerji, *Pryhody Khorobroho Holuba* (Gay Neck, the Story of a Pigeon) (Ukrainian – American Publishing and Printing Co., Chicago, 1961)
12. L.H. Boussenard, *Bortsi za Voliu* (The Fighters for Freedom). (OKO, New York, 1965).

Implementation of the competence idea into the primary education in Ukraine – the reflective analysis based on the European Union benchmarks

Oleksandra Savchenko[1], and *Olena* Lokshyna[2,*]

[1]Institute of Pedagogy of the NAES of Ukraine, Primary Education Department, Kyiv, 04053, Ukraine
[2]Institute of Pedagogy of the NAES of Ukraine, Comparative Education Department, Kyiv, 04053, Ukraine

Abstract. The generation of the competence education idea in Ukraine based on the EU benchmarks is studied. It is underlined that the transformation of the national education into the competence base is a component of the European integration of Ukraine. The authors made conclusion that the Ukrainian education has gone from proclaiming the idea of the competence education, its conceptual justification to implementation. The research results evidence that the support of the competence idea at the level of educational policy is the most successful in Ukraine. At the same time, at the implementation level the competence idea meets a number of challenges due to the novelty of a clear correlation between strategic, didactic and methodological levels. The authors conclude that the problem of the formation of key competences, firstly of transversal or cross-curricular ones, remains underdeveloped; the aspects of cross-curricular links and integration of the education content requires further development; the level of assessing younger students in competencies is even more underdeveloped. On the other hand, the authors generalise that common globalization challenges, powerful integration processes in Europe, Europeanization affecting the European countries outside the EU borders contribute greatly to the synchronization of the education in Ukraine with EU standards.

1 The competence education as the response of the European Union to the challenges of the globalised world

Globalization and digital revolution have determined the need to transform the conceptual foundations of education – a knowledge model that had dominated in Europe for centuries could no longer satisfy the rapidly changing demands of the European societies. The competence-based education became an answer. "The shift to a digital, knowledge-based economy, prompted by new goods and services, will be a powerful engine for growth, competitiveness and jobs. In addition, it will be capable of improving citizens' quality of life and the environment… Every citizen must be equipped with the skills needed to live and work in this new information society" was stated in the Lisbon European Council Presidency Conclusions as of March 23 and 24, 2000. The Lisbon European Council proclaimed the necessity to develop a European framework of basic skills to be provided through lifelong learning as a key measure in Europe's response to globalisation and the shift to knowledge-based economies [1].

The process of the implementation of the competence idea in the European Union (EU) is long and multidimensional. The adoption of the European framework was proceeded by thorough research on the origin [2] and the essential characteristics of competences [3].

The discussions on the list of key competences needed for citizens of the modern world have been led by the leading international organizations. They are the Council of Europe with W. Hutmacher's Symposium Report "Key competencies in Europe" (1996) within "Secondary Education for Europe" Project; Organisation for Economic Cooperation and Development with "Definition and Selection of Competences (DeSeCo)" Project (1997-2003); the European Union "Key Competencies. A developing concept in general compulsory education" Eurydice Report (2002) [4], [5], [3].

The research and discussion period was completed with the adoption of the European Reference Framework of Key Competences for Lifelong Learning in 2006, which transformed the conceptual foundations of the European education. It is about changing the understanding of the mission of learning from the knowledge transition to the formation of individuals' personal characteristics of universal nature to be applied in a variety of rapidly changing contexts – individual, professional, social. To this end, not only knowledge but also skills / attitudes, as well as the ability to apply them in real life, are proclaimed as the components of the key competences.

* Corresponding author: luve2001@hotmail.com

The 2006 Reference Framework sets out eight key competences: communication in the mother tongue; communication in foreign languages; mathematical competence and basic competences in science and technology; digital competence; learning to learn; social and civic competences; sense of initiative and entrepreneurship; and cultural awareness and expression [6].

The adoption of the 2006 Reference Framework initiated the transformation of education into a competence basis in the EU Member States. In 2012 the monitoring results of this transformation were presented in the Eurydice Report "Developing Key Competences at School in Europe: Challenges and Opportunities for Policy" [3]. The methodological framework applied for the analysis can be regarded as valuable for the Ukrainian education that is under the same transformation.

In a highly competitive environment of the globalized world, the EU is making efforts to preserve its role as an economic leader, considering education as a key tool for building the knowledge economy. An updated European Reference Framework of Key Competences for Lifelong Learning was approved by a Council Recommendation as of May 22, 2018. Reflecting the new needs of the European societies, new emphasis is placed on the list of key competences and their interpretation. The 2018 Framework sets out the following eight key competences with a view to respond education to new political, economic and social realities: literacy competence; multilingual competence; mathematical competence and competence in science, technology and engineering; digital competence; personal, social and learning to learn competence; citizenship competence; entrepreneurship competence, cultural awareness and expression competence [7].

When analysing the 2018 Framework, you should pay attention to the following important points:

- multidimensionality of competences as efficient tool for modern life: the EU's position on understanding competence as a combination of knowledge, skills and attitudes was reaffirmed.
- lifelong and life wide perspective for competences acquisition: under the importance of lifelong learning, the idea of acquisition of key competences in a lifelong learning perspective, from early childhood throughout adult life, and through formal, non-formal and informal learning in all contexts, including family, school, workplace, neighbourhood and other communities is important;
- continuity in competences acquisition: a competence is defined as an open entity that an individual can develop within lifelong learning.

Thus, the adoption of the updated European Reference Framework for Key Competences for Lifelong Learning (2018) confirms the position on the invariability of competency education for EU – "In the knowledge economy, memorisation of facts and procedures is key, but not enough for progress and success. Skills, such as problem solving, critical thinking, ability to cooperate, creativity, computational thinking, and self-regulation are more essential than ever before in our quickly changing society. They are the tools to make what has been learned work in real time, in order to generate new ideas, new theories, new products, and new knowledge" stated in the Council Recommendation of 22 May, 2018 on key competences for lifelong learning [7].

The impact of this position on other European countries, including Ukraine, is equally important.

2 Methodology and literature review

The problem of competences has been essential for the Ukrainian education for almost twenty years, since the introduction of the competence idea at the European level. During this time, the Ukrainian scholars have made a solid input to the development of the competence education theory. In particular, the works of N. Bibik [8], I. Ermakov [9], L. Khoruzha [10], V. Lugovyi [11], O. Pometun, I. Rodigina, O. Savchenko, L. Vashchenko and others have enriched the understanding of the essential characteristics of competences, the nature of their origin.

The Ukrainian comparative educators have worked extensively on the dissemination of the European experience of the transformation of education into the competence base. For the first time O. Ovcharuk presented the European vision of the competence education in 2004 [13].

A long period of reflection and discussion has contributed to the systematization of the views of scholars, enriched the initial vision of the essence of the competences in the Ukrainian realities, and initiated the development of the technologies of the transformation of the national education on a competence basis.

The aim of the article is the reflective analysis of the competence education development in Ukraine from the perspective of the EU approach. The study was conducted through the lens of the EU analytical framework, based on the benchmarks proposed in the Eurydice Report "Developing Key Competences at School in Europe: Challenges and Opportunities for Policy" (2012) [14]. In particular, the following benchmarks have been applied:

(1) Competence idea support at the education policy level,

(2) New competence-based curriculum and

(3) Assessment of students in key competences.

The benchmark (1) "competence idea support at the education policy level" has been used for the analysis of the strategic/legislative/regulatory competence innovations; and the benchmarks; the benchmarks (2) "the new competence-based curriculum" and (3) "assessment of students in key competences" – to analyse the degree of the competence idea implementation into the educational practice in Ukraine. Together, all three benchmarks made it possible to view an aggregate picture of the implementation of the competence approach in Ukraine. Besides, the degree of synchronization of this process in Ukraine with the EU was analysed by the authors.

The data of the Eurydice Report "Developing Key Competences at School in Europe: Challenges and Opportunities for Policy" (2012) represent the state of implementation of the competence idea in the EU Member States after the European Framework adoption in 2006. In Ukraine, the systematic transformation of the national education into the competence base was initiated after the adoption of the Law of Ukraine "On Education" in 2017. Against this background, the benchmarks of the 2012 Eurydice Report appear to be valid for Ukraine.

The choice of the level of primary education in Ukraine is due to the fact that it is from the primary education level that the cardinal reform "New Ukrainian School" has been initiated after the adoption of the Law of Ukraine "On Education" (2017) [15].

The sources used for the analysis provide qualitative evidence in the forms of strategic benchmarks and legislative acts, existing national/international studies and reports. The selection of the sources depended on their appropriateness to the research idea of the paper.

In order to reach the aim of the paper the qualitative methods were used. The method of the literature review was used to find how scholars addressed the research problem. The analytical framework method was used to limit the scope of the relevant data by focusing on the selected benchmarks. The comparative method provided a base for the comparison of the competence idea implementation in EU and in Ukraine as well as of evaluation of the degree of synchronization of this process in Ukraine with the EU.

3 The transformation of the primary education in Ukraine into the competence base under the European integration

Ukraine has chosen the path of the European integration since independence declaration in 1991. Accordingly, the national education has been developing in the direction of synchronization with the European standards for almost thirty years. A competence idea has become one of the benchmarks of such synchronization.

Discussion on the competences for the national education started within the UNDP Ukraine projects "Education Innovation and Renewal for Improved Wellbeing and Poverty Reduction" (2002-2003) and "Education Policy and Peer Education" (2004). In the books "Reform Strategy for Education in Ukraine: Educational Policy Recommendations" (2004) and "Competence-based Approach in the Modern Education: World Experience and the Ukrainian Prospects (Education Policy Library)" (2004) the Ukrainian scholars under the umbrella of the Ministry of Education and Science of Ukraine launched the process of definition and selection of competences for the Ukrainian school [13], [8].

The Law of Ukraine "On Education" (2017) finalized the long-standing movement of the Ukrainian education towards the implementation of a competency idea. The Law declares that the goal of the secondary education is a comprehensive development, upbringing and socialization of a person capable to live in a society and interact with nature, to have a desire for self-improvement and life-long learning, ready for conscious life choices and self-realization, responsibility, work activity and civic activity. The achievement of this goal is ensured through the formation of key competencies [15].

The primary school in Ukraine, as an integral part of the general secondary education, is also under transformation into a competence base. The State Standard of Primary General Education 2000 (being active for more than ten years) can be considered as the first normative document proclaiming the competence idea. At the same time, the traditional definition of the outcomes of school education in the knowledge & skills format due to the underdeveloped theory of the competence-based approach did not contribute to it proper implementation [16].

Hard work of the Ukrainian scholars supported the development of the national competence education theory. The competence education has been proclaimed as a personal & activity one that shifts the focus to a person's ability to take practical action in a particular context. The traditional "know what..." education result has been proposed to change into the "know how..." result. A new vision of the goal of education – from "formation of a comprehensively developed, harmonious personality" to "formation of an active, competent personality" – has been substantiated. The approaches to defining the content of learning – not "from the goal" but "from the result" – have been grounded. The development of a subject competence concept has been another valuable contribution of the Ukrainian scholars into the national theory of the competence education.

Based on the European experience (primarily the findings of the DeSeCo project) and the needs of the Ukrainian school, the following key competencies were identified: ability to learn (educational); civic; general cultural; competence in information and communication technologies; social; entrepreneurial; health saving [8]. It was the first attempt of the Ukrainian scholars in this area. They tried to adapt the European vision to the Ukrainian realities, i.e. the health saving competence was proposed in view of the ill-health indicators of the Ukrainian students.

In 2010, under the further modernization of the primary education in Ukraine, the Concept of the Second Generation of the State Standard of Primary General Education was developed. It became the theoretical basis for the development of the State Standard of Primary General Education (approved by the Cabinet of Ministers of Ukraine as of April 20, 2011).

The State Standard (2011) proclaims that the development of a child's personality is the priority of each educational field, taking into account the specifics of the subjects, the realization of the developmental and educational potential of the content and methods in the context of the competence approach. The document grounded on the European vision of the competence idea. In particular, the notions of "competence approach" in education, "key competence" and "cross-curricular competence" were officially defined taking into consideration the European approach.

It was first announced at the official level a list of cross-curricular and subject key competences that younger students should master: civic, communicative, social, aesthetic, mathematical and natural sciences competences [17].

The start of the systematic transformation of the primary education in Ukraine on a competence basis can be regarded as the most important positive that have been achieved with the adoption of the second generation of the State Standard of Primary General Education. The state of the Ukrainian education theory under the European integration provided the substantiation of the following fundamental principles of the competence education:

- the use of a competence always occurs in a particular context (in a particular educational situation);
- a competence is always a result that characterizes what an individual can do, rather than describing the process in which he/she acquires that competence (for example, a student does not say "I read, I wrote ..." but shows what exactly he/she can);
- clearly measured and approved standards are required to measure an individual's ability to exercise competence, taking into account what an individual can do at a specific time.

The new Law of Ukraine "On Education" (2017) enshrined the idea of the competence education. The Law declares that the goal of the secondary education is a comprehensive development, upbringing and socialization of a person capable to live in a society and interact with nature, to have a desire for self-development and life-long learning, be ready for conscious life choices and self-realization, responsibility, work and community activities.

The achievement of this goal is ensured through the formation of the following key competencies: fluency in the state language; ability to communicate in mother tongue (in case it differs from the state language) and foreign languages; mathematical competence; competence in the area of natural sciences, engineering and technology; innovation; ecological competence; information and communication competence; life-long learning; civic and social competences related to the ideas of democracy, justice, equality, human rights, well-being and a healthy lifestyle, with the awareness of equal rights and opportunities; cultural competence; entrepreneurship and financial literacy [15].

The Law on Education (2017) has become the regulatory framework for further competence-oriented reforming primary education in Ukraine. During the 2016-18 years, a new State Standard of Primary Education and two Typical Educational Programmes, in which key and subject competences have been defined as educational outcomes, have been developed and approved.

The State Standard of Primary Education (2018) proclaims that the goal of primary education is the comprehensive development of a child, his/her talents, abilities, competences and transversal skills in accordance with age, individual psychophysiological characteristics and needs, formation of values, development of autonomy, creativity and curiosity [18].

The list of key competences named in the Standard is identical with the list in the Law. At the same time, the State Standard of Primary Education specifies the content of competences, taking into account the specific tasks of this level of education. In particular,

- Fluency in the state language competence for primary school implies the ability to express verbally and in writing thoughts, feelings, clearly and reasonably explain the facts, as well as a love of reading, a sense of the beauty of the word, an awareness of the role of the language for effective communication and cultural expression, Ukrainian readiness language as a mother tongue in different situations of life;
- Ability to communicate in mother tongue (in case it differs from the state language) and foreign languages involves active use of the mother tongue in various communication situations, including in everyday life, educational process, cultural life of the community, the ability to understand simple expressions in a foreign language, to communicate in appropriate situations intercultural communication skills;
- Mathematical competence involves the identification of simple mathematical dependencies in the surrounding world, modelling processes and situations using mathematical relations and measurements, awareness of the role of mathematical knowledge and skills in the personal and public life of man;
- Competences in the area of sciences, engineering and technology. They presuppose the formation of curiosity, the desire to seek and propose new ideas, independently or in a group to observe and explore, formulate assumptions and draw conclusions from the experience, to know themselves and the world through observation and research; Innovation involves being open to new ideas, initiating change in a close environment (class, school, community, etc.), forming knowledge, skills, attitudes, which are the basis of a competency-based approach, provide further ability to successfully learn, pursue a professional activity, feel part of community and take participation in community affairs;
- Ecological competence implies awareness of the basics of environmental management, adherence to the rules of environmental behaviour, saving the use of natural resources with an understanding of the importance of nature conservation for the sustainable development of society;
- Information and communication competence involves mastering the digital literacy framework for development and communication, the ability to safely and ethically use the means of information and communication competence in learning and other life situations;
- Lifelong learning involves mastering the skills needed for further learning, organizing one's own learning environment, obtaining new information to use to assess learning needs, defining one's own learning goals and ways to achieve them, learning to work independently and in a group;
- Civic and social competences related to the ideas of democracy, justice, equality, human rights, well-being and a healthy lifestyle, awareness of equal rights and opportunities include: the ability to act as a responsible citizen and participate fully in public and social life, including schools and class, based on an understanding of social, economic and political concepts and

sustainable development, a critical appraisal of major events in national, European and world history, and respect for human rights us and the rule of law, the appreciation of the cultural diversity of different peoples, and the identification of oneself as a citizen of Ukraine, showing respect for others and tolerance, being able to cooperate constructively, sympathize and act in conflict situations, including in relation to various forms of discrimination. Careful attitude towards personal, social and physical well-being and health, maintaining a healthy lifestyle; understanding the rules of conduct and communication that are generally accepted in different communities and environments, the ability to act in the face of uncertainty and multitasking;

- Cultural competence involves involvement in various types of artistic creativity (fine arts, music and other arts) through the discovery and development of natural abilities, creative expression of personality;
- Entrepreneurship and financial literacy imply initiative, a willingness to take responsibility for their own decisions, the ability to organize their activities to achieve goals, an awareness of the ethical values of effective cooperation, a willingness to put into practice the ideas that have been initiated, to make their own decisions [18].

The experience of the recipients of education, their needs, which motivate learning, knowledge and skills, which are formed in different educational environment (school, family), different social situations determining formation of attitude to them are proclaimed as the basis for the formation of key competences.

The Standard defines the requirements to the mandatory learning outcomes and competencies for students in the following educational areas: language & literature (Ukrainian language and literature, languages and literature of relevant indigenous and national minorities; foreign language education); mathematical; science; engineering; informative; social and health saving; civic and history; art; physical education.

It should be emphasized that introduction of new educational terminology of the education results reflects a change of the goal of education, which becomes activity-efficient. This necessitates rethinking of selection of content, ensuring its closeness to the vital needs of children, reconciling the complexity and volume of the learning material with the predicted results.

The new educational outcomes is another innovation of the State Standard of Primary Education (2018). The content was the source category of the State Standard of Primary Education 2011. It defined the instruction results according to the themes and scope of content.

The content design of the 2018 Standard is fundamentally different, i.e. the instrumental knowledge being the basis to master the methods of action, prevails in teaching. If educational outcomes are the starting point in defining the content of primary education, then the competence approach becomes a means of achieving new quality of education. Under these conditions, teachers must objectively distinguish between learning process and learning outcomes, and students gradually learn to distinguish between knowledge of facts, phenomena, and knowledge of action.

The focus on different types of integration we consider as another advantage of the competence-based State Standard of Primary Education 2018. It provides the ability to combine subject learning with integrated courses in typical educational programmes.

Therefore, the new 2018 State Standard of Primary Education is based on new conceptual framework that is in line with the European competence approach. It opens opportunities for the development of young personalities able to successfully live, work, self-develop in a fast-paced and open world.

The Typical Primary Education Programmes with the lead of O. Savchenko are clearly synchronized with State Standard in terms of the competence idea. Specifically, the "logical consistency and sufficiency of students' acquisition of subject competencies" and "interconnected development of key and subject competences in each educational area" are defined among the principles of Programme design.

In the 2018 State Standard of Primary Education the general mandatory results are defined in the key competences and educational area format. In the Typical Education Programmes the specific expected results are defined and presented along the content lines, taking into account the opportunities of each educational area to form key and subject competencies [19], [20].

4 Synchronization of the education in Ukraine with the EU benchmarks

Evaluation of the degree of the synchronization of the education in Ukraine with the EU benchmarks made it possible to reach a number of important conclusions.

The competence education is recognized as a priority in Ukraine like in the EU. Ukraine, as a European state, is under the influence of the similar with the EU globalisation challenges, which makes competence education relevant. Besides, the implementation of the European competence idea in Ukraine is facilitated by the European integration, which is enshrined in the Constitution of Ukraine.

At the same time, it should be noted that there is no separate strategy for implementation of the competence education in Ukraine – the competence idea is a part of the comprehensive strategy of school education reforming "New Ukrainian School".

It is necessary to state the synchronization of the conceptual vision of the competence education in Ukraine with the EU. The competence idea adopted in Ukraine is in line with the European vision.

The provided in the Table 1 comparison of the key competences proclaimed at the policy level in EU and in Ukraine confirms this statement.

At the level of the primary education key documents in Ukraine the list of key competences is identical to the list approved by the Law "On Education" (2017). This effectively aligns the competences of the primary education in Ukraine with the European standards.

However, it should be noted that the adoption of the renewed 2018 European Reference Framework of Key

Competences for Lifelong Learning aimed at responding to the new demands of the European societies opens prospects for Ukraine in the area of conceptual architecture.

Table 1. Key competences for lifelong learning proclaimed at the policy level in the EU (2006 vision) and in Ukraine.

List of key competences indicated in the European Reference Framework of Key Competences for Lifelong Learning (2006) [6]	List of key competences indicated in the Law of Ukraine "On Education" (2017) [15]
Communication in the mother tongue	Fluency in the state language
Communication in foreign languages	Ability to communicate in mother tongue (in case it differs from the state language) and foreign languages
Mathematical competence and basic competences in science and technology	Mathematical competence
	Competence in the area of natural sciences, engineering and technology
	Innovation; ecological competence
Digital competence	Information and communication competence
Learning to learn	Life-long learning
Social and civic competences	Civic and social competences related to the ideas of democracy, justice, equality, human rights, well-being and a healthy lifestyle, with the awareness of equal rights and opportunities
Sense of initiative and entrepreneurship	Entrepreneurship and financial literacy
Cultural awareness and expression	Cultural competence

As can be seen from Table 2, it is about further synchronizing both the list and the essence of the key competences that are necessary for life in Europe in the 21st century.

Is it important that Ukraine applies the European implementation algorithm. It comprises the comprehending the essence of competencies; development of methodological/strategic benchmarks and legislative/normative base; technologies for practical implementation.

The monitoring of the implementation of the announced initiatives to track benchmarks is an important component of this algorithm in the EU. In particular, the results of such monitoring in the Eurydice Report "Developing Key Competences at School in Europe: Challenges and Opportunities for Policy" (2012) made it possible to assess the degree of the process, see the systemic challenges that need to be addressed. Unfortunately, the practice of monitoring the implementation of the announced/started reforms has not yet become mandatory in Ukraine.

We consider broad consultations (on-line consultations, position papers, consultation meetings, expert seminars) initiated by the European Commission on the review of the Recommendations on Key Competences for Lifelong Learning (2006) as a promising tool for the educational policy in Ukraine.

Table 2. Key competences for lifelong learning proclaimed at the policy level in Ukraine and in the EU (2018 vision).

List of key competences indicated in the Law of Ukraine "On Education" (2017) [15]	List of key competences indicated in the European Reference Framework of Key Competences for Lifelong Learning (2018) [7]
Fluency in the state language	Literacy competence
Ability to communicate in mother tongue (in case it differs from the state language) and foreign languages	Multilingual competence
Mathematical competence	Mathematical competence and competence in science, technology and engineering
Competence in the area of natural sciences, engineering and technology	
Innovation; ecological competence	
Information and communication competence	Digital competence
Life-long learning	Personal, social and learning to learn competence
Civic and social competences related to the ideas of democracy, justice, equality, human rights, well-being and a healthy lifestyle, with the awareness of equal rights and opportunities	Citizenship competence
Entrepreneurship and financial literacy	Entrepreneurship competence
Cultural competence	Cultural awareness and expression competence

The practice of application of the national-based approaches to implement the competence education idea by the EU Member States within the framework of EU Framework is valuable for Ukraine as well. A number of countries (Austria, Lithuania, Poland and Spain) launched national strategies across all eight key competences proposed by the European Reference Framework of Key Competences for Lifelong Learning 2006. Others (as on 2012) in their national strategies focused on several key competencies, including digital and entrepreneurial ones [14]. From the very beginning, Ukraine has taken a holistic approach, introducing a full list of key competencies identified by legislation [15]. It opens up new possibilities and at the same time creates additional difficulties.

The competence-based curriculum is a new reality for both EU and Ukraine. The challenge here is a new content that would provide the formation of key competences, primarily of the transversal ones.

The problem of integrating the content of education to form these competences exists both in the EU Member States and in Ukraine. In particular, the Eurydice Report "Developing Key Competences at School in Europe: Challenges and Opportunities for Policy" (2012) notes low levels of integration with respect to digital competencies in teaching mathematics, science and languages, even in countries where computer availability is high [14]. For Ukraine, the problem of the formation of transversal/cross-curricular competences by means of the integrated content is a new one. Proper teacher training is equally important.

The assessment of students in the transversal competences is one of the most unresolved issues for Ukraine. The EU Member States started the process of developing tests for assessing transversal competences much earlier. Therefore, their experience requires careful study.

5 Conclusions

Speaking about the transformation of the primary education in Ukraine into the competence base, we should state its dynamic movement from the stage of proclaiming the idea of the competence education, its conceptual justification to implementation.

There is a clear consistency in the implementation of the competence approach at the policy level (benchmark 1). The competence idea is reflected in the key state documents – in the Law of Ukraine "On Education" (2017), the State Standard for Primary Education (2018), and Typical Programmes. This testifies to the intention of the state to implement of the competence idea, the priority of the competence education for Ukraine.

It should be emphasized that the support of the competence idea at the level of educational policy is the most successful in Ukraine.

The approval of the new competence-based curriculum – the National Standard of Primary Education (2018), as well as Typical Primary Education Programmes is the achievement at the implementation level (benchmark 2). The prioritization of the outcome component in the structure of these documents is the defining feature.

At the same time, since the implementation of the new competency-based curriculum has been just started, this creates a number of challenges. Primarily, they are due to the novelty of a clear correlation between strategic, didactic and methodological levels.

The question of the need of the presence of the whole list of key competences identified by the Law "On Education" in the primary school curriculum requires special reflection – should all of them be the result of learning at this education stage?

The essence of a key competence as an integrated entity involves reaching a cross-curricular level. A competence approach to education requires the use of different types of integration of the educational content. In particular, in the typical educational programs for grades 1-2 and 3-4 of general secondary education with the lead of O. Savchenko, the subject education is combined with the integrated courses. For example, the new course "I'm Exploring the World" combines five educational areas and is due for instruction in grades 1-4 [21], [22].

At the same time, the problem of the formation of key competences, firstly of transversal or cross-curricular ones, remains at the methodological level; the aspects of cross-curricular links and integration of the education content requires further development.

The level of assessing younger students in competencies is even more underdeveloped (benchmark 3). Traditionally, the learning outcomes have always been recorded at the subject level. Today, the development of valid tools for assessing students in transversal competences is a serious challenge.

Additionally, the following questions remain relevant:

- When key competences mastered by younger students can be subject of the national level assessments?
- Which evaluation tool is best to assess the transversal competences of the younger students?
- How to assess students' mastering of value orientations that are integral components of key competences?

Finally, an effective feedback tool has not yet been worked out, the process is still discrete.

Thus, in the context of the European integration, the nature of the development of the Ukrainian education is determined by the strategic benchmarks of the EU. The transformation of the Ukrainian primary education into a competence base is a prime example.

Of course, we should note the lag in the implementation of the competence idea in Ukraine in time compared to the EU countries (the competence education in Ukraine was enshrined at the legislative level by the Law of Ukraine "On Education" only in 2017). Ukraine is not a member of the EU and therefore does not directly fall within the requirements / mechanisms / methods applied within the EU.

However, common globalization challenges, powerful integration processes in Europe, Europeanization affecting the European countries outside the EU borders contribute greatly to the synchronization of the education in Ukraine with EU standards.

References

1. The Lisbon European Council 23-24 March 2000 Presidency Conclusions (European Council, Lisbon, 2000), https://www.europarl.europa.eu/summits/lis1_en.htm. Accessed 29 Dec 2019
2. M.K. Smith, Competence and Competency (infed, 1996), http://www.infed.org/biblio/b-comp.htm. Accessed 29 Dec 2019
3. Key Competencies. A developing concept in general compulsory education. Eurydice Survey 5. (European Commission, Directorate-General for Education and Culture, Eurydice, Brussels, 2002)
4. W. Hutmacher, Key Competences for Europe. Report presented at the Symposium "Key Competences in Europe", Council for Cultural Cooperation, Council of Europe, Berne, Switzerland, 27-30 March 1996
5. Definition and Selection of Competencies: Theoretical and Conceptual Foundations (DeSeCo). Background Paper (OECD, DeSeCo, Switzerland, 2001)
6. Recommendation of the European Parliament and of the Council of 18 December 2006 on key competences for lifelong learning, OJ. **L 394**, 10–18 (2006)

7. Council Recommendation of 22 May 2018 on key competences for lifelong learning (Text with EEA relevance) ST/9009/2018/INIT, OJ. **C 189**, 1–13 (2018)
8. N. Bibik et al., *Kompetentnisnyi pidkhid u suchasnii osviti: svitovyi dosvid ta ukrainski perspektyvy. Biblioteka z osvitnoi polityky* (Competency-based approach in modern education: world experience and Ukrainian prospects: Educational policy series). (K.I.S., Kyiv, 2004)
9. I.G. Yermakov (ed.), *Zhy`ttyeva kompetentnist` osoby`stosti: vid teoriyi do prakty`ky`* (The life competence of a personality: from theory to practice). (Tsentrion, Zaporizhzhia, 2005)
10. L.L. Khoruzha, Pedagogichna osvita: teoriya i prakty`ka **7**, 178–183 (2007)
11. V. Lugovyi, Pedagogika i psy`xologiya **2**, 13–26 (2009)
12. I. Rodygina, Kompetentnisno spryamovany`j pedagogichny`j proces (Competence-oriented pedagogical process) (2008), https://osvita.ua/school/method/1963/. Accessed 8 Jan 2020)
13. O. Ovcharuk, in *Reform Strategy for Education in Ukraine: Educational Policy Recommendations* (K.I.S., Kyiv, 2003), pp. 13–41
14. Developing Key Competences at School in Europe: Challenges and Opportunities for Policy. Eurydice Report (European Commission, EACEA, Eurydice, Brussels, 2012)
15. Zakon Ukrainy Pro osvitu (The Law of Ukraine On Education) (2017), https://zakon.rada.gov.ua/laws/show/2145-19. Accessed 28 Nov 2019
16. Postanova Kabinetu Ministriv Ukrainy vid 16 ly`stopada 2000 r. № 1717 Pro perexid zagal`noosvitnix navchal`ny`x zakladiv na novy`j zmist, strukturu i 12-richny`j termin navchannya (Resolution of the Cabinet of Ministers of Ukraine as of November 16, 2000, No. 1717 On the Transition of the Secondary Education Institutions into a New Content, Structure and 12-years Duration of Instruction) (2000), https://zakon.rada.gov.ua/laws/show/1717-2000-%D0%BF. Accessed 9 Jan 2020
17. Postanova Kabinetu Ministriv Ukrainy vid 20 kvitnia 2011 r. № 462 Pro zatverdzhennia Derzhavnoho standartu pochatkovoi zahalnoi osvity (Resolution of the Cabinet of Ministers of Ukraine as of April 20, 2011, No. 462 On the Adoption of the State Standard of Primary General Education (2011), https://zakon.rada.gov.ua/laws/show/462-2011-%D0%BF. Accessed 8 Jan 2020
18. Postanova Kabinetu Ministriv Ukrainy vid 21 liutoho 2018 r. № 87 Pro zatverdzhennia Derzhavnoho standartu pochatkovoi osvity (Resolution of the Cabinet of Ministers of Ukraine as of February 21, 2018, No. 87 On the Adoption of the State Standard of Primary General Education (2018), https://zakon.rada.gov.ua/laws/show/87-2018-%D0%BF. Accessed 8 Jan 2020
19. Typova osvitnia prohrama dlia 1-2 klasiv zakladiv zahalnoi serednoi osvity pid kerivnytstvom Savchenko O.Ia. (The typical educational program for grades 1-2 of general secondary education with the lead of Savchenko O.Ya.) (2019), https://osvita.ua/doc/files/news/659/65920/5d9d8feb1ce4f316467013_1.pdf. Accessed 8 Jan 2020
20. Typova osvitnia prohrama dlia 3-4 klasiv zakladiv zahalnoi serednoi osvity pid kerivnytstvom Savchenko O.Ia. (The typical educational program for grades 3-4 of general secondary education with the lead of Savchenko O.Ya.) (2019), https://mon.gov.ua/storage/app/media/zagalna%20serednya/programy-1-4-klas/2019/11/3-4-dodatki.pdf. Accessed 3 Jan 2020

Development of multilingual education in Ukraine

Nina Nikolska[1,*], and *Oksana* Pershukova[2]

[1]National University of life and environmental sciences of Ukraine, 15 Heroiv Oborony Str., Kyiv, Ukraine
[2]National Aviation University, 1 Lubomyra Guzara Ave., Kyiv, Ukraine

Abstract. Language education is an important tool that forms the consciousness of the individual and its ability to be socially mobile in society, contributes to a dialogue of cultures in a globalizing world. New realities require changes in the definition of the level of knowledge of foreign languages, the new approaches definition to the selection of content and organization of materials, the use of adequate forms and types of control. Of particular relevance is multilingual education on the principle of Ukrainian + foreign in the light of Common-European polymeric trends, the integration of our state in the European educational space, the spread of student exchange practices and the openness of our country's educational system for foreign students. The spread of the foreign languages studying fits into the context of European directives, which recommend that the Member States teach two other languages of the European Union in educational institutions, in addition to the state language. Multilingual education in Ukraine is an adequate answer to the question of linguistic and cultural diversification in Europe. An important feature of multilingual education in Ukraine is to take into account the regional needs of local ethnic communities, national characteristics and Common European trends.

1 Introduction

1.1 The problem statement

A characteristic feature of modernity is the increase in the scale and pace of change due to the information society formation. Priority is currently given to information, high technology and humanistic principles of governance, based on democratic public decision-making. The information community is characterized by the globalization of processes and phenomena that form a single world and a single information space. This contributes to the establishment of a single global culture of mankind, as well as the interpenetration and enrichment of national cultures [1].

The interconnectedness and interdependence of countries and nations of the modern world, the intensive development of globalization processes determine the growth of interest in studying international experience in public life's various areas, including the activation of international comparative studies in the educational sphere. Education reforms constitute an important aspect of European countries' social policies. The main trends analysis in the development of the theory and practice of education in the modern world to search for new educational models that meet the requirements of our time is the study object of comparative pedagogy. In the context of this scientific direction, there are many views of scientists on the substantiation of trends in the development of the world and European education in the 21st century [2], however, they are very cautious in making forecasts and warn that their assumptions may be erroneous, since the future is multivariate, and social and pedagogical processes are largely unpredictable, and only time will adequately demonstrate the truth or falsehood of one or another prognostic the script.

1.2 Literature review

According to the Ukrainian researcher O. Lokshyna, the trend is the basic category of comparative pedagogy [3]. Therefore, we dwell on the essential characteristics of this term in more detail. The trend - (from the Latin "tendo" – "I direct, I want") means the direction of development, a manifest opportunity, which under certain conditions turns into reality [4]; Collins English Dictionary suggests viewing the trend as a tendency to a specific characteristic, or a general course [5], and the Macmillan electronic dictionary provides such a translation: 'something will likely happen'. A trend as a category of comparative pedagogy (according to N. Lavrychenko) serves as a form for determining patterns that, on the whole, have no other reality than approaching something in a trend. O. Lokshyna interprets the concept of tendency not only from the position of revealing the direction of movement but also in the context of the essence of qualitative changes occurring during this movement [3]. A. Sbruyeva sees the conditioning of trends in education by its functions, the implementation of which can have the result of both favourable and unfavourable external socio-cultural conditions. At the same time, A. Sbruyeva draws attention to the contradictions that arise in educational systems: between global and local problems, between traditions and modernism, between competition in achieving success and aspirations for equal

* Corresponding author: nina777-07@hotmail.com

opportunities. Therefore, awareness in the areas of educational trends provides the basis for forecasting and updating, increasing the level of integrity. A. Ohienko refers to the implementation of educational trends 'as a result of the contradictory interaction of the determining force of laws of the internal structure, functioning and development of the educational sphere on the one hand, and external factors due to the impact on the educational integrity of other cultural subsystems – from the others' [6, p. 380]. Consequently, the trend in comparative pedagogy is the educational changes direction, due to the numerous political, social, cultural and scientific factors that interact in society, often causing contradictions, and result in the emergence of a new quality of educational processes. The trend is based on a set of ideas and aspirations, and therefore its manifestations can cause a confrontation of interests, which results in the strengthening or weakening of some processes, changes in educational policy or practice at a certain level depending on the direction and intensity of society development. The work of the comparativists and a detailed analysis of the directions of educational changes development in foreign countries, their intensity levels, and the interaction results, is a prerequisite for predicting the further development of the educational direction and developing proposals for updating the domestic school. In the multilingual education of schoolchildren in Western European countries, the trend personifies the direction of pedagogical phenomena development aimed at developing students' ability to communicate in several languages, as well as possession of a complex of knowledge, skills, experience, and attitudes that are part of intercultural competence. This development direction is the result of political, economic, socio-cultural and ethnolinguistic processes in Europe and the world, as well as in any of the Western European countries where training takes place, but regional, local and individual development levels are also significant. Because of the above mentioned, the European and global trend in the development of multilingual education has largely influenced the development of multilingual education in Ukraine.

1.3 Methods

The purpose of the article is to highlight the main directions of pedagogical research in the field of multilingual and bilingual education through the analysis of the achievements of Ukrainian researchers in the field of pedagogy. The methodological basis of the study was an analysis of published texts, a comparison of their content, the synergistic approach to their interpretation, combined to make possible conclusions.

2 Multilinguaism forming

2.1 Bilingual education in Ukraine

The concept of 'native language' is not unambiguous, as it seems at first glance. For many people, the mother tongue is one that a person has the best of and which is fundamental to her/his communication. This is the language of her 'kind', having deep roots in her native land. Such a model of linguistic, personality cultural development is common, but it is not the only one. In the ethnic reference book, which refers to linguistic, sociological terms, it is stated that 'atypical options for the formation of the child's speech in the mass migration context, high territorial and social mobility of people are becoming more common and require a clear definition of their native, maternal, ethnic, functionally first language' [7, p. 15–17].

Local scientist I. Hydzuk on this occasion notes that the reality for our country is a huge variety of conditions, situations that do not fall into any categorical formula. So, in Ukraine there are various population groups:

- Ukrainians, who learned their language from their father and mother, actively use it in all spheres of life and consider it undoubtedly native, and speak Russian poorly;
- Ukrainians, for whom the Ukrainian language is the mother language, but they studied in a Russian-language school, after graduation spoke Russian in many fields as official, and speak it better than Ukrainian;
- bilingual immigrants from families where parents are representatives of different nationalities, but for one of the parents Ukrainian is their native language;
- Russians who have accepted the Ukrainian idea and speak two languages;
- Russians who, according to national identity, remain Russian and speak mostly Russian, although they are patriots of Ukraine.

The researcher also mentions the facts about known cases of Ukrainian native language rejection, which is perceived as non-prestigious [7, p. 17–18].

The development of Ukraine as a multinational state, the formation of a civil democratic society opened to other countries, peoples, and cultures, as well as the process of general secondary education transformation in the context of society globalization, requires the new approaches development to teaching the younger generation in the context of the revival of ideas of national and multicultural education. Among the main tasks set for modern education in Ukraine is the preparation of young people for life in a multicultural, globalizing society. Of particular importance in educational activities is the formation of the ability to conduct a productive dialogue, including a foreign language, with representatives of other countries, cultures, nations [8, p. 9–10].

The present of our country is characterized by two interconnected development trends: the strengthening of Ukraine as an independent state and its integration into global processes, in particular in the European educational space. For Ukraine, entering the mentioned space means focusing on state-public management principles; increasing the role of alternative systems and innovative processes; widespread introduction into all links of new, first of all, information technologies; ecological and cultural orientation of education and science; providing citizens with the right to choose equal conditions for general education, access to science, regardless of their socio-genetic capabilities; moral and spiritual improvement of citizens; ensuring the continuity of personality education throughout life [8, p. 9–10].

The process of Russification of Ukrainian population lasted for several centuries. Therefore, the process of the state Ukrainian language strengthening is developing so slowly and not easily. Local specialist in the field of bilingual education L. Tovchyhrechka identifies eight main stages of the development of 'Ukrainization' in Ukraine.

The first stage (1922-1929) went down in history as the policy of 'Ukrainization'. The beginning of Ukrainization in the publication of the following documents: Decree of the Council of People's Commissars of the Ukrainian SSR of July 27, 1923 'On Measures in the Case of Ukrainization of Schools and Teaching and Cultural-Educational Institutions' and Decree of August 1, 1923 'On Measures to Ensure Equality of Languages and on Assisting the Ukrainian Language Development'. In 1929, training in the Ukrainian language was carried out by more than 80% of secondary schools, over 75% of technical schools and 30% of institutes [9].

The second stage (1930-1952). Assimilation policy in the practice of language planning in the USSR. The crowding out of national languages into Russian took place by spreading it in the status of a language of interethnic communication with a gradual increase in National-Russian bilingualism within national collectives; in the future, monolingualism should end in Russian.

The third stage (1953-1958). The Soviet government, under the leadership of Khrushchev, developed a program to accelerate the Russification of non-Russian peoples of the Soviet Union. The law on the free choice by parents of the instruction language and on permission to refuse to study the Ukrainian language.

The fourth stage (1958-1962) was characterized by the emergence of the theory of 'the promising and unpromising language'. In the early 60s of XX century, a scientific council was created in the system of the USSR Academy of Sciences on the problem of 'Patterns of the Development of National Languages in Connection with the Socialist Nations Development,' the orientation of which was clearly illustrated by its program published in the journal Questions of Linguistics (1962). The main task of the council was to create a theoretical base for the forced Russification of several republics. On this basis, the peoples' languages of the USSR were divided into promising and unpromising. The first ones included, of course, 'the great Russian language', Armenian, Georgian, Latvian, Lithuanian and Estonian. Other languages fell into the category of unpromising. Thus, the Ukrainian and Belarusian languages, as well as the Azerbaijani language and the speeches of the Central Asian republics, were excluded from the promising group.

The fifth stage (1963-1975). Dissemination of the theory of 'objective rapprochement of nations', which provided for the infusion of closely related languages into the Russian language. During this stage, the Russification of all life spheres - education, science, production, the media, culture – was noted with particular intensity.

The sixth stage (1975-1985) is determined by the substantiation of the concept of a single Soviet people. During this period, the crowding out of the Ukrainian language of higher education and science intensified. The language of instruction in vocational schools and higher educational institutions has become Russian. The Russification linguistic and cultural policy in all areas of public life has led to the fact that in the east and south of the Ukrainian SSR the Russian language has finally changed its native language in the field of scientific communication. All this led to a sharp narrowing of the functioning of the Ukrainian language in most of Ukraine and the decline in its prestige. At the end of the 70s, the assimilating influence of the Russian language in eastern and southern Ukraine turned over from city to village.

The seventh stage (1985-1990) is characterized by the idea of harmonious National-Russian bilingualism. At this time, the development of the scientific program 'The Role of the Russian Language as a Means of Interethnic Communication' significantly increased.

The eighth stage (1991-2015) – the formation of language policy in Ukraine [9, p. 157–160].

Multilingualism is a feature of the national structure of the population of Ukraine. According to the All-Ukrainian Population Census of 2001. Representatives of more than 130 nationalities and ethnic groups live in Ukraine. The number of representatives of the titular nation is 77.8%. The second largest group is the Russians, their number is 17.3%. Other ethnic minorities: Belarusians (0.6%), Moldavians (0.5%), Crimean Tatars (0.5%), Bulgarians (0.4%), Hungarians (0.3%), Romanians (0.3 %), Poles (0.3%), Jews (0.2%), Greeks (0.2%), Tatars (0.2%), Georgians (0.1%), Gypsies (0.1%) , Azerbaijanis (0.1%), Germans (0.1%), Gagauz (0.1%) [10]. These data indicate a difficult language situation in the country, which is further complicated by the specific sociolinguistic status of the Ukrainian language. It is fixed at the legislative level as the state one, the Ukrainian language is actually in a weak position, especially in large cities in the east of the country – it is the Russian language that dominates. Ukrainian-Russian bilingualism is observed on the prevailing territory of the country, and in some regions, the state language is a minority language, because most ethnic Ukrainian and other nationalities are Russian-speaking. An attempt to determine the language balance, or the real ratio of the two most common languages - Ukrainian and Russian - in various spheres of public and private life based on statistical and sociological data, was made by A. Medvedev, vice president of the Ukrainian League for Public Relations. The main conclusions are as follows: Ukraine as a whole is a bilingual state, in which the largest group are exclusively Ukrainian-speaking citizens, and somewhat smaller are two approximately identical groups of bilingual and exclusively Russian-speaking. However, bilingual is mainly ethnic Ukrainians and those who consider Ukrainian their native language. Ethnic Russians and those who call Russian their native language, are part of the monolingual. Besides, 15% of ethnic Ukrainian consider their native language Russian, and 20% of ethnic Ukrainian communicate exclusively in Russian. A significant number of citizens of non-Ukrainian and non-Russian nationalities, and among some ethnic groups - the vast majority, consider the native language Russian, and not the language of their nationality. The fact that the ratio of the speakers of Ukrainian and Russian languages does not correspond to the ratio of ethnic Ukrainian and Russian indicates the deformation of the linguistic situation in Ukraine [11].

Since Independence gaining multilingual education in Ukraine was developing according to two main models: 1) secondary education for children from families of the majority of the population and 2) secondary education for children from families of national minorities. The initial conceptual provisions that determine the content and organization of language education in Ukraine and the legal basis for the formation of state policy to ensure the rights of national minorities are the Declaration on State Sovereignty of Ukraine, the Constitution of Ukraine, Laws of Ukraine 'Based on State Language Policy' (2012), 'on the Concept of State Language Policy' (2010), 'On national minorities in Ukraine' (No. 2494-XII of 06/25/1992), 'On ratification of the European Charter for Regional languages or minority languages' (No. 804-IV of 05/15/2003), 'On Education' (No. 1144-XII of 06/04/1991), 'On General Secondary Education' (No. 1642-III of 04/06/2000) and other legislative acts of Ukraine. These documents ensure the right of national minorities to meet educational needs in their native language, preservation, and ethnic culture development. At the same time, the need to create conditions for the proper mastery of the state language in secondary schools with instruction in the languages of national minorities is emphasized [12]. The Law of Ukraine 'On National Minorities in Ukraine' regulates the right of all national minorities to use and study their mother tongue in state education all institutions or through national cultural societies [13].

The Law of Ukraine 'Based on State Language Policy' (2012) applies to 18 languages: Russian, Belarusian, Bulgarian, Armenian, Gagauz, Yiddish, Crimean Tatar, Moldavian, German, modern Greek, Polish, Roma (Gypsy), Romanian, Slovak, Hungarian, Ruthenian, Karaime and Crimean. The law contains an ambiguous definition of the term 'mother tongue', interpreting it as follows: 'mother tongue is the first language that a person mastered in early childhood'. According to the conclusions of the Main Scientific and Expert Directorate of the Verkhovna Rada of Ukraine, the definition of the term 'native language' in the law is inaccurate and incorrect, since it follows from the conclusion that the native language can be someone else's speech if, in certain circumstances, a person has mastered it first. It underlines that free choice of the language instruction is an inalienable right of citizens of Ukraine, which is implemented under this Law, subject to the mandatory study of the state language in an amount sufficient to integrate into Ukrainian society. Citizens of Ukraine are guaranteed the right to receive education in the state language and regional or minority languages. This right is ensured through a network of preschool child care facilities, general secondary, extracurricular, vocational and higher state and communal educational institutions with Ukrainian or other languages of instruction, which is created following the needs of citizens and under the legislation of Ukraine on education [11]. The law establishes that the state language is Ukrainian, but significantly expands the use of regional languages if the number of speakers of these languages is at least 10% of the population in a certain region and some cases less than 10% [11].

Introduction of a sectoral programme on improving the Ukrainian language teaching in secondary schools (Ministry of Education and Science of Ukraine Order No. 461 of 28 May 2008) began a step-by-step transition to teaching in Ukrainian from the fifth grade onwards the following subjects: history, geography, labour training and physical culture. The elements of bilingual education have been decided to apply by combining the possibilities of the invariant component (in the language of the national minority) and the variable component (in the state language). The specifics of such training are set out in the methodological recommendations. Based on the analysis, we have identified the characteristic features of bilingual education in Ukraine: 1) preliminary acquaintance with separate language units (separate words, word combinations, sentences) and on such basis to offer work with a fragment of the text or the whole text; 2) considerable attention is paid to terminology mastering, words of terminological and general meaning used in various syntactic constructions, word combinations in the subject content - formation of skills to use terms and describe processes; 3) attraction of pupils to all kinds of speech activity in the target (Ukrainian) language in subject lessons, development of both receptive and productive skills, attention is paid to the development of both monological and dialogic speech; 4) regular independent work, work in pairs and groups is provided; 5) in the content of teaching the Ukrainian language improvements are made, as attention is paid to mastering the special vocabulary used in the lessons of the natural and mathematical cycle (numerals, adjectives, adverbs, participles), the formation of skills of terminology using in subject contents for the development of language knowledge (e.g. sound analysis).

According to the Act 'On general secondary education' and Cabinet of Ministers Decision No. 1392 of 23 November, 2011 on the approval of the State standard for basic and complete general secondary education, new model curricula for general education institutions of the second degree have been developed (Ministry Order No. 664 of 29 May 2014). Therefore, in 2014/15 there was a return to the system whereby the institution and parents have the right to choose whether or not to study a second foreign language. According to the new rules, the compulsory study of a second foreign language has been introduced. It can be any foreign, Russian or any other language of a national minority. The school administration became deprived of the right to decide whether to introduce or discontinue the study of a particular foreign language alone. All proposals must be discussed and agreed upon with the parents and approved by the teachers' council, provided that appropriate staffing is available. The above mentioned caused the appearance of a *third specific model of multilingual education* in Ukraine. According to this model the Russian language, usually, the language used in the family of a schoolboy or a schoolgirl is studied as a second foreign language. As the analysis shows, there is no such model in Western European countries. The introduction of such a model in Ukraine corrected the situation in which schoolchildren (Ukrainians by nationality) communicate in everyday life and family environment in Russian, and remain semi-literate because of the lack of this language in the curriculum. The reasons for this situation are caused by, first of all, the deformation of the language situation in Ukraine, the weak position of the Ukrainian language in the society of the country, the relatively close kinship of the Ukrainian and Russian languages, the prevalence of Russian language in Europe and the world. Common problems for all these models are lack of a sufficient number of quality educational programmes, properly developed teaching and methodological materials and methods for consistent teaching of several languages; lack of systematic professional training and professional development of teachers who are ready to teach several languages and cultures or subject content in two languages (state + foreign, national + state); limited number of topics in the curricula dedicated to introducing the languages and cultures of national minorities to the public.

As a result of situation analysis, we have concluded that multilingual education in Ukraine develops according to three main models: *the first* - multilingual education for children from the families of the majority of the population; *the second* - multilingual education for children from families of national minorities; *the third* model of multilingual education is specific; a similar model does not exist in Western Europe. In its context, the Russian language, usually the language used in the student's family, is studied as a second foreign language. The reasons for this state of affairs lie in historical and socio-cultural aspects. Among them is the policy of Russification for a long period of Ukrainian history [14].

So, language education in Ukraine went the hard way, which was significantly influenced by historical events, ideological settings, socio-cultural conditions and many more. Today in Ukraine there is a system of continuous language education, supported by the language policy of the state, aimed at mastering the mother tongue of every speaker, the compulsory mastery of the state Ukrainian language by all citizens (regardless of nationality, religion, social production sphere) and knowledge of foreign languages, as well as language tolerance education.

2.2 Multilingual education in Ukraine

The development of secondary education for the first model includes teaching the native students of the Ukrainian language and teaching one or two foreign languages. This model has extensive experience in functioning in the context of specialized schools with the training of several subjects in a foreign language, which is now implemented in the organization of teaching foreign languages to specialized subjects. As emphasized in the White Paper ..., among the shortcomings of this model is the tendency to make the Ukrainian-speaking part of the population monolingual, since in many schools with the Ukrainian language of instruction, only Ukrainian and one foreign language are studied, the level of knowledge of which is low and does not give reason to speak about bilingualism in the Ukrainian system education [15, p. 28-29]. In many schools, until recently, only one foreign language was studied. Therefore, the foreign-language proposal was quite disproportionate - the mass learning of English, a small amount of German and a rather insignificant share of French and Spanish. Some specialized schools suggest two foreign languages (English-German, English-French). Cases of learning other foreign languages in schools (Japanese, Chinese, Polish, Hungarian, Czech) are single and often experimental. Besides, training sometimes is hampered by an insufficient level of methodological training of teachers, a poor focus on modern communicative techniques, a lack of real practice and modern teaching aids, in particular, multimedia.

Starting from the academic year 2012–13, compulsory foreign language learning from the first grade and compulsory second foreign language learning from the fifth grade were introduced in the following language combinations: English (IM1) + German (IM2) English (IM1) + French (IM2) German (IM1) + English (IM2) French (IM1) + English (IM2), "language of the neighbouring country" (IM1) + English (IM2). So far, such training is largely unorganized, as V. Hamanyuk notes, the correct formation of multilingual competence requires intensification of the educational process, a high level of systematization and abstraction, which should result in a more conscious and deeper understanding of linguistic phenomena, transfer of knowledge and skills in the field of study a new language and the linguistic abilities development: verbal intelligence, language flexibility, language memory, and linguistic intuition. This means that the implementation of the conceptual foundations of multilingual education in Ukraine, defined in the White Paper ... as 'developing models of language education taking into account the sequence of content and teaching methods at certain stages, the correlation of different language courses ... the introduction of bilingual education' is slow enough.

The study of the state language by representatives of national minorities is a prerequisite for the inclusion of ethnic community representatives in the public life of the state since the educational, socio-cultural and economic prospects of the individual depend to a large extent on this kind of competency. The right to learn the official language of a country is part of the language of human rights. Ukrainian is studied as the state language in all educational institutions of Ukraine without exception, and the right of national minorities to meet the educational needs in their native language, the preservation, and development of ethnic and cultural characteristics of representatives of national minorities is ensured. In general, in Ukraine, only 31 languages are studied at schools, including 19 languages of national minorities, and Ukraine is one of the few countries in the world where teaching is conducted in eight languages of national minorities. Education of students from families of national minorities is the second model of secondary education in Ukraine. This system was created over the years of independence from scratch. Our state as a whole fulfils its obligations under the Council of Europe, in particular, are provided for by the European Charter for Regional Languages and the Framework Convention for the Protection of National Minorities [16]. However, the intentions outlined in the documents do not always coincide with real opportunities in ensuring educational rights and often have a declarative character. Although during the years of independence in Ukraine in general educational institutions with instruction in the national minorities languages, conditions have been mainly created for the proper mastery of the state language, however, there are certain problems in teaching the Ukrainian language at schools, in particular with the Hungarian and Romanian languages of instruction on the organization of the educational process and methodological support. Since the introduction of teaching the Ukrainian language was begun only after Ukraine gained independence, this process was accompanied by the urgent creation of scientific and methodological centres, the primary tasks of which were the programs and textbooks creation. The authors' lack of theoretical training on the problems of selecting the content and methodology of language teaching affected the quality of instruction: the programs were imperfect, the textbooks were created without taking into account the sociolinguistic situation and the sociocultural characteristics of minorities, and therefore often had a compilation character. Also, the textbooks for primary and high school contained too much theoretical material and were similar to textbooks for schools with the Ukrainian language of instruction. Another problem was the ratio of the components of education in the mother tongue and the state language - it was disproportionate at all levels of education: in the elementary school, out of the total weekly load of 85 hours, only 14 are devoted to mastering the state language, in the primary school, 155 versus 22 in the senior school [17].

In recent years, a tendency toward mastering foreign languages has been observed in Ukraine more and more transparently, they are actively acquiring the status of a socio-economic and political means of understanding between various representatives of the world community in various spheres of life. These realities objectively determine the expansion of the functions of foreign languages and the renewal of the tasks of mastering them in modern society. First of all, it is:

- formation of readiness for social interaction among pupils/students to jointly solve various problems and to achieve mutual understanding compromises;
- tolerant attitude towards the peoples whose language is studied, familiarization with the dialogue of cultures as an important phenomenon of peaceful coexistence and mutual enrichment;
- mastering the techniques of independent work with the language, learning strategies, the ability to use compensatory techniques in case of language resources shortage;
- creative use of a foreign language for their expression, awareness of language differences with their native language, knowledge, and ability to use the features of its use in typical social communication situations;
- the formation by pupils/students of an individual learning style, self-control of their level of mastery of foreign language communicative competence[14].

The result of language training should be personal multilingualism (plurilingualism), which provides for a deliberate differentiation of language systems and a relatively free transition from one language to another, depending on the changing situation and needs.

3 Conclusions

As a result of our research, we have concluded that historical events, ideological settings, socio-cultural conditions and many other factors influenced greatly on the language situation and language education in Ukraine.

The development of multilingual education in different countries of the world and Ukraine as well has become a requirement of time. Multilingual education in Ukraine is developing according to three main models: *the first* - multilingual education for children from the families of the majority of the population; *the second* - multilingual education for children from families of national minorities; *the third* model of multilingual education is specific; a similar model does not exist in Western Europe. In its context, the Russian language, usually the language used in the student's family, is studied as a second foreign language. The reasons for this state of affairs lie in historical and socio-cultural aspects. Among them was the policy of Russification for a long period of Ukrainian history and which resulted in the deformation of the language situation in modern Ukraine.

Based on the study results, proposals on the development of multilingual education of schoolchildren in Ukraine are substantiated. Their main directions can be represented at several levels. At the *conceptual level*: to improve the conceptual foundations of multilingual education, taking into account Western European experience and national characteristics, the requirements of the labour market and Ukrainian society. At the *educational and administrative level*: to provide support for educational initiatives to develop multilingualism at the regional and local levels to preserve the existing linguistic and cultural heritage and subsequent integration into the European and world educational space; to develop and ensure the implementation of the state language policy in the education system, taking into account regional characteristics; create a sufficient number of high-quality educational programs for various levels of education, taking into account the kinship / non-recognition of languages; to contribute to the implementation of the teaching English experience as a language of worldwide communication; to improve the system of professional training and teachers' professional development, to provide training for bilingual specialists or those who are ready to teach not only two languages as systems, but also subject content in two languages (state + foreign, national + state) with experience in a multicultural environment. At the *scientific and methodological* level: to study the foreign experience of immersion, receptive and other technologies of multilingual education and to test them experimentally in the conditions of the Ukrainian school. At the *school level:* encourage the diversification of language learning in schools; introduce the process of integration of the curriculum in language subjects as a way to avoid competition and rivalry, increase the cooperation level, collegiality and mutual assistance among teachers of language subjects; provide methodological support from school administration and academic institutions. At the *individual level*: to promote the development of language abilities based on individualization, learning process differentiation; to apply the students' experience of learning languages, use the already existing metalinguistic consciousness; to learn to apply various strategies, to form a habit and need for students to master the languages throughout their lives [14].

Implementation of the proposals will contribute to the formation of responsible citizens of Ukraine, capable of thinking globally, ready to live and work next to representatives of different peoples and cultures. Improving the process of forming the individual multilingualism (plurilingualism) of young Ukrainians will contribute to the preservation of their cultural identity and the development of their creative potential, provide an opportunity for professional mobility, will be an effective means of supporting political stability in society.

References

1. T. Petrova, Tendencii razvitija sovremennogo obrazovanija (Trends in the development of modern education). Observer **2**(193) (2006)
2. Z. Kolokol'nikova, Vedushhie tendencii razvitija sovremennogo mirovogo obrazovatel'nogo processa (Leading trends in the development of the global educational process). Dinamika sovremennoj nauki (2012), http://www.rusnauka.com/21_DSN_2012/Pedagogica/4_114364.doc.htm. Accessed 16 Sept 2019
3. O. Lokshyna, Tendenciya yak katehoriya porivnyal'noyi pedahohiky (Trend as a category of comparative pedagogy). Comp. ped. stud **8** (2011), http://lib.iitta.gov.ua/7118/1/18585-27831-1-SM.pdf. Accessed 27 Jan 2019
4. Ukrainian Dictionary: Academic Interpretative Dictionary (1970-1980), http://sum.in.ua/s/tendencija. Accessed 26 Feb 2015
5. Collins English Dictionary, http://www.collinsdictionary.com/dictionary/english/tendency. Accessed 26 Feb 2015
6. O. Ohienko, *Trends in Adult Education in Scandinavian Countries* (Ellada, Sumy, 2008).
7. I.F. Hudzyk, *Competently-oriented teaching of the Russian language in elementary grades (in schools with the Ukrainian language of instruction).* (Bukrech, Chernivtsi, 2007).
8. S.M. Nikolaenko, *Education and Science: Legislative and Methodological Foundations* (Politekhnika, Kyiv, 2004), p. 280
9. L. V. Tovchyhrechka, Dissertation, Sumy Makarenko State Pedagogical University, 2012
10. Proceedings of the 2001 All-Ukrainian Census (2001), http://2001.ukrcensus.gov.ua/results/general/language. Accessed on 10 Feb 2012.
11. Law of Ukraine "On the Principles of State Language Policy" (2012), http://zakon4.rada.gov.ua/laws/show/5029-17. Accessed 16 Sept 2014
12. Law of Ukraine "On Education" (1991), http://zakon1.rada.gov.ua/laws/show/1060. Accessed 16 Sept 2014
13. Law of Ukraine "On National Minorities in Ukraine" (1992). Accessed 16 Sept 2014
14. O.P. Pershukova, Dissertation, Ivan Ziaziun Institute of Pedagogical and Adult Education of the National Academy of Educational Sciences of Ukraine, 2016
15. V.G. Kremen, *White Paper on National Education of Ukraine* (Acad. ped. of Sciences of Ukraine, Kyiv, 2009)
16. On the Methodological Recommendations of the Ministry of Education and Science on the Application of Elements of Bilingual Learning, Regulatory Letter No. 1/9-581 of 28.08.2009, Information Bulletin of the Ministry of Education and Science of Ukraine (2009). Accessed 16 Sept 2014
17. T. Bodnarchuk, S. Tsura, Osoblyvosti orhanizaciyi bilinhval'noyi osvity dlya ditej nacional'nyh menshyn v Ukrayini (Peculiarities of balancing education organization for children of national minorities in Ukraine). Bulletin of the University of Lviv **28**, 111–121 (2012)

Ivanna Blazhkevych about the development of pre-school education as a factor in building a strong state

Hryhorii Tereshchuk[1], *Oleksandra* Yankovych[2], *Kateryna* Binytska[3,*], *Olena* Binytska[3], and *Iryna* Kuzma[4]

[1]Ternopil Volodymyr Hnatiuk National Pedagogical University, Department of Service Industry, Technologies, Occupational Health and Safety, Ternopil, 46027, Ukraine
[2]Kujawy and Pomorze University in Bydgoshch, Department of Early Education, Special Pedagogy, Resocialization, 85-023 Bydgoshch, Poland
[3]Khmelnytskyi Humanitarian-Pedagogical Academy, Department of Pedagogy, Khmelnytskyi, 29013, Ukraine
[4]Ternopil Regional Communal Institute of Postgraduate Pedagogical Education, Department of Pedagogy and Psychology and Inclusive Education, Ternopil, 46027, Ukraine

Abstract. The activity and creative work of Ukrainian activist, writer, teacher Ivanna Blazhkevych (1886–1977) in the context of the development of pre-school education as a factor of strengthening the state have been thoroughly investigated in the article. For the first time the periodization of main stages of life and development of cultural-educational and scientific ideas of Ivanna Blazhkevych were substantiated in the article. The basic directions of pedagogical and cultural-educational activity and creativity of I. Blazhkevych have been clarified: pre-school-pedagogical (organization of pre-school institutions; substantiation of methods of pre-school education; determination of requirements for kindergarteners-educators); educational (participation in the elimination of illiteracy), ethnographic (collection of ethnographic material: recording folk songs, legends, sayings); national-patriotic (preservation of national traditions, mother tongue); guardianship (assistance to local war victims, prisoners of war, establishing orphanages for children); economic (founding cooperatives; working as a director of the district union of cooperatives in Ternopil), literary (writing books for children).

1 Introduction

1.1 Problem statement

The need for modernization of the pre-school education system in Ukraine in the context of European integration processes prompts a comprehensive study of the genesis of foreign educational systems and technologies in order to implement positive ideas in modern institutions. At the same time, the history of Ukrainian pre-schooling, when the genesis of kindergartens was influenced by progressive foreign reform pedagogy and domestic concepts and approaches to education, is worth studying. These processes took place in Eastern Galicia at the end of the 19th – beginning of the 20th century. On the one hand, the kindergarten folk traditions of pre-school pedagogy were implemented in children's "*okhoronky*" (until 1903, so-called kindergartens in the region; later – *zakhoronky*); on the other hand, European achievements of pre-school education were actively used.

Among the famous figures of Ukrainian pre-school pedagogy, which is not sufficiently popularized in modern educational books for students, it is necessary to mention the name of the famous educator and writer, the founder of children's "zakhoronky" Ivanna Blazhkevych, who almost a century ago expressed the view that "the path to "clear gate" lies through the education of the people, through "the construction of national life, and it must be started from the youngest generation, and this process is continuous" [1, p. 2].

Although Ivanna Blazhkevych is the founder of children's institutions, she substantiated the methodology of pre-school education, defined the requirements for kindergarten teachers, but her activity and creativity is reflected only in a few publications. At the same time, the study and popularization of the figure of Ivanna Blazhkevych in the context of pre-school education development is one of the important factors for building quality pre-school education in our country and hence its economic progress.

1.2 Analysis of recent researches and publications

Analysis of scientific works, textbooks, reading books on the history of pre-school education, monographs, dissertation researches showed that to date in the national historical-pedagogical science there are only single attempts to investigate and summarize the scientific-pedagogical and literary legacy of Ivanna Blazhkevych despite her significant role not only in the development of pre-schooling, but also in the Ukrainian region.

Ivanna Blazhkevych, as well as Nataliia Kobrynska and Kostiantyna Malytska, are the founders of pre-

* Corresponding author: rfn.yz87@gmail.com

school education in Eastern Galicia, these activists started the national pre-school education.

Studying historical-pedagogical sources shows that some theoretical and practical problems of pre-school education in the pedagogical thought of Ukraine in the second half of the 19-20th centuries have been analyzed in a number of researches (N. Salyha [2], I. Uliukaieva [3], B. Melnychuk [4], V. Khoma [4], O. Hanusyn [5]). However, the monograph of Z. Nahachevska "The Ukrainian Pre-Schooling in the Educational Space of East Galicia" (end of the 19th century – 1939) is especially distinguished by the soundness of representation of this issue. Among a number of prominent educational figures of that time Z. Nahachevska singled out the figure of I. Blazhkevych [6], described her recommendations on the establishment and organization of seasonal kindergartens in the early 1930-s, in particular, stating: "The book, identified by the author as a manual, consists of the introduction and 13 chapters. The basis of the manual was the experience of working with pre-school children by I. Blazhkevych herself" [7, p. 136].

The works of I. Rutylo [8], M. Chorna [9] deal with highlighting of pedagogical achievements of the Ukrainian activist, writer, teacher I. Blazhkevych. The scientific researches of I. Kuzma, who revealed the role of I. Blazhkevych in the history of media education, are characterized by the novelty, because through the development of aesthetic tastes the writer influenced the formation of media literacy of pre-schoolers at the time when there were no special media educational terms [10, p. 30; 11, p. 108].

Therefore, there are only a few scientific works in which the pedagogical heritage and contribution to the genesis of pre-school pedagogy by I. Blazhkevych are fragmentally reflected. However, we have not identified a comprehensive scientific study of the outstanding figure of ***Ivanna Blazhkevych in the context of the development of pre-school education and pedagogy, both past and present; her ideas that are worth reviving.*** At the same time, the pedagogical articles and books of the writer of the early 20th century of pre-school education are still quite interesting in the study of the historical development of pre-schooling, and remain relevant in pre-school theory and practice.

Objective of the research: on the basis of retrospective analysis to thoroughly study the creativity of Ivanna Blazhkevych as for the development of pre-school education as a factor in building a strong state and to determine the possibilities of realization of perspective ideas in the modern Ukrainian pre-schooling.

2 Material and methods

Research-bibliographic (for searching and systematizing sources), historical-retrospective (for analyzing the content and directions of pedagogical and cultural-educational activity I. Blazhkevych in historical retrospective), comparative (for comparative analysis of pre-school education in Eastern Galicia, creativity of the founders of pre-school education), biographical (for study the views and ideas of Ukrainian public figures, scientists, writers as for the development of theory and practice of pre-school education in specific socio-political conditions); prosopographic (for creating comprehensive characteristic of a prominent activist I. Blazhkevych), prognostic (for determining the possibilities of implementing of I. Blazhkevych's ideas in modern pre-school institutions).

3 Results

In Galicia, the ideas as for the social upbringing of small children emerged at the same time as the development of public pre-school education in Western European countries (the prerogative belongs to the German educator F. Froebel), and the emergence of appropriate prerequisites in the industrial production, which required new workers, including women, so a woman could no longer take care for the children who needed it. In addition to manufacturing, there have been social changes. We mean the emancipation of society. With the expansion of ideas about the outside world, access to education, women came from their private family world into the public world: they sought to get education, to acquire a profession. Of all the professional categories, women were most likely to pursue a career in schooling and literature [5, p. 128].

Therefore, it was precisely against the background of socio-economic factors that proposals were made to create pre-school establishments.

It is known that in 1848 the peasant Ivan Zaluzhnyi from Kamianka Volynska at the meeting of the Main Russian Council for the first time proposed the creation of Ukrainian pre-school establishments, but at that time this proposal was not supported [12, p. 19].

At that time, Polish and German charity communities were already developing in the Ukrainian territories, which together with the church, private individuals founded the gardens of angels, pre-school establishments, elementary pre-schools, zakhoronky, a day nurseries, Froebel's nurseries for children from rich families. Already in 1888 in Galicia there were 25 "zakhoronky" and Froebel's nurseries [12, p. 25].

Galician intelligentsia, rural communities, advocating the teaching of children in the Ukrainian language, considered the opening of national pre-school establishments and primary schools the first step in addressing this issue.

Theoretical and methodological basis for the development of pre-schooling in Eastern Galicia was the educational ideals, views and cultural-educational activity of Mykhailo Hrushevskyi, Ivan Franko, Oleksandr Dukhnovych, Nataliia Kobrynska, Kostiantyna Malytska, Omelian Partytskyi, Hryhorii Vretsiona, Ivan Petryshyn and other progressive figures.

The pedagogical foundations of the educational work of the first kindergartens were developed under the influence of Western Ukrainian theory and practice of social education, which mainly absorbed the achievements of Friedrich Froebel, Maria Montessori, Ovid Decroly and used their provisions on the purpose of pre-school education. At the same time, the achievements and acquisitions of the national experience of family education, which was based on Ukrainian ethno-pedagogy and moral principles of Christianity, were realized.

Considerable contribution to the development of pre-school education in Galicia made educators whose names and activity are little known to the contemporaries: I. Bartoshevskyi, I. Blazhkevych, M. Hrushevska, N. Kobrynska, V. Kotsovska, K. Malytska, I. Petryshyn, D. Shukhevych-Starosolska and others.

Due to their activities in the field of theory and practice of pre-school education, the purpose of the first kindergartens (1909) was defined as the protection of children from the effects and dangers of the environment; education of children, implementation of national patriotic and religious education [12, p. 59].

Ivanna Blazhkevych's figure is characterized by the diversity of activities and achievements among the theorists and practitioners of pre-school education. She is known in the Ternopil region primarily because she is a native and has spent there all her life. The regional archive of Ternopil region has a fund that holds documents related to work for the Ukrainian people by I. Blazhkevych. It is there that stores Manuscripts of Essays: About Children and Parents [13].

On the basis of the holistic and systematic analysis of pedagogical, cultural and educational activity and creativity of I. Blazhkevych, the main stages of which are shown in Table 1.

Table 1. Main stages of life and development of cultural-educational and scientific ideas of Ivanna Blazhkevych

No	Name of the stage, years	Main events
1.	***Educational-ethnographic (1886–1912)***	Born on October 9, 1886 in the village of Denysiv Kozlivskyi (Kozivskyi district) in Ternopil region in the family of the village teacher Omelian Borodiievych. There were 4 children in the family. Ivanna's sister's name is Melaniia. Brothers – Ostap and Yevhen. Receiving primary education in the village of Denysiv. Personal acquaintance with I. Franko in 1894.
2.	***Educational-guardian (1912–1928)***	Studying at the seven-year high school named after Queen Jadwiga in Ternopil. Active collection of ethnographic material: recording of folk songs, legends, proverbs. Combining self-education with active social activity. Organization in 1904 in the village of Denysiv of the women's section of the "Circle of Ukrainian Girls" (1904-1911) (cooking courses, systematic lectures of hygiene for all segments of the population in order to eliminate illiteracy in the countryside). The first literary attempts (1908). In 1909, at the First All-Ukrainian Agricultural Exhibition in the town of Stryi, she was awarded a diploma and a silver medal for her ethnographic collection. Perception of I. Franko's idea of the economic uplift of peasants through agricultural unions. Teaching activity with her husband teacher in the village of Subotiv, village of Zalukvy in Stanislav region. Participation in the elimination of illiteracy, enhancement of the general culture in the countryside. 1914 – acting as the headmaster of school in the village of Zalukva. Assisting local war victims, prisoners of war, establishing orphanage shelters for children. Description of military misery in the book "Woman on the Battle Line". Studying at the Lviv Teacher's Seminary. Activity in the society of child protection and youth care. Teaching at school. Establishment of reading rooms, dramatic circles.
3.	***Integration (1929-1977) – combination of pre-school-pedagogical socio-economic and literary activity***	***Stage I (1929-1939)*** *Creation of kindergartens and substantiation of methods of pre-school education of children. Orientation towards the achievement of European pedagogy.* ***Stage II (1939–1977)*** *Was dominated by active literary creativity and the concern for the creation of the new institutions for pre-school children. Orientation to the ideas of Soviet pedagogy.* Organization of the first kindergarten in 1929. Justification of the methods of pre-school education. Writing books for children "St. Nicholas" (1920); "Nativity Scene" (1924); "Cute Little Book" (1928), "The Fluffy King" (1929). Economic education of children and young people. Establishment of cooperatives. Work as a director of the territorial district cooperative union in Ternopil. Organization of agricultural courses. Writing an article "At the 50th Anniversary of the First Ukrainian zakhoronky" (1942). Organization of pre-school establishments. Since 1950 active literary creativity. 1963 – a member of the Union of Writers of Ukraine.

So, the main directions of pedagogical and cultural-educational activity and creativity of I. Blazhkevych are the following: pre-school-pedagogical (organization of pre-school institutions; substantiation of methods of pre-school education; determination of requirements for kindergarten educators); educational (participation in the elimination of illiteracy), ethnographic (collection of ethnographic material: recording folk songs, legends, sayings); national patriotic (preservation of national traditions, mother tongue); guardianship (assistance to local war victims, prisoners of war, establishing orphanages for children); economic (founding cooperatives; work as a director of the Union of cooperatives in Ternopil), literary (writing books for children).

Thus, Ivanna Blazhkevych is known as a children's writer. Her poems, stories, and nonfiction articles of her time were published by progressive Galician newspapers and magazines. The creative activity and public work of Ivanna Blazhkevych were supported by Olha Kobylianska, Uliana Kravchenko, Tymofii Borduliak, with whom the writer was on friendly terms. For children, Ivanna Blazhkevych began writing in the 1920-s. In the publishing house "Child's World" were published one by one her plays, stories, poems: "St. Nicholas in 1920", "Taras at the Deacon", "Deed in Honor of Taras", "Vertep", "Sweet Little Book", "Fluffy King", "On Mother's Day", "Ivas-Kharacternyk", "Narration" and others. After 1939, Ivanna Blazhkevych was actively involved in the creation of the Denysiv seven-year school, kindergarten, worked as a teacher, and took care of the cultural transformations in the village. New children's books by the writer "Podolianochka" (1958), "Arrival of a Stork" (1971), "Is There Something Brighter in the World" (1977), "Swallows" (1986), "In Kindergarten" (1993) and others [14].

Today, I. Blazhkevych's poems for children are often retold by children and parents, in particular, "Little Ukrainian Girl" "Is There Any Light in the World?" I. Blazhkevych spread the idea that the power of state formation is in education, and it is necessary to start education from the youngest age. The prominent educator has made other important steps towards improving the lives of Ukrainians: she is the founder of the cooperative movement in Ternopil and Podillia region.

At the same time, the figure of Ivanna Blazhkevych as a theoretician and practitioner of pre-school education in Eastern Galicia (beginning of the twentieth century) deserves attention, taking into account the urgency and the low level of popularization. In 1929, she organized in Denysiv, her native village, the first pre-school institution. Anna Zharska was the headmistress; Ivanna Blazhkevich was the curator. The medical care was given by Dr. Ivan Yavorskyi, a doctor from neighboring Kupchyntsi. The singing was taught by Ivan Blazhkevych, a retired teacher. There were 48 children who paid for their education. 6 children were educated free of charge, based on the decision of the department. The average number of children attending the kindergarten was 30-33 people [8, p. 107].

The payment was 5 zloty per month. The children's mothers gave 1 kg. of cereals or flour a week, 1 loaf of bread for 10 days, and each child brought with them a daily bottle of milk (1 liter) [8, p. 107].

Work in the kindergarten began at 6 o'clock. a.m. The children came without breakfast. At 8 o'clock they drank milk, at 10 o'clock they ate bread with butter or honey, at 12 o'clock they had lunch: thin gruel or dough with milk. Once a week there was lunch with meat.

From 14^{00} to 15^{00} there was a "dead" hour in the kindergarten, with full silence. The children slept in separate beds [8, p. 107].

At 16^{00} there was light lunch, consisting of milk and bread. There was self-service in the kindergarten. A person on duty appointed "matchmakers", "bridesmen" and "headman", who supervised the order, making sure that no one was hurt.

In 1929 the kindergarten was headed by Nusia Zharska, who became a favorite of the whole village. She loved all the children as relatives, taught them the games, fun and songs that were heard throughout the village. Such care of the pupils was appreciated by the mothers of children from the kindergarten and, as a gift, they bought her a wristwatch with the words: "From grateful mothers". Funds for the purchase of the watch were collected by the mothers of children [8, p. 108].

At the end of the education in kindergarten there was a children's holiday, with a program performed by the children: they sang, recited poems, and danced in the round, taught by Nusia Zharska. The dwellers of the village Denysiv gathered at the solemn event and welcomed every performance of the holiday program with the warm applause [8, p. 107].

Following the example of the village Denysiv the kindergartens in the nearby villages were founded. For kindergartens, the Society for the Protection of Children and Youth Welfare in Lviv trained headmistress in courses in the city of Lviv and in other cities [8, p. 108].

Recommendations for the founding and organization of seasonal kindergartens in the early 1930-s were reflected in the already mentioned work of Ivanna Blazhkevych, "Kindergarten in the Village during Summer Work".

The most valuable are the recommendations approved by the author and analyzed by Z. Nahachevska on the methods of lessons with children in kindergarten. Given that "service to the child is service to God", I. Blazhkevych considered interest as the main impetus for the activities and entertainment of children. She rejected the detailed work planning for each day: in her opinion, "... the teacher teaches the children what they are interested in" [6, p. 137].

The pedagogue and the children's writer considered the innate tendency to speak as the basis for the spiritual development of the child. That is why a lot of attention was given to the use of the native word – poems, stories, fables, dramatizations. "The science of poems" is the most elaborated in the manual. Not only scientifically justified requirements for the selection and educational orientation, methods and techniques of work with poetic works, but also created by I. Blazhkevych or skillfully selected texts of poems for formation of high patriotic feelings of a child are of great importance ("Prayer of the Ukrainian Child", "Who am I?") [13; 12, p. 137].

For the development of sensations I. Blazhkevych offered exercises to distinguish between colors, size, condition, body temperature, etc. The special attention was paid to games, motors, fun, and the author presented their models from S. Rusova's work "Theory and Practice of Pre-School Education", collections of A. Zhyvotyk, manual of O. Sukhoverska and others. At the same time, I. Blazhkevych considered it as ill-advised decision to introduce Ukrainian children into the alien and distant nature of the games offered by the Polish textbooks [11; 12, p. 138].

The guide contains tips for conducting "free" conversations, viewing and describing pictures, teaching children singing, music. In our opinion, it can still be used today in the work with pre-school children.

The logical conclusion of the work on the creation of organizational-methodological foundations of seasonal kindergartens was the publication by the General Board of the Native School Society of the Instruction entitled "How to Found and Maintain Kindergartens", which appeared in Lviv in 1932. Prepared based on the accumulated theoretical and practical experience of Ukrainian seasonal kindergartens, this instruction was intended as an official instruction from the leading organization that took over the leadership and control over the activities of all types of Ukrainian pre-school institutions. It revealed the main reasons and needs for opening kindergartens, the procedure for their establishment, contained instructions for the kindergarten teacher, requirements for kindergarten hygiene, nutrition for children. The instruction also offered a detailed plan for classes with children on hot and rainy days [12, p. 138-139].

It is noteworthy that, in recommending the indicative lesson plan, the creators of the instruction did not blindly follow it, but only sought to give the guide "approximate concept of how to deal". The instruction required to conduct classes taking into account the age characteristics of children, without coercion, on the basis of free choice, interest. It was emphasized that "… there are no precise education plans in kindergarten, because kindergarten is not a school in which children are taught, and we can never change the kindergarten to school". The best method of education was determined by the method of projects. It was believed that the methods of M. Montessori and O. Decroly, which were used in permanent zakhoronky, "will not be used in the kindergarten" [12, p. 139].

The instruction also had directions for organizing performances, holidays in kindergartens, and the like.

It was the first program of activity of Ukrainian seasonal kindergartens aimed at organizing a unified system of educational work with pre-school children. Despite some omissions inherent in the general pedagogical thought of the time, it generally corresponded to the purpose and tasks of comprehensive education of pre-school children, facilitated the transformation of this type of pre-school institutions into important centers of the physical and spiritual development of children and their ethnization.

I. Blazhkevych was interested in the problems of parenting both as a mother and as a teacher. She published notes and short correspondence on the pages of "Woman's Fate", in which she advised to teach children the simplest homework, to entrust them with the care of rabbits and poultry [15, p. 20].

In "Cultural Family", published in "Self-Education" in 1934, the author raises important problems of social upbringing – about the relationships between a man and a woman in the family, about raising children – worthy citizens of their homeland: "It is a very important thing to have a solid line in the upbringing of [children]. What has been commanded – must be fulfilled, what was once forbidden – must be so. The mother who prefers to punish the child in the first place, and in a minute is sad herself to caress and fondle the child. For the second time, such a child does not know what is free and what is not free, what is being punished and what is praised for, and will do what he/she likes, often to the own and others' detriment. Such a mother loses respect of the child, because the child sees that the mother herself does not know what is free and what is not free". The author immediately makes an instructive conclusion: "As we can see, being a smart mother is not such an easy thing, so a mother must read a lot and take care of her education in general" [8, p. 106].

The above points out that in the 1920-s – early 1930-s the appropriate organizational, methodological-theoretical basis for the activities of Ukrainian zakhoronky and kindergartens was established in Eastern Galicia. The most characteristic features of this process were: orientation of the theory and practice of pre-schooling of the region on the content dominants of the creative heritage of prominent representatives of national pedagogical thought, natives of the Dnieper Ukraine; active participation of women in the formation of organizational and theoretical-methodological foundations of national pre-school education; stable tendency to take into account the possibilities of Ukrainian institutions of guardianship, education and preparation of children for school as well as the peculiarities of functioning of society in the face of increasing oppression by the official Polish authorities.

The article "In the 50th Anniversary of the First Ukrainian Zakhoronky" (1942) stated: "By the efforts of the "Native School" seasonal kindergartens were organized during summer work in the villages and towns. In that work, the Ukrainian national teachers took active part, directing the new kindergartens to the right way of educating and teaching the village pre-school children. Also, reading rooms of "Prosvita" and branches of "Union of Ukrainians" were keenly interested in the kindergarten business and with their cooperation the number of kindergartens increased from year to year [8, p. 106].

I. Blazhkevych paid attention to the competent selection of pedagogical staff for kindergartens. Among the recruitment criteria she has in the first place the love that is embedded in every woman. It "encourages the kindergarten teacher to become more familiar with pre-school education; she will understand its importance, gain the will to know every child entrusted with her tutelage, will love these little children as relatives, will learn with them new games every time, and invent new games by herself", – wrote Ivanna Blazhkevych's granddaughter Iryna Rutylo [8, p. 107].

It should be noted that Ivanna Blazhkevych was convinced that the prospects of state building depend on the development of schooling: "bright future" lies through the quality education and it is necessary to start from the preschool education, and this process is continuous" [1, p. 2].

I. Blazhkevych, being convinced that it is education that would enable to build bright future in Ukraine, paid considerable attention to the economic education of pre-schoolers. At the same time, she considered economic activity of citizens as a significant factor in strengthening the state. She did not only promote cooperative ideas in Eastern Galicia, she was also the founder and the first chairperson of the cooperative commission at the Union of Ukrainian women in Lviv, the chairperson of the supervisory board of the district dairy and chief director of "Raifaisenka" in Denysiv, a long-time member of the board "Audit Union of Cooperatives in Lviv", a member of the supervisory board and director of the Podillia Union of Cooperatives in Ternopil [1; 16, p. 177].

I. Blazhkevych predicted that the cooperative movement would enable people to achieve well-being. In her autobiography, she stated, "based on cooperative movement we will get better destiny, the better destiny for the whole nation, we will bring happiness and well-being" [16, p. 178; 8].

Becoming a member of the Socialist Party in 1963 and a freelance correspondent I. Blazhkevych continued to assist in the setting up of kindergartens in the villages. In the archive of the house-museum many evidences of her petitions are stored – reviews from district committees, regional committees, from editorial boards about consideration of her letters, till the requirements for opening of permanent kindergartens, construction of premises for them in Denysiv and other surrounding villages were satisfied [8, p. 108]. In addition, she cared about the high quality of teaching staff in them, was indignant that there were educators who, avoiding work in the beet fields, went to work in kindergartens, having neither proper qualification, nor pedagogical abilities, nor the desire and love for children. She was also worried about the poor material base of the kindergartens: there were not enough toys, furniture, good food. In addressing these issues, I. Blazhkevych wrote articles in the newspaper. In addition, she took care of the aesthetics of the illustrations in the children's magazines "Baby" and "Barvinok".

Unfortunately, such a prominent figure has no proper respect in the history of pre-school pedagogy.

The study of modern textbooks, manuals (Z. Borysova, Z. Nahachevska, N. Salyha, I. Uliukaieva) for the students the history of pre-school pedagogy showed that the figure of Ivanna Blazhkevych, her educational and cultural-educational activities are not studied. And this is at a time when significant attention is paid to the contribution to the pre-school pedagogy of figures whose creativity is of no interest to contemporaries. Thus, there are articles about V. Belinskyi, M. Dobroliubov, N. Krupska and others in the textbook on the history of pre-school pedagogy ("Readings on the History of Pre-School Pedagogy (2004)"), however, there is not a single word about those people who founded Ukrainian pre-school institutions in Eastern Galicia [17].

Some educational and methodical works on the background of the development of Ukrainian pre-schooling cover the issues of activity of Galician educated women. We refer to the educational and methodical manual, authored and compiled by N. Salyha. It mentions I. Blazhkevych's activities. The author notes in particular: "At the origins of the theory and practice of public pre-school education were N. Kobrynska, K. Malytska, and I. Blazhkevych" [2, p. 58].

However, I. Blazhkevych's views on pre-school education should be reflected in more detail, since the interest to them is not only a reflection of the spirituality of society, but also the belief that it is only through the development of education that the level of economic growth and well-being of people can be enhanced.

The current state of pre-school education in Ukraine is defined as the stage of development between the old system of values, attitudes, traditions and the new, which is emerging, and is aims at providing optimal conditions for the education and upbringing of creative, active personality. In the 21st century, new requirements have been put forward regarding the goals, objectives, structure and content of pre-school education, related to the current general globalization processes, changing worldview and values, Ukraine's desire to integrate into the European social-humanitarian and administrative space, preserving own achievements. Implementation of the pedagogical, cultural and creative achievements of Ukrainian educators of Eastern Galicia second half of the 19th – beginning of the 20th century is also relevant nowadays because we can trace clear parallels to contemporary realities: the need for changes that would ensure the development of a child, the upbringing of national-patriotic values. In this way, it is advisable to improve and expand the networks of institutions of different types, change the mode of their operation; implementation of modern technologies for active socialization of children, correction of their physical and mental development; newest innovative models of integrated education. Modernizing the content of pre-school education requires defining clear methodological principles that take into account current world and European trends in its development and, at the same time, are based on the heritage of Ukrainian culture [18, p. 1].

4 Conclusions

Forming a new educational paradigm, it is important to use the following pedagogical ideas by I. Blazhkevych:
- expansion of the network of pre-school institutions, creation of institutions of various types, including part-time;
- strengthening of national-patriotic upbringing in pre-school education institutions (formation of love for the native land);
- formation of aesthetic tastes of children as a direction of media education by means of Ukrainian periodicals;

- defining requirements for kindergarten teachers. We believe that it is expedient to develop Professional Standard "Kindergarten Teacher";
- early involvement of pre-school children in self-government, in particular the distribution of tasks among pre-school children, in particular; this method of self-government is advisable to use as a way of preventing bullying.

At the same time, the use of books authored by the Ukrainian writer Ivanna Blazhkevych, who wrote popular poetry for young children is relevant in the educational process of pre-school institutions.

In our opinion, the renewal of interest in the activity and creativity of I. Blazhkevych as one of the leading national figures on the part of pedagogues, parents, children, students-future specialists of pre-school education, reflection of the system of her work in textbooks and manuals on the history of pre-school pedagogy will become another important indicator of positive changes in the educational field of modern Ukraine. However, the idea of strengthening the Ukrainian state through the development of education, especially for the youngest children.

References

1. I. Blazhkevych, *Dytiachyi sadok na seli v chasi litnikh robit* (Kindergarten in the village during summer work). (Lviv, 1930)
2. N. Salyha, *Istoriia doshkilnoi pedahohiky* (History of Pre-School Pedagogy). (NAIR, Ivano-Frankivsk, 2014)
3. I.H. Uliukaieva, *Istoriia doshkilnoi pedahohiky* (History of Pre-school Pedagogy). (Disa plius, Kharkiv, 2015)
4. B. Melnychuk, V. Khoma, Blazhkevych Ivanna Omelianivna, in *Ternopilskyi entsyklopedychnyi slovnyk*, ed. by H. Yavorskyi et al. (Zbruch, Ternopil, 2004), p. 142
5. O.B. Hanusyn, Dissertation, Lviv Ivan Franko National University, 2018
6. Z.I. Nahachevska, Dissertation, Kyiv Taras Shevchenko University, 1995
7. I.Blazhkevych, *Opovidannia* (The story). (Svit dytyny, Lviv, 1937)
8. I. Rutylo, Ivanna Blazhkevych i doshkillia (Ivanna Blazhkevych and Preschooling). Scientific notes of TDPU, Series "Pedagogy" 11, 104–108 (2002)
9. M. Chorna, Diialnist Ivanny Blazhkevych u Podilskomu soiuzi kooperatyviv uprodovzh 20-30-kh rokiv XX stolittia (Ivanna Blazhkevych's activity in Podillia union of cooperatives during 20-30-s years of the XXth century). Visnyk Lvivskoi komertsiinoi akademii, Seriia: Humanitarni nauky 11 (2013)
10. I. Kuzma, Formuvannia mediahramotnosti ditei starshoho doshkilnoho viku: teoriia i tekhnolohiia (Formation of media literacy of older preschoolers: theory and technology). (Vector, Ternopil, 2019)
11. S. Rusova, *Teoriia i praktyka doshkilnoho vykhovannia* (Theory and practice of preschool education). (Ukr. hromad. vyd. fond, Praha, 1924), p. 123
12. Z. I. Nahachevska, *Ukrainske doshkillia v osvitnomu prostori Halychyny (kinets XIX st. – 1939 r.)* (Ukrainian Preschool in the Educational Space of Halychyna (end of 19th century – 1939)). (NAIR, Ivano-Frankivsk, 2015)
13. I. Blazhkevych, *Rukopysy narysiv: Pro ditei ta batkiv* (Manuscripts of Essays: About Children and Parents), State archive of Ternopil region, f. P 3205, op. 1, cf. 32, 4 sheets
14. U Ternopoli vshanuvaly vidomu ukrainsku dytiachu pysmennytsiu Ivannu Blazhkevych (The famous Ukrainian children's writer Ivanna Blazhkevych was honored in Ternopil) (2016), https://day.kyiv.ua/uk/news/081016-u-ternopoli-vshanuvaly-vidomu-ukrayinsku-dytyachu-pysmennycyu-ivannu-blazhkevych. Accessed 28 Nov 2019.
15. *Knyha pamiati Ivanny Blazhkevych* (Book of memory of Ivanna Blazhkevych). Halych 32 (1991)
16. M. Chorna, Kooperatyvna diialnist Ivanny Blazhkevych u Skhidnii Halychyni (Cooperative Activity of Ivanna Blazhkevych in Eastern Galicia). Scientific Notes of Ternopil Volodymyr Hnatiuk National Pedagogical University, Series "History" 1, 96–103 (2012)
17. Z.N. Borysova (ed.), *Khrestomatiia z istorii doshkilnoi pedahohiky* (Readings on the History of Pre-School Pedagogy). (Higher School, Kyiv, 2004)
18. N.V. Huk, *Derzhavna polityka rozvytku doshkilnoho vykhovannia v Ukraini* (State policy for development of kindergarten in Ukraine) (2014), http://www.kbuapa.kharkov.ua/e-book/conf/2014-3/doc/3/06.pdf. Accessed 28 Nov 2019

Mathematical education at Kryvyi Rih State Pedagogical University: history, analysis of achievements and prospects of development

Tetiana Kramarenko*, *Volodymyr* Korolskyi, and *Dmytro* Bobyliev

Kryvyi Rih State Pedagogical University, 54 Gagarina Ave, Kryvyi Rih, 50086, Ukraine

Abstract. The research deals with development of mathematical education at Kryvyi Rih State Pedagogical University (KSPU). The goal and objectives of the research include distinguishing and characterizing basic stages of formation and development of KSPU's mathematical education, informing about the state of teaching, methodological, scientific and research activity, and defining prospects for developing the Department of Mathematics and Methods of its Teaching in future. By stages of developing mathematical education at KSPU, the authors mean periods of its 90-year development noted for certain peculiarities of organization and methods of training Mathematics teachers. Teaching, research and life dedicated to people are criteria that form the basis for determining basic stages in the Department of Mathematics' development. A special edition Mathematical Education in Kryvyi Rih Pedagogical University: Personality Dimension was issued to honour the 90th anniversary of the Department of Mathematics after analyzing information from the University archive, museum and accessible personal data of Mathematics teachers – University teachers, graduates and researchers. The data also concern their publications in ORCID and Google Scholar databases, the University repository, etc. There are characterized basic stages of developing mathematical education in the educational institution and its research trends.

1 Problem statement

From 60 % to 90 % of Mathematics teachers from Dnipropetrovsk region schools are graduates of Kryvyi Rih State Pedagogical Institute/University. During 90 years of its existence, a corps of highly professional teachers has been formed developing children education aimed at building the foundation for their further life success. As Ya. V. Shramko points out, teaching research, and life dedicated to people are priorities of a tertiary pedagogical institution [1]. Considering the fact that the teacher's role implies not only transmitting knowledge, but also forming the pupil's personality, the significance of a pedagogical university for those who used to go and go to region schools is evident.

2 Analysis of publications

V. N. Soloviev highlights some issues of development history of the Faculty of Physics and Mathematics in his paper dedicated to the 70th anniversary of Kryvyi Rih State Pedagogical Institute [2]. V. P. Rzhepetskyi presents some important stages of forming the Department of Physics and Methods of its Teaching as well as characteristics of main research trends on the University site [3]. Works by I. S. Mintii, S. O. Semerikov, V. N. Soloviev reflect the 25-year history of the Department of Computer Science and Applied Mathematics [4]. The given paper deals with scientific, teaching and guiding work, the teaching staff, students' training and job placement and the Department's prospects.

To honour the 85th anniversary of the University establishment, the scientific library has prepared the bibliographic list *Academic and Intellectual Centre of Dnipropetrovsk Region* [5] that presents materials from the library fund documents, the historical museum of Kryvyi Rih State Pedagogical Institute and accessible electronic resources.

Yet, the problem of developing mathematical education at Kryvyi Rih State Pedagogical University is not studied to the fullest in the mentioned papers.

3 The research goal and objectives

The research aims at singling out and characterizing basic stages of forming and developing mathematical education at Kryvyi Rih State Pedagogical Institute / University in order to inform about the state of teaching, guiding and research activity and prospects of the Department of Mathematics and Methods of its Teaching in future.

4 Main material

T. H. Kramarenko prepared a specialized edition *Mathematical Education at Kryvyi Rih Pedagogical University: Personal Dimension* dedicated to the 90th anniversary of the University and the Department of Mathematics [6]. The book contains materials from the University archive, accessible data from University

* Corresponding author: kramarenko.tetyana@kdpu.edu.ua

teachers', graduates', and researchers' personal pages as well as data on their research papers according to ORCID and Google Scholar databases, the University repository, the National Library of Ukraine named after V. I. Vernadskyi, etc. There are data on present-day teachers of the Department of Mathematics and Methods of its Teaching as well as on the pioneers of mathematical education at this educational institution. Those are well-known graduates who excelled at pedagogical activity including researchers, Mathematics and Computer Science teachers, excellent workers of Ukrainian education, the best school principals of Dnipropetrovsk region, and gifted young people who have chosen to be teachers. The edition is aimed at popularizing scientific and pedagogical experience of Kryvyi Rih region's educators – graduates and workers of the Department of Mathematics and Methods of its Teaching, enhancing the prestige of teaching Mathematics, Physics and Computer Science, spreading ideas of self-education and self-improvement in profession, theoretical and methodological foundations of implementing innovative technologies into Mathematics training. Short biographical reports, data on basic scientific and methodological activity trends are to orient readers towards choosing their own personal trajectory of professional growth.

Let us provide a short description of basic stages of developing mathematical education at the educational institution.

4.1 Stage 1, 1930-1941

In 1930, the Department of Physics and Mathematics was founded at Kryvyi Rih Institute of Vocational Education to teach a series of physical and mathematical subjects. Since then, history of the Faculty of Physics and Mathematics has started. In 1933, the Institute was transformed into the Pedagogical one and the first Mathematics teachers graduated in 1934. They acquired a wide range of knowledge and skills. 29 subjects including 18 field-related ones were taught at that time.

Such well-known mathematicians as L. A. Kareta, V. Boryshkevych and V. V. Sakk had been teaching in the early years of the Institute functioning [7]. Little is known about that period as part of the archive was lost in the war years. Professor L. V. Kareta taught in about 1933-1936. Besides, he was one of the founders of higher vocational education in Kryvyi Rih region working at the research sector of Kryvyi Rih Ore Mining Institute, the activity of which dealt with ore mining systems and search for efficient ways of mineral mining under certain conditions [7].

In 1936, after graduating from Dnipropetrovsk State University with an honour, B. M. Yakhnin (Speciality *Mathematics Teacher*) was placed to work at Kryvyi Rih State Pedagogical Institute [6, p. 133]. He started as a department assistant with a 480-hour yearly time sheet, but from 1938, he had been working as Dean of the Faculty of Physics and Mathematics and a senior teacher of the Department of Mathematics.

V. P. Zapolskyi worked as a Mathematics University teacher from September 1934 to 1935 and later on after a short pause until August 1941 [6, p. 62]. Having a 10-year experience of working at school, Zapolskyi taught methods of teaching Mathematics.

Fig. 1. Professor L. V. Kareta.

Fig. 2. Associate Professor B. M. Yakhnin.

Fig. 3. V. P. Zapolskyi.

Fig. 4. Associate Professor F. K. Kosyk.

In 1939-1941, F. K. Kosyk was Head of the Department of Mathematics [6, p. 78]. In April 1941, Fedir Kalenykovych defended his Candidate's thesis in Physical and Mathematical Sciences *Green Functions for Higher Order Operators* at Kharkiv State University and was awarded a scientific degree of Associate Professor of Mathematics confirmed by the Higher Attestation Commission on March 27, 1943. At that time he had six research, papers two of which were published.

Teaching and devotion were priorities of Stage 1 of the Department development.

4.2 Stage 2, 1944-1972

In 1944-1945, 1947-1951, and 1953-1955, Candidate of Physical and Mathematical Sciences F. K. Kosyk, was Head of the Department of Mathematics. In 1945-1947, Candidate of Physical and Mathematical Sciences A. A. Markov, was temporarily in charge of the Department of Mathematics. In 1955-1959, V. P. Zapolskyi, a senior teacher, presided as Head of the Department.

B. M. Yakhnin became his successor. In 1959-1972, he was acting as Head of the Department and in 1962 he was appointed as such on a competitive basis. Working at the Institute, Yakhnin taught courses in Mathematical Analysis and Theory of Functions, conducted special seminars and was in charge of a students' Mathematics Circle. In 1963, Yakhnin defended his Candidate's thesis and became Candidate of Physical and Mathematical Sciences. As a researcher, he worked on the issue *Investigation into Approximation of Functions on Classes by Means of Orthogonal Function Systems* in the field of the constructive theory of functions. He had a number of research papers on Mathematics and teaching and guiding manuals for students, some of which being published in *Journal of Higher Educational Institutions* and *Achievements of Mathematical Sciences* including *Studying the Upper Limit of Fourier-Valee-Poussin Totals by Means of Orthogonal Polynomials; Investigation into Some Processes of Function Approximation on Classes of Fourier-Valee-Poussin Totals According to Jacobi Polynomials; On Fourier-Valee-Poussin Totals for an Aperiodical Case* and *Poussin-Abel Totals as an Approximation Toolkit.*

In 1950, after graduating from the Faculty of Physics and Mathematics with distinction with a degree in Mathematics, M. P. Khoroshko [6, p. 130] was appointed Assistant of the Department of Mathematics and from 1954, he worked as a senior teacher. Later, he was chosen Dean of the Faculty of Physics and Mathematics. He upgraded his skills at Moscow State University and studied at Dnipropetrovsk State University as a postgraduate student. In 1969, Khoroshko defended his Candidate's thesis and got a Candidate degree in Physical and Mathematical Sciences. In 1974, he became Associate Professor of the Department of Mathematics (the course of *Theory of Functions and Functional Analysis*). Khoroshko is an author of almost 20 published research papers in Mathematics including those on uniform approximation of continuous functions by polynomials according to the Haar system; some experimental problems associated with approximation; assessment of the upper limit of Fourier coefficients on some classes of Haar, Walsh and Rademacher functions.

On the stage until 1972, along with the University restoration, we can mention *researches into Mathematics* as a leading type of activity of the Department.

Fig. 5. Associate Professor M. P. Khoroshko.

It is worth noting that in this period, the Department was engaged into issues of approximating functions by polynomials in various spaces. B. M. Yakhnin started that work in close cooperation with scholars from Dnipropetrovsk State University of the Order of the Red Banner named after the 300th Anniversary of Ukraine-Russia Reunion. The Department worked closely with Mykola Korneichuk (1920-2003), Academician of the National Academy of Sciences of Ukraine, a well-known Ukrainian mathematician, a pupil of Serhii Nikolskyi, Academician of the Academy of Sciences of the USSR. At that time, much work was done including the first results immediately published in a respectful Soviet scientific journal *Achievements of Mathematical Sciences* [8]. The paper proved dependency between sequences of Lebesgue functions $L_n^{(\alpha,\beta)}(x)$, $n \in \mathbb{N}_0$ that determined the exact upper limit of expanding $f(x)$ functions from $C_{[-1,1]}$ space according to Jacobi sequences $J_n^{(\alpha,\beta)}(x)$, $n \in \mathbb{N}_0$:

$$\begin{cases} L_n^{\left(\frac{1}{2},-\frac{1}{2}\right)}(x) = L_n^{\left(-\frac{1}{2},\frac{1}{2}\right)}(x) + O\left\{\dfrac{\left|\sin\left(n+\frac{1}{2}\right)\arccos x\right|}{\sqrt{1-x}} + 1\right\}, \\ L_n^{\left(\frac{1}{2},\frac{1}{2}\right)}(x) = L_n^{\left(-\frac{1}{2},-\frac{1}{2}\right)}(x) + O\left\{\dfrac{\left|\sin n \arccos x\right|}{\sqrt{1-x^2}} + 1\right\}, \end{cases}$$

where $n \to \infty$ i $O(1)$ is a value uniformly limited as to x and n.

This work is still relevant today as it is referenced in articles from well-known mathematical journals in 2015 and 2016. This paper has not become passed today as it was referred to in many famous mathematical journals in 2015 and 2016.

Yakhnin established similar regularities for Lebesgue functions while assessing a Jacobi polynomial series expansion with weight of $P(x) = \sqrt{\frac{1+x}{1-x}}$ and that of P. L. Chebyshov polynomials [9].

That work was performed in cooperation with the Department Associate Professors M. P. Khoroshko and O. I. Polovyna.

Khoroshko [10] built effective approximations for class functions H_ω^L, H_ω i H_V , by Haar polynomials by means of L matrix and obtained evaluation of built approximations. This paper is still topical nowadays as it was referred to in 2006 and 2016.

Polovyna was engaged in approximating functions by algebraic polynomials [11]. The mentioned research was highly appreciated and supported by A. M. Kolmohorov, Academician of the Academy of Sciences of the USSR.

Fig. 6. Associate Professor O. I. Polovyna.

4.3 Stage 3, 1972-1979

As V. N. Soloviev indicates [2], the Faculty reached its climax in the 1970s-1980s. At that time, Specialities *Physics and Mathematics* and *Mathematics* both involved five student groups, each including students taking correspondence courses. In 1972, the 1st *All-Union Conference on Problems of Diffusion and Defect Formation in Solid Bodies* was held at the Department. In 1975, a new research trend appeared – *Computer Modeling of Defects in Crystals* – followed up with regular conferences of All-Union scientific status. Professor A. E. Kiv, Doctor of Physical and Mathematical Sciences, the research supervisor of the seminar, became a laureate of State prizes of the USSR and the Ukrainian SSR. The Faculty departments conducted theoretical researches into physics of semiconductors, physics of solid bodies, theoretical and applied mechanics, methods of teaching Mathematics and Physics in cooperation with research institutes of Kryvyi Rih, Dnipropetrovsk, Kyiv and Moscow.

In 1975, the Department of Mathematical Analysis separated from the Department of Mathematics. Such Candidates of Physical and Mathematical Sciences, Associate Professors as B. M. Yakhnin, M. P. Khoroshko and O. I. Polovyna (Head of the Department in 1972-1982) worked there. Those researchers were engaged into various investigations including those in cooperation with the research school of M. P. Korneichuk, Candidate and later Doctor of Physical and Mathematical Sciences.

I. O. Polovyna was an author of a number of research papers in Mathematics, took an active part in scientific conferences and was in charge of scientific seminars on the theory of function approximation, variation calculus, linear programming and wrote several textbooks [6, p. 96]. In 1972-1977, he worked on the issue *Approximation of Functions Continuous over the Interval by Polynomials* considered as his Doctor's thesis.

In the same years, the Department of Mathematics comprising B. H. Druz (Head of the Department till 1976), S. V. Utkina, L. S. Naryshkina, N. V. Bohatynska and others enhanced their researches into the problems of pedagogics and methods of teaching Mathematics. In 1972, B. H. Druz defended his Candidate's thesis and got his Candidate of Pedagogical Sciences degree in methods of teaching Mathematics. He published his papers in journals specialized in methods of teaching Mathematics and, in 1976, got a degree of Associate Professor of the Department of Mathematics (the course of *Methods of Teaching Mathematics*).

Fig. 7. Associate Professor B. H. Druz.

From April 1976, S. V. Utkina acted as Head of the Department of Mathematics, and from September of 1977 till September 1982, she was appointed such on a competitive basis. In that period, the Department staff was working on the integrated multi-authored theme *Teaching Mathematics in Forms 4-10 of the Secondary School under the New Programme*.

Fig. 8. Associate Professor S. V. Utkina.

Besides teaching and research in the field of Mathematics in particular, the Department's activity was focused on life dedicated to people. O. I. Polovyna used to deliver lectures on scientific and popular topics for secondary school teachers and pupils including the issues of applying Mathematics to Physics. The Department established a theme-based workshop to provide methodological support to school teachers of Kryvyi Rih.

4.4 Stage 4, 1979-1992

In 1979, a well-known Mathematics methodologist H. P. Bevz came to Kryvyi Rih State Pedagogical Institute to deliver lectures on methods of teaching Mathematics for Mathematics students. He was the Institute graduate and after finishing the postgraduate course, he worked as a senior teacher of the Department of Mathematics. Bevz was a renowned personality in the field of teaching Mathematics and a theme-based workshop was arranged at the Institute for the Institute staff from both departments of Mathematics as well as those of Pedagogy and Physics. In 1986, the assistant of the Department, L. O. Chernykh defended her Candidate's thesis and got a Candidate of Pedagogical Sciences degree under the supervision of H. P. Bevz (*Improvement of Methods of Explaining Geometrical Concepts and Theories for School Plane Geometry Course*). The scientific school of H. P. Bevz continued enhancing the Department's potential through conducting researches into methods of teaching Mathematics. A great number of the Department teachers including S. H. Shyperko and O. V. Vikhrova were the pupils of Bevz while studying at the post-graduate course.

When P. I. Shevchenko became Rector of Kryvyi Rih State Institute, the teaching staff's research potential greatly increased. Shevchenko got his Candidate of Pedagogical Sciences degree when he was Head of the Education Department in Kryvyi Rih Council. Young professionals of the Institute became more engaged into researches while studying at post-graduate courses in Kyiv and other scientific centres, that fact increasing the number of Candidates and Doctors of Sciences working at the Institute. P. I. Shevchenko focused professionals' attention on application of Mathematics to inter-subject links of field-specific subjects, especially those studied by future teachers of production-type service work [8].

In 1981, the Department senior teacher S. V. Utkina defended a Candidate's thesis in Pedagogical Sciences (*Methods of Forming Geometrical Concepts at 8-year School through the Element System-Based Approach*).

Candidate of Pedagogical Sciences, Associate Professor A. L. Zhokhov contributed much to enhancing investigations into methods of teaching Mathematics. He was appointed Head of the Department of Mathematics on a competitive basis in 1982-1990. Among researchers of Kryvyi Rih State Pedagogical Institute who successfully defended their Candidates' theses supported by A. L. Zhokhov, one can mention I. F. Safir (*Enhanced Efficiency of Students' Mastering Algebra on the Basis of Structuring Training Materials (Incomplete Secondary Education*), 1984), A. M. Drozdov (indicated in the minutes of the Department meeting concerning A. L. Zhokhov), A. M. Kapinosov (*Methods of Forming Skills of Providing Arguments while Studying Mathematics in Forms 5-6*, 1988), etc. A. L. Zhokhov delivered lectures and conducted seminars on methods of teaching Mathematics, Algebra and Theory of Numbers, Mathematical Analysis, Mathematical Logics and History of Mathematics.

The mentioned stage of the Department history is noted for *enhanced researches into methods of teaching Mathematics*.

From 1980, Candidate of Physical and Mathematical Sciences (Speciality *Functional Analysis and Theory of Functions*) V. H. Filipenko started working as Associate Professor of the Department.

In 1982-1992, Associate Professor, a 1961 University graduate P. I. Ulshyn headed the Department of Mathematical Analysis. In 1971-1974, P. I. Ulshyn was a post-graduate student of the Institute of Mechanics of the Academy of Sciences of the Ukrainian SSR. In 1981, he defended a Candidate's thesis *Peculiarities of Movement of a Thread of Varied Length Conditioned by Non-Linear Properties*, (Speciality *Dynamics, Strength of Machines, Devices and Apparatuses)* at the Institute of Machine Building Problems of the Academy of Sciences of the Ukrainian SSR.

Fig. 9. Professor A. L. Zhokhov.

Fig. 10. Associate Professor P. I. Ulshyn.

1979-1992 are characterized by enhanced potentials of the Department of Mathematics due to the staff transferred from Kryvyi Rih Research Ore Mining Institute and the Laboratory of Mathematical Research Methods at the Research Institute R*udventyliatsiia* which was transformed into the All-Union Research Institute of Labour Safety in Mining in 1972. In 1976, V. V. Korolskyi began working as an assistant and then a senior teacher and Associate Professor of the Department of Mathematical Analysis [12]. Candidate of Technical Sciences H. M. Shevtsova, started working as a senior teacher and Associate Professor in 1987. In 1987, V. V. Petrov, after working as an assistant (Speciality *Computer Science and Computing Machines*), took the post of the leading specialist of this course, Associate Professor V. N. Soloviev, who was sent away to take a year internship. In September 1988, the Department senior teacher chosen on a competitive basis, Candidate of Technical Sciences M. M. Hudymenko taught Algebra and Theory of Numbers and later became Associate Professor of the Department.

In 1986, O. I. Oleinikov began working as a senior teacher at the Department of Mathematical Analysis. While working at the Department, he accomplished his Candidate's thesis *Rheological Ratios for Isotropic Media with Dilatancy and Internal Friction and Solution of some Problems*. He successfully defended his thesis for a degree of Candidate of Physical and Mathematical

Sciences in 1989 and developed his own courses – *Convex Analysis, Geometrical Transformations and Symmetry and Mathematical Modeling*.

Fig. 11. Professor O. I. Oleinikov.

O. I. Oleinikov [13] proved the theory of unity of a solution for the model of heterogeneous-elastic material for geometrically linear arrangement of the boundary value problem of the elasticity theory on the basis of precise solution of inequalities of local convexity of elastic potential and stableness and established necessary and sufficient conditions for the stableness postulate to build the model. He analyzed unity and stableness conditions for an individual case – the model of loose material – and built their geometrical interpretation. O. I. Oleinikov was engaged into implementing the method of boundary elements into calculating the stress-strain state of rocks.

Analysis indicates that the Department's potential increased due to contribution of Candidates of Technical Sciences whose scientific interests were associated with application of mathematical modeling to mining and adjacent sciences. In particular, while working as an assistant of the Department of Mathematical Analysis from 1979 and a senior teacher from 1982, V. N. Soloviev prepared and, in 1981, defended his Candidate's thesis and got a Candidate of Physical and Mathematical Sciences degree. At that time, he participated in arranging All-Union workshops on computer defects in crystals (Kryvyi Rih, 1975, 1977, 1982), supervised students' research work, developed curricula on Numerical Mathematics and worked out instructions for doing laboratory works.

4.5 Stage 5, 1992-1997

As rapid informatization of society conditioned the need for highly-qualified Computer Science teachers and engineers, the Institute introduced such new specialities as *Mathematics and Computer Science* (at the end of the 1980s), *Physics and Computer Science* (the mid-1990s) and *Computer Science* (2001). In 1992, the Department of Computer Science and Applied Mathematics was created. V. N. Soloviev became the Department Head and defended his Doctor's thesis in Physical and Mathematical Sciences in 1993. S. O. Semerikov, I. O. Teplytskyi, N. A. Kharadzhian and others defended their Candidates' theses in Pedagogical Sciences under the supervision of Doctor of Physical and Mathematical Sciences V. N. Soloviev.

In 1992-1994, the Department of Mathematical Analysis was headed by Candidate of Physical and Mathematical Sciences, Associate Professor O. I. Polovyna. From 1990 to 1994, Candidate of Pedagogical Sciences, Associate Professor L. O. Chernykh was the Department Head. In 1994, the Department of Mathematical Analysis and that of Mathematics merged under the supervision of Candidate of Physical and Mathematical Sciences O. I. Olienikov, who defended his Doctor's thesis in Physical and Mathematical Sciences at the Institute of Mechanics of the National Academy of Sciences of Ukraine in 1995. He also published a monograph based on his research in 1999. His scientific contribution accounted for 58 papers including 16 published abroad.

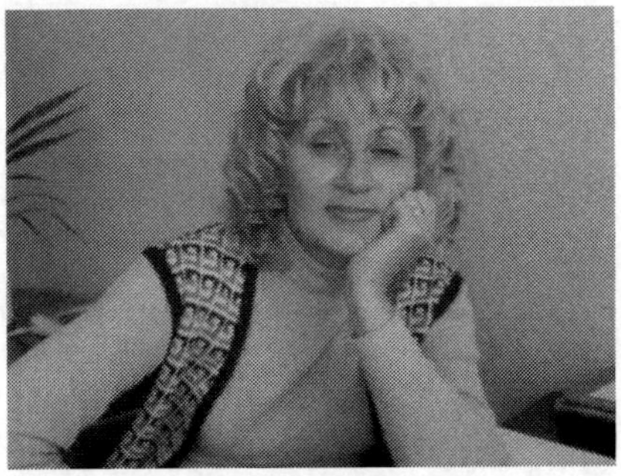

Fig. 12. Associate Professor L. O. Chernykh.

M. O. Rashevskyi, who graduated from Kryvyi Rih State Pedagogical Institute in 1988, conducted his researches at the Department of Mathematics followed by a successfully defended Candidate's thesis in Physical and Mathematical Sciences (Speciality *Differential Equations*) in 1995. His further scientific interests were associated with the theory of probability and other fields of Mathematics.

In 1997, O. V. Vikhrova defended her Candidate's thesis *Systematization and Generalization of Pupils' Mathematical Knowledge while studying Algebraic Structures* and got her Candidate of Pedagogical Sciences degree (13.00.02 *Theory and Methods of Teaching. Mathematics*).

4.6 Stage 6, 1997-2011

In 1997-2011, Candidate of Technical Sciences, Professor V. V. Korolskyi was Head of the Department of Mathematics at Kryvyi Rih State Pedagogical Institute/University.

In 1997-2002, there were three research trends in the Department activity – mathematical models and methods of heterogeneous elasticity (supervised by Doctor of Physical and Mathematical Sciences O. I. Oleinikov); methods of enhanced teaching of Mathematical Analysis, Geometry and Higher Mathematics (supervised by Candidate of Technical Sciences, Professor V. V. Korolskyi); development of methods for training Mathematics teachers working under educational humanitarization conditions (supervised by Candidate of Pedagogical Sciences, Associate Professor L. O. Chernykh).

Fig. 14. Professor V. V. Korolskyi.

In this period, there were two defended Candidates' theses including that by I. V. Lovianova – *Formation of Senior Schoolchildren's Intellectual Skills through Studying Natural Sciences* in 2006 (13.00.09 – *Theory of Teaching*) supervised by Doctor of Pedagogical Sciences, L. V. Kondrashova.

In 2003-2010, the Department of Mathematics staff worked on the integrated theme *Students' Independent Work in Studying Mathematical Subjects under Credit-Based Module Learning* supervised by Candidate of Technical Sciences, Professor V. V. Korolskyi.

In 2010, Candidate of Pedagogical Sciences, Associate Professor L. O. Chernykh, Candidate of Technical Sciences, Associate Professor P. I. Ulshyn, Candidate of Pedagogical Sciences, Associate Professor I. V. Lovianova, the senior teacher L. P. Korolska, the senior teacher S. H. Shyperko worked on the integrated theme of the Department, in particular, Stage 4 *Control and Evaluation of Students' Knowledge in Mathematics under Credit-Based Module Learning*.

Topicality. Under Credit-Based Module Learning, the problem of control and relevant evaluation of academic results of individual students is of great importance. After studying practices of similar departments of universities from Ukraine, Belarus and Russia accompanied by accumulation of the Department staff's experience, there appeared a necessity to systematize scientific and practical results of organizing, controlling and assessing mathematical knowledge under current conditions of higher pedagogical education.

Module and rating control and evaluation aimed at introducing students' independent work through developing their self-control and self-evaluation skills are becoming urgent.

The research tasks include:
- studying and introducing various mathematical models of assessing students' knowledge in Mathematical subjects into the Department activity;
- arranging and expanding potentials of various types and forms of control over training quality under the rating evaluation system to assess students' results in Mathematics;
- studying potentials and efficiency of conventional and modern means of making control and measuring devices for each mathematical subject;
- developing general requirements to form rating evaluation for the Department subjects considering varied methods of accumulation of grades in different mathematical subjects.

Research methods: theoretical analysis of scientific, psychological, pedagogical, teaching and guiding literature on issues under investigation, analysis and generalization of pedagogical experience; synthesis of achievements in didactics and methods of teaching Mathematics at higher educational institutions.

Theoretical research results:

1. There are developed general principles of forming rating for the Department subjects (informative content, systematicity, consistency, transparency, consideration of various activity types).
2. The role and functions of the rating evaluation system of academic results in fundamental mathematical subjects are determined:
- possible application of various control types and forms;
- encouragement of students' training and cognitive activity due to step-by-step evaluation of various work types;
- motivation of students for systematic work during the term;
- increased objectivity of final (examination) evaluation;
- reduced significance of random factors;
- uniform distribution of student-teacher load.
3. The system (scheme) of regulating control steps to form students' rating evaluation based on the concept of *training and rating units* is developed.

Practical research results:

1. Under the rating-based module learning, teachers implement various assessment modes to evaluate students' knowledge in mathematical subjects – models considering task parameters, performance time, acquisition levels; models based on probability criteria; models based on the theory of fuzzy sets.
2. There is created methodological and computer-assisted support to arrange and control students' results in mastering the Department's subjects.
3. The elaborated structure of rating evaluation of students' mathematical results enables monitoring students' individual academic trajectories (discussion at the Department meetings twice a term).

Testing of research results:

1. The research results were implemented into the Department activity: knowledge control is based on a weighted total of students' grades by various methods.
2. The research results of the integrated theme were reflected in some scientific and methodological papers and the Department teachers' reports at scientific conferences.

It is the Department's tradition to participate in organizing the international scientific and practical conference *Theory and Practice of Fundamental Subjects at Higher School* that has been conducted since 2001 [14]. The Department of Mathematics staff works closely with the Department of Computer Science and Applied Mathematics. In this respect it is worth mentioning Candidate of Pedagogical Sciences I. O. Teplytskyi and Doctor of Pedagogical Sciences S. O. Semerikov (*Theoretical and Methodological Basics of Fundamentalization of Teaching Informatics Subjects at Higher Educational Institutions*, 2009). Cooperation concerns issues of introducing Information Technologies into Mathematics training, participation in scientific conferences on problems of ICT implementation in education, All-Ukrainian scientific and methodological workshops *Computer Modeling in Education* and international scientific and practical conferences *Innovative Computer Technologies*. Being an organizer of a number of international scientific and practical conferences, S. O. Semerikov focuses on creating high-quality platforms for scientific discussions.

As a result of cooperation, a team of authors including V. V. Korolskyi, T. H. Kramarenko, S. O. Semerikov and S. V. Shokaliuk prepared a teaching manual *Innovative Information and Communication Technologies in Teaching Mathematics* (2009). In 2008, T. H. Kramarenko defended his Candidate's thesis (Formation of Schoolchildren' Personal Features of Under Computer-Oriented Learning of Mathematics) and got a Candidate's of Pedagogical Sciences degree (13.00.02 – Theory and Methods of Teaching. Mathematics) under the supervision of Doctor of Pedagogical Sciences, Professor, Active Member of the National Academy of Pedagogical Sciences of Ukraine M. I. Zhaldak. Teaching aids visualization during lectures and practical classes, in particular in Mathematics, allows students to understand the learning material better, to increase the applied orientation of learning and the communication competence both learners and teachers. One of the ways to improve the abstractions visualization in Mathematics is a pedagogically sound and appropriate application in the teaching the modern ICT. Report cleared out the integral computer-oriented approach. It in detail its use in the study of school course of Mathematics, directed on forming of personality qualities of student such as organizational, cognitive, creative; ones; criteria and levels of formed of personality qualities are specified.

4.7 Stage 7, 2011-2019

In 2011, the Department of Mathematics was transformed into the Department of Mathematics and its Teaching Methods. From that time till April 2019, the Department was headed by Candidate of Technical Sciences, Professor V. V. Korolskyi who was also an Honorary Figure of People's Education of Ukraine.

From January 2020, Candidate of Pedagogical Sciences, Associate Professor D. Ye. Bobyliev was chosen Head of the Department of Mathematics and Methods of Teaching on a competitive basis.

In the mentioned period, the Department staff conducted a series of researches, while one Doctor's thesis and five Candidates' theses were defended.

There were defended four theses in methods of teaching Mathematics (Speciality 13.00.02 – Theory and Teaching Methods. Mathematics), their results being implemented into future Mathematics teachers' training. In 2015, the Department Associate Professor I. V. Lovianova defended her Doctor's thesis in Pedagogical Sciences (*Theoretical and Methodological Foundations of Teaching Mathematics at Profession-Oriented School*) supervised by Doctor of Pedagogical Sciences, Professor N. A. Tarasenkova) [15].

The dissertation [15] is devoted to the problem of scientific substantiation of the theoretical-methodological bases of formation of professional orientation of the individual students in the senior profile school during their mathematical training, creating and implementing real learning process of methodical system of professionally oriented teaching mathematics. The dissertation is offered the concept of mathematical training of pupils of profile school which basic provisions are based on: the role of mathematical training in education; on the group of the principles, among which the classical didactic principles; the principles of profile training; the principles of mathematical training of seniors; the principles of design of process of training in mathematics at the profile school. Methodological tools of research of problems of profile differentiation of training are chosen: in historical aspect – the comparative approach (for comparison in the sphere of a diachrony and comparison of the phenomena removed spatially); general scientific approaches (system-structural, operational, axiological, semiotic, competence) to provide training, which ensures the formation and development of the individual pupil; task approach to teaching mathematics, which provides the organization of mastering the content of professionally directed training of mathematics profile school pupils through the introduction of a learning content professionally designed tasks; the state documents on updating and improvement of the content of mathematical education; the historical and modern tendencies of development of school mathematical education in the context of its profiling.

In 2017, the Department part-time assistant O. V. Ambroziak defended his Candidate's thesis (*Methodological Foundations of Implementing a Competence-Based Approach in Teaching Linear Algebra for Future Informatics Teachers*) supervised by Doctor of Pedagogical Sciences, Professor I. V. Lovianova.

In 2018, D. Ye. Bobyliev defended his Candidate's thesis (*Methods of Teaching Functional Analysis for Future Mathematics Teachers*) supervised by Doctor of Pedagogical Sciences, Professor I. V. Lovianova. Insufficient attention to the practice of applying numerical methods in the existing functional analysis manuals compiled for applied specialties can be explained by several circumstances. Firstly, the very ideology of functional analysis is tuned to the high abstractness of this section of mathematics. Secondly, the training trajectories of this discipline were structured at a time when computer technologies were still far from the leading role in education, and therefore, their connection to the educational process was not perceived as something natural and not burdensome. Thirdly, numerical methods are traditionally presented in a separate course of computational mathematics (or course of numerical methods) [16].

The approximation of the course of functional analysis to computational mathematics contributes to the continuity and coherence of vocational training. Perhaps this is even the only way to fully implement functional analysis in a pedagogical university. The convergence with computational mathematics should be such as to fully prove the theoretical fact to the number: to trace the projection of abstract ideas into the plane of numerical methods and to give an opportunity to immediately test methods in computational practice. Of course, the measure of this convergence should be reasonable, so that functional analysis does not lose its identity and is not substituted by the course of computational mathematics. To solve these problems, a scientific methodological research was conducted and a set of two textbooks was developed: a summary of lectures and a collection of tasks on functional analysis for pedagogical universities [16].

In 2014, I. S. Dereza defended her Candidate's thesis in Pedagogical Sciences (*Organization and Pedagogical Conditions of Controlling Mathematics Students' Independent Work at Pedagogical University*), Speciality 13.00.06 – Theory and Methods of Educational Management, under the supervision of Doctor of Pedagogical Sciences, Professor L. V. Kondrashova.

In 2017, K. V. Polhun defended her Candidate's thesis in Pedagogical Sciences (*Organization of Inclusive Teaching of Physical and Mathematical Subjects for Handicapped Students at Higher Educational Institutions*), Speciality 13.00.09 – Theory of Teaching) under the supervision of Doctor of Pedagogical Sciences, Professor Z. P. Bakum [17]. The report contains the results of theoretical and experimental research of the organization process of inclusive learning of physical and mathematical disciplines by students with disabilities in technical institutions of higher education. Psychological factors of organizing inclusive learning of physical and mathematical disciplines by students with disabilities in technical institutions of higher education are identified.

Dominant methodological approaches (individual-oriented, competence-oriented, and systemic one) have been identified that contribute to the efficient organization of inclusive learning of physical and mathematical disciplines by students of the category under investigation in technical institutions of higher education, specific teaching principles have been determined. Forms, methods, techniques of teaching processes have been examined that ensure the formation of mathematical competence in students with special needs. The expediency and the necessity of using the research method of learning, the heuristic method, the method of problem statement, the method of projects, the method of learning by cooperation, the method of simulation of professional situations, as well as information and communication technologies along with adaptive ones have been proven.

It has been found that the mathematical competence of the student of a technical institution of higher education can be represented by a synthesis of the following components: motivating-values component, cognitive component, action-dominated component, and reflective one. In accordance with the structure of the mathematical competence of students, the relevant criteria (valued-orientation criterion, cognitive criterion, procedural criterion, evaluative and regulative one) and their indicators have been identified to determine the formation levels (low, medium, sufficient, high) of each of the components of that competence.

A functional and structural model of inclusive learning of physical and mathematical disciplines by students with disabilities in technical institutions of higher education has been developed, which requires creating a set of certain teaching conditions and contains four interrelated components: the target-oriented component, the content-oriented component, the operation-and-action-oriented component and the control-and-evaluation one.

The teaching and methodological materials developed during the study have been represented in the author's Guide called "Inclusive Learning of Mathematics in Institutions of Higher Education" and in the Methodology Recommendation on the use of electronic instructional and methodological package for higher mathematics under the conditions of inclusive learning [17].

In 2011, the Department of Mathematics and Methods of its Teaching in cooperation with the Department of Computer Science and Applied Mathematics conducted the All-Ukrainian scientific and methodological conference of young scholars *Innovative Information and Communication Technologies for Teaching Mathematics, Physics and Informatics at Secondary and Higher Educational Institutions* with the corresponding collection of papers published [18].

In 2009-2014, the Department staff worked on two integrated themes. One of them is *Technology of Differential Teaching of Mathematics at Schools of General Education* supervised by Candidate of Technical Sciences, Professor V. V. Korolskyi, Candidate of Pedagogical Sciences, Associate Professor A. M. Kapinosov.

The *project executives* were senior teachers N. V. Bohatynska, L. R. Korolska, H. M. Bilousova, Candidate of Technical Sciences, Associate Professor P. I. Ulshyn, Candidates of Pedagogical Sciences, Associate Professors T. H. Kramarenko, L. O. Chernykh and O. V. Vikhrova.

The *research is aimed* at theoretical substantiation of technology of theme-based level teaching of Mathematics, developing teaching and guiding support of the technology and testing efficiency of the implemented technology in teaching Mathematics at secondary schools.

In 2014, Stage 4 (finalizing and correcting) of the theme development was performed.

The research final results were three packages.

Package 1 *Theoretical Substantiation of Differential Technology of Teaching Mathematics*: there were substantiated content and organization-methodological principles and rules of designing, arrangement and implementation of theme-based level differential teaching of Mathematics; models for teaching multi-level pupils; benchmark characteristics of training levels in Mathematics in the form of multi-level criterion tasks and problems.

Package 2 *Theoretical Procedures of Theme-Based Levelled Differential Teaching of Mathematics*: there was description of the technology as a series of solving didactic problems, local technologies and the technology of basic lesson types.

Package 3 *Methodological Toolkit of Theme-Based Level Differential Teaching of Mathematics*: there were developed blocks of training tasks, control tasks, lesson plans, course books and teaching manuals for level teaching.

The developed products were in electronic format. There were 3 published course books, five teaching manuals, 30 scientific and methodological papers eight of which were published in dedicated journals. There were also 25 qualification and 40 course papers written and defended.

Testing of the research results was performed through delivering lectures and conducting practical classes at advanced training courses for Mathematics teachers at Kryvyi Rih State Pedagogical University comprising 50 lectures at the University and 60 lectures at regional institutes of post-graduate education. There were conducted several scientific and methodological workshops for Mathematics teachers.

In 2015-2019, the Department of Mathematics and Methods of its Teaching staff worked on the integrated theme *Theoretical and Methodological Foundations and Technologies of Implementing a Competence-Based Approach into Training of Mathematics Teachers* supervised by Candidate of Technical Sciences, Professor V. V. Korolskyi. The theme was approved at the Department of Mathematics and Methods of its Teaching meeting, Minutes 11, 20 May, 2015. The research character was fundamental and applied.

I-II. Theoretical and Methodological Stage (January 2016-December 2017).

The stage was aimed at developing theoretical and methodological foundations of the technology of forming professional competences (basic, specific, partial) while studying content modules of mathematical subjects and courses of methods for teaching Mathematics.

The research tasks included:

1. Determining the role and the place of content modules of mathematical subjects (Mathematical Analysis, Algebra and Theory of Numbers, Geometry, Discrete Mathematics, Theory of Probability and Mathematical Statistics, Functional Analysis, Methods of Teaching Mathematics) in forming (developing, studying in depth) professional competences of future Mathematics teachers.
2. Determining tasks on the basis of content modules of subjects (see 1) relevant to developed professional competences.
3. Revealing peculiarities of methodological systems of teaching content modules being guided by the competence-based approach (result-oriented, organization- and content-based).

Theoretical results:

1. The role and the place of content modules of subjects in the framework of new curricula for Speciality *Mathematics* (Terms 1 and 2: Mathematical Analysis, Algebra and Theory of Numbers, Geometry) were determined.
2. Professional competences and basic professional tasks for forming field-specific competences were determined.
3. Peculiarities of methods for teaching content modules guided by the competence-based approach were partially ascertained.

The structure of teaching manuals to study content modules for the Department subjects during students' independent work was elaborated.

The research was in accordance with the aim and the tasks of the corresponding stage – development of theoretical and methodological foundations of forming professional competences in studying content modules of mathematical subjects and courses of methods of teaching Mathematics.

The research resulted in setting the role and the place of content modules studied in Term 1 while forming (developing studying in depth) professional competences and defining basic professional tasks relevant to competences, ascertaining peculiarities of methods of studying content modules guided by the competence-based approach.

Stage III: Applied (technological) (September 2017-December 2018).

The research tasks included description of formation of professional competences, development of training and monitoring materials.

The research results included teaching and guiding manuals and reference books.

Stage IV. Reflexive (controlling and concluding): January – December 2019. Finalizing the Department theme. The research was aimed at developing methodological, theoretical and method-based foundations of the competence-based approach in Mathematics teachers training while studying mathematical subjects and courses of methods of teaching Mathematics.

The research topicality. The models of professional competences (scientific-subject and method-based) of Mathematics teachers were crude and underdeveloped as well as the method systems (objectives, content, organization and methods, monitoring) of module learning for mathematical courses and methods of teaching Mathematics based on the competence-based approach both on the theoretical and practical (technological) levels.

The research is aimed at analyzing the research results, correcting methods, developing recommendations, generalizing, systematizing data and preparing a monograph and teaching manuals.

In 2019, there were three monographs two of which were published in Ukraine and one – abroad, eight course books and teaching manuals, eight papers four of which were published in Ukraine and four – abroad The results of 2019 included three monographs, one of which published abroad, eight course books and teaching manuals; eight papers, four of which published abroad (Scopus and Web of Science databases) [19-24].

The published course books included *Geometry (specialized level), Textbook for Form 10 of Secondary Schools* by Professor I. V. Lovianova [25]; the revised edition of the teaching manual *Mathematics. Integrated Edition for State Final Attestation and External Independent Evaluation* by H. M. Bilousova, A. M. Kapinosov, and P. I. Ulshyn [26], the revised edition of the teaching manual for training future Mathematics and Computer Science teachers *Innovative Information and Communication Technologies for Teaching Mathematics* by T. H. Kramarenko, V. V. Korolskyi, S. O. Semerikov, S. V. Shokaliuk and M. I. Zhaldak (ed.) [27].

It was carried out the maintename of mathematics, proper theoretical and task material, in particular, for the deep study of mathematics, pedagogical programmatic facilities, computer-oriented methods and forms of studies, were considered principles of construction of the system of developing tasks. The methodical recommendations are developed in relation to the use GRAN1, GRAN-2D, GRAN-3D, DG, GeoGebra in an educational process. The results of pedagogical experiments confirm the efficiency of the offered components of the computer-oriented system methodical of studies [27].

S. O. Semerikov attributes Artificial Intelligence Education Applications, Applications in Education, Conversational User Interfaces, Blockchain in Education, Immersive Technology Design Thinking, Competency-Based Education Platforms and Adaptive Learning Platforms to the main tendencies of using ICT in education. Since augmented reality technology already has an important place in innovative development, it can also have significant potential for implementation in Mathematics learning. That is why this technology needs more detailed study [28].

Because augmented reality is intrinsically linked to 3D-construction, its usage in conjunction with Dynamic Mathematics systems like GeoGebra, can significantly increase the level of visualization in Mathematics and enhance students learning. In addition, Augmented Reality can become a tool for enhancing STEM-based learning for students majoring in Mathematics and Computer Science [29-30]. At present, the use of augmented reality technology in teaching, including Mathematics, requires development, research, and testing.

Electronic academic courses based on the electronic management system MOODLE are widely applied to training Mathematics teachers.

The Department of Mathematics and Methods of its Teaching has been a vital and integral part of the Faculty of Physics and Mathematics of Kryvyi Rih State Pedagogical University since its founding. The department arranges training work at the Faculty of Physics and Mathematics as majors and at other departments as minors.

Annually the faculty welcomes school leavers and encourages them to apply for the area of expertise "Mathematics", as well as the following supplementary specialties: "Computer Science".

According to the staff division in the beginning of academic year 2019-2020 the department has 1 Doctor of Pedagogical Sciences, 8 Candidates of Sciences; 1 Professor, 4 Associate Professors.

Teachers of the Department of Mathematics and Methods of its Teaching carry out research work in accordance to the general plan of the department. The system of preparation of scientific and pedagogical staff actively functions at the department through postgraduate studies.

The department offers postgraduate studies in the following areas of expertise 13.00.02 – Theory and Methodology of Teaching the Mathematics.

The bachelor's degree program is designed to take four years to complete. The following courses are available at the faculty to obtain the Bachelor's degree: "Mathematical analysis" [22], "Complex analysis", "Functional analysis"[16], "Financial Mathematics", "Calculus", "Differential equations" [&&], "The basis of further Mathematics", "Methodology for teaching mathematics", "Analytical geometry", "Differential geometry", "Mathematical Logic", "Linear Algebra" [23], "Algebra and Number Theory", "Elementary Mathematics", "Discrete Mathematics", "Mathematic and statistic", "Information and communication facilities for teaching mathematics", "Probability theory and mathematical statistics", "Intel Training for the Future Course" seminars and special courses on Mathematics. The Master's degree requires completing an undergraduate degree. The faculty offers the following academic courses:, "Methodology for teaching Mathematics in the profile school", "Selected geometry questions", "History of development of mathematical education Higher Mathematics", "Methods of mathematical statistics in scientific research", etc. completion of these courses is awarded with the Master's.

The department provides a solid foundation for sustainable scientific development.

Faculty graduates become teachers of Mathematics and Computer Science. They are involved into scientific activities not only in Kryvyi Rih and Ukraine research establishments, but also in the leading foreign institutions abroad.

5 Prospects of the Department's research activity

The short-term prospect of the department involves improving scientific, methodological and information provision of training future Mathematics teachers as well as improving their qualification.

Nowadays, a university teacher's professional activity is shifting towards the informational space. There are many factors causing this including introduction of education assisting systems like Moodle. Innovative technologies and new training methods are being implemented into the educational system. Digitalization of education has become urgent.

Blended learning embracing digital distance technologies and contact teacher-student communication is becoming a well established practice. Blended learning is an educational technology, which combines traditional (face-to-face) training and elements of distance technologies (on-line training). Combination of these two forms of training provides for their equal significance in the training process. It should be noted that blended learning envisages considerable amount of students' independent work and their participation in building their own academic trajectories. It acquires specific significance in arranging future Mathematics teachers' training characterized by students' high governance and required development of independent training skills. For this reason, the Department is planning to conduct the research called *Blended Learning Technology in Fundamental and Professional Training and Improvement of Qualification of Mathematics Teachers* in 2020-2024. The research aims at developing, substantiating and testing the model of implementing the blended learning technology into Mathematics teachers' fundamental and professional training and improvement of their qualifications as well as finding ways of implementing the technology.

In the immediate future, the Department should adopt a systematic assessment of students' knowledge according to content-based modules for the relative subjects through using the system of controlling electronic academic courses in MOODLE and publish accumulated experience of work with under- and postgraduates in the form of teaching, guiding and research materials.

The mid-term prospect involves improvement of the academic and methodological activity of the department and gradual transition to the training and technological activity. In other words, the Department teachers should master the technological approach to training, develop models of professional knowledge and present training materials of the Department-related subjects as adaptive knowledge-oriented training technologies. At this stage of working on the research, the Department staff's activity should be aimed at developing theoretical principles of using MOODLE and investigating into opportunities to increase the Department's virtual intellectual potential through applying models of highly qualified specialists' professional knowledge. Scientific substantiation and experimental testing of economic efficiency of MOODLE application to training Mathematics teachers and improving their qualification is an important trend of the Department's activity. Basic mid-term tasks of the Department development include further improvement of the scientific school of methods for Mathematics training by increasing the number of postgraduates for Speciality *014.04 Secondary Education (Mathematics)*. Teaching and guiding aids developed at this stage should reflect results of testing blended learning technologies accompanied by experimental results published in research papers and monographs.

The final stage of working on the multi-authored research provides substantiation of application of the blended learning technology in the full cycle to training Mathematics teachers. While working at the mentioned research, the Department staff should focus on improving the material and technical support, increasing the number of printed and electronic papers (monographs, textbooks, teaching guides on subjects of fundamental and professional training of Mathematics teachers), improving the image of the Department in the Ukrainian scientific space, expanding contacts with foreign scientific organizations and educational institutions, and taking an active part in international scientific and research projects on educational problems.

6 Conclusions

The research considers basic stages of history of the Department, analyzes its main achievements. The Department's activity is aimed at implementing academic curricula of specialists training, re-training and enhancing their qualification, training scientific and pedagogical staff, elaborating a single concept of the academic content for the subjects taught at the Department, developing requirements and suggestions for upgrading the content and volume of curricula and preparing scientific and methodological support of the subjects. The Department is engaged into developing and implementing innovative academic technologies into the training process, conducting researches aligned with research trends of the University. Much attention is paid to cooperation with academic and research institutions of Ukraine and other countries. The Department staff participates in training, re-training and upgrading secondary school and University teachers' qualification.

References

1. Ya.V. Shramko, Misiya pedahohichnoho universytetu (Mission of the Pedagogical University) (2017), https://kdpu.edu.ua/blogs/2017/10/03/misiya-pedagogichnogo-universytetu/#more-8. Accessed 28 Nov 2019
2. V.M. Soloviov, in *Computer simulation and information technology in the natural sciences* (KSPU, Kryvyi Rih, 2000), pp. 3–6
3. V.P. Rzhepetskyi, Istoriya kafedry fizyky ta metodyky yiyi navchannya KDPU (History of the Department of Physics and methods of its teaching KSPU) (2017), https://kdpu.edu.ua/fizyky-ta-metodyky-ii-navchannia/zahalna-informatsiia/pro-kafedru/102-istoriia-kafedry-fizyky-ta-metodyky-ii-navchannia.html. Accessed 28 Nov 2019
4. I.S. Mintii, S.O. Semerikov, V.M. Soloviov, in *Proceedings of the V All-Ukrainian Scientific-Practical Conference of Young Scientists "Scientific Youth-2017"*, Kiev, 2017, ed. by O.M. Spirin, A.V. Yatsyshyn, pp. 167–171
5. O.M. Kravchenko (ed.), *Osvitn'o-intelektual'nyy tsentr Dnipropetrovshchyny* (do 85-richchya zasnuvannya Kryvoriz'koho pedahohichnoho instytutu DVNZ "Kryvoriz'kyy natsional'nyy universytet") (Kryvyi Rih, 2015), http://elibrary.kdpu.edu.ua/handle/0564/299. Accessed 28 Nov 2019
6. T.H. Kramarenko (ed.), *Matematychna osvita u Kryvoriz'komu pedahohichnomu: osobystisnyy vymir* (Mathematical Education at Kryvyi Rih Pedagogical University: Personal Dimension) (Kryvyi Rih State. Ped. Univ., Kryvyi Rih, 2020), http://elibrary.kdpu.edu.ua/xmlui/handle/123456789/3617. Accessed 08.02.2020
7. Ya.V. Shramko (ed.), *Profesory Kryvoriz'koho pedahohichnoho* (Professors of Kryvyi Rih Pedagogical) (Kryvyi Rih, 2015)
8. B.M. Yakhnin, Uspehi Mat. Nauk **13**, 6(84), 207–211 (1958)
9. B.M. Yakhnin, Izv. Vysš. Učebn. Zaved. Matematika 1(2), 202–207 (1958)
10. M.P. Khoroshko, Mat. Zametki **6**, 47–54 (1969)
11. N.P. Korneichuk, O.I. Polovyna, Dokl. Akad. Nauk SSSR **166**, 281–283 (1966)

12. H.M. Vinyar (ed.), *Volodymyr Viktorovych Korol's'kyy: Docendo discimus (Navchayuchy, navchayusya): (do yuvileyu profesora, kandydata tekhnichnykh nauk, zaviduvacha kafedry matematyky ta metodyky yiyi navchannya Kryvoriz'koho derzhavnoho pedahohichnoho universytetu, zasluzhenoho pratsivnyka osvity Ukrayiny): biobibliohrafichnyy pokazhchyk* (Volodymyr Korolskyi: Docendo discimus (Teaching, studying): (to the anniversary of professor, candidate of technical sciences, head of the department of mathematics and teaching methods of Krivoy Rog State Pedagogical University, Honored Worker of Education of Ukraine): biobibliographic index) (Kryvyi Rih State. Ped. Univ., Kryvyi Rih, 2020), http://elibrary.kdpu.edu.ua/xmlui/handle/123456789/3045. Accessed 28 Nov 2019
13. O.I. Oleinikov, E.V. Mogilnikov, Dalnevost. matem. Journal **3**, 2, 242–253 (2002)
14. *Teoriya ta metodyka navchannya matematyky, fizyky, informatyky: zbirnyk naukovykh prats* (Theory and methods of teaching mathematics, physics, computer science: a collection of scientific works) **IX** (NMetAU Publishing Department, Kryvyi Rih, 2011), http://elibrary.kdpu.edu.ua/handle/0564/737. Accessed 28 Nov 2019.
15. I.V. Lovianova, *Profesiyno spryamovane navchannya matematyky u profil'niy shkoli: teoretychnyy aspekt* (Theoretical and Methodological Foundations of Teaching Mathematics at Profession-Oriented School). (Chabanenko Yu. A., Cherkasy, 2014)
16. I.V. Lovyanova, D.Y. Bobylyev, in *Current Status and Prospects of Mathematical Educational* (SCASPEE, Budapest, 2018), pp. 37–45
17. K.V. Polgun, Dissertation, Ternopil Volodymyr Hnatiuk National Pedagogical University, 2017
18. *Innovatsiyni informatsiyno-komunikatsiyni tekhnolohiyi navchannya matematyky, fizyky, informatyky u serednikh ta vyshchykh navchal'nykh zakladakh* (Innovative information and communication technologies of teaching mathematics, physics, computer science in secondary and higher education institutions) (Kryvyi Rih State. ped. Univ., Kryvyi Rih, 2011). http://elibrary.kdpu.edu.ua/handle/0564/579. Accessed 28 Nov 2019.
19. N.O. Zinonos, E.V. Vihrova, A.V. Pikilnyak, Prospects of Using the Augmented Reality for Training Foreign Students at the Preparatory Departments of Universities in Ukraine. CEUR Workshop Proceedings **2257**, 87–92 (2018), http://ceur-ws.org/Vol-2257/paper10.pdf. Accessed 28 Nov 2019
20. I.V. Lovyanova, D.Y. Bobylyev, in *Current issues in ensuring the quality of mathematical education* (SCASPEE, Budapest, 2019), pp. 39–50
21. K. Vlasenko, O. Chumak, I. Sitak, O. Chashechnikova, I. Lovianova, Espacios, **40**(31), 11 (2019)
22. I. Lovianova, D. Bobyliev, A. Uchitel, Cloud calculations within the optional course Optimization Problems for 10th-11th graders. CEUR Workshop Proceedings **2433**, 459-471 (2019), http://ceur-ws.org/Vol-2433/paper31.pdf. Accessed 28 Nov 2019
23. M.M. Volikova, T.S. Armash, Y.V. Yechkalo, V.I. Zaselskiy, Practical use of cloud services for organization of future specialists professional training Cloud Technologies in Education. CEUR Workshop Proceedings **2433**, 486–498 (2019), http://ceur-ws.org/Vol-2433/paper33.pdf. Accessed 28 Nov 2019
24. K. Vlasenko, I. Lovianova, I. Sitak, O. Chumak, O. Kondratyeva, Universal Journal of Educational Research **7(9)**, 1892–1900 (2019)
25. M.I. Burda, N.A. Tarasenkova, O.M. Kolomiyets', I.V. Lovianova, Z.O. Serdyuk, *Heometriya (profil'nyy riven') pidruchnyk dlya 10 klasu zakladiv zahal'noyi seredn'oyi osvity* (Geometry (profile level) textbook for the 10th grade of institutions of general secondary education) (Orion, Cherkasy, 2019)
26. H.M. Bilousova, A.M. Kapinosov, P.I. Ul'shyn, O.Y. Chyzh, Ya.F. Hap'yuk, L.I. Kondrat'yeva, O.M. Martynyuk, S.V. Martynyuk, L.I. Oliynyk, *Matematyka. Kompleksne vydannya dlya pidhotovky do DPA i ZNO* (Mathematics. Complex edition for preparation for DPA i ZNO) (Pidruchnyky i posibnyky, Ternopil, 2019)
27. T.H. Kramarenko, V.V. Korolskyi, S.O. Semerikov, S.V. Shokalyuk, *Innovatsiyni informatsiyno-komunikatsiyni tekhnolohiyi navchannya matematyky* (Innovative information and communication technologies of mathematics teaching) (KSPU, Kryvyi Rih, 2019). http://elibrary.kdpu.edu.ua/xmlui/handle/123456789/3315. Accessed 28 Nov 2019.
28. A.E. Kiv, M.P. Shyshkina, S.O. Semerikov, A.M. Striuk, Y.V. Yechkalo, AREdu 2019 - How augmented reality transforms to augmented learning. CEUR Workshop Proceedings **2547**, 1–12 (2020), http://ceur-ws.org/Vol-2547/paper00.pdf. Accessed 09 Feb 2020
29. T.H. Kramarenko, O.S. Pylypenko, V.I. Zaselskiy, Prospects of using the augmented reality application in STEM-based Mathematics teaching. CEUR Workshop Proceedings **2547**, 130–144 (2020). http://ceur-ws.org/Vol-2547/paper10.pdf. Accessed 09 Feb 2020
30. T. G. Kramarenko, O. S. Pylypenko, Problemy pidhotovky uchytelia do vprovadzhennia elementiv STEM-navchannia matematyky (Problems of preparation of teacher for implementation of elements STEM-teaching mathematics). Physical and Mathematical Education 4(18), 90–95 (2018). doi:10.31110/2413-1571-2018-018-4-014

The development of the dialogic culture as a "centaur problem" of the professionalization of the future educators

Natalya Tokareva*

Kryvyi Rih State Pedagogical University, Kryvyi Rih, 50086, Ukraine

Abstract. The article deals with the analysis of the problem of developing the dialogic culture of the future educators in terms of their professionalization. The author has shown that the congruent coverage of the specifics of forming an individual as an entity of professional activity in the continuum of the modern culture can be carried out only within the dialogic mode. It has been found that de-valuing the psychological component and accumulating systems of algorithmic tasks of the test type in the practice of the professional training makes it impossible to develop the communicative (and especially dialogic) culture of a specialist. The dialogic culture is interpreted by the author as an integrative construct of an individual, i.e. a set of communicative and technological indicators of the systemic behaviour (and professional-role one in particular) of the activity entity. The empirical research procedure had as its objective to outline the psychological tendencies in the development of the indicators for the communicative culture of the professionalization entities (bachelor degree and master degree students) in the modern educational space. The dominance of the monologic model of communication in the respondents' behavioural scenarios and the absence of the self-management development dynamics in communication have been statistically corroborated. The findings of the research are interpreted as the theoretical foundation of a modern model for the dialog-phenomenological interaction that is adequate to the new challenges of the time and the educational reform principles.

Introduction

The cosmopolitan tendencies of the post-industrial era of the civilization development, the continual character of the renewal of the living space in the 21st century lead to permanent entropy, chaotic nature, dramatization of the scenarios of being and worldview positions of the present-day "average person". The multimodality and the accessibility of informational content under the unstable conditions of the life's activities exacerbate the contradiction of the existential personal choices of the cultural and historical continuum of the self-fulfilment of an individual.

The involvement of a person in the various aspects of social structures and relationships is accompanied by considerable stress on the psyche. The "destruction" of the inner integrity and the affective cognitive emotions of a person under the unstable conditions of the life's activities reach such a level that the threat of psychological maladjustment, mental and somatic diseases of the psychogenic spectrum becomes real. Social apathy, stereotyping of thinking strategies, conformism, avoiding (or even fearing) the manifestation of creativity in constructing one's own life concepts are often human reactions to the social-psychological situation of uncertainty. The constructive life creation by the individual in today's paradoxical world is largely related to adaptation scenarios that are interpreted as active processes of interaction with the social environment, one of the most important forms of that interaction being professional activity.

Professionalism of the individual is recognized an important constant of the efficient implementation of the professional activity [1], which conditions not only the possibility of skilled work, but also the optimization of the process of modelling personal constructs, ensuring integrity, the constructive orientation [2-4] and the resource potential [5-7] of the activity entity. A significant vector of an individual's professionalization is undoubtedly high-quality professional education, whose development potential is marked by the orientation to a new level of the development of a person capable of making responsible choices and inclined to move in a socially constructive direction of self-fulfilment. And, first of all, this concerns pedagogical education, whose quality determines the efficiency of the formation of worldview mindset of the person of the new generation, the so-called generation-next that is filled with the potential resources for the affirmation of their "otherness", capable of independently comprehending a new experience of the efficient interpersonal interaction under the modern conditions of life creation.

In the context of the above, the issues of rethinking the paradigmatic aspects of the socio-cultural and professional identity, its aesthetics under the conditions of an unstable information society are becoming

* Corresponding author: tokareva152681@gmail.com

particularly relevant. At the same time, the congruent coverage of the specificity of an individual's formation as an entity of self-development in the continuum of the dynamic worlds of the modern culture can undoubtedly be effected within the dialogic mode [8-11, 13-17]. After all, the assimilation of the individual's social experience, the accumulation of social and personal behavioural patterns (including professional and role-playing ones) are possible only under the conditions of communication, which is characterized by entity-to-entity links and relationships [4; 13; 14; 18; 19]. "Forming a culture of dialogue is one of the real ways of overcoming the depth of contradictions of the modern culture and a prerequisite for the proper communication between civilized people, – N.G. Uskova emphasizes, – because a dialogue is the foundation of being and worldview, of the development of cognitive and mental activity of a person" [10, p. 3]. In a dialogue, a person constitutes and reveals himself/herself as an individual, as an entity of self-determination, free self-definition in relation to reality thus realizing his/her own "identity that acquires its full and universal expression precisely in the process of dialogue. One voice completes nothing and resolves nothing. Two voices are the minimum of life, the minimum of being" [19, p. 121].

However, according to the analysis of the scientific literature [10, 20-22] and our own experience of pedagogical activities, professional training of future specialists (and educators in particular) takes place mainly outside the continuum of the dialogue of cultures, without taking into account the dynamic tendencies and the paradoxical nature of the modern socio-cultural space. The digitalization of educational systems and the widespread introduction of the knowledge control tests do not contribute to the development of the dialogic culture of the education entities either, which creates a paradox of the social consciousness, – the dimension of the "centaur problem".

Zh.T. Toshchenko's research convincingly proves that "centaur problems" are generated by the contradictions of life experience and testify not only to the existence of incompatible values, attitudes in particular social groups, but also cultivate such contradictions, which leads to incompatible statements and actions [23, p. 11]. A "centaur problem" manifests itself when individually diverse actions are rational, but combined into a system reveal not only incoherence, but also total contrariety, incompatibility with each other [23, p. 12].

In this context, to the best of our belief, devaluing the psychological component and accumulating systems of algorithmic tasks of the test type in the practice of the professional training makes axiomatic the impossibility to develop the communicative (and especially dialogic) culture of a specialist. After all, the assimilation by a person of the socio-cultural and professional experience, the verification of subjective behavioural patterns in the modern information environment are possible only on the basis of the active communication [4, 10, 16, 17, 20-22, 24, 25] that implies the intensification of the existing communicative intentions, their translation into a more technological form (involving the use of a linguistic code apparatus and other semiotic systems), which ensures the achievement of a predicted result as opposed to a random process.

In order to prove the above, a theoretical and empirical research was carried out with the **purpose** of outlining the psychological tendencies of the development of the communicative culture indicators for professionalization entities (bachelor degree and master degree students) in the modern educational space.

The conceptual idea of this research was the proposition that the development of the communicative culture (and dialogic one in particular) of the future educators is an invariant indicator of an efficient pedagogical process determined by the requirements of the modern information space and the tendencies of the individual development of the professionalization entities.

Research methods and techniques

The contradictory nature of the problematic field of studying the psychological tendencies of the development of the professionalization entities' communicative (and in particular dialogic) culture determines the requirements for the correct formation of the sample cohort for the empirical component of the research. Given the sensitivity of adolescents to the transient nature of the socio-cultural space of the modern information environment and the sensitivity of this age period persons to the formation of the experience of the professional and role-playing behaviour, representatives of the student community were selected as the entities of the research program.

The research sample consisted of 226 persons: 123 second-year bachelor degree students (17-18 years of age) and 103 master degree students (20-22 years of age) of Kryviy Rih State Pedagogical University (Ukraine).

In order to more fully define the general dynamics of the phenomenon under research, the target sample cohort was supplemented by the juvenile respondents (43 5th and 6th grade pupils (11-12 years of age) of one of the city of Kryviy Rih secondary schools), whose answers were interpreted by us as a model for the presentation of the communicative culture in the genesis of childhood (younger juvenile age).

The systemic-integrative content of the subject field of the students' communicative culture indicators research determined the structuring of the logic of the research program in terms of the implementation of the genetic method of systemic research, which states the need to analyze the patterns of functioning and transformation of mental phenomena, the correlation of the actual and the potential in the mental development.

In order to achieve our goal and to implement the program of studying the development features of the communicative culture of students in the period of their professionalization, we used the following psycho-diagnostic methods that were adequate to the research subject:

- S. L. Bratchenko's "Individual's orientation in communication" method that makes it possible to assess the importance of the dialogic and monologic vectors of

an individual's orientation in the "I – the Other one" coordinates;
- "The study of the capability for self-management in communication" method (questionnaire) intended to determine the mobility and adaptability of an individual in different situations of communication [26, p. 164-166].

The primary data processing was carried out by calculating the average values of the material obtained. The IBM SPSS Statistics 19 ("Statistical Package for the Social Science") computer program was used to compile and analyze the empirical material. The variables were checked for the standardized character of the distribution of a feature.

Findings and discussions

The theoretical and methodological review of the issue of the development of the communicative culture of students in the period of their professionalization [20, 21, 24] makes it possible to interpret the communicative culture of students as a system-forming component of the general and professional culture of the future specialists.

In general, we share the conceptual approach of L. A. Aukhadeyeva, who views the communicative culture as the ability of an individual to integrate his/her own consciousness into the global (not narrowly professional) space of social consciousness and thus to achieve significant results in his/her personal and professional development [20, p. 228]. One of the forms of implementing the communicative culture of an individual is dialogic culture [10, 16, 17], which can be interpreted as an integrative construct of an individual, i.e. a set of communicative and technological indicators of the systemic behaviour (and professional-role one in particular) of the activity entity.

In the context of the philosophical and anthropological worldview, a recourse to a dialogue as an explanatory principle of constructing subjective reality is traced. M. M. Bakhtin [18, 19], for example, interprets dialogue as a special form of interaction between equal and equivalent consciousness entities: everything vital is inherently dialogic [18, p. 327]. The existence of man, as the philosopher points out, is directly related to "the other" [11, 18], which conditions the formation of the worldview positions of an individual (the value-meaning intentions), the assimilation of the repertoire of the comprehensible roles and behavioural strategies in terms of the interpersonal interaction (those roles and strategies denoting the interpersonal intention capacity of the dialogue). The event-related content of the dialogue (the implementation of the "event of being") conditions its definition as the space of spiritual-semantic and moral-ethical dimensions of the human life, which transform the abstract-theoretical and conditionally potential attitudes to the discourse of life into real actions and practical deeds of a person as a responsible entity of being and co-being. So consequently, the development of the dialogic culture of an individual facilitates the existential self-affirmation, which in turn promotes, supports (and sometimes significantly influences the process of) decision-making and making choices in the context of the entity's mental field.

In the context of the aforementioned, we tested individual personality indicators for the maturity of the dialogic culture of schoolchildren and students within the sample cohort.

The analysis of the dialogic and monologic vectors of an individual's orientation in the "I – the other one" coordinates (data obtained using S. L. Bratchenko's "Individual's orientation in communication" method) made it possible to identify the development tendencies of the communicative orientation of the respondents at different stages of their professionalization (Table 1).

Table 1. The dynamics of the identification of the types of orientation in communication among the juveniles and adolescents of the sample cohort.

Age groups / Types of the individual's orientation in communication	11–12 years of age (*N*=43)	17–18 years of age (*N*=123)	20–22 years of age (*N*=103)
	Average values (Mx)		
Dialogic orientation	**2.37**	0.71	0.33
Monologic orientation	3.47	3.85	**3.92**
Authoritarian	2.44	3.05	**3.55**
Manipulative	2.69	2.93	**3.96**
Alterocentric	3.75	**5.54**	4.50
Conformist	**6.00**	5.17	4.19
Indifferent	2.50	2.54	**3.38**

The analysis of the average values (Mx) of an individual's orientation tendencies in the communication among schoolchildren and among students recorded the dominance of the monologic model (the maximum values are presented in the master degree students' group (Mx = 3.92)).

The dialogic orientation of the students' communicative behavior is pronounced at the minimal level (in particular if compared with the indicators for younger juveniles).

The differential analysis of the monologic orientation in the genesis of the students' professionalization makes it possible to state:
- an increase in the students' propensity for the manipulative, authoritarian and indifferent (a behavioural pattern oriented to the rigid nature of communicative scenarios) models of the communicative behaviour (the maximum Mx values are presented in the master degree students' profile);
- a decrease in the level of the tendency to the conformist behaviour (the maximum Mx values are detected for the teenage schoolchildren, which is related to the childlike dependence of the juvenile respondents on the people of authority);
- the instability of the development of the alterocentric model of the communicative behaviour in the process of the respondents' professionalization (the maximum Mx values are presented in the bachelor degree students' profile); this model is characterized by the refusal of the students of this group (II year study) from the equality in

communication in favour of the interlocutor, i.e. the orientation to the "object" position for themselves.

The Spearman rank correlation method was used to corroborate the significance of the data obtained. The method made it possible to clarify the direction and the strength of the correlation between the types of the respondents' communicative behaviour in the age dynamics. The following was found:

- during the process of the higher education, the respondents' dialogic orientation in communication decreases statistically significantly (ρ = –0,209) at the average level of statistical significance ($p \leq 0,01$);
- the level of the conformist model of the monologic orientation decreases statistically significantly (ρ = -0,180) at the average level of statistical significance ($p \leq 0,01$), which indicates the efforts of the students (and especially those of the master degree course) to assert their own interests in the process of communication while ignoring the needs of the others;
- the manipulative type of the monologic orientation increases with age statistically significantly (ρ = 0,252) at the average level of statistical significance ($p \leq 0,01$);
- the level of the authoritarian (ρ = 0.106) and indifferent (ρ = 0,128) types of the communicative behaviour increases at a sufficient level of statistical significance ($p \leq 0,05$).

Although the relationships found during the correlation analysis process are weak ($0,01 < \rho \leq 0,29$ and $-0,29 < \rho \leq -0,01$), they signify the potential vectors for the deployment of the key communicative tendencies of an individual's orientation in communication during the controversial period of professionalization.

The significance of the differences in the identification of the types of the monologic orientation in communication among students of the different academic periods was also corroborated by means of the variance analysis (the correctness of its use was proved by the Levene homogeneity-of-variance criterion at $p > 0,05$). The one-way variance analysis revealed statistically credible meaningful differences between the Mx values in different age groups of the respondents by such types of the monologic orientation of an individual in communication as manipulative (F = 4,5 at $0,001 = p \leq 0,05$), alterocentric (F = 3,3 at $0,012 = p \leq 0,05$) and conformist (F = 3,8 at $0,005 = p \leq 0,05$). Communication are statistically insignificant (at $p > 0,05$), Mx values compared differ only quantitatively.

The differences between the empirical research indicators for the manifestation of the authoritarian and indifferent types of orientation in the juveniles'.

We have also explored such an indicator of the communicative culture of an individual as the capability for self-management, which outlines the dynamics of the respondents' adaptability in different situations of the social interaction.

"The study of the capability for self-management in communication" questionnaire was used [26, p. 164-166]. The generalized results of the psycho-diagnostic measurement of that feature in the respondents of different age groups are shown in Table 2.

Table 2. Self-management capability dynamics for the juveniles and adolescents of the sample cohort.

Age groups / Analysis parameters	11-12 years of age (N=43)	17-18 years of age (N=123)	20-22 years of age (N=103)
	Average values (Mx)		
Level of self-management in communication	**14.32**	14.23	13.54

The analysis of the data obtained makes it possible to ascertain the dominance in the respondents of the sample cohort of the medium level of the capability for the self-management in communication (with a standard deviation of 2,78 on the questionnaire scale within the range of 8–17 points), which denotes the impulsive rebelliousness and the renunciation of responsibility that is inherent in the respondents of the older age group.

Checking the significance of the dynamics indicators for the level of the development of self-management in communication in the respondents of different age groups using the one-way ANOVA method (the correctness of the use of this method is corroborated by the Levene homogeneity-of-variance criterion at $p > 0,05$) made it possible to ascertain the absence of statistically credible meaningful differences between the groups of juveniles and adolescents (F = 0,134 at $0,87 = p > 0,05$).

The Mx values of the level of self-management in communication between the respondents of different ages (and therefore of different stages of professionalization) vary only quantitatively, which indicates that there is no constructive development of self-management skills in communication during the process of the higher education.

Conclusions

It should be acknowledged that the absolutization of certain cognitive positions (and in particular the technologization of the core pedagogical training of specialists) cannot in principle ensure the unambiguity of the solution of the problem of the development of dialogic culture of future specialists, and, accordingly, leads to the emergence of a "centaur situation", whereby the altered living environment conditions the non-adaptive logic of behavioural scenarios of a person.

The results of empirical investigation certify about the absence of object-oriented influence in the educational institutions regarding the development of dialogical culture of pupils and dialogical model of professional interaction in the model "Student – Teacher" without violation of professionally-communicative balance.

The mastering by a life creation's entity of the dynamic function of interpretation (in the form of multi-entity interpretive constructs: narration, dialogue, reasoning, judgment) makes it possible to solve a number of problems of personal development: it eliminates the semantic uncertainty of the social existence situation and in particular in the field of professional activity; it explains the need to correlate (or not to correlate) one's own versions, thoughts and constructs with other people's versions; it implements the reflection of one's own assessments and judgments thus conditioning the certainty and constancy of the criteria for such a reflection & etc. The communicative (and above all dialogic) culture of an individual signifies the deepening of the individual's narratives, intensifies the overall development of the individual under the conditions of the polylogue of cultures.

Dialogization of educational system creates the conditions to re-analyze integral "I" of the subjects of teaching and educational process as inter-discursive activity.

The above idea nourishes the concept of tolerance, forms the theoretical foundation of a modern model for the dialog-phenomenological interaction that is adequate to the new challenges of the time and the educational reform principles.

The format of reflexive-dialogical professional interaction has been selected as a strategic vector of modelling process organization of rational personality constructs.

Functional modelling of communicative modelling of teachers' personality profile under the conditions of the present-day educational environment is interpreted as an act of continuous formatting of the subjective psychic reality, i. e. of a unique model of personal dimensions of a human being's socio-cultural environment.

The communicative aspects of the flexible introduction of functional programs for educational environment humanization is a necessary condition to ensure the long-term psychological support for multivariate extension and refinement of the system of personality constructs of professional activity agents in the context of their modelling. Activation of the agent's potential (the internal determination of behaviour) and goal-oriented social and psychological motivation to create meaningful scenarios of personality's creative life (the external determination of adaptive behaviour) have been identified as relevant areas for modelling personal constructs of teachers.

Constructive dialogue strategies of communication are the beginning of the adolescent productive life creativity and demand aimed modeling in the active person becoming period.

The introduction of dialogue system in the education is a crucial factor for development of personal potential of each participant during the teaching and educational process.

The attractors of the process of communicative modelling of personal constructions are the mechanisms of context making of communicative contour of phrases in the dialogue situation.

We are convinced that the research findings presented in this article will contribute to solving a wide range of theoretical and practical problems related to understanding the dynamic context of the semantics of the psychological support for professionalization.

We see the prospects of the research in studying the semantic features of the "image of one's own changes" of the professionalization entities, which will make it possible to build guidelines for the work of psychologists with students in the context of preventing the difficulties of the development of the dialogic culture as a whole.

Reference

1. S.A. Druzhilov, *Psihologiya vyizhivaniya v sovremennom mire* (The psychology of survival in the modern world). (Gumanitarnyiy Tsentr, Harkov, 2012)
2. E.E. Sapogova, Scientific Pap. of The Witelon St. Univer. of App. Scien. in Legnica. **11**(2), 29–43 (2014)
3. V. Tatenko, *Metodologiya sub'yektno-vchinkovogo pidhodu: socialno-psihologichnij vimir* (Methodology of subject-action approach: socio-psychological dimension). (Milenium, Kiyiv 2017).
4. N.M. Tokareva, Dissertation, Institute of Psychology, Kyiv, 2015
5. S.B. Kuzlkova, S.V. Puhno, Probl. suchasn. psiholog. 1(15), 72–77. (2019). doi:10.26661/2310-4368-2019-1-15-11
6. O.V. Lishin, Prakt. psiholog. ta soc. robota 12, 23–29 (2010)
7. S.A. Merzakanov, Istor. i soc.-obrazovat. mysl. 6(22), 128–132 (2013)
8. G.V. Dyakonov, *Osnovyi dialogicheskogo podhoda v psihologicheskoy nauke i praktike* (The basics of the dialogical approach in psychological science and practice). (RIO KGPU im. V. Vinnichenka, Kirovograd, 2007)
9. O.G. Ivanovskaya, *Semanticheskij rezonans i ponimanie tekstov* (Semantic resonance and understanding of texts). (Rech, Sankt-Peterburg, 2011)
10. N.G. Uskova, Dissertation, MSU, Moskva, 2003
11. N. Tokareva, Psycholog. Zesz. Nauk. (Zielona Gora), 1, 35–48 (2018)
12. T.N. Ushakova, *Rozhdenie slova: Problemy psihologii rechi i psiholingvistiki* (Birth of a word: Problems of speech psychology and psycholinguistics). (Institut psihologii RAN, Moskva, 2011)
13. S. L. Frank, Vopr. filosof., 6, 68–131 (1990)
14. T.D. Sherban, Dissertation, Institute of Psychology, Kyiv, 2005
15. N. Fairclough, *Discourse and Social Change.* (Polity Press, Cambridge, 1992).
16. H.J.M. Hermans, Cult. & Psychol., 7(3), 243–281 (2001)
17. H.J.M. Hermans, J. of Construct. Psycholog. 16(2), 89–130 (2003)
18. M.M. Bahtin, *Estetika slovesnogo tvorchestva* (Aesthetics of verbal creativity). (Iskusstvo, Moskva, 1979)
19. M.M. Bahtin, *Avtor i geroy: K filosofskim osnovam gumanitarnyih nauk* (Author and hero: To the philosophical foundations of the humanities). (Azbuka, Sankt-Peterburg, 2000)

20. L.A. Auhadeeva, Filolog. & Cult. 1(27), 226–231 (2012)
21. A.Yu. Sutugin, T.G. Muhina, Mir nauki, kult., obrazov. 2(51), 121–124 (2015)
22. L. Nystedt, in *Toward a psychology of situations: an interactional perspective*, ed. by D.Magnusson (Lawrence Erlbaum, Hillsdale, 1981), pp. 375–391
23. Zh.T. Toshenko, Sociolog. vlasti 2, 6–21 (2011)
24. E.M. Karimulaeva, A.M. Kurbanova, Mir nauki, kult., obrazov. 6 (67), 52–54 (2017)
25. V.A. Yasvin, *Obrazovatelnaya sreda: ot modelirovaniya k proektirovaniyu.* (Educational environment: from modeling to design). (Smysl, Moskva, 2001)
26. N.P. Fetiskin, V.V. Kozlov, G.M. Manujlov, *Socialno-psihologicheskaya diagnostika razvitiya lichnosti i malyh grupp* (Socio-psychological diagnosis of the development of personality and small groups). (Izd-vo Instituta Psihoterapii, Moskva, 2002)

SHS Web of Conferences **75**, 02002 (2020) https://doi.org/10.1051/shsconf/20207502002
ICHTML 2020

Entrepreneurship education as a factor of society's modernization

Hennadii Korzhov[1,*], and *Yaroslav* Pasko[2]

[1]National Technical University of Ukraine 'Igor Sikorsky Kyiv Polytechnic Institute', Department of Sociology, 37 Peremogy Ave., Kyiv, 03056, Ukraine
[2]Borys Grinchenko Kyiv University, Department of Philosophy, 13b Tymoshenko Str., Kyiv, 04212, Ukraine

Abstract. Entrepreneurship education (EE) is often regarded as one of the most important elements of the policy of stimulating small business and self-employment, creating start-ups and innovative business projects. The implementation of EE in the system of professional training at a new stage in the development of society could significantly increase the potential of using universities in the process of economic growth. The article sheds light on the nature and features of EE, analyzes the prospects and difficulties of its integration with other areas of training. The study demonstrates great prospects for the implementation of EE into the program of higher education training. EE has to be concentrated around several important characteristics of entrepreneurship and entrepreneurs themselves. The implementation of EE at the curriculum and extra-curriculum activities at the institutions of higher education of Ukraine, the combination of theoretical, problem- and project-oriented learning with the immersion into practical activities concerned with business planning, creating and running, training of students in groups together with nascent and experienced entrepreneurs would be extremely useful and productive in terms of motivating more and more young people to take an entrepreneurial path and achieve prosperity and self-actualization. The practical realization of EE will contribute to the formation of more resilient and competitive local communities as well as a more inclusive, just, equal, and happy society.

1 Introduction

Private enterprise is not only the most important economic phenomenon of a market society, but also a key socio-cultural phenomenon that has made a significant contribution to the processes of modernization, the establishment of institutions, structures, value-motivational and behavioral models of a modern society. Although entrepreneurship has deeply embedded in the life of modern Ukrainian society, it has not to a full extent become the driving force behind the development of a civilized market economy and a highly developed social sphere. Private enterprise is influenced by many negative factors related to corruption, raiding, violent actions, non-compliance with contractual obligations, connections with the criminal world and corrupt officials, tax evasion, neglect of moral standards and demonstrative consumption of 'new Ukrainians' - all these occur against the background of social polarization and low quality of life of wide strata of society. Entrepreneurs themselves often demonstrate a low level of professionalism and profound anomic demoralization. In this sense, business in Ukraine is more inherent in the attributes of adventurous capitalism rather than a modern, rational variation of market relations. What can contribute to overcoming the above-mentioned maladies of Ukrainian entrepreneurship is entrepreneurship education (EE), in particular at the level of higher professional education.

In recent years, many countries in the post-Soviet space have been discussing the need to introduce new organizational and managerial forms of interaction between the industrial and scientific-educational sectors, in particular scientific and industrial research centers based in leading universities. Such centers, which combine the financial, material, and technical resources with qualified personnel of universities, could become the driving forces of the innovative model of the country's industrial development. So, on the basis of the National Technical University of Ukraine 'Igor Sikorsky Kyiv Polytechnic Institute', the first in Ukraine innovative ecosystem Sikorsky Challenge, which is designed to attract creative youth to innovative entrepreneurship and which includes a start-up school, business incubator, venture fund and other elements of innovative infrastructure, was created and already has achieved resounding success. There innovative technological ideas are successfully nurtured, startup companies are launched and developed. Thus, Ukraine has certain successes – though local ones – in creating innovative infrastructure based on research universities.

In recent years one can observe the growth of interest in entrepreneurship education. It is often considered as one of the most significant elements of the policy of

* Corresponding author: korzhovgena@yahoo.com

stimulating small entrepreneurship and self-employment, creating start-ups and innovative business projects. However, even in the developed countries, the most advanced in terms of introducing education in the field of business administration, the issue of EE implementation remains quite relevant. As a matter of fact, there is an essential difference between conventional business administration and entrepreneurship education. Indeed, EE implies a broader approach, sets broader goals than business education that is more narrowly focused on creating and managing a business based on a scientific approach to management. The EE is aimed at the formation of entrepreneurial skills, approaches and models for solving problems in various fields of life, and not just in commercial activities.

Up until recently post-Soviet system of education displayed no signs of integrating this innovative and promising approach into the curriculum of contemporary higher education institutions. However, lately in a number of leading universities in Ukraine and other post-Soviet countries, an innovative environment has been created for potential entrepreneurs, for those who are willing and ready to start their own business. However, the degree of involvement of engineering students in this activity leaves much to be desired for a number of reasons, in particular, the lack of intellectual and socio-psychological preparedness as well as the lack of necessary knowledge and skills.

Despite its growing popularity, EE has not been sufficiently studied in the context of its integration with other specialized training models. In the framework of the ongoing paradigm shifts in modern higher education, a special place is occupied by the idea of combining business training with engineering education in order to form a broad social base for the development of industrial entrepreneurship. The development of private initiative in industrial sectors of the economy is particularly acute in post-Soviet countries that have experienced the de-industrialization phase – a sharp reduction in industrial production, the mass closure of plants and factories, the washing out of skilled personnel, and a catastrophic decline of the share of industry in total production. The implementation of industrial education in the system of professional training of engineering personnel at a new stage in the development of society could significantly increase the potential for using technical universities in the process of economic revival.

The major goal of this article is to reveal the prospects and difficulties of entrepreneurship education implementation in the Ukraine's system of higher education. In theoretical terms the article is inspired by life-enhancing philosophy and learning paradigm to be discussed below. The combination of these approaches allows for analyzing the peculiarities of EE as well as the prospects and difficulties of its implementation in the Ukraine's system of higher education. Methodologically, the paper is based upon the analysis of recent years publications on the issue of entrepreneurship education published in specialized Western academic journals mostly in the 2000-2010s. The publications were sorted out according to their relevance: those papers which were directly devoted to the issue under examination and contained EE in the title or list of keywords were selected. The major bulk of publications analyzed in the given article come from the following journals: Education and Training, Journal of Small Business Management, Entrepreneurship: Theory and Practice, Journal of European Industrial Training, European Journal of Engineering Education, Industry & Higher Education, Journal of Small Business and Entrepreneurship, The Journal of Entrepreneurship, Entrepreneurship Theory and Practice, International Small Business Journal, International Journal of Entrepreneurial Behaviour & Research. The analysis of recent publications is provided in the relevant sections of the article.

The structure of the paper is determined by the following logic of explication of the presented problems. The first part of the work reveals the essence and specificity of entrepreneurial education, its functions, potential and limitations. The second part discusses the role and importance of EE in modern Europe, primarily in the EU countries, where over the past few years a policy has been implemented to expand this type of education at various levels of training. Based on several large-scale studies, the achievements and difficulties of implementing EE in higher education are analyzed. The third section of the article raises the question of the productive and creative potential of integrating entrepreneurship and engineering education.

2 Entrepreneurship education: concept, functions, opportunities and limitations

The preparation of young people for the creation and development of their own business projects as a path to personal self-actualization and a driver of social development has a long and deeply enrooted place in the system of university training in all highly developed countries of the world.

In the literature one can find various definitions of the concept of 'entrepreneurship education'. The whole range of various definitions of the phenomenon can be divided into two categories – narrow and broad ones. The former put emphasis on the learning specific skills and knowledge pertinent to the initiation and successful running of business. The latter center on broader implications of this type of education, its usability in various spheres of the individual's life, both public and private. According to one, a rather narrow definition, EE should be understood as 'a conscious activity, the aim of which is to develop a student's readiness to observe and understand entrepreneurial activity and to be aware of its connections in the development of one's own personality' [1, p. 175].

EE is a focused activity that contributes to the formation of a certain way of students' thinking (mental models, ways of perceiving reality, worldviews) and relevant skills that can be utilized in the framework of entrepreneurial activity. An entrepreneurial way of thinking is understood as aspiration, ability, and willingness to translate ideas into social practice thanks to the resources available to the individual. In brief, an entrepreneurial way of thinking promotes not just the creation of a business, but the implementation of sustainable development models. However, such an understanding of EE should be recognized as quite narrow, aimed mainly at stimulating entrepreneurship – whether it is the creation of small business or the formation of leaders of large companies.

A broad interpretation of the concept means the formation of entrepreneurially oriented and independent people capable of rational goal-setting and active solving of complex and difficult life situations. These abilities are required not only in business, but also in many other areas of life and circumstances – both ordinary and critical.

In this context it is noteworthy to put emphasis on dynamic and creative/productive force of knowledge making use of a life-enhancing philosophy put forward by Mikhail Bakhtin and Gilles Deleuze. The philosophy affirms the openness towards potentialities and transformation of life, possibility to create life beyond present experiences. Learning when considered in light of life-enhancing philosophy can also be seen as a process of self-creation, of becoming other [2, p. 60], and a road to self-actualization. They accentuate a sharp contrast between management and entrepreneurship education. The former is about various methods of control and governing in established organization settings, while the latter is about creating new ways of doing things, about re-creating old things in a new manner. People approach 'life as a multiplicity of becomings' [2, p. 63]. However, the authors stress that one can observe a certain contradiction in Swedish society between, on one hand, a cultivation of entrepreneurial approach in schools and, on the other, dominant values, norms, and discourse of society which promote large-scale high-technology driven growth and professional careers in major corporations [2, p. 68].

On the basis of the philosophy of becoming in two Swedish universities – Stockholm and Malmö – a master program in entrepreneurship was launched open to all students with diverse educational background. The specially designed program invited students to approach entrepreneurship as part of society rather than simply part of business. Students were expected to learn as much as possible from each other and would-be entrepreneurs. They worked together in small groups with business development projects in regard to real-life cases. An important conclusion has been made: heterogeneity drives creativity [2, p. 72]. The organizers also learnt that openness was important for learning to happen.

Eventually, the expansion of EE contributes to economic growth, the creation of new firms and jobs, and to the growth in the welfare of the whole society. At the same time, it should be emphasized that the benefits of developing entrepreneurial competencies go beyond purely financial profits, creation of new economic and commercial enterprises, the spread of start-up activities. In fact, it extends to other spheres of the life of individuals and society.

Additionally, EE can be considered not only at the individual, but also at the organizational and systemic level. In this case, it is understood as a comprehensive concept that modifies and explains the functioning of the entire educational institution and even educational system as a whole. In this context the purpose of education is regarded to promote the development of entrepreneurship, social activity, innovation, and creativity. In the framework of such an integrated approach, entrepreneurship is not just being introduced as a separate academic discipline or set of disciplines, but it becomes the main educational principle that guides the entire training model, an idea that combines the content and forms of all educational activities.

Numerous publications of the last two decades have emphasized the special role of the university in the development of entrepreneurial attitudes, intentions, and talents, in increasing the level of motivation for creating businesses. Scientists attempt to answer the questions 'What? How? When to teach?' within this type of education [3]. It is not easy to answer these questions, given the fact that efforts to introduce EE into university academic and extracurricular, basic and additional activities have yielded mixed results. Some studies have shown that under the influence of EE students demonstrated increased motivation and intentions to go into business [3, 4]. At the same time, there are studies that indicate that formal training in itself can reduce entrepreneurial motivation. The matter is that instead of cultivating an entrepreneurial spirit modern universities formally teach business and management and weakly affect the entrepreneurial culture itself.

Many authors call for a radical revision of the traditional approach adopted in pedagogy where well-known, classical methods and forms of teaching are used. It is not only about how important it is to apply active teaching methods, case studies, focus group discussions, disputes, debates, etc. It is about how to focus on students' requests and expectations, as well as to actively involve their everyday experience and cognitive models in the learning process. This issue has long been actively discussed in the context of the transition from the pedagogy of teaching, instruction (instruction paradigm) to the pedagogy of teaching, learning (learning paradigm) [5, p. 14]. The question is raised of an even deeper revision of the usual educational paradigm: the transition from purely functionalist ideas about what an entrepreneur should know and be able to, to comprehend the living experience of what it means to be an entrepreneur, to think and live like an entrepreneur, and to achieve success and to fail in business. The idea is to encourage current and potential entrepreneurs to reflect on the topic of their behavior and practices, the meaning and goals of their activities, and thereby stimulate creative and self-critical perception, awareness of the complexity and ambiguity of their activities, and evaluating the effectiveness of their current activities. Such an approach requires the active involvement of scientists, teachers, and practitioners (entrepreneurs, officials, representatives of business associations) in the study process, in order to conduct a meaningful dialogue and jointly develop a model of practical learning [6].

Such a profound shift in the educational paradigm is dictated by the characteristics of entrepreneurial activity. Bird [7] describes entrepreneurial behavior as the result of a combination of rational reasoning and intuitive choice, thereby emphasizing that entrepreneurship includes creation and reflection, action and reflection; immediate response to environmental changes as well as intentional, pre-planned actions. This image of entrepreneurship as a specific activity to create new profitable enterprises requires an action-oriented approach. An entrepreneur is distinguished by his ability to take, in collaboration with others, the necessary measures to allocate resources in accordance with emerging opportunities.

3 Entrepreneurship education as 'learning by doing' and entrepreneurial socialization

Recently, there have been a growing number of publications that substantiate the empirical, experiential nature of the process of learning entrepreneurship [8, 9], or 'learning by doing' [10]. It is understood that starting and doing business itself creates a learning environment in which business people search for answers to the questions of survival and development of their enterprises, decide how to act and with whom to do it. Business becomes a training organization, a sort of 'university' for entrepreneur. This does not at all preclude seeking advice, taking courses, attracting consultants, and other forms of training. However, they are all subject to the same goal, namely: the successful implementation of a business idea and business plan in life. Such training is a feedback or response to the actions of customers, suppliers, regulatory authorities, professional consultants, competitors, and other stakeholders. This training is carried out in the process of personal communication, and not by means of participating in formal forms of training [10, p. 19]. In general, this approach hinges on the results of scientific studies showing that entrepreneurs derive the lion's share of new knowledge and skills from practical experience [11]. Those with experience in building their own business are more successful at subsequent entrepreneurial activity: in particular, they more often set up a second and third start-up [12].

At present, there is no well-grounded and developed in detail theory of entrepreneurial socialization that allows researchers and practitioners to identify those factors and mechanisms that lead to the establishment of an entrepreneurial career, to the choice and successful conduct of business as a life project. However, a number of works shed light on the role certain elements of such a socialization play in guiding the individual to entrepreneurial path. For example, a study of the history of the formation and career development of the panel of entrepreneurs revealed the significant role of experience in one's youth, namely, important responsibilities in various spheres of daily life and early experience in doing business, for further choosing a definite life path. Diverse labor and educational experiences also favorably affect the entrepreneurial motivation of youth. Specialized courses in entrepreneurship or business fundamentals can increase people's confidence in their own abilities, which is essential for starting a business career. A study of the top managers of the 500 largest American companies showed that they had solid experience – both in business and in other areas of life – before embarking on the path of a successful entrepreneur. Negative work experience in large organizations convinced some of them to choose an entrepreneurial career trajectory. Although there does not appear to be a single sequence of life events or experience patterns that lead to an entrepreneurial path, previous studies seem to indicate that early family and childhood experiences, education and training, and some work experience contribute to entrepreneurial behavior [13, p. 11]. Further studies of entrepreneurial socialization are likely to lead to the emergence of models that will better predict entrepreneurial behavior than models based solely on individual factors. Such models can underpin the development of more effective entrepreneurial training.

At the same time, the benefits of EE should not be evaluated solely in accordance with economic criteria by the direct results of training, in particular by the number of newly created companies among graduates of relevant educational programs. In addition to the immediate benefits, there are indirect positive outcomes of EE implementation that are long-term [14]. The frequency of founding new firms by university graduates depends on the type and content of EE as well as on students' profile. In some universities due to the introduction of EE the level of the founding of startups by graduates was extremely high (for example, Babson College in the USA, Twente University in the Netherlands).

4 Value of entrepreneurship education for modern society

Let us now turn to the discussion of those important positive contributions that EE makes to the modernization of society. What is the significance of EE for modern society? The following key positive functions should be highlighted:

1) The formation of such a social environment that would be favorable for the initiation and development of business, for increasing the level of socio-cultural legitimacy of entrepreneurship and, thereby, for the growing attractiveness of business career for the younger generation. The fulfillment of this task should be based on scientifically verified facts obtained as a result of studying the barriers to involvement in entrepreneurship of both individuals and entire social groups. Of vital importance for countries with economies in transition is stimulating the development of entrepreneurial potential in productive areas of activity and deterrence in non-productive and especially destructive forms [15] and, thereby, contributing to the improvement of the business environment. The 'entrepreneurial spirit', about which Max Weber wrote so vividly, has a tendency to periodic fluctuations - in some periods of history it flares up,

whilst in others it fades out. EE is designed to give it a new impetus.

2) The distribution of entrepreneurs between productive and unproductive activities can have a significant impact on the innovativeness of the economy, the rate of economic growth, technical and technological progress, and the level of economic development of society. The widespread presence in society of incentives for unproductive entrepreneurship holds back competition, investment in industry, the growth of labor productivity, the introduction of advanced technologies, and the modernization of the economy. However, it should be emphasized that not only the modification in the structure of remuneration is important, but also the system of preferences of the entrepreneurs themselves, their goals and cultural factors, which can be modified under the influence of education.

3) The attraction of entrepreneurial talents to those areas of economic activity, which often remain of little demand, but they are the ones who make an indispensable contribution to the development of economic potential and improving the well-being of the whole society. In particular, we are talking about industrial production.

4) The development of entrepreneurial motivation among students by creating a high need for achievement, along with other components of entrepreneurial motivation complex, encouraging interest in business and the desire to create and develop their own innovative projects.

5) Giving encouragement for the formation among students of attitudes, approaches to solving problems, ways of thinking and patterns of behavior conducive to successful entrepreneurial activities.

6) Teaching specific entrepreneurial competencies for students, in particular the ability to see hidden opportunities, put forward innovative ideas, generate insights, find niches, and to fill in the 'structural holes' referred to in the structural theory of social networks [16]. An entrepreneur occupies a structural void when he unites previously unconnected individuals, groups, and organizations, contributing to the creation of a more integrated social structure.

7) Development of competencies demanded in the conditions of the post-industrial economy and highly mobile labor market: innovativeness, creativity, enterprise, ability to translate ideas into reality, rationality and the ability to reasonable risk, responsibility, as well as ability to conduct successful negotiations.

Despite the fact that EE is gaining popularity in recent years, there are a very limited number of publications in which the effectiveness of this type of training is empirically verified. One of the rare examples of this kind of research reveals whether the level of involvement in business is different among graduates of a master's program aimed at developing entrepreneurial competencies compared with graduates who studied in related programs. Graduates from different years (1987 – 1994) of one of the Norwegian business schools were studied. It turned out that graduates who specialized in training in the field of entrepreneurship showed better results for all the characteristics that have entrepreneurial content (the number of self-employed, the number of firms founded after graduating from a business school, the number of people who founded firms and owned firms at the time of the survey, as well as the number of graduates who preferred self-employment). The study demonstrated the close relationship between specialized entrepreneurship training and subsequent entrepreneurial behavior – both actual and potential in the form of intentions to start a business or become self-employed in the future [17, p. 158]. At the same time, the authors noted a number of limitations of this study. In particular, the goal of entrepreneurship education is not only to help create more firms, but also to improve the quality of the business. The study did not in any way evaluate the qualitative characteristics of the established enterprises. Anyway, the data obtained speaks in favor of EE. Both the behavior and intentions of those who specialized in entrepreneurship differ from the behavior patterns of those who choose other areas of specialization. Based on this and a number of other studies, it can be concluded that entrepreneurship, at least in part, is determined by factors that lend themselves to formation and change in the learning process.

There have been repeated and yet unsuccessful attempts to develop diagnostic tests, procedures, questionnaires that would help in the process of entrepreneurial training. In the late 1990-s, on the basis of Schön's theory of practical training [18] and Kolb's model of action education [9], the 'entrepreneurial action ability' test was developed [19]. Such a learning model is akin to the so-called action research [20, 21, p. 126], in which an individual or group of people can learn from their own experience and make this experience accessible to others. The purpose of such training is to solve a practical problem, improve the activity, and increase its effectiveness. In the case of EE, this strategy implies an attempt to answer the question 'how?'. The emphasis is on enhancing the rationality of action, in contrast to the traditional approach, based on the priority of rationality of decision [19, p. 158]. A test of entrepreneurial action ability can serve as an important diagnostic tool to identify the relationship between the ability to act and the actual actions to create a business.

Another important aspect of EE is the learning model itself. Gibb [10, p. 24] distinguishes between traditional ("didactic") and entrepreneurial learning models, which are fundamentally different in their approaches to mastering the material, the role of teacher and student, the nature of educational tasks, and modeling of learning environment.

Thus, in the process of implementing EE the very paradigm of education in higher education is gradually changing, becoming more and more problem and project-oriented, aimed at a more harmonious combination of theoretical knowledge and practical skills, more flexible and open to new experience, requiring greater return and reflexivity by students and teachers.

5 Entrepreneurial education in Europe as a factor in the transformation of society

Since the 80s of the XX century in the USA and many countries of Western Europe EE programs are actively developed with the involvement of the state, business and civil society organizations. Since then, EE has become quite widespread on the European continent. Training programs have been introduced and adapted to the needs of different target groups. Various institutions have been involved in the development of EE curriculum. In recent years, the concept of EE, which became widely accepted and almost universally accepted, was used primarily in the USA in the 1990s. In the UK they prefer to use the phrase 'enterprise education' [10, p. 12]. In the British educational system, the emphasis was on the formation of the personal qualities of entrepreneur.

According to a large-scale study covering 31 countries, including 27 EU member states, approximately 5 out of 21 million students studying in European universities, are directly involved in certain types of EE. At the same time, 11 million students do not have access to any form of classroom or extracurricular activity that seeks to cultivate an entrepreneurial spirit [22, p. 22]. In this regard, a united Europe lags significantly behind North America, where the cultivating of entrepreneurial attitudes and skills has been lasting for a long period of time.

The implementation of EE is carried out on the basis of the Entrepreneurship Action Plan adopted in the EU in 2004, which created a strategic framework and set five major goals in the field of policies aimed at stimulating entrepreneurial dynamism. One of these goals concerns the development of entrepreneurial attitudes and style of thinking that would help EU countries successfully cope with the challenges of the modern era and attract a sufficient number of people in the field of private entrepreneurship [22, p. 13]. Despite the course officially adopted by the EU, discussions are ongoing regarding the need and specific forms of introducing EE at universities.

Until now, entrepreneurship courses have not been sufficiently integrated into the curricula of European universities, especially those countries that joined the EU in 2004. Entrepreneurship is taught mainly in economic specialties and in business schools. At the same time, the lion's share of professions does not include entrepreneurial disciplines in the curricula.

However, over time, the situation is changing: EE is gaining a wider field in the training of highly qualified specialists. One of the pioneers in the widespread introduction of entrepreneurial and business education is Finland. In the late 1990-s this small Scandinavian country began a comprehensive reform of the entire education system, one of the key components of which was the implementation of entrepreneurial modules at all levels of training. Gradually, entrepreneurial approaches to educational activities were introduced at all levels and now form a continuum from elementary school to college and university.

The pan-European policy of stimulating entrepreneurship sets the task of improving EE and more actively involving universities in activities to stimulate economic development. The documents of the European Commission note the importance of EE in the modern world: 'In addition to equipping young people with the skills needed for the 21st century, entrepreneurship education is a means to increase social inclusion; it can increase the number of entrepreneurs – social and commercial, and it can be a gateway for a greater integration of the framework for key competences for lifelong learning' [22, p. 7].

A large-scale study supported by the European Commission in 2012 revealed the effectiveness of EE programs offered by universities in four main areas: the development of key entrepreneurial competencies, intentions in relation to entrepreneurship, the impact on individual employment prospects, and the impact on the economy and society overall [22, p. 8]. The study was based on a comparison of various characteristics of graduates of two types of educational programs: those on which, along with the main specialization, training was also carried out within the framework of EE, and those that did not undergo entrepreneurial training. Those who underwent such training demonstrated more clearly expressed entrepreneurial attitudes and intentions, quickly found work after graduation, showed a greater inclination and ability to innovate, even being in the role of employees, and more often established their own companies. As the results of the study show, training in entrepreneurial programs positively affects almost all characteristics that are important both for entrepreneurs themselves and for the whole modern society. It is these characteristics that form the basis of a productive economic culture of society, contribute to its dynamism, innovativeness and competitiveness. In addition to improving entrepreneurial knowledge, the EE contributes to the development of entrepreneurial attitudes: it increases the level of initiative, enterprise, risk propensity and the need to achieve [23]. No EE effect is observed only in the sphere of self-efficacy and structured behavior. Entrepreneurship training also develops entrepreneurial skills – creativity, adaptability, analytic abilities, social networking, and motivation. Thus, the vast majority of key entrepreneurial competencies can be formed in the process of properly organized EE.

Hirsch et al. [24] accentuate various specific forms of entrepreneurial thinking that may well be utilized in the process of EE. They include structural thinking, bricolage, effectuation, and cognitive adaptability. 'Entrepreneurs think differently from nonentrepreneurs. Moreover, an entrepreneur in a particular situation may think differently from when faced with some other task or decision environment. Entrepreneurs must often make decisions in highly uncertain environments where the stakes are high, time pressures are immense, and there is considerable emotional investment. We all think differently in these strained environments than we do when the nature of a problem is well understood and we have time and rational procedures at hand to solve it. Given the nature of an entrepreneur's decision-making environment, he or she must sometimes (1) think structurally, (2) engage in bricolage, (3) effectuate, and (4) cognitively adapt' [24, p. 7].

6 Implementation of entrepreneurial education in a training program for engineers

Of particular interest are those directions of introducing entrepreneurship education, which are concerned with its integration into the training program for engineers and other technical specialists and professionals. The potentially high prospects of this area for the stimulation of business activity are explained, first of all, by deep scientific and technical knowledge among engineers and the possibility of their application in creating high-tech startups in the knowledge economy. In a market economy, the development of small and medium-sized enterprises in the field of industrial production, which can give a significant impetus to the country's socio-economic development, is particularly relevant. The implementation of entrepreneurial skills training among engineering specialists can noticeably contribute to the development of small and medium enterprises in the manufacturing sector. Demand for this kind of education also exists among engineering students themselves, a sizeable part of which take interest in entrepreneurial activity in the future.

The discussion on the introduction of EE in the training program of engineers is part of a broader debate on the issue of the modernization of engineering education, its goals and forms, the need to overcome a narrow mono-professional approach and introduce interdisciplinary training programs. Many scientists support the idea of combining engineering knowledge with the competencies necessary to create and run a business organization. They emphasize the importance for engineers to possess also managerial abilities and skills as well as to be able and ready to create their own business structures that will concentrate advanced scientific knowledge and technologies, and to become the cores of innovation. In current conditions of the formation and rapid development of the knowledge economy, engineers are faced with increased requirements – the ability to design, create and manage complex technical and economic systems, solve creative problems that require not only high scientific and technical training, but also competencies in the field of the so-called 'soft skills', namely: leadership and team work, communication and time-management, problem-solving and creativity, adaptability, interpersonal skills, and productive work ethic. However, soft skills are expected to characterize employees at different positions and of various occupations. Potential entrepreneurs have to be trained in a specific set of qualities, attitudes, cognitive and non-cognitive abilities which altogether can be called entreprenerual skills.

It is the development of these skills and competencies that will determine the success of professional activities in the 21st century, which was highlighted by the World Economic Forum in 2015. Among the key competencies were the ability to solve problems comprehensively, critical thinking, creativity, people management, coordination with others, emotional intelligence, ability to assess the situation and make decisions, ability to negotiate, cognitive flexibility [25]. Apparently, modern education should also focus on the formation of these competencies. It is worthwhile to draw attention to the fact that a significant part of the competencies highlighted above is absolutely necessary for entrepreneurs.

Studies on the impact of entrepreneurial education on the level of business development give mixed results in different countries, due to the presence of a number of cultural, institutional and structural factors that mediate this influence. Nevertheless, all the data speak in favor of EE as a positive motivator for starting own business. An interesting experience was gained in Canada, especially among students of engineering specialties. It was found that 40% of engineering graduates who completed entrepreneurial training at universities established their own firms upon graduation [26]. A comparative study of the United States and South Korea demonstrated the importance of EE in creating a supportive business culture, especially for countries with not so deep-rooted entrepreneurial traditions [27].

A number of publications discuss examples of introducing EE in individual countries or universities in diverse socio-cultural contexts, in particular, in Greece [28], Spain [29], Holland [30], Belgium [31], Sweden [32], France and the USA [4], Finland [1], Canada [26] and several other countries.

The integration of engineering and entrepreneurship education is one of the possible and promising options for a harmonious, balanced combination of the classical (liberal) and utilitarian (professional) paradigms within the framework of a single direction of professional training. In fact, we are talking about symbiosis, the interpenetration of the traditional approach to training an engineering specialty and a new interpretation of the social role and mission of an engineer not only as a carrier of advanced scientific knowledge and technical expertise, but also as a socio-economic innovator, leader, intellectual, and holder of productive motivational qualities, transformer and integrator of society based on progressive and productive values. Moreover, the dissemination of such values among the general public, including those who are not directly involved in business activities, increases entrepreneurial potential and creates a favorable breeding ground for nurturing business talents, both commercial and social. Thus, the rest of society also benefits from the growth and expansion of carriers of productive values, attitudes and patterns of economic behavior.

An analysis of existing approaches to the training of entrepreneurs allows us to highlight the key limitations and shortcomings in the established model of EE:

1) It is inapplicable due to the low efficiency of classical teaching methods and forms that ignore the complex and diverse nature of the challenges, problems, risks and uncertainties that entrepreneurs face in their activities.

2) The historically established individualistic approach focuses on the personality of the entrepreneur, pushing into the background macro-social conditions and factors

of an institutional and structural nature that affect the development of business activity. Entrepreneur operates in conditions of high uncertainty. Learning to act in an uncertain business environment is one of the key tasks of the EE.

3) There is a narrow understanding in educational and academic circles of EE as learning how to establish and develop a business rather than how to efficiently act in ever-changing environment with limited information.

4) Rationalized systematic approach to teaching entrepreneurship by applying traditional theories and management techniques to entrepreneurial situations is poorly adapted to the social reality business people have to deal with – complex, unique, insufficiently defined, risky, and emotional conditions [6, p. 137]. This approach does not allow to plunge into the real world of entrepreneurship and to feel all the specifics of this activity. A number of studies show the effectiveness of situational entrepreneurship training, in particular through active inclusion in the family business. EE appears as a contextually determined, embedded in everyday activities, immersed in a system of social relations and obligations, as well as value and regulatory determined. Family and business act as two mutually overlapping communities of practice, fields of practice-based knowledge. EE proceeds through learning from intergenerational exchange, reproduction, and transformation [33].

The implementation of EE requires a thorough and comprehensive analysis of the macrosocial environment in which such changes will be introduced. There are several significant trends and factors that must be taken into account. First, the mass-scale expansion of higher education, which in the 1990-2000s acquired unprecedented proportions in most post-socialist countries. For example, if in the 1990/91 academic year 174.5 thousand people entered universities, institutes and academies in Ukraine, and 136.9 thousand graduates graduated, then in 2006 these indicators were 507.7 and 413.6 thousand respectively [34]. In other words, the higher education system instructs 3 times more specialists than in the late Soviet period. Second, the expansion of higher education took place in conditions of a deep economic crisis and sharp reduction of the real sector of the economy, de-industrialization of the country and a catastrophic drop in production volumes. Objectively, the national economy could not absorb such a large number of highly qualified specialists. Third, in parallel there was a sharp reduction in the training of specialists of the post-secondary vocational level, skilled technicians, craftsmen, and workers. Fourth, the increase in quantitative indicators in higher education was accompanied by a significant decrease in the qualitative characteristics of both students and graduates. Fifth, there have been radical changes in the structure of training of specialists, their distribution by industry. There was a mass decline in the sphere of technical, exact, and engineering sciences, a peculiar "flight from science" associated with the devaluation of professionalism and the pursuit of formal qualifications.

In addition to the general trends in vocational education noted above, the development of EE is also influenced by specific factors associated with the historical processes of the formation and development of entrepreneurship itself. First and foremost, this area of independent economic activity has been banned for a long time. In the Soviet Union it was impossible to legally do business, in contrast to the socialist countries, where political regimes in most cases showed great tolerance for the carriers of entrepreneurial functions, allowing private initiative with varying degrees of freedom. The rigidity of the Soviet system in relation to businessmen negatively affected the processes of the revival of entrepreneurship in the new historical conditions: there were no generations directly related to entrepreneurial activity, and it had to be revived from scratch.

Second, the lost positive traditions of doing business and the deep economic crisis in which the formation of the private sector of the economy took place, led to the dominance of destructive socio-economic practices, the spread of unproductive forms of activity, in particular intermediary, speculative, financial operations, and the closure of industrial and manufacturing industries, the desire for quick and easy profit, narrowing the temporal horizon of economic activity, abandoning long-term investment.

Third, citizens of the newly independent states had a very vague and often wary and hostile idea of entrepreneurship, due to the lengthy process of indoctrination and massive state propaganda, which aimed to achieve the full delegitimization of private enterprise. Thus, there were no scientific and pedagogical staff, nor real entrepreneurs who could contribute to the development of EE and entrepreneurship itself.

Implementation of EE can bring both direct positive results (revitalizing entrepreneurial activity, improving the quality structure of small businesses, economic growth and improving living standards), and indirect (changes in the structure of training and employment: reducing the proportion of humanitarian, legal and economic students not related with business, profile).

At the same time, the implementation of EE is faced with a number of specific difficulties arising at the institutional level due to the conservatism of the educational institutions themselves. Among them, special attention should be paid to the following difficulties:

1) The concentration of business disciplines in business schools raises serious doubts and leads to negative consequences. As experience shows, the most innovative and viable business ideas arise in an environment related to technical, scientific, cultural creativity, among specialists who think extraordinary and are looking for ways to commercialize their creative ideas. The best option could be the introduction of EE in the framework of specialized training programs with a focus on entrepreneurship development opportunities within the framework of this program. For example, the implementation of diverse, development-oriented entrepreneurial competencies, disciplines and modules within the framework of engineering training to

stimulate business activity in the field of innovative industrial production.

2) The rigid structure of educational institutions creates difficulties for the implementation of multidisciplinary approaches. The integration of EE and education in a specific area - natural science, technical, social and humanitarian - requires overcoming a highly specialized framework. The question arises whether universities are ready to introduce such innovations. To what extent are educational institutions themselves entrepreneurial in essence, in organizational culture and everyday practices, or do they remain bastions of bureaucracy, formalism? [35]

3) Programs should be adapted to the needs of students of specific specialties, be professionally oriented. So, EE for engineering specialists should focus on innovative and technological models of entrepreneurship in the industrial sphere, on the possibility of creating and developing their own business using engineering knowledge, based on the commercialization of technical developments.

7 Conclusions

Thus, in the context of modernization of the higher education system, the implementation of EE may encounter serious obstacles at the systemic – structural and institutional – level. In other words, an objectively existing demand in society for the development of a productive entrepreneurial culture and the formation of relevant competencies on a large scale can be met with resistance. At this stage, hopes should be pinned on individual innovative universities that are ready to become pioneers in the implementation of EE, which is achievable thanks to wide university autonomy in accordance with the law of Ukraine 'On Higher Education'. On the other hand, it is important to understand to what extent students and teachers are ready for this kind of innovation.

There are several formidable obstacles the implementation of EE in Ukraine will encounter. Some of them are explained by conservatism typical of post-soviet system of education with mimicry and pseudo-reforming. Others are generated by factors external in relation to the institute of education.

EE has to be concentrated around several important characteristics of entrepreneurship and entrepreneurs themselves. In our opinion, EE is expected to help overcome or at least minimize the following negative or even toxic aspects of current entrepreneurship in Ukraine: unproductive and counter-productive (e.g., speculative) forms of activity, tendency towards violence, authoritarian forms of management, close interconnections with political sphere, the spread of 'shadow economy' practices, and lack of sustainability (fast and easy profit-making, lack of ecological consciousness, no labor rights guarantee, etc.).

Regional authorities may well promote training for entrepreneurship and thus contribute to financially resilient local communities. This is very topical in the context of de-centralization process taking place in Ukraine during last several years. On the whole, the implementation of EE at the curriculum and extra-curriculum activities at the institutions of higher education of Ukraine, the combination of theoretical, problem- and project-oriented learning with the immersion into practical activities concerned with business planning, creating and running together with nascent and experienced entrepreneurs would be extremely useful and productive in terms of motivating more and more young people to take an entrepreneurial path and achieve prosperity and self-actualization. The practical realization of EE will contribute to more resilient and competitive local communities as well as a more inclusive, just, equal, and happy society.

References

1. T. Lähdeniemi, Industry & Higher Education **11**(3), 174–175 (1997)
2. D. Hjorth, B. Johannisson, Revue de l'Entrepreneuriat **8**(2), 57–78 (2009)
3. N. Arranz, F. Ubierna, M. Arroyabe, C. Perez, J. de Arroyabe, Studies in Higher Education **42**(11), 1979–2008 (2016)
4. J.Boissin, B. Branchet, S. Emin, J. Herbert, J. of Small Business and Entrepreneurship **22**(2), 101–122 (2009)
5. R. Barr, J. Tagg, Change: The Magazine of Higher Learning **27**(6), 12–26 (1995)
6. D. Higgins, K. Smith, M. Mirza, The J. of Entrepreneurship **22**(2), 135–160 (2013)
7. B. Bird, *Entrepreneurial Behavior* (Scott, Foresman and Company, Glenview, 1989)
8. D. Politis, Entrepreneurship theory and practice **29**(4), 399–424 (2005)
9. D. Kolb, *Experiential Learning: Experience as the Source of Learning and Development* (Prentice Hall, Englewood Cliffs, 1984)
10. A. Gibb, Int. Small Business J. **11**(3), 11–34 (1993)
11. R. Sullivan, Int. J. of Entrepren. Behaviour & Res. **6**(3), 160–175 (2000)
12. L. Lamont, J. of Small Business Management **10**(3), 36–41 (1972)
13. W. Dyer, Entrepreneurship, Theory and Practice **19**(2), 7-21 (1994)
14. L. Galloway, W. Brown, Education and Training **44**(8/9), 398–405 (2002)
15. W. Baumol, J. of Pol. Economy **98**(5), 893–921 (1990)
16. M. Adams, M. Makramalla, W. Miron, Technology Innovation Management Rev. **9** (4), 19–27 (2014)
17. L. Kolvereid, Ø. Moen, J. of Eur. Industrial Training **21**(4), 154–160 (1997)
18. D. Schön, *Educating the Reflective Practitioner* (JesseyBass, San Fransisco, 1986)
19. B. Johannisson, H. Landström, J. Rosenberg, Eur. J. of Engineering Educ. **23**(4), 477–496 (1998)
20. A. Touraine, *Vozvrashcheniye cheloveka deystvuyushchego. Ocherk sotsiologii* (The return of an acting human. A sociological sketch) (Nauchny mir, Moscow, 1998)
21. V.A. Yadov, *Sotsiologicheskoye issledovaniye: metodologiya, programma, metody* (Sociological research: Methodology, programme, methods) (Samara, 1995)

22. European Commission, Effects and impact of entrepreneurship programmes in higher education: report of the expert group (European Union, Brussels, 2012)
23. D. McClelland, *Human Motivation* (Cambridge University Press, Cambridge, 1973)
24. R. Hisrich, M. Peters, D. Shepherd, *Entrepreneurship*, 10th edn (McGraw-Hill Education, New York, 2017)
25. E. Mykhailenko, T. Bleyon, Zerkalo nedeli 43-44, (2016), https://zn.ua/EDUCATION/postindustrialnoe-obrazovanie-kuda-idem-rezultaty-ukraino-kanadskogo-mezhuniversitetskogo-pilotnogo-proekta-okazalis-dovolno-neozhidannymi-_.html. Accessed 21 Mar 2017
26. T. Menzies, Y. Gasse, *Entrepreneurship and the Canadian Universities: A Report of a National Study of Entrepreneurship Education* (St. Catharines, Brock University, 1999)
27. S. Lee, D. Chang, S.-B. Lim, Int. Entrepreneurship and Management J. **1**(1), 27–43 (2005)
28. L. Papayannakis, I. Kastelli, D. Damigos, G. Mavrotas, Eur. J. of Engineering Educ. **33**(2), 199–210 (2008)
29. N. Arranz, M. Arroyabe, J. de Arroyabe, Studies in Higher Education **44**(11), 2011–2024 (2019)
30. H. Bonnet, J. Quist, D. Hoogwater, J. Spaans, C. Wehrmann, Eur. J. of Engineering Education **31**(2), 155–167 (2006)
31. R. Donckels, J. of Small Business & Entrepreneurship **9**(1), 35–42 (1991)
32. B. Johannisson, Entrepreneurship & Regional Development **3**(1), 67–82 (1991)
33. E. Hamilton, J. of Small Business and Enterprise Development **18**(1), 8-26 (2011)
34. Derzhavna sluzhba statystyky Ukrayiny. Zaklady vyshchoyi osvity (State service of statistics of Ukraine. Institutions of higher education) (2019), http://www.ukrstat.gov.ua/operativ/operativ2005/osv_rik/ osv_u/vuz_u.html. Accessed 28 Nov 2019
35. P. Coyle, Industry & Higher Education **28**(4), 263–269 (2014)

The research on educational environment of institution of higher education involving experts: results and analysis

Mariia Bratko*, and *Liudmyla* Khoruzha

Borys Grinchenko Kyiv University, 18/2 Bulvarno-Kudriavska Str., Kyiv, 04053, Ukraine

Abstract. Based on the analysis of scientific resources, the role of educational environment in personal building, development and education is actualized. It is stated that educational environment is one of the most essential factors of those processes. The author's definitions of notions "educational environment", "educational environment of institution of higher education" are given. The authors advocate the point that educational environment is a system of facilities for personal education. It is emphasized that the environmental approach in education is changing accents in teaching which is aimed at creating and developing educational environment, which has to meet educational needs of a student. The author's technique of evaluating the state of educational environment of institution of higher education is grounded and approved, the rating scale of this technique is introduced. The technique suggests making expert evaluation of educational environment of institution of higher education according to qualitative (modality) and quantitative (professional breadth, professional saturation, sociocultural intensity, congruence, openness, mobility, informativeness) parameters. The typology of educational environment of institution of higher education is presented (innovative-professional, formal-professional, pragmatically-oriented, formal general cultural educational environment). The analysis and interpretation of the results obtained from the experiment at University College of Borys Grinchenko Kyiv University are given. The evaluation of educational environment of the college has been made by three groups of experts, which have included students, teachers and parents. Despite the fact that evaluation by different groups of experts are similar, it is revealed that it has essential differences, which has been proved with appropriate statistical methods. It is determined that the substantiation of the whole monitoring system of educational environment of institution of higher education and corresponding diagnostic tools has great potential.

1 Introduction

One of the most significant factors of the quality of education due to its specifics is educational environment of institution of higher education. Interest in researching it as a component of integral social and living environment of an individual has been growing for the last few decades. At the same time the paradigm of educational environment in personal building and development has changed from the necessary condition to the active contributor that must be created and realized in order to provide high-quality education. The scientific discourse presents different views on the phenomenon of educational environment which is considered as a part of sociocultural space (N. Krylova [1]); a sociocultural system (V. Kozyriev [2]); a system of impacts and factors of personal developvent (V. Yasvin [3]); a complex of opportunities for personal learning and developing (S. Deryabo [4]); a system of key factors that determine personal education and development (V. Lebedeva [5]); a complex of human practices and material systems (T. Warger [6]); a complex of conditions in which instruction takes place and which influence students' performance and relationships (B. Fraser [7]); a product of a mutual activity of subjects of education (V. Slobodchikov [8]); a complex of facilities for personal education (M. Bratko [9]). We prefer to regard educational environment as a field of opportunities for personality. If conditions characterize 'space', environment is characterized by the notion 'opportunities', i.e. space becomes environment when conditions become opportunities. English-speaking publications extensively demonstrate the results of the study on the impact of educational environment on personality, which researchers interpret as educational environment (J. Salmi [10], S. Cotterill [11]); educational climate (R. Hiemstra [12]); academic environment (A. Lizzio [13]), study/learning environment (I. Abulrub [14], R. Moos [15], T. Wagner [6], B. Fraser [7]). The analysis of the resource base enables to state the development of the environmental approach in education that provides the subject of management with techniques and technologies of using educational environment for personal development and self-development, ensuring the quality of education. To achieve the objective of our research we appeal to approaches to examination, diagnostics and evaluation of educational environment including

* Corresponding author: m.bratko@kubg.edu.ua

educational environment of institution of higher education presented in works of S. Deryabo [4], R. Moos [15], V. Rubtsov [16], B. Fraser [7], V. Yasvin, S. Rybinskai [17], which has become the theoretical basis for developing the author's approach to evaluation of educational environment of institution of higher education. Monitoring the state of educational environment of institution of higher education may turn into a significant component of the internal system of ensuring the quality of education and its results may serve to make good managerial decisions.

2 The objective of research

The purpose of our research paper is substantiation and approval of the author's technique of evaluation of a current state of educational environment of institution of higher education by the example of University College of Borys Grinchenko Kyiv University according to qualitative and quantitative parameters in order to take managerial decisions concerning its development.

3 Research methodology

The purpose of research has made us use the complex of the relevant methods: scientific literature analysis in order to establish the state of the problem development, the definition of the categorical and conceptual apparatus of investigation; synthesis, generalization, systematization for theoretical substantiation and practical development of approaches to diagnostics of the current state of educational environment of institution of higher education; empirical: expert evaluation, vector modelling; mathematical and statistical methods (Mann-Whitney U-test, Kolmogorov-Smirnov test, Friedman test) to assess the experimental work results.

4 Results and discussion

The analysis of the phenomenon of environment, the thesaurus of environmental approach in education, scientific resources dealing with issues about examining educational environment as a factor of personal development and the author's point presented in our previous publications [18] enable to determine the content of fundamental notions of the research. On basis of methodology of the system and environmental approaches, it is defined that the educational environment is a multilevel system of conditions / circumstances / factors / opportunities that provides optimal parameters of the educational activity of certain educational individual in all aspects - target, content, process, result, resource. The educational environment is characterized by: systematic, integrity, unity, emergence, variability, organization, structural, plasticity, communicativeness, eventfulness, configurability, saturation, vectority, sphericity, openness, dialog / polychology, organization, stability, adaptability, functionality ability for development and self-development.The educational environment of a higher educational institution is an integrity that encompasses a set of conditions, impacts, opportunities for vocational training, personal development and socialization of the future specialist. The component structure (personal, axiological-semantic, information-content, organizational-active, spatial and substantive components) and functions (polystructural educational in the unity of educational-professional, educational-socializing, educational-cultural functions and personalized-developing function) of the educational environment of the university college have been determined as well. On the whole, realization of the environmental approach in higher education, professional training takes into account guaranteeing quality education by means of mediation of educational environment of an institution. The environmental approach includes increase in activities directed at designing, modeling and creating educational environment that have to meet educational needs of a student efficiently.

Analyzed approaches to evaluation of activities of educational institutions, effectiveness of educational process, quality of educational environment (L. Vashchenko [19], S. Deryabo [4], V. Yasvin [20] and ets.) enable to define the main qualitative and quantitative parameters of evaluation of educational environment of institution of higher education. As a rule, in researches of environment of institutions of secondary education presence or absence of conditions and opportunities for development of activity or inactivity of the subject of education and their personal freedom is taken as a criterion indicator of modality. In our study we take into consideration educational environment of institution of higher education. Therefore, the qualitative parameter (modality) of educational environment of institution of higher education is adjusted with its main functions (polystructural educational one which consists of educational-professional, educational-socializing, educational-cultural and personalized-developing functions) and is characterized from the typological point of view. Defining the type of educational environment of institution of higher education, the degree of its orientation to profession, socializing, inculturation, general personal development is taken into account. Such an approach enables to determine the types of educational environment of institution of higher education considering existing approaches to typology of educational environment: *innovative-professional* (modern, filled with professional content and activities with well-developed practice-oriented component of educational process, prioritizing development of professional competences of future specialists, creative thinking, 21st century skills, open-mindness and adaptability to sociocultural context); *formal-professional* (filled with professional context but preference is given to theoretical instruction, practice-oriented component of educational process is underdeveloped, characterized by low saturation of socio-humanitarian life of an educational institution); *pragmatically-oriented* (environment aimed at meeting individual educational needs of students and their parents concerning organization of professional training and socio-humanitarian life of an educational institution); *formal general cultural* (first of all aimed at general cultural and personal development of students, occasionally harmful to professional training). Researchers of educational environment suggest different lists of its quantitative parameters. The most frequently mentioned ones are the following: breadth, intensity, awareness, resilience, community, dominance, social activity, mobility, information, formality, emotionality, orientation, structure, coherence, professionalism, security. We have made a list of characteristics for evaluation of educational environment of institution of higher education taking into consideration researchers' discoveries, our own position, peculiarities of an institution of higher education as an institution that provides professional training. Hence, in our case, the quantitative parameters feature filling the environment of an institution of higher education with various factors of professional training, namely: *professional breadth* (quantity and quality of subjects and objects, processes, phenomena with professional orientation), *professional saturation* (the degree of saturation with professional resources, technologies, contacts with people in profession, educational projects, clubs and studios), *sociocultural intensity* (the degree of saturation with conditions, impacts, opportunities for personal socializing and inculturation), *congruence* (the degree of adjustment of functioning all components), *openness* (demonstrates the degree of social orientation and involvement of educational environment in the surrounding world, availability of ways for social partnership to perform multiaspect educational function of educational environment), *mobility* (the degree of ability of educational environment to meet demands of outside context concerning the content, forms, methods and technologies of professional training), *informativeness* (features the degree of saturation and availability of information resources).

The approbation of determined criteria for evaluation of educational environment of institution of higher education by means of the expert method was carried out at University College of Borys Grinchenko Kyiv University (further on College) during 2012-2018. This publication presents the results of the research conducted in 2018. The participants of educational process – students, teachers, parents – performed a function of experts.

Expert groups were offered to evaluate the types of educational environment (its modality) according to 5-point scale, where 5 points corresponds to the highest level of the certain type manifestation and 1 point – to the lowest one. It is important to remember that there cannot be 'pure' type of educational environment. Any educational environment of an educational institution possesses characteristics of every type. However, characteristics of a peculiar type can dominate, thus they determine general orientation of environment.

Evaluation of educational environment of institution of higher education according to modality was made by 294 students (22,7% out of overall number of the student body at the time of the experiment, all courses and majors which are provided at College were represented), 83 teachers (49% out of general number of pedagogical staff) with different work experience in institutions of higher education (10 people with work experience of less than 5 years; 19 people – from 5 to 10 years; 23 people – from 10 to 20 years; 31 people – from 20 to 30 years), 250 parents. A special google-form was created to carry out the survey. Identification of nature of received results and their verification were made by means of statistical methods. All calculations were made by means of digital tables Excel, statistical package SPSS [21].

Generalized results of evaluation by experts (students, teachers, parents) of manifestation of types of educational environment (modality) at University College in 2018 are presented in Table 1.

Table 1. The results of evaluation of manifestation of types of educational environment by experts at College, 2018.

No	Type of educational environment	Expert group		
		Students	Teachers	Parents
1	innovative-professional	4,22	4,13	4,14
2	formal-professional	3,31	3,34	3,22
3	pragmatically-oriented	3,72	3,51	3,71
4	formal general cultural	3,71	3,23	3,22

Generalized results demonstrate that the highest points by each group of experts were given to innovative-professional environment. However, other types of environment have also a high degree of manifestation. It must be stated that evaluation of parent experts and teacher experts are almost identical in all types. Evaluation of innovative-professional and pragmatically-oriented types of environment is similar by all types of experts. Points by student experts are significantly different from those by other experts concerning the formal general cultural type of environment. In comments students wrote that they were excessively involved in different entertaining events and concerts and they didn't always understand their connection with the content and goals of professional training. The answers of experts show that educational environment of College corresponds to its aim – providing professional training. In our opinion, it is a bit advance evaluation. In future it is necessary to enhance opportunities for students to obtain professional knowledge and skills at College. We are going to demonstrate approaches to evaluation of reliability of received results by means of statistical methods by the example of the group of student experts. The qualitative and quantitative composition of the group of student experts (according to specialty) and the results of evaluation by them (average indicators in every specialty) of types of educational environment at University College are presented in Table 2. Kolmogorov-Smirnov test has shown that students' evaluation data are not arranged appropriately, therefore for their comparison we apply non-parametric tests. Since they were dependent groups, the same students evaluated educational environment according to four types. Table 3 and Table 4 show the comparative results of evaluation of types of educational environment of University College by student experts and statistics according to Friedman test. As we see, a posteriori knowledge of statistics χ^2 equals 156,721, its significance is less than 0,05, so the differences between 4 lines of points are considerable.

Table 2. The results of evaluation of manifestation of types of educational environment at College by student experts, 2018.

No	Specialty	Amount of students	Type of educational environment			
			innovative-professional	formal-professional	pragmatically-oriented	pragmatically-oriented
1	journalism	33	4,15	3,18	3,48	3,52
2	social work	11	4,00	2,55	3,36	3,64
3	design	22	3,59	3,41	3,41	3,91
4	choreography	5	4,20	3,80	3,60	3,40
5	science of law	8	4,13	3,25	3,25	3,75
6	secondary education (music)	14	3,71	3,00	3,50	3,36
7	information, library and archives	23	4,35	3,26	4,00	3,83
8	primary education	70	4,43	3,33	3,87	4,01
9	pre-school education	27	4,44	3,67	3,85	3,78
10	secondary education (arts)	16	4,00	3,31	4,06	3,19
11	finances	24	4,29	3,50	3,83	3,67
12	secondary education (physical education)	12	4,58	3,17	3,75	3,58
13	management	29	4,28	3,28	3,62	3,55
14	total	294	4,22	3,31	3,72	3,71

Table 3. The comparative results of evaluation of types of environment at College by student experts (Friedman test), 2018.

No	Type of educational environment	Average ranks
1	innovative-professional	3,09
2	formal-professional	2,02
3	pragmatically-oriented	2,46
4	formal general cultural	2,44

Table 4. Statistics (Friedman test) for student experts.

1	N	294
2	Chi-Square	156,721
3	df	3
4	Asymp. Sig.	0,000

Calculations for other groups of experts are made in an analogical way. Table 5 and Table 6 show the comparative results of evaluation of types of educational environment at University College by teacher experts and statistics according to Friedman test. As we see, a posteriori knowledge of statistics χ^2 equals 30,172, its significance is less than 0,05, so the differences between 4 lines of points are considerable. Table 7 and Table 8 show the comparative results of evaluation of types of educational environment of University College by parent experts and statistics according to Friedman test. As we see, a posteriori knowledge of statistics χ^2 equals 183,803, its significance is less than 0,05, so the differences between 4 lines of points are considerable.

Expert evaluation of educational environment of College according to the quantitative parameters (professional breadth, professional saturation, sociocultural intensity, congruence, openness, mobility, informativeness) in 2018 was made by three groups of experts consisting of students, teachers and parents of about 50 people in each group. The survey was conducted by means of google-form. General data about the results of evaluation of educational environment of College according to specific characteristics, namely – average value in expert groups according to the quantitative parameters (professional breadth, professional saturation, sociocultural intensity, congruence, openness, mobility, informativeness) is presented in Table 9. The graphic image of the results of evaluation of educational environment of College according to the parameters is presented in Fig. 1.

Table 5. The comparative results of evaluation of types of environment at College by teacher experts (Friedman test), 2018.

No	Type of educational environment	Average ranks
1	innovative-professional	3,08
2	formal-professional	2,37
3	pragmatically-oriented	2,37
4	formal general cultural	2,17

Table 6. Statistics (Friedman test) for teacher experts.

1	N	83
2	Chi-Square	30,172
3	df	3
4	Asymp. Sig.	0,000

Table 7. The comparative results of evaluation of types of environment of College by parent experts (Friedman test), 2018.

No	Type of educational environment	Average ranks
1	innovative-professional	3,18
2	formal-professional	2,10
3	pragmatically-oriented	2,65
4	formal general cultural	2,06

Table 8. Statistics of Friedman test for parent experts.

1	N	250
2	Chi-Square	183,803
3	df	3
4	Asymp. Sig.	0,000

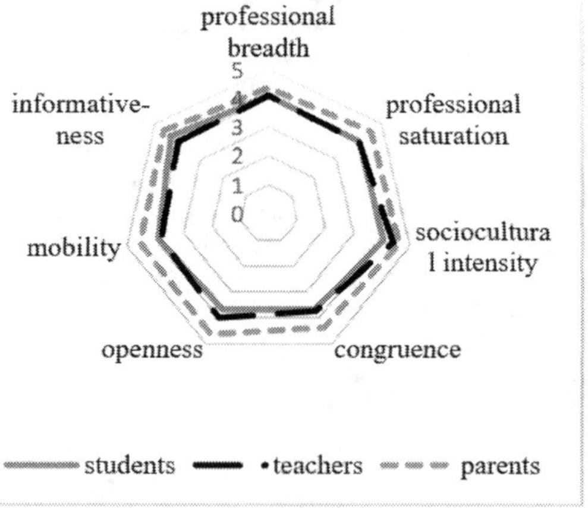

Fig. 1 The graphic image of the results of evaluation of educational environment of College according to the parameters, 2018.

Table 9. The results of evaluation of educational environment of college according to the parameters (average values, in expert groups), 2018.

No	Parameter	Expert group: Students	Teachers	Parents	Average value
1	professional breadth	4,10	4,10	4,37	4,19
2	professional saturation	4,00	3,98	4,54	4,17
3	sociocultural intensity	4,06	4,42	4,63	4,37
4	congruence	3,67	3,74	4,37	3,93
5	openness	3,67	4,00	4,63	4,10
6	mobility	3,90	3,80	4,54	4,08
7	informativeness	4,29	3,98	4,61	4,30

Let us analyze obtained results. First of all, we are going to identify how different are points of certain groups of experts in pairs according to Mann-Whitney U-test. Let us present the analysis results and give examples of calculations in Table 10 and 11. The analysis of the calculation results in pair of students – teachers shows that points by students and teachers differ considerably according to the following parameters as sociocultural intensity, informativeness (significance is less than 0,05).

Table 10. The comparison of evaluation of educational environment of College according to the quantitative parameter of Mann-Whitney U-test for 2 expert groups of students and parents, 2018.

No	Parameter	Status	N	Mean rank	Sum of ranks
1	professional breadth	students	50	42,17	2108,5
		parents	41	50,67	2077,5
		Total	91		
2	professional saturation	students	50	39,65	1982,5
		parents	41	53,74	2203,5
		Total	91		
3	sociocultural intensity	students	50	37,62	1881,0
		parents	41	56,22	2305,0
		Total	91		
4	congruence	students	50	36,93	1846,5
		parents	41	57,06	2339,5
		Total	91		
5	openness	students	5	35,85	1792,5
		parents	41	58,38	2393,5
		Total	91		
6	mobility	students	50	38,12	1906,00
		parents	41	55,61	2280,0
		Total	91		
7	informativeness	students	50	40,97	2048,5
		parents	41	52,13	2137,5
		Total	91		

Student experts evaluate informativeness higher (+0,31) than teacher experts do. Teacher experts evaluate sociocultural intensity higher (+0,36) than student experts do. Other parameters do not have significant differences. Evaluation by students and parents differ in all parameters except professional breadth (significance is less than 0,05). In particular, parents evaluate educational environment according to parameters – professional saturation (+0,54), sociocultural intensity (+0,57), congruence (+0,7), openness (+0,96), mobility (+0,55), informativeness (+0,32). Evaluation by teachers and parents differ considerably in all parameters except professional breadth (significance is less than 0,05). We should state that parents evaluate educational environment higher than teachers in all parameters. In particular: professional saturation (+0,56), sociocultural intensity (+0,21), congruence (+0,63), openness (+0,63), mobility (+0,74), informativeness (+0,63). On the whole, points by parent experts are higher than those of other experts in every parameter. Although student experts and teacher experts evaluate educational environment differently, their points are very similar. Such results were unexpected. Researches by V. Yasvin [20] show that the highest points are usually given by teacher experts and administration board, the lowest ones – by students, and point by parents are between them. From interviewing parents, it is stated that they idealize the place where their children are studying. Partially, parents explain that by the fact that they took responsibility in choosing an educational institution taking into consideration their children's age and they did not choose the worst one.

Table 11. Statistics of Mann-Whitney U-test for 2 expert groups of students and parents, 2018.

No	Parameter	Mann-Whitney U	Wilcoxon W	Z	Asymp. Sig. (2-tailed)
1	professional breadth	833,5	2108,5	-1,699	0,089
2	professional saturation	707,5	1982,5	-2,742	0,006
3	sociocultural intensity	606,0	1881,0	-3,715	0,000
4	congruence	571,5	1846,5	-3,890	0,000
5	openness	517,5	1792,5	-4,328	0,000
6	mobility	631,0	1906,0	-3,395	0,001
7	informativeness	773,5	2048,5	-2,255	0,024

5 Conclusions

1. The role of educational environment as a factor of personal education and development, which not only has to be taken into consideration in educational process but also has to be created, is emphasized. Based on existing approaches to interpreting the notions 'educational environment', 'educational environment of institution of higher education', it is determined that educational environment is a multilevel system of conditions / circumstances / factors / opportunities that provides optimal parameters of the educational activity of certain educational individual in all aspects - target, content, process, result, resource. The educational environment of a higher educational institution is an integrity that encompasses a set of conditions, impacts, opportunities for vocational training, personal development and socialization of the future specialist. The component structure and functions of the educational environment of the university college have been determined.

2. The author's technique of evaluating the state of

educational environment of institution of higher education is suggested. It makes expert evaluation of educational environment of institution of higher education according to the qualitative (modality) and quantitative (*professional breadth, professional saturation, sociocultural intensity, congruence, openness, mobility, informativeness*) parameters. The typology of educational environment of institution of higher education (innovative-professional, formal-professional, pragmatically-oriented, formal general cultural educational environment) is presented.

3. The results of approbation of technique of evaluating the state of educational environment of institution of higher education by the example of educational environment of University College of Borys Grinchenko Kyiv University are presented. Reliability of obtained results is proved by means of Mann-Whitney U-test, Kolmogorov-Smirnov test, Friedman test. It is shown that expert groups consisting of students, teachers and students' parents evaluate educational environment, where the experiment took place, in a different way but their points are very similar.

Further research perspectives are connected with development of the integral system of monitoring educational environment of institution of higher education and corresponding diagnostic tools. In particular, research interest deals with development of criteria of evaluating the components of educational environment of institution of higher education.

The research was carried out within the framework of the complex scientific theme of the Department of Theory and History of Pedagogy of Borys Grinchenko Kyiv University "The content and technologies of ensuring the quality of life long pedagogical education in the context of European integration"), SR No 0116U003295.

References

1. N. Krylova, *Antropologicheskiy, deyatelnostnyiy i kulturologicheskiy podhodyi* (Anthropological, activity and cultural studies approaches). (Scientific Research Institute of School Technologies, Moscow, 2005)
2. V. Kozyrev, Educator: Science, technology, practice 7, 26–32 (1999)
3. V. Yasvin, *Obrazovatelnaya sreda: ot modelirovaniya k proektirovaniyu* (Educational environment: from modeling to design), 2nd edn. (Smyisl, Moscow, 2001)
4. S. Deryabo, *Uchitelyu o diagnostike effektivnosti obrazovatelnoy sredyi* (For teacher about the diagnosis of the effectiveness of the educational environment). (Molodaya gvardiya, Moscow, 1997)
5. V. Lebedeva, V. Orlov, V. Panov, in *Abstracts of the 1rd Russian Conference on Environmental Psychology*, Psychological Institute Russian Academy of Education, Moscow, 3–5 Dec. 1996
6. T. Warger, G. Dobbin, Learning Environments: Where Space, Technology, and Culture Converge (2009), https://aect.org/docs/ELI3021.pdf. Assessed 22 Oct 2019
7. B. Fraser, in *Studies in educational learning environment: An international perspective*, ed. by S.C. Goh, M.S. Khine (World Scientific Publishing, Singapore, 2002), pp. 1–25
8. V. Slobodchikov, The psychology of learning 1, 4-24 (2010)
9. M. Bratko, Higher education environment: functional aspect. Pedagogical process: theory and practice 1–2, 11-16 (2015)
10. J. Salmi, *The Challenge Of Establishing World-Class Universities* (World Bank, Washington, 2009)
11. S. Cotterill, Capture 4, 47-54 (2013)
12. R. Hiemstra, Aspects of effective learning environments. New Directions For Adult And Continuing Education **50**, 5–12 (1991). doi:10.1002/ace.36719915003
13. A. Lizzio, K. Wilson, R. Simons, University students' perceptions of the learning environment and academic outcomes: implications for theory and practice. Studies In Higher Education **27**, 27–52 (2002). doi:10.1080/03075070120099359
14. I. Abualrub, B. Karseth, B. Stensaker, The various understandings of learning environment in higher education and its quality implications. Quality In Higher Education **19**, 90–110 (2013). doi:10.1080/13538322.2013.772464
15. R. Moos, *Evaluating educational environment: procedures, measures, findings and policy implications* (Jossey-Bass, San Francisco, 1979)
16. V. Rubtsov, T. Ivoshina, *Proektirovanie razvivayuschey obrazovatelnoy sredyi shkolyi* (Designing the developing educational environment of the school). (Izd-vo MGPPU, Moscow, 2002)
17. V. Yasvin, S. Rybinskaya, *Kompleksnaya otsenka obrazovatelnyih organizatsiy* (Integrated assessment of educational organizations). (Izd-vo MGPPU, Moscow, 2015)
18. M. Bratko, *Upravlinnia profesiinoiu pidhotovkoiu fakhivtsiv v osvitnomu seredovyshchi universytetskoho koledzhu: teoriia i praktyka* (Management of Professional Training of Specialists in the Educational Environment of University College: Theory and Practice) (Axiom, Kamianets-Podilskyi, 2017)
19. L. Vashchenko, *Upravlinnia innovatsiinymy protsesamy v zahalnii serednii osviti rehionu* (Management of innovation processes in general secondary education of the region). (Tyrazh, Kyiv, 2005)
20. V. Yasvin, Research Educational Environment in Russian Psychology: from Methodological Discussion by the Empirical Results. Philosophy. Psychology. Pedagogy **18**, 80–90 (2018). doi:10.18500/1819-7671-2018-18-1-80-90.
21. L. Panchenko, E. Adamenko, *Kompyuternyiy analiz dannyih* (Computer data analysis). (Publishing House of the State Institution of Taras Shevchenko State University, Lugansk, 2010)

Managing the development of training of competitive professionals in vocational educational institutions

Larisa Sergeeva[1,*], and *Tatiana* Stoychik[2]

[1]University of Management Education, 52-A Sichovyh Striltsiv Str., Kyiv, 04053, Ukraine
[2]Kryvyi Rih Professional Mining and Technology Lyceum, 1 Elistynska Str., Kryvyi Rih, 50053, Ukraine

Abstract. Significant transformation processes in Ukraine, inspired by globalization, require a rethinking of educational goals, objectives, and educational imperatives. Within the educational process in vocational (vocational-technical) educational institutions, the training system is forced to take into account the global context of actions and consequences in the preparation of a competitive specialist. The purpose of this article is to analyze the current state of development of vocational (vocational-technical) education in Ukraine and the network of institutions of vocational (vocational-technical) education; identifying factors that affect the quality of training of a competitive specialist and controversies that need to be resolved in the training of professionals. To analyze the current state of preparation, a set of theoretical and empirical methods was used, with the help of which a number of problems that inhibit the process of ensuring the quality of training of a competitive specialist were identified. In order to study the current state of preparation, a set of questionnaires was developed and, within the framework of a pedagogical experiment, questionnaires were conducted among entrants, educators, graduates, pedagogical workers, heads of institutions of vocational (vocational-technical) education in the regions of Ukraine and employers, and the results of the questionnaire were analyzed.

1 Introduction

The global market over the past two decades has given rise to a series of negative social phenomena that characterized "wild" capitalism until the middle of the twentieth century: sharp differentiation of incomes, the formation of centers of poverty and hunger, and the undermining of social security systems. Low wages, alienation and marginalization are seen as a direct consequence of putting into practice the logic of the global free market.

In the context of the challenges of globalization, which actualize the search for new paradigms of social development, the problems of professional development of the individual, development of his creative potential, constant self-improvement and vocational training throughout life are of particular importance. The problems of professional development in the period of complex transformations of the labor sphere of modern society are of particular importance. This is due to changes in the value-meaning self-actualization of the individual and deformation of its formation in the conditions of increasing requirements for professional mobility and competitiveness of the specialist [1].

The Ukrainian system of vocational (technical and vocational) education (herein after referred to as V(VT)E has faced a number of challenges, first of all: there is a discrepancy between the quality and directions of vocational training, the levels of qualification of V(VT)E applicants on labor market needs and personality demands; low level of prestige of working professions and professional qualifications [2, p. 2]; threatening insufficient level of quality of V(VT)E, mobility and competitiveness of graduates of institutions of vocational (vocational-technical) education (herein after – V(VT)E; shortage of qualified personnel, lack of qualification level to the real needs of the labor market; ensuring availability of V(VT)E, degree of obtaining professional qualifications; outdated material and technical base of V(VT)E; the need to expand the professional competencies of specialists; new requirements for the content of education and training process.

Content education requires a holistic system of cultural content. This system should cover the balanced development of the mental, emotional, value, volitional and physical spheres [3].

The challenges are due to the following factors: centralized and bureaucratized system of governance in the area of V(VT)E; insufficient amount of funding for V(VT)E; low level of wages of pedagogical and other employees of V(VT)E; lack of co-operation between the V(VT)E with employers and business partners; lack of data on the state of the labor market; the imperfection of the

* Corresponding author: lase2508@gmail.com

vocational guidance and career counseling system for young people and adults [4].

In the process of training highly qualified specialists, there are contradictions that need to be resolved, in particular, between: the need for training of highly qualified specialists and the existing system of education and scientific and methodological support in the educational process; the need to train competitive professionals and the existing V(VT)E system, which does not adequately ensure the mobility and competitiveness of professionals in the labor market; personal and professional development of the specialist in the conditions of the educational institution and insufficient opportunities of this aspect in his future competitiveness. It should be added that a competitive specialist is a subject of professional activity, which achieves the best performance in competitive working conditions due to the maximum integration of the competences required for this purpose [5, p. 7]. It is competition that acts as a specific situation, a phenomenon in which the competitiveness of any subject is formed and developed on the basis of the psychological interaction of the participants in the production process.

According to the analysis of scientific works, in recent years scientists have increasingly raised the problem of the competitiveness of the specialist and is often considered in the pedagogical aspect associated with the formation and management of the process of becoming a person in the process of professional training (Alexander G. Romanovsky [6], Sergey V. Smirnov [7]), revealing the essence of future specialists' competitiveness (Lyudmila A. Dudko [8], Elena L. Kholodtseva [9], Sergey M. Shyrokobokov [10], Rais A. Fathutdinov [11] and other.), structure and characteristics of competitive personality traits (Tatiana Y. Andriyako [12], Igor Ansoff [13], Evgeniy O. Klimov [14], Larisa M. Mitina [15] Tatiana A. Slivina [16], Olena A. Fil [17] and others), features of development of competitiveness of future specialists of different specialties in the process of vocational training in higher education (Irina P. Sarattseva [18], Svetlana O. Hazova [19], Andy Adcroft [20], Abrahim Bakar [21], Rassa Viederyte [22] and others).

In pedagogy, according to the definitions of scientists in different fields, the concept of a specialist's competitiveness must be interpreted as a specialist's ability to compete, that is, his or her corresponding competencies and personal preferences that allow him / her to win the competition.

The development of competitiveness is of a systemic nature and requires the creation of certain organizational and pedagogical conditions, educational and information environment in which the educational process takes place, educational and production activities that create the conditions for personal and professional formation of personality.

Developing the competitiveness of the future specialist is a complex, step-by-step, multi-faceted process that must take into account the individual's experience of creative professional activity and experience of constructive professional communication, that is, modeling a competitive educational environment that guarantees the quality of professions.

In view of the above, the purpose of the article is to analyze the current state of the balanced development of V(VT)E; identification of components that affect the quality of vocational training and education of a competitive specialist.

2 Research methodology

According to the goal, a set of interrelated research methods has been applied. *Theoretical methods:* systematic analysis of pedagogical and methodological literature, aimed at studying the current state of the problem of research, which made it possible to reveal the concept of "effectiveness of the activities of V(VT)E" and identify its main components. Based on the study of materials related to the concept of quality of education, such as "Regulation of the quality of education as a philosophical and educational problem" (V. G. Viktorov), "Quality of education. Resulting qualities of education. Methodical provision of the state standard of initial vocational education" (V. G. Kazakov), "Management of the development of a professional educational institution: praxeological foundations" (L. M. Sergeeva et al.), "Professional development. Encyclopedia of Education" (V. V. Rybalka), "Theoretical and methodological foundations of managing the development of a vocational school" (L. M. Sergeeva) have clarified the essence of the concept of "quality of education" and defined the basic approaches to determining quality.

Empirical methods: interviews and questionnaires of employers; testing of GV(VT)E applicants for the purpose of socio-psychological selection for training; questionnaire of pedagogical staff (teachers and practice leaders) – questionnaire developed by the authors; *methods of mathematical statistics:* in order to verify the accuracy of the data, quantitative and qualitative processing of the obtained results by means of computer equipment was carried out.

It was interviewed: 14 institutions, 9 of which are V(VT)E and 5 institutions of professional higher education; 175 participants; 175 educators; 86 employers; 84 management and 280 teaching staff; 548 graduates of the V(VT)E about ten districts of Ukraine.

3 Results and discussion

At the beginning of the study, we considered it necessary to consider the overall state of development of V(VT)E in Ukraine.

As of 01.01.2018, there were 754 GV(VT)Is in Ukraine. Compared to 1990, the network of institutions decreased by 492 institutions (table 1, fig. 1) [23].

An important problem that today has a negative impact on reducing the contingent in the country's GV(VT)E is

https://doi.org/10.1051/shsconf/20207502004

that it is perceived as a panacea for social inclusion and integration of vulnerable populations. But this is not about competitiveness in the global marketplace, lifelong learning and innovation.

Table 1. Network GV(VT)E and the Dynamics of the Contingent of Educators in Ukraine

Indicator	Amount of GV(VT)E	Reduction of GV(VT)E	Average Filling GV(VT)E, persons	Reduction of the contingent, persons
1990	1246		516	
1995	1179	– 67	471	– 45
2000	970	– 209	541	+72
2005	1023	+ 53	485	– 56
2010	976	– 47	444	– 41
2014	814	– 162	388	– 56
2015	798	– 16	381	– 7
2016	787	– 11	363	– 18
2017	756	– 31	356	– 7
2018	754	– 2	253	– 106

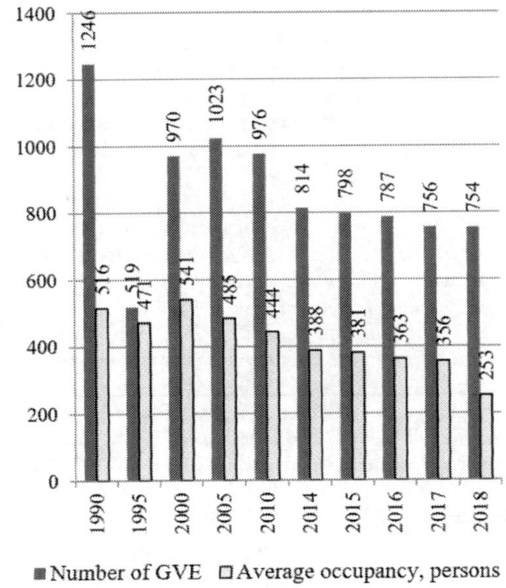

Fig. 1. General Vocational(VT)Education Network and the Dynamics of the Contingent of Educators in Ukraine.

In many countries, vocational education is considered a vital component of improving production efficiency, stimulating the development and competitiveness of the economy, reducing unemployment, as well as a tool to improve the situation of the poor.

The Europe 2020 Strategy: Strategies for Intellectual and Balanced Development Promoting Community Engagement prioritize three interconnected elements of the classic concept of balanced development: intellectual development: that is, the development of a knowledge and innovation based economy; balanced development, that is, development of the economy, with efficient use of resources, and development that promotes social activity: that is, the development of an economy with a high level of employment [24].

According to Eurostat, almost half of all second-year secondary school graduates in 28 European Union countries have gone through vocational training programs. And in the case of the Republic of Austria, the Czech Republic, the Republic of Croatia and the Finnish Republic, this proportion was 70% and above.

In Ukraine, the contingent of GV(VT)E applicants for 2018 is 253.9 thousand people, including: on the basis of complete general secondary education (11 classes) – 62.0 thousand (24.5%), on the basis of basic secondary education (9 classes) – 178.1 thousand people (70.1%), on the basis of incomplete basic secondary education (up to 9 classes) – 2.2 thousand people (0.5%), trainees – 12.5 thousand persons (4.9%) (fig. 2) [23].

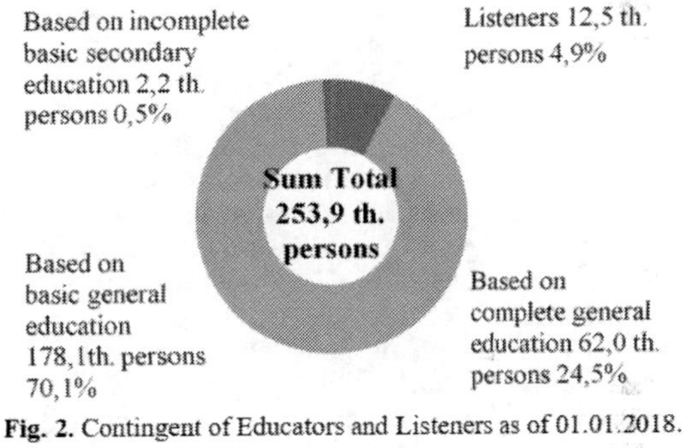

Fig. 2. Contingent of Educators and Listeners as of 01.01.2018.

The analysis indicates a significant reduction in the contingent of the GV(VT)E. There is a tendency of not fulfilling the plan of the order for training of qualified specialists, so in 2016 at the planned indicators 127 thousand persons (100%) were enrolled for training of 118,5 thousand persons (93%), thus the regional order for 112,5 was not fulfilled thousand people, state – 6 thousand people. The formation of a contingent of GV(VT)E on a residual principle significantly affects the quality of education of the educational service provider. According to V. G. Viktorov, the quality of education is a category that by its nature reflects various aspects of the educational process – philosophical, social, pedagogical, demographic, economic and others [25, p. 8].

According to United Nations Secretary-General Ban Ki-moon, quality education is not just a content delivery system; rather, it is a system designed to help job seekers fully unlock their potential and enter society as full and productive citizens ready to lead the future [26].

The quality of education is defined as the basis for the political, economic independence of the state, as a major factor in ensuring the competitiveness of both the individual graduate and the country as a whole. There are

two main approaches to defining the essence of the quality of education, which is considered from the point of view:
- achievement of certain norms, standards, goals and needs (individuals, societies, states). This approach is based on establishing a match between the planned and achieved educational quality indicators, but does not focus on identifying the reasons for the violation of this compliance, finding out the factors of influence, measuring their weight, intensity, etc., which does not allow to develop an effective management strategy aimed at improving quality of education and development of educational institution;
- modern theory and practice of quality management as a theoretical category. Within this approach, the quality of education is a clearly defined dynamic object that can change in the right direction in accordance with managerial influence [27, p. 34].

Therefore, the quality of education can be interpreted as: the social ideal of human education; as a result of her educational activities; as a process of organizing education and training; as a criterion for quality educational activities.

The quality of educational activity characterizes the level of organization of the educational process in an educational institution, provided it meets the standards of V(VT)E and the professional development of future specialists. The purpose of this process is to develop an employee's professional culture, which is manifested in a high level of professional education, competence and skill, in realizing them the value of their profession and themselves as a professional, dedication to their profession, achieving on this basis socially and personally significant creative results [28, p. 733].

It is important to include the notion of "efficiency" in the "quality of education" characteristic and be considered as a criterion for evaluating the activities of the teaching staff in any field, including management. Ensuring high management effectiveness is part of the overall problem of enhancing the effectiveness of GV(VT)E activities. The effectiveness of V(VT)E can be determined in the areas that should be measured, plausible and predetermined in the clear results of their achievement, in particular: the content of vocational training and its comprehensive methodological support; availability of V(VT)E worlds; professional success of graduates; personnel potential of the educational institution; material and technical base; financing [27, p.5].

The results of the level of quality of education and its effectiveness are likely to be ensured within the framework of pedagogical research. The leading method of pedagogical research is experiment. In science, an experiment is understood as a way of studying phenomena in regulated conditions, which allows you to reproduce, observe and record phenomena by hardware methods or by using appropriate scientific documentation. In an experiment, any phenomenon can be investigated under different conditions, repeated several times under the same and changed circumstances. The essence of the experimental method is that it is aimed at the study of cause and effect relations between the objects studied, which allows you to truly master the laws of pedagogy for practical purposes [29, p. 13].

The limits of using pedagogical experiment as a method of scientific cognition in pedagogy are considerably broadened due to the fact that the training of future specialists in the GV(VT)E combines all the necessary components of education. The process of professional training of specialists takes place both in training and in production conditions. The experiment helps to identify the links between theoretical and industrial learning, the quality of the acquired knowledge, skills and abilities, to establish the ambiguity and complexity of actions of many pedagogical factors. This, of course, increases the possibilities of pedagogical experiment before other methods.

The tasks of further improvement of pedagogical science, especially in the field of V(VT)E, require from a pedagogical experiment multifaceted and at the same time – holistic consideration of the object of study, system-structural analysis of interacting factors.

I stage the experiment stage in our study was aimed at substantiating the choice of the number of respondents, determining the actual level of quality of training of competitive specialists (incoming diagnosis).

II stage – approbation of the results of the research of the quality management system of training of competitive specialists.

III stage – determining the final level of competitiveness of the educational institution as a component of the system of management of the development of GV(VT)E (initial diagnosis). Comparison and statistical processing of the results obtained.

The ascertainment (stage I) of the pedagogical experiment is such that it establishes only the real state of affairs in the process of research. The purpose of the ascertaining stage of the pedagogical experiment was to determine the actual level of quality of the training of competitive specialists in the conditions of GV(VT)E.

14 institutions were involved in the ascertaining stage of the pedagogical experiment, 9 of which were GV(VT)E and 5 professional higher education; 175 participants; 175 educators; 86 employers; 84 management and 280 teaching staff; 548 graduates of the GV(VT)E about 10 regions of Ukraine, including:

GV(VT)E: Kryvyi Rih Professional Mining and Technology Lyceum of Dnipropetrovsk region (27 graduates), DPTNZ "Pershotravensky Mining Lyceum" of Dnipropetrovsk Region, (29 graduates), Ternovsky Professional Mining Lyceum of Dnipropetrovsk Region (56 graduates), DPT "Kyiv" (30 graduates), Bilopilsky Higher Vocational School of Sumy Region, (47 graduates), State Educational Center" Odessa Center of Vocational Education "of Odessa region (30 graduates), State Professional Lyceum of Ukrainka Kharkiv region (32 graduates), Regulations

"Dniprorudnenskyy professional college" Zaporizhzhya region (47 graduates), Higher professional mining school (Gorishni Plavni, Poltava region (26 graduates).

Higher Education Institutions of 1st and 2nd levels of accreditation: Lviv State College of Food and Processing Industry of the National University of Food Technologies (47 graduates), Ivano-Frankivsk College of Restaurant Service and Tourism of the National University of Food Technologies (27 graduates), Ternopil National Technical College Ivan Puluj University (64 graduates), Zhytomyr Trade and Economic College of Kyiv National University of Trade and Economics (30 graduates), Kyiv College of Information National Aviation University (32 graduates).

In the program of ascertaining experiment a special place is occupied by our survey of employers, which involved 86 employers from 10 regions of Ukraine, which is 30% of their total number, provided an opportunity to study the structure of employment of the population of the region by types of economic activity. 7280 employees (58.8% of the total number of employees in selected types of economic activity) work in the enterprises selected for the survey, Dnipro, Zaporizhzhya, Zhytomyr, Ivano-Frankivsk, Kyiv, Lviv, Odesa, Poltava, Sumy, Ternopil region and Kyiv city.

Target groups: Group I: Entrants to educational institutions, applicants, graduates of the GV(VT)E; Group II GV(VT)E heads (directors and deputy directors, heads of departments, senior masters), representatives of regional education departments, teachers and masters of production training of GV(VT)E; III – stakeholders, representatives of local employers and businesses; local employment centers.

The survey of 175 entrants aimed to identify the motives for choosing a future profession.

According to the results of the survey, it is found that the preferred age of respondents is 16-19 years; education - basic general secondary education (110 persons (63%)).

The survey identified problems related to the selection and formation of motivation for choosing professions, in particular 84 respondents (48%) entered the vocational training institution, having learned about the chosen profession from friends, from parents and relatives – 19 people (10.9%), independently through the Internet – 21 people (12%) and from other sources – 17 people (9.7%), and only a fifth of all interviewees received professional information from teaching staff (34 persons (19.4%)) (fig. 3).

The possibility of quick employment attracts 97 (55.4%) respondents in the chosen profession; 82 persons (47%) – high salary after graduation; 56 people (32%) – prospects for professional growth; and only 29 people (16.5%) have the ability to work in their chosen profession (fig. 4).

At the end of the GT(VT)E 167 (95.4%) respondents have high salary expectations, including UAH 3-5 thousand – 36 people (21.6%); 7-9 thousand UAH – 89 persons (53.3%); 9-12 thousand UAH – 42 (25.1%); 142 persons (81.1%) – employment by profession; and only 21 people (12%) expect the opportunity to continue their studies (fig. 5).

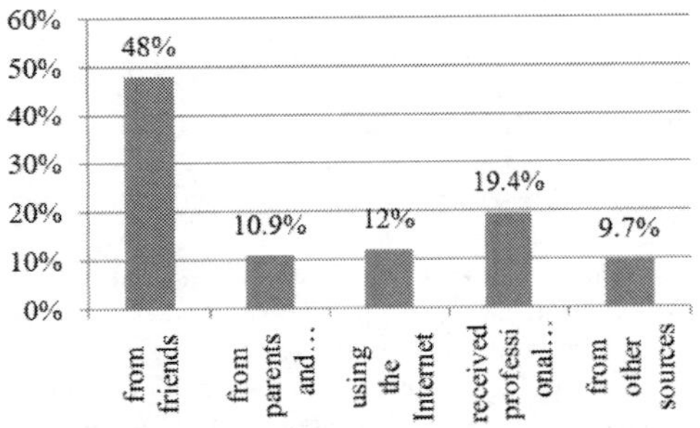

Fig. 3. Formation of motivation for choosing professions based on the results of the survey of entrants.

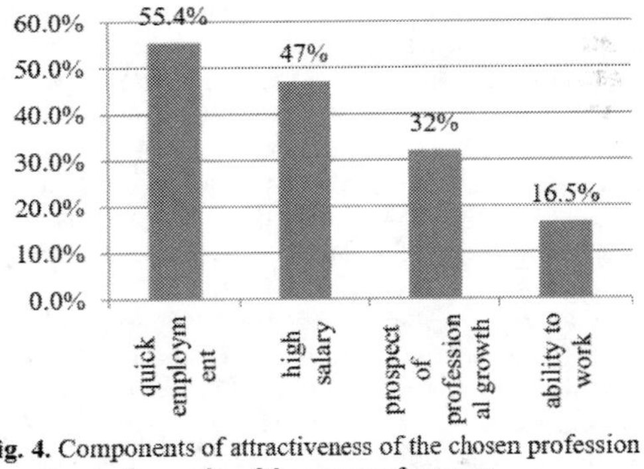

Fig. 4. Components of attractiveness of the chosen profession according to the results of the survey of entrants.

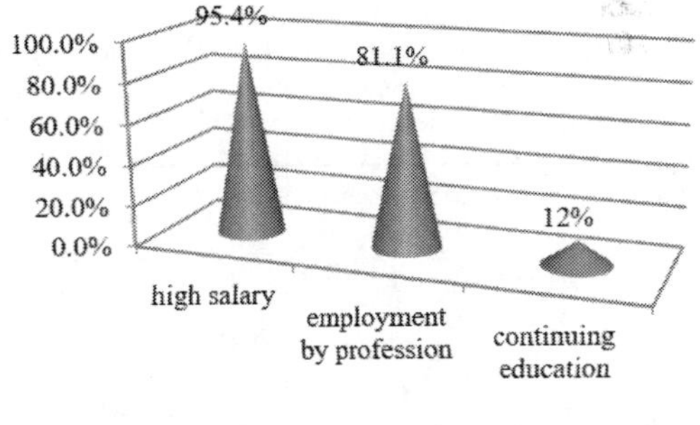

Fig. 5. Expectations of entrants after the end of the GV(VT)E (by the results of the survey).

According to the results of the survey of entrants on the first achievements after graduation, 149 (85%) of respondents consider it necessary to have a prestigious job and 123 persons (70,2%) to buy their own housing.

In the study of the current state of the quality of education in the GV(VT)E, 175 applicants participated, of which a considerable part (137 persons (78.3%)) state that the knowledge, skills they receive in the educational establishment are sufficient for future professional activity. 135 (77%) of the interviewed interviewees believe that in the process of vocational training, the Lyceum staff adhere to a democratic style of communication. 45 (25.7%) of the surveyed respondents said that the most attractive for them is the organizational form of conducting classes, as the speeches of pedagogical workers. Only 33 (18.8%) respondents expressed a desire to participate in the development of various projects and their implementation. For 168 (96.1%) interviewed applicants, the criteria for assessing the knowledge, skills used by teaching staff are clear.

For qualitative mastering of subjects of theoretical and theoretical training, the applicants lack the educational literature and access to the Internet (92 persons (52.6%); Insufficient applicants consider educational support and point to the lack of multimedia training facilities (64 persons (36.6%). The main components in preparation are the formation of professional knowledge, skills and competences necessary for competition for a job in the labor market, and the formation of the qualities necessary for self-realization in professional activity (153 persons (87.4%).

140 (80%) of the respondents believe that pedagogical staff of the Lyceum pay due attention to the development of independent work skills; 95 (54.3%) of applicants claim that their interests, abilities and psychological characteristics are taken into account in the training process.

All respondents noted the positive role of the use of multimedia educational facilities by pedagogical staff. Respondents state that it is possible to improve the quality of theoretical and vocational training by increasing the level of professional competence and quality of work of pedagogical staff – 49 persons (28%)), changes in the content of disciplines – 36 persons (20,6%)), development of a culture of communication between pedagogical employees and students – 90 people (51.4%) and improvement of the material and technical base of GV(VT)E – 64 people (36.6%) (fig. 6).

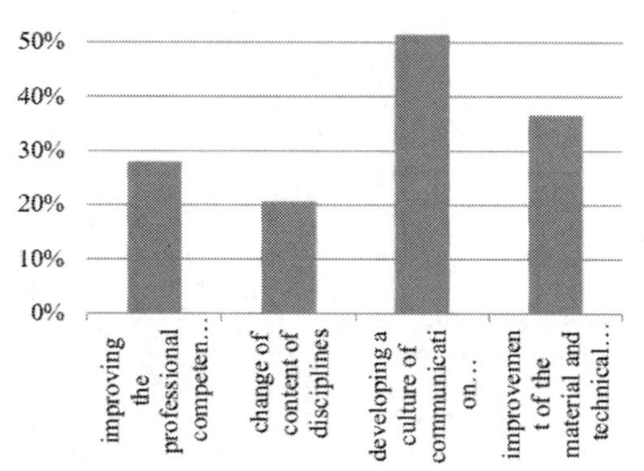

Fig. 6. Ways to increase the level of vocational training in the GV(VT)E on the results of a survey of applicants.

135 (76.9%) of the interviewed applicants recommend to the friends of the GV(VT)E for admission and further education. 280 respondents took part in the questionnaire to determine the level of professional competence and self-education of pedagogical staff.

Their qualitative composition is represented in the main I qualification category and 13 tariff category (fig. 7).

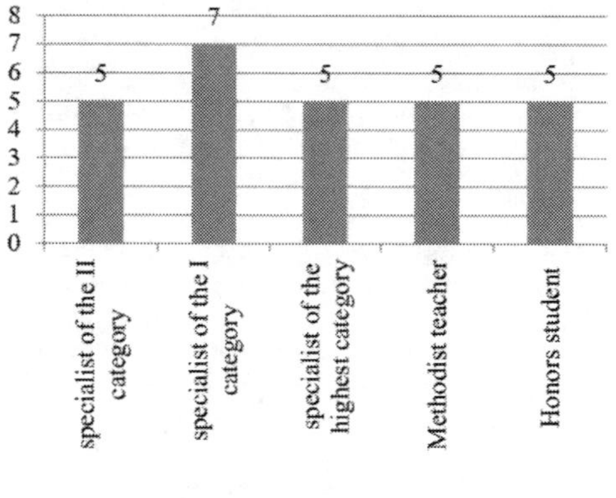

Fig. 7. Qualitative composition of teaching staff.

The study found that 210 people (75%) had completed higher education. 85 people (30,3%) have up to 3 years of pedagogical experience in V(VT)E; 61 persons (21,7%) from 3 to 10 years; 61 persons (21,7%) from 10 to 25 years; 73 persons (26,3%) – 25 years and over (fig. 8).

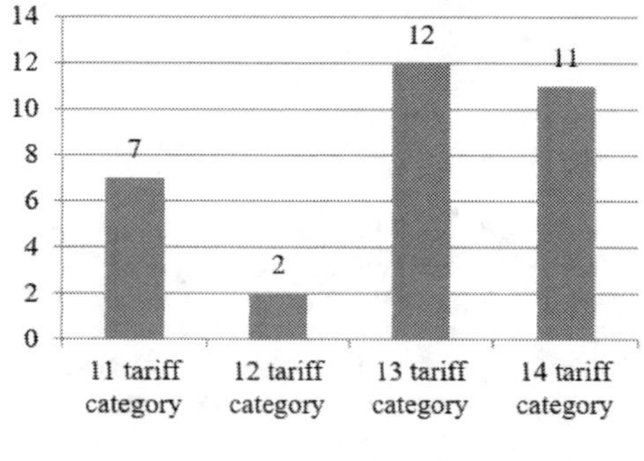

Fig. 8. Qualitative composition of masters of industrial training.

One-third of teaching staff (85 people (30.3%) introduce innovative technologies into the educational and production process.

210 (75%) of respondents consider lack of time, 42 (15%) of persons – poor awareness of innovations and only 28 (10%) of persons – lack of necessary theoretical knowledge (fig. 9).

210 (75%) of employees evaluate the quality of conditions for the development of innovative activity in the educational institution as favorable, 70 (25%) consider the quality of conditions satisfactory.

 https://doi.org/10.1051/shsconf/20207502004

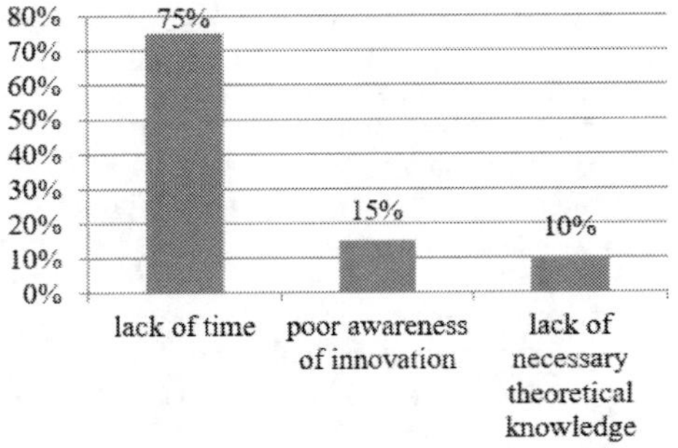

Fig. 9. Obstacles in the development of innovations by pedagogical workers and their introduction into the process of specialist training.

All pedagogical workers (100%) carry out self-educational activities, most of which (230 persons (82.1%)) assess their own level of readiness to study independently and acquire knowledge as average and carry out self-education occasionally, without a system.

Self-education 196 (70%) of pedagogical workers are hampered by lack of free time, and 84 (30%) lack of computer literacy and lack of necessary reference and methodological literature. Today, there are teachers who do not have computer equipment at all, and therefore do not use Internet resources in teaching activities (25 people (8.9%)).

224 (80%) teachers consider their own level of readiness to learn and acquire knowledge sufficient, 42 workers (15%) consider their average level of readiness and 14 pedagogical workers (5%) define their own level as high (fig. 10).

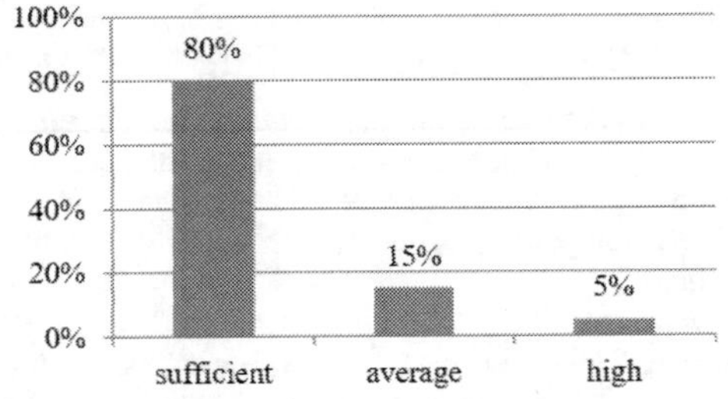

Fig. 10. Levels of readiness to study and acquire knowledge by pedagogical staff.

The most effective forms of self-education 168 (60%) of pedagogical workers choose individual work (visiting libraries, studying scientific and methodical and educational literature) and participation in conferences, seminars, training courses. At the same time, 12 teachers (4.2%) have no desire to engage in exploratory activities.

Most pedagogical workers (168 persons (60%)) consider the level of organization of professional training of competitive specialists in their educational institution sufficient, 84 (30%) believe that the level of organization of vocational training is average and only 28 (10%) – high (fig. 11).

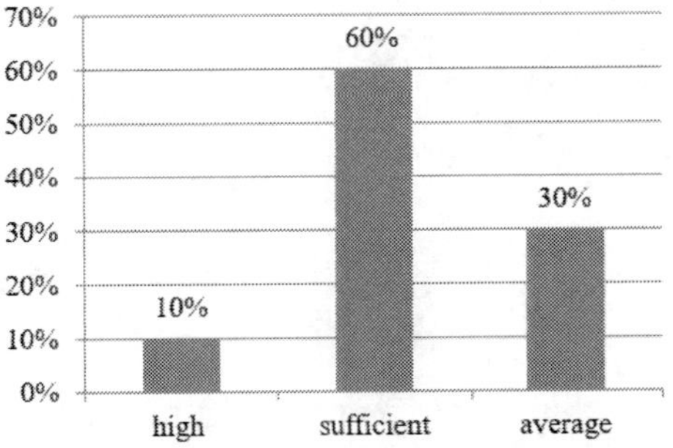

Fig. 11. Level of organization of professional training of competitive specialists (according to the survey of pedagogical staff).

During the survey of employers it was found that out of 86 enterprises 63 (73%) are private, 59 enterprises (68%) have been working for more than 10 years. At 26 enterprises (30%), the number of employees reaches more than 500 people. 32 employers (37%) are satisfied with the existing system of co-operation between the PA (OA) and the enterprise in the preparation of a competitive specialist. The quality of professional training of future specialists was rated as "partially satisfactory" (68 people (79.2%)). Almost equal number of employers gave a positive assessment of the quality of training of specialists (6 people (6.9%)) and not satisfied with the quality of training (12 people (13.9%)) (fig. 12).

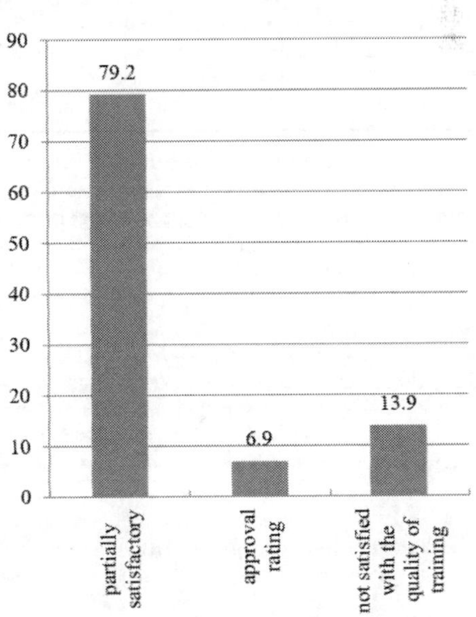

Fig. 12. Assessing the quality of professional training of a future specialist based on the results of a survey of employers

Employers consider qualification, ability to study, responsibility, ability to work in a team, additional education or related profession, mobility as components that affect the level of a specialist's competitiveness.

The employer assesses the quality of the graduate in certain areas as high (5 points) – 11,1%, sufficient (4 points) 62,3%, medium (3 points) 12,2% and low 14,4% (fig. 13).

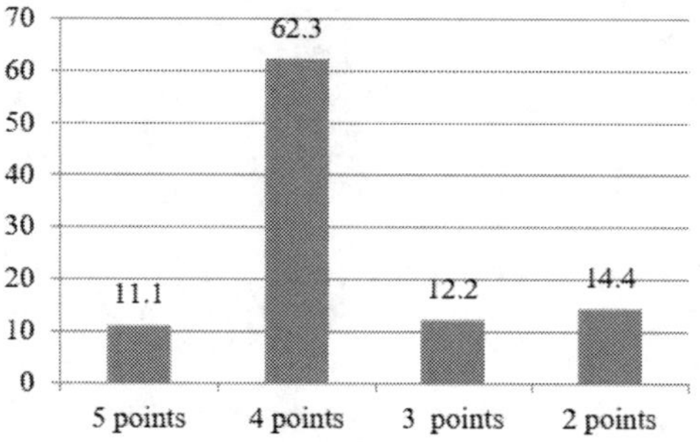

Fig. 13. Assessment of the quality of professional training of a specialist according to the directions according to the results of the survey of employers.

At the same time, employers believe that to improve the quality of training of competitive professionals it is necessary:

- to carry out internships of pedagogical workers of vocational-theoretical and vocational-practical training in the conditions of production, not once every five years, but in accordance with changes in production technologies;
- to carry out internships of pedagogical workers of general education (compulsory mathematics, physics, chemistry) in real production conditions, in order to provide a professional direction in teaching;
- to adapt the State Standards of V(VT)E on the changes in production technologies.

According to the results of the survey of senior staff 84 V(VT)E About 10 regions of Ukraine, according to the program of advanced training courses (category of deputy directors of GV(VT)E on educational and production work) at the Central Institute of Postgraduate Education of the State University of Education Management of the National Academy of Pedagogical Sciences of Ukraine that the main challenges faced by the GV(VT)E is the lack of adequate training facilities, outdated equipment, or none at all (39 educational institutions (47%). 11 (13%) of respondents believe that the competitiveness of competitive specialists is the personal qualities of the educational acquirer (unwillingness to study, small vocabulary, etc.); Interviews with entrants are conducted as a method of qualitative selection at admission.

In the survey, 61 respondents (73%) said that in the GV(VT) E, in the professional training of future specialists, information and communication technologies and distance learning technologies are constantly used. 23 (27%) respondents believe that cooperation with employers, international partners, holding workshops and competitions of professional skill is the most appropriate in improving the quality of training of competitive specialists.

The most common forms of social partnership in the regions are traineeships of pedagogical workers on the basis of enterprises 62 (74%) respondents; introduction of elements of dual training system 11 (13%); development of educational literature with the involvement of employers, conducting production training on the basis of enterprises, excursions and seminars with the participation of enterprise technologists – 11 (13%) persons (fig. 14).

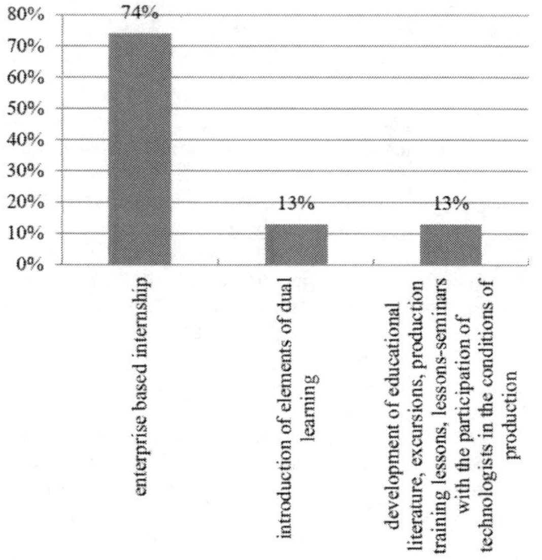

Fig. 14. Forms of social partnership of GV(VT)E and employers of different regions of Ukraine (according to the survey results).

The survey also found that 44 (53%) educational institutions do not cooperate with scientific institutions, the remaining 40 (47%) of the surveyed respondents cooperate with the Institute of Vocational Education of the National Academy of Pedagogical Sciences of Ukraine, the Central Institute of Postgraduate Education and the Bilotserkivsky Institute "University of Management Education", Ukrainian Engineering-Pedagogical Academy, Kharkiv Academy of Continuing Education.

23 (27%) of the surveyed GV(VT)E executives believe that the problem of training competitive professionals should be addressed at the state level, in particular by revising the State Standards of V(VT)E in accordance with the requirements of professional competences in the professions; 28 (33%) of the respondents consider that the updating of the educational and material and technical base is the most effective tool in the training of competitive specialists; 17 (20%) of the respondents indicate the need to increase the professional competence of teaching staff; 17 (20%) refused the answer (fig. 15).

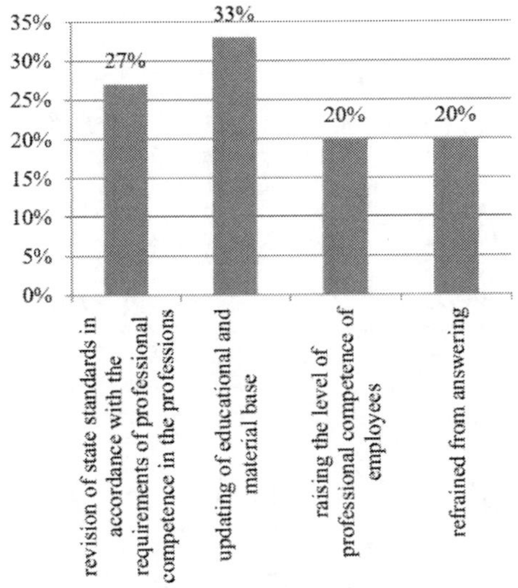

Fig. 15. Ways to Improve the Training of Competent Professionals of GV(VT)E.

4 Conclusions and prospects for further research

In the course of the theoretical analysis we have come to the conclusion that the concept of "competitiveness" is considered as the ability of a specialist to compete, that is, he has the appropriate competencies and personal preferences that allow to win the competition.

The main factor in ensuring competitiveness is the quality of education, which is defined as the social ideal of human education; as a result of her educational activities; as a process of organizing education and training; as an effective criterion for the functioning of the educational system.

The results of this study are a pedagogical experiment, in which the quality and effectiveness of V(VT)E in the following areas were determined: the content of vocational training and its comprehensive methodological support; motivation and availability of V(VT)E; professional success of graduates; personnel potential of educational institution; material-technical base.

The survey conducted by the author's questionnaires gives grounds to claim that the majority of the graduates of the GT(VT)E have developed a satisfactory level of satisfaction with professional training; according to employers, the content of vocational training programs needs significant changes; teaching staff do not have experience in the manufacturing sector; lack of funding and an outdated material base remains a problem.

This article aims to realize the idea (prospect of further research) of improving the quality of education, capable of ensuring a high level of competitiveness of specialists, by creating a holistic system of cultural content that will cover the balanced development of the mental, emotional, value, volitional and physical spheres.

References

1. P. Sauh, *Alhorytm realizatsii Ya-kontseptsii osobystosti u protsesi profesiinoho samovyznachennia* (An algorithm for realizing the self-concept of personality in the process of professional self-determination). Ya-kontseptsiia akademika Nelli Nychkalo u vymiri profesiinoho rozvytku osobystosti (The self-concept of Academician Nelly Nichcalo in measuring the professional development of the individual). (Instytut pedahohichnoi osvity i osvity doroslykh NAPN Ukrainy, Kyiv, 2014), pp. 62–69
2. Proekt kontseptualnykh zasad reformuvannia profesiinoi osvity Ukrainy "Suchasna profesiina osvita" (Draft Conceptual Framework for the Reform of Vocational Education in Ukraine "Contemporary Vocational Education") (2019), https://mon.gov.ua/ua/news/mon-rozrobilo-proekt-konceptualnih-zasad-reformuvannya-profesijnoyi-osviti-ukrayini-suchasna-profesijna-osvita. Accessed 20 Dec 2019
3. B.M. Bim-Bad, Pedagogiko-antropologicheskoe osnovanie teorii i praktiki sovremennogo obrazovaniia (Pedagogical and anthropological basis of the theory and practice of modern education). http://www.bim-bad.ru/biblioteka/article_full.php?aid=95. Accessed 20 Dec 2019
4. Kontseptsiia realizatsii derzhavnoi polityky u sferi profesiinoi (profesiino-tekhnichnoi) osvity "Suchasna profesiina (profesiino-tekhnichna) osvita" na period do 2027 roku (The concept of state policy in the field of vocational (vocational-technical) education "Modern vocational (vocational-technical) education" for the period up to 2027) (2019), https://zakon.rada.gov.ua/laws/show/419-2019-%D1%80#n8. Accessed 20 Dec 2019
5. G.O. Kozlakova (ed.), *Kontseptualni zasady psykholoho-pedahohichnoi pidhotovky konkurentozdatnoho fakhivtsia v umovakh pisliadyplomnoi pedahohichnoi osvity: na vybirtsi naukovo-pedahohichnykh pratsivnykiv* (Conceptual bases of psychological and pedagogical training of competitive specialist in the conditions of postgraduate pedagogical education: on a sample of scientific and pedagogical workers). (Universytet menedzhmentu osvity, Kyiv, 2015), p. 40
6. A.G. Romanovsky, O.S Ponomaryov (ed.), *Rozvytok liderskoho potentsialu natsionalnoi humanitarno-tekhnichnoi ta upravlinskoi elity* (Development of the leadership potential of the national humanitarian-technical and administrative elite). (Politekhnichnyi instytut, Kharkiv, 2017), p. 290
7. S.V. Smirnov, Dissertation, Pivdennoukrainskyi natsionalnyi pedahohichnyi universytet im. K.D. Ushynskoho, 2018

8. L.A. Dudko, Dissertation, Instytut sotsiolohii, 2004
9. E.L. Kholodtseva, *Osobennosti podgotovki spetcialistov, konkurentosposobnykh na rynke truda* (Features of training specialists competitive in the labor market). Sbornik trudov studentov i molodykh uchenykh Kemerovskogo gosudarstvennogo universiteta (Collection of works of students and young scientists of Kemerovo State University). (Poligraf, Kemerovo, 2004), pp. 88–90
10. S.N. Shirobokov, Dissertation, Omskii gosudarstvennyi pedagogicheskii universitet, 2000
11. R.A. Fathutdinov, *Globalnaia konkurentosposobnost* (Global competitiveness). (Ekonomika, Moscow, 2009)
12. T.Y. Andriyako, Pedahohichna sutnist i struktura konkurentospromozhnosti fakhivtsia (Pedagogical essence and structure of specialist competitiveness). Visnyk Cherkaskoho universytetu 189(2), 116 (2010)
13. A. Ansoff, *Strategicheskoe upravlenie* (Strategic Management). (Ekonomika, Moscow, 1989)
14. E.A Klimov, *O stanovlenii professionala: priblizhenie k idealam kultury i sotvorenie ikh (psikhologicheskii vzgliad)* (On becoming a professional: approaching to the ideals of culture and their creation (psychological view)). (IPPS, Moskva, 2006)
15. L.M. Mitina, *Psikhologiia razvitiia konkurentosposobnoi lichnosti* (Psychology of competitive personality development). (Voronezh, Moskovskii psikhologo-sotcialnyi institut, 2002)
16. T.A. Slivina, Dissertation, Sibirskii gosudarstvennii universitet, 2008
17. O.A. Fil, Dissertation, Instytut psykholohii im. H. S. Kostiuka APN Ukrainy, 2006
18. I.P. Sarattseva, Dissertation, Dalnevostochnyi gosudarstvennyi universitet putei i soobshchenii, 2005
19. S.A. Hazova, Dissertation, Adygeiskii gosudarstvennyi universitet, 2011
20. A. Adcroft, J. Teckma, R. Willis, International Journal of Public Sector Management **23** (**6**), 588 (2019)
21. A.R. Bakar, I. Hanafi, Assessing employability skills of technical-vocational students in Malaysia. J. Soc. Sci **3**, 202–207 (2007)
22. R. Viederyte, Globalization influence on competitive education and knowledge spreading activity. Economics & Management **14**, 1032–1036 (2009)
23. Three elephants of vocational education. UGP Education Expert Group (2019), http://education-ua.org/images/uploads/285/00012.png. Accessed 20 Dec 2019
24. Europa 2020, Strategia na rzecz inteligentnego i zrównoważonego rozwoju sprzyjającego włączeniu społecznemu (Komunikat Komisji, Bruksela, 2010), https://ec.europa.eu/eu2020/pdf/1_PL_ACT_part1_v1.pdf. Accessed 20 Dec 2019
25. V.G. Viktorov, Dissertation, Instytut vyshchoi osvity APN Ukrainy, 2006
26. S. Slade, What Do We Mean by a Quality Education? (2017), https://www.huffpost.com/entry/what-do-we-mean-by-a-qual_b_9284130. Accessed 20 Dec 2019
27. L.M. Sergeeva (ed.), *Upravlinnia rozvytkom profesiino-tekhnichnoho navchalnoho zakladu: teoriia i metodyka* (Development Management of a Professional School: theory and methodology). (Lira-K, Kyiv, 2017)
28. V.V. Rybalka, Profesiinyi rozvytok (Professional development), in *Entsyklopediia osvity*. (Yurincom Inter, Kyiv, 2008)
29. A.P. Belyaeva (ed.), *Prognosticheskie i metodologicheskie funktcii pedagogicheskogo eksperimenta. Organizatciia i provedenie pedagogicheskogo eksperimenta v uchebnykh zavedeniiakh.* (Prognostic and methodological functions of pedagogical experiment. Organization and conduct of pedagogical experiment in vocational education institutions). (Editus, Saint Petersburg, 1992)

The theoretical foundations of the regional training program for teachers' professional development "Education in conflict"

Iryna Trubavina[1,*], *Liudmyla* Tsybulko[2], *Natalia* Mohyliova[3], and *Volodymyr* Ivanchuk[4]

[1]H. S. Skovoroda Kharkiv National Pedagogical University, Department of General Pedagogics and Pedagogy of the Higher School, 2, Valentynivska Str., Kharkiv, 61168, Ukraine
[2]Donbas State Pedagogical University, Department of Pedagogy, 19 Batiuk Str., Sloviansk, 84116 Ukraine
[3]Donbas State Pedagogical University, Department of Speech Therapy and Special Psychology, 19 Batiuk Str., 84116, Sloviansk, Ukraine
[4]Donbas State Pedagogical University, Department of Philosophy, Social-political and Legal Sciences, 19 Batiuk Str., Sloviansk, 84116. Ukraine

Abstract. The article is devoted to the issue of developing professional skills of teachers working in a situation of a conflict. The purpose of the research is to substantiate the theoretical foundations of the course and to develop the content of a new training program for Ukrainian teachers' professional development. The scientific novelty of the research lies in the content of the teachers' professional development training program, which is grounded on a set of scientific approaches of different levels: philosophic, general scientific, specifically scientific, pedagogical approaches. The importance of children's security assurance and warranty of their right to education in conflict is under our special consideration. The content of the program includes the following modules: 1) "Children's problems caused by the armed conflict"; 2) "Education in conflict: the essence, standards (on the UNISEF and UNESCO information)"; 3) "Ways of implementation education in conflict in Ukraine". The practical significance of the research lies in the multidisciplinary character of the regional training program "Education in conflict" and the recommendations to the methods of teaching it.

1 Introduction

The topicality of the research is stipulated by a search for new opportunities to alleviate the aftermaths of the conflict in the eastern region of Ukraine in order to realize the child's right to education, which is related with the necessity to form special professional competences in teachers to be applied in the context of the conflict. The system of teachers' professional development under the current educational reform and in the context of the conflict in eastern Ukraine faces some contradictions: 1) between the numerous scientific researches in pedagogy considering teacher training and professional development targeted in most cases at forming professional skills for work in the renovated Ukrainian school [1, 2, 3], for dealing with conflicts in non-belligerent education [4, 5, 6, 7; 8], some certain researches considering teacher training for successful activities in a situation of a military conflict [9, 10, 11] and a need for a scientifically grounded content of a teachers' professional development program to be applied in the situation of the conflict in eastern Ukraine; 2) between an array of facts in Ukraine and in the world considering violence against the child's right to education, the impact of the conflict on realization of this right (UNESCO, UNISEF, INEE, Ukrainian Institute for the Study of Extremism) [12, 13, 14, 15, 16, 17, 18, 19, 20, 21, 22] and lack of competence for work in education in conflict; 3) between an array of international researches and recommendations available in accordance with the problem of education in regions of military conflicts [12, 13, 14, 15, 18, 19, 20, 21, 22] and lack of theoretical foundations of the content of professional development programs for Ukrainian teachers as well as the programs themselves. The online training course "Conflict Sensitive Education" is rather short (2-3 hours), presented in English and does not reveal all Ukrainian realia. The UNISEF recommendations (Conflict Sensitivity and Peace-building Programming Guide, November, 2016 [24]) are not a training course but a set of self-training material. This groundwork gives the fundamental conception of the problem and its importance. However, they are not theoretically grounded and are not aimed at forming specific competences.

Thus, the purpose of the article is to substantiate the theoretical foundations of the teachers' professional development training program, which can successfully be implemented into work in the context of the conflict in eastern Ukraine. The purpose stipulates the following tasks: to set up theoretical foundations which can make grounds for a new content of the program and to work out recommendations for its application.

* Corresponding author: trubavina@gmail.com

2 Methodology

The methods of the research are theoretical analysis of books and articles on the issue, induction, deduction, synthesis and modeling, summarizing. The research is grounded on the information and reports of UNESCO, UNISEF, INEE and other organizations in relevance with the problem, the list of which is given in the References [10-25]. A thorough theoretical analysis of the material mentioned proves that formation of professional competences for teaching in conflict requires knowledge in different fields of human activities: pedagogy, didactics, security, management, sociology, conflict resolution, social work, jurisprudence, political science, etc. It is the cause of the fact that school becomes the centre of the area in conflict and teachers have to perform additional functions in their work: increasing concern for the life and safety of children and the pedagogical staff in educational establishments; taking measures to provide children's and adults' security; giving social assistance; conflict management and team-building in groups with children, who have experienced psychological trauma, have witnessed hostilities; detection of violation of children's rights; development of the curricula and programs that can be applied in conflict; selecting forms and methods appropriate for teaching in conflict, e-learning is used whenever possible; providing parents with information considering public assistance; creating necessary conditions to implement the child's right to education in conflict; identifying the needs of the families; teachers have to perform functions of peace-builders and protectors in the community and in the educational establishment, are forced to deal with cases they are not experienced in. These aspects of pedagogical activity are new to pedagogical practitioners and have not been taught in pedagogical universities. Therefore, changes in working conditions extend teachers' duties and increase responsibility, require better competences, higher qualification and appropriate actions (ability to resist stress, work in extreme conditions, organize adults, parents, the community to provide children's security in educational establishments, creating necessary conditions to implement the child's right to education, etc.). The necessity of considering all these in the content of the new training program stipulates the importance of providing fundamentally new grounds and a set of approaches to the solution of this problem. Thus, there exists an urgent necessity in an integrated multidisciplinary program. It means that the content of the training program for the professional development for teaching in conflict should involve new theoretical foundations in the form of a set of scientific approaches of different levels and different fields relevant to the program's content, methods and forms of its application. It should be noted that this problem is urgent both for teachers in eastern Ukraine and in the rest territory of the country since all children, members of their families suffer from the aftermaths of the conflict. This presupposes the double character of the program: its considering the regional features on the one hand and its flexibility in applying on the other hand (for instance, different amount of time to study the educational material in dependence with educational purposes of a certain region, which provides a regional approach of management to the program's content). In fact, a training seminar for teachers of the Donetsk Regional Institute of Continuous Education has so far been held in Ukraine on the UNISEF material [13, 21, 24], in which international experience and recommendations are described. It has raised the issue, outlined the ways of its solution in the world. Nevertheless, there is neither any theoretical foundation of the program's content nor practical recommendations as to the training methods. We claim that theoretical foundations of the contents of the regional educational programs on the issue must be grounded both on the universal values, the experience set out in the UNISEF materials and on the Ukrainian values and features that indicate the need to rely on:

1) the axiological approach, the principal values of which are human life and security;
2) the pragmatic approach to meet specific issues concerning children's and teachers' safety in the country;
3) the regional approach, which considers the nature, the context of the conflict and its manifestation in a certain region, specific features of life in the region and the level of the population security and is aimed at developing flexible educational standards, curricula, forms of teaching taking into account the level of safety of children and population, along with the state approach, which guarantees the right to education for all children;
4) the judicial approach which requires respect the child's right to education and creation of conditions to implement it in the conflict; organization of the community to provide measures for children's security;
5) the political approach focused on welfare policy, democracy and human rights regardless the military conflict;
6) the innovative approach, which involves teachers innovating in the organization and implementation of the educational process targeted at the best possible realization of the child's right to education and their safety in conflict; their authority to make and choose curricula and programs, which will best meet the requirements of education in conflict;
7) the interdisciplinary approach, which integrates all the professional specialists and teachers of the community to create conditions necessary to implement the child's right to education; ensures the safety of children;
8) the synergetic approach explaining teachers' self-organization in extreme conditions, their ability to identify the risk and danger for children. UNISEF [13, 21, 24] claims the necessity to meet special standards of education in conflict.

So, the content of the program must be based on the risk-oriented approach (a theory of risks and hazards) for the sake of preserving lives and health of all the participants of the educational process and implementation of the child's right to education; considering specific features and resources of the community in realizing every child's right to education and interaction of the participants of the educational process for this purpose, which testifies to the need for reliance on the theory of the UN participation, the resource approach.

Teachers work with different categories of children: internally displaced children, the war veterans' children, children whose parents were killed in the armed conflict, children living in local communities in ordinary, not always prosperous families and experiencing social and economic aftermaths of the war. It is necessary to know the whole complex of children's problems in the region, community and school, to be able to identify them, to comprehensively and systematically react to them, to be able to find an individual approach to each child and to act as a peace-builder, to form in children the peace culture. Therefore, the child-oriented, problem-oriented, complex, gnoseological (for identifying and studying the child's problems), systematic, diagnostic, cultural, personality-oriented, differential approaches are also the basis of the content of our program. Teachers require practical recommendations and tips regarding their actions in extreme situations. Such material is partly presented in UNESCO materials [16, 18, 19, 20], INEE [17] but it is of general character, not specific to Ukrainian realia and Ukrainian legislation in education, organization of educational process, is not an educational program aimed at developing teacher competences. Thus, it is important in the content of teacher training to use such approaches as the pragmatic, situational, crisis interference, self-assistance, optimization, algorithmic, interdisciplinary approaches as well as to create frameworks and conditions of their application for teachers' effective coordinative actions aimed at combining efforts and optimizing the use of resources. These approaches are important to develop the ability to act on a ready-made algorithm or compile it quickly in the conditions of limited time and information concerning the life and health safety of children and adults, their right to education. Regardless of the conditions of the conflict the human and child's rights must always be abode (respected, protected, implemented [26]). Therefore, the teacher must comprehend and appreciate the human rights approaches [26] to be able to select the best strategy to protect them in a particular situation: informational, power, economic, political, balanced, non-resistance to evil by violence, exposure of violence and hatred. Particularly noteworthy is the informational approach to denote its importance for teaching activities in extreme situations and for preventing conflict-related problems. The teacher should be aware of how to use information, find it, check it, report it and distribute it in the conflict situation to prevent panic, to perform effective activity. Providing teacher training requires considering the andragogic approach for learning from experience, accounting teachers' competences, their motivation to study. The next among the essential approaches are the competence-oriented and the action-oriented ones which lead to the ability to act under certain circumstances. This proves the necessity of training, individual work, exercising and putting it into practice. Developing the content of the curriculum as an educational discipline requires appropriate theoretical foundation and its application into the teaching process is not possible without relevant approaches – andragogic, personality-oriented, e-learning.

Taking into account that school in the conflict zone becomes the centre of the area and provides help to the entire community, it is advisable to apply the social-work-approaches to the program development: diagnostic (functional), crisis-interference, client-oriented, model-creating-oriented, self-care support. Special attention should be paid to scientific approaches regarding the behaviour of a person in extreme situations aimed at gaining skills in controlling oneself and making influence on the others: sociological, psychological, social-psychological approaches, which train to treat the conflict objectively, to behave adequately in conflict and find out solutions.

On this basis, it is possible to identify a set of theoretical foundations for teachers' professional development training program for their successful activities in conflict in Ukraine (considering our classification and levels of scientific approaches [27], classification and levels of scientific approaches by I. Lypsky [28]):

1) philosophic approaches: gnoseological, child-centrism, pragmatism, theory of participation;
2) general scientific approaches: axiological, synergetic, systemic, activity, interdisciplinary, complex, optimization, resource, innovative;
3) specifically scientific approaches: approaches to the protection of human rights (informational, power, economic, political, balanced, non-resistance to evil by violence, exposure of violence and hatred), the risk-oriented approach to the security theory, the situational approach to the management theory, the political approach, the regional and state approaches to public administration, the legal approach, the approaches to social work: problem-oriented, diagnostic, crisis interference, self-care support; the conflictological approaches: the approach to psychology of behavior in extreme situations, the cultural approach; the algorithmic approach through the theory of resolution, occupational safety approaches to work in extreme situations;
4) the pedagogical approaches: andragogic, competence-oriented, differentiated, educational, personality-oriented, e-learning. Thus, the development of the program content, selection of the forms and methods of teaching it involves integration of a set of approaches as the important, complex and multidisciplinary character of the problem cannot be restricted to pure pedagogy or psychology. It draws to the necessity of interactive teaching it by experts in different fields. It should be noted that the content of the teachers' professional development training program "Education in conflict" is grounded on the subjects, which have been taught in peacetime (occupational safety, life safety, medical training, general, age and pedagogical psychology), which eliminates their repetition in the content of the new program. This program is supposed to be applied for providing successful education in conflict only, though the previously acquired knowledge and competencies may

be used by teachers as an update of the material learned to teach new. For example, group discussions on the issues of injury prevention and traffic rules are a common practice for all teachers. However, they have not been taught to explain children how to act in cases of hostage-taking, terrorist act, mine danger, etc. This definition of risks and dangers for children in conflict creates serious challenges for educators who have lived, studied and worked in peacetime. These aspects make up the content of the program proposed.

3 The results of the research

For all those reasons, we have developed the content of the regional program "Education in conflict" for professional training of teachers and secondary school administration, which is recommended for H. S. Skovoroda Kharkiv National Pedagogical University and Donbas State Pedagogical University. The program contains an explanatory note, a list of purposes and tasks, the syllabus, the content through the modules and themes, the description of the structure of the discipline, explanation of the methods and forms of teaching, methodological support, a list of references and information resources. The program is first and foremost intended for teachers of Kharkiv and Donetsk regions (which are the areas of conflict), where this knowledge is in urgent need.

The explanatory note indicates the topicality of the issue discussed regarding the risks and hazards for children, their families and teachers caused by the armed conflict in eastern Ukraine. The new working conditions require considerate innovation of the content of teachers' professional training. The purpose of the program is relevant with the development of teachers' general and professional competences for their activity in the conflict zone. The objectives of the course are as following: development of professional and general competences in teachers and heads of institutions of general secondary education in accordance with the job requirements and the conflict conditions; formation of teachers' availability to create conditions to realize the child's right to education, detection of cases of its violence, assurance of children's life and health protection, assistance to the families in conflict.

It should be noted that the professional development training program is based on the Declaration of School Safety, UNISEF and UNESCO reports, the Law of Ukraine "On Education" (2017). The program contains three modules: "Children's problems caused by the armed conflict", "Education in conflict: the essence, standards (on the UNISEF and UNESCO information)", "Ways of education implementation in conflict in Ukraine". The program offers a 30-hour course of full-time, part-time and distance learning: 10 hours of classroom work; 20 hours of individual work (part-time learning).

The subject matter of the module "Children's problems caused by the armed conflict" provides school teachers and administration for profound knowledge as to the problems, which children of different categories have in the result of the armed conflict in eastern Ukraine as well as the aftermaths of this conflict and problems concerning realization of the child's right to education: social, legal, psychological, pedagogical, social-pedagogical, finance, economic, health problems. The program reveals the nature of child centrism, which is proclaimed an ideology of education in conflict. A separate category of children and their families are those living in compact places of residence. The problems of Ukrainian children are compared to the problems of children from abroad, who are also in conflict zones. The module purpose is to cultivate in teachers condemn against any violation of the child's right to education and availability to their detection, protection and creation of all the conditions for its realization, which shall be grounded on a thorough learning of the child-problem-statistics, considering visions and recommendations of distinguished people, treatment of the problem by different human rights organizations, regarding Ukrainian statistics as to the killed and injured children, internally displaced children, children in the "gray zone", orphans in the result of the armed conflict, cases of bullying in educational establishments as a result of the culture of war in children instead of the culture of peace at school, etc. The module reveals the risks and dangers for children in the conflict as well as the role of education in their prevention on the basis of the risk-oriented approach.

Module "Education in conflict: the essence, standards (on the UNISEF and UNESCO information)" provides innovation and development of teacher's skills and knowledge for teaching process organization at secondary school in the conflict zone, in accordance with the international recommendations, standards and experience in teaching process improvements. The experience of Ukrainian and foreign education in conflict has been generalized. The module introduces UNESCO and UNISEF reports on the issue of education in military conflict, home recommendations on children protection in conflict through education, reveals the standards of education in conflict in Ukraine, the judicial and political approaches to such education. Furthermore, the module explains Ukrainian educational policy, the subjects and objects of education in conflict.

Module "Ways of education implementation in conflict in Ukraine" is aimed at teachers' acquiring specific skills for teaching process implementation in the conflict zones, including the "gray zone", creating the secure educational environment, which implies a friendly-to-child school, non-violence pedagogy, mediation, understanding services, school social psychological and pedagogical services, subject-subject relations between teachers and children, cooperation with parents, providing security for children and teachers in the conflict conditions, preventive work and crisis intervention in order to assure children's life and health protection. The module explains the organization of secondary education in conflict.

The curriculum contemplates providing online lectures. Part-time and distance learning contains individual self-studying of some educational material, reading literature on speciality and doing practical tasks. Estimation of the skills and knowledge acquired in the professional training course is provided through

https://doi.org/10.1051/shsconf/20207502005

individual tasks and exercises and on the results of the written test on speciality. Teachers can optionally describe their own experience.

Furthermore, the content of the regional training program for teachers' professional development "Education in conflict" is explained through the modules and themes:

Module 1. "Children's problems caused by the armed conflict".

Theme 1.1. Children's problems caused by the armed conflict in Ukraine and abroad: problems of internally displaced children, problems of children of soldiers of war in Donbass, problems of children from the occupied territories, from the "gray zone", from places of compact residence, problems of all Ukrainian children caused by the conflict in eastern Ukraine. Psychological, pedagogical, social, economic, finance, legal, social-pedagogical, health problems of children in conflict education. Statistics on the education of children in conflict. Child-centrism in education.

Theme 1.2. International organizations about aftermaths of a military conflict for children: UNESCO, UNISEF, other organizations and distinguished people regarding the aftermaths of the conflict for children and adults, the impact of the conflict on the behaviour and children's rights realization, the right to education in particular.

Theme 1.3. Risks and dangers for children in conflict: the table of risks and dangers for children which are caused by conflict. The role of education in their prevention. Protection of teachers' work and prevention of traumatic behavior of children in conflict.

Module 2. "Education in conflict: the essence, standards (on the UNISEF and UNESCO information)"

Theme 2.1. The notion "education in conflict". The essence and meaning of education in conflict. Legal and political approaches to its application. Ukrainian regulations regarding education in conflict.

Theme 2.2. Standards of education in conflict: The subject matter of the UNISEF standards of education in conflict. The ways of their application in Ukraine. The terms of implementation of education in conflict.

Theme 2.3. Ukrainian and foreign experience of education in conflict: The experience of education in conflict in the Donetsk and Luhansk regions. The experience of education implementation in the countries with military conflicts (on UNISEF material).

Module 3. "Ways of education implementation in conflict in Ukraine"

Theme 3.1. Access to education in conflict: The subject matter of access to education and cases and reasons of its restriction. The ways of providing access to education. Ukrainian experience. International recommendations.

Theme 3.2. Security of children and their families, teachers in the conflict: Understanding of security and danger. Security theory. School security declaration. Security measures in secondary schools. School security plan. Regulations of teachers' actions in extreme situations at school. School as the centre of the area in the conflict. Friendly-to-child school. Bullying prevention. Non-violence pedagogy. Mediation. Social psychological and pedagogical services. Teachers' ability to resist stress, ways of its developing.

Theme 3.3. Organization of educational process in secondary education in conflict: The curriculum and the programs for secondary comprehensive schools in conflict. The forms of education available in conflict. Specific features of education in conflict. Recommendations regarding the educational process in conflict.

Table 1 presents the structure and hour distribution of the discipline.

Table 1. The structure of the discipline "Education in conflict"

Module themes	**Classroom hours**		**Individual work**
	lectures	***tutorials***	
Module 1. "Children's problems caused by the armed conflict"			
Theme 1.1. Children's problems caused by the armed conflict in Ukraine and abroad	1		
Theme 1.2. International organizations about aftermaths of a military conflict for children	1		4
Theme 1.3. Risks and dangers for children in conflict	1	1	2
Module 2. "Education in conflict: the essence, standards (on the UNISEF and UNESCO information)"			
Theme 2.1. The notion "education in conflict"	0,5		2
Theme 2.2. Standards of education in conflict	1		4
Theme 2.3. Ukrainian and foreign experience of education in conflict	0,5	1	
Module 3. "Ways of education implementation in conflict in Ukraine"			
Theme 3.1. Access to education in conflict	0,5		2
Theme 3.2. Security of children and their families, teachers in conflict	1	1	4
Theme 3.3. Organization of educational process in secondary education in conflict	0,5		4
Total	7	3	20

Implementation of the regional training program "Education in conflict" involves teaching methods which are different in the character of cognitive activity, in the ways of pedagogical guidance (with emphasis on individual work using the internet resources), in the resources of knowledge (teachers' professional experience included).

The forms of learning recommended are classroom work in micro-groups, lectures, tutorials and training workshops. Application of the program implies using some methodological support: e-version of lecture, tutorial and individual work material; problem questions to the material studied at the tutorials; tests for self-control; tests checking the level of mastering the material studied; methodological material, professional literature for learners' individual work; the list of references recommended to the learning course.

We have found out that scientific pedagogy has dealt only with fragmentary researches on the problem of substantiating the content of teachers' professional development. It has been proved that the problem is not purely pedagogical, its solution requires accounting the heritage of different fields of knowledge and professional activities, which implies: new functions (peace-builder, right-protector, social work) and risks and hazards for children and their families, which are possible only in conflict conditions; an array of problems in realization the child's right to education in conflict and their impact on children.

Therefore, the theoretical foundation of the training program for teachers' professional development "Education in conflict" is a set of scientific approaches of different fields and levels. The content of the program must be of multidisciplinary character and requires reading by experts in different fields (educators, psychologists, social workers, security specialists, lawyers). It has been determined that development of the content of the regional training program for teachers' professional training "Education in conflict" is based on a set of scientific approaches:

1) philosophic approaches: gnoseological, child-centrism, pragmatism, theory of participation;
2) general scientific approaches: axiological, synergetic, systemic, activity, interdisciplinary, complex, optimization, resource, innovative;
3) specifically scientific approaches: approaches to the protection of human rights (informational, power, economic, political, balanced, non-resistance to evil by violence, exposure of violence and hatred), a risk-oriented approach to the security theory, a situational approach to the management theory, political approach, regional and state approaches to public administration, legal approach, approaches to social work: problem-oriented, diagnostic, crisis interference, self-care support; conflictological approaches: an approach to psychology of action in extreme situations, cultural approach; an algorithmic approach through the theory of resolution, approaches to occupational safety in extreme situations;
4) pedagogical approaches: andragogic, theories of education content, competence-oriented, differentiated, personality-oriented, e-learning.

These theoretical foundations provided grounds for the regional training program for teachers' professional development "Education in conflict" that offers a 30-hour course on 3 modules: "Children's problems caused by the armed conflict", "Education in conflict: the essence, standards (on the UNISEF and UNESCO information)", "Ways of education implementation in conflict in Ukraine".

It should be noted that we conducted a training seminar on the program discussed in December 2019 upon the request of educators of comprehensive secondary school № 17 in Kharkiv, which was attended by the whole teaching and administrative staff (40 members). The teachers have admitted necessity and urgency of the program "Education in conflict", lack of professional skills and experience for dealing with children in conflict and have ordered the professional development course on the basis of this program in H. S. Skovoroda Kharkiv National Pedagogical University. Special appreciation has been expressed to practical guidance and exercises for teachers on dealing in conflict. Being well-aware of peace-building and mediation problems, the teachers claimed unsystematic character of their professional activity in conflict. Therefore, the first training seminar has confirmed a non-systemic set of professional competences in teachers for providing education in conflict and has proved the topicality of the content of The regional training program for teachers' professional development "Education in conflict".

4 Conclusion

To sum up, we would state that upgrading a teacher to work in conflict is an urgent issue that cannot be restricted to the program provided, which is limited in time (30 hours) and themes to study. But this program is the first attempt to develop both theoretical foundations (a set of scientific approaches of different levels) for determining its content in new teacher environment in Ukraine (in conflict) and its content as a multidisciplinary one. It considers international recommendations, standards and experience, current scientific approaches. Unfortunately, there are no analogues of the program in Ukraine for now. It emphasizes the integrated character of the content and efforts in solving the complexity of the problem. It trains teachers to deal with the problems concerning realization of the child's right to education and assurance of children's life and health protection, to fulfill new functions in the context of the conflict in eastern Ukraine. Our further research concerns the experimental appraisal of the program in order to adjust its content considering the results of the experiment and teachers' experience.

References

1. M.V. Elkin, T.F. Bielcheva, M.M. Oksa, M.I. Vorovka, M.M. Holovkova, *Modern pedagogical technologies in future teachers' training* (Vydavnychyi Budynok Melitopolskoi Miskoi Drukarni, Melitopol, 2014)
2. H.F. Ponomariova, *Educating the future teacher: theory and practice* (Ranok, Kharkiv, 2014)
3. V.M. Chaika, *Teacher's intellectual culture as a factor of innovations in pedagogical activity* (TNPU, Ternopil, 2014)
4. N.M. Bilyk, Sc. and Ed. a New Dim. Ped. and Ps. III (33), **66**, 23-26 (2015)
5. N.M. Ilyina, *Conflicts in a comprehensive school* (Universytetska Knyha, Sumy, 2015)
6. M.A. Trukhan, *Formation of the readiness of future social educators to work on the resolution of professional conflicts in educational institutions* (Khmelnytskyi, 2016)
7. K.V. Rozum, *Forming the readiness of future teachers of the humanities profile to prevent and*

resolve conflicts in educational work with students (Odessa, 2015)

8. A.M. Pynzenyk, L.A. Perminova, T.I. Molnar, V.V. Liapina, I.V. Voroniuk, N.V. Kabelnikov, *Professional pedagogical training of future teachers of primary general educational institution in the new Ukrainian schools* (part 1) (Ilant, Kherson, 2018)
9. S.A. Kroshka, in *Proceedings of the All-Ukrainian scientific and practical conference on Education of Ukraine in the conditions of military conflict in the Donbass,* Separate Subdivision "Lysychansk Pedagogical College as part of Luhansk Taras Shevchenko National University", Lysychansk, 2017
10. I.M. Trubavina, O.L. Mikhailenko, K.A. Yurieva, *Methodological recommendations for the organization of optional events in schools for the purpose of instilling dignity, peace and social consensus* (Planeta-Print, Kharkiv, 2017)
11. K.B. Levchenko, V.H. Panok, I.M. Trubavina, *Peacebuilding. Conflict Prevention and Resolution Using Mediation: A Socio-Pedagogical Aspect* (FOP Stetsenko V.V., Kyiv, 2016)
12. A. Ivashchenko, Human Rights Workshop: Education and Protection against Gender Violence in Armed Conflict. (Ukrainian Helsinki Human Rights Union, 5.04.2018), https://helsinki.org.ua/articles/pravozahysnyj-praktykum-osvita-ta-zahyst-vid-hendernoho-nasylstva-v-umovah-zbrojnoho-konfliktu. Accessed 10 Dec 2019
13. Education in Fragile, Conflict & Violence Contexts (Brief, November 9, 2018), https://www.worldbank.org/en/topic/education/brief/education-in-fragile-conflict-violence-contexts. Accessed 10 Dec 2019
14. J. English, World has failed to protect children in conflict in 2018. Widespread violations against children in conflict continue in shocking year-on-year trend (2018), https://www.unicef.org/eca/press-releases/world-has-failed-protect-children-conflict-2018. Accessed 11 Dec 2019
15. Children of war: the study of childhood problems in Ukraine in the face of military aggression (Ukrainian Institute for the Study of Extremism, 2015), http://uire.org.ua/doslidzhennya/1738. Accessed 11 Dec 2019
16. Training under shells. Police improve student safety system in ATO zone (Kramatorsk Police Department, 2016), http://kramatorsk-police.dn.ua/news/view/1089. Accessed 11 Dec 2019
17. Education and Armed Conflict: A Call to Action (2011), http://www.unesco.org/new/ru/goodwill-ambassadors/news-single view/news/education_in_armed_conflict_a_call_to_action. Accessed 12 Dec 2019
18. International network for Education in Emergencies, INEE Guidance Note on Conflict Sensitive Education (2013), https://inee.org/resources/inee-guidance-note-conflict-sensitive-education. Accessed 12 Dec 2019
19. International network for Education in Emergencies, Three Steps to Conflict Sensitive Education (2013), http://www.iiep.unesco.org/sites/default/files/inee_cse_graphics.pdf. Accessed 12 Dec 2019
20. M. Bouquet, Conflict-sensitive education – why and how? (2019), https://en.unesco.org/news/conflict-sensitive-education-%E2%80%93-why-and-how. Accessed 11 Dec 2019
21. J. Sany, A. Smiley, What is conflict-sensitive education and why is it necessary? (2018), https://www.fhi360.org/news/what-conflict-sensitive-education-and-why-it-necessary. Accessed 12 Dec 2019
22. UNICEF, Peacebuilding, Education and Advocacy Programme (2014), https://www.unicef.org/sites/default/files/2018-10/F_UNICEF1006_LFP_R3-online_single.pdf. Accessed 12 Dec 2019
23. Conflict-sensitive education policy. A Preliminary Review (Education Above All, Doha, 2012), http://protectingeducation.org/sites/default/files/documents/eaa_conflict_sensitive_education_policy.pdf. Accessed 10 Dec 2019
24. International network for Education in Emergencies, Conflict Sensitive Education, online course (Sphere, 2018), https://spherestandards.org/online-courses/conflict-sensitive-education-online-course/ Accessed 12 Dec 2019
25. UNICEF, Conflict Sensitivity and Peacebuilding Programming Guide (2016), https://www.unicef.org/media/59156/file. Accessed 12 Dec 2019
26. L.M. Karamushka, T.M. Dziuba. *Psychology of conflict management in the organization (on the material of educational organizations)* (Naykovyi svit, Kyiv, 2009)
27. Introduction to Human Rights (Programm Office, Vienna, 1999)
28. I.M. Trubavina, *Theoretical and methodological foundations of social and pedagogical work with the family* (T. Shevchenko LNU, Luhansk, 2009)
29. I.A Lipskiy, *Social pedagogy: a methodological analysis* (TC Sfera, Moscow, 2004)

Analysis of the state of formation of future teacher's communicative competence as a condition of providing the quality of general secondary education

Iryna Trubavina[1,*], *Svitlana* Vasylieva[1], and *Anna* Martyniuk[2]

[1]H. S. Skovoroda Kharkiv National Pedagogical University, Department of General Pedagogy and Pedagogy of Higher School, Kharkiv, 61102, Ukraine
[2]H. S. Skovoroda Kharkiv National Pedagogical University, Department of English language, Kharkiv, 61102, Ukraine

Abstract. The scientific novelty of the article is that the authors have revealed the average level of the state of formation of communicative competence (CC) of the third-year students. The cognitive and emotional components dominate among all CC components (it can be observed among students of the Faculty of Natural Sciences, Faculty of Arts). The behavioral component of the CC is better formed among students of the Faculty of Foreign Philology and the Faculty of Arts and Graphics. The communicative tolerance prevails among students of the Faculty of Physical Education as well as the Faculty of Arts and Graphics. Moreover, the two factors influencing CC were revealed: the external factors (organization of students' subject leisure; professional-oriented educational work with students; content of education at institutions of higher pedagogical education: specialty of the chosen profession, increased cycle of humanities and psychological and pedagogical disciplines, active pedagogical practice of students; age of students to work with, different categories of children and adults to communicate with, pedagogical skills and CC of teachers); the internal factors (students' attitude towards self-development, self-education, students' life and social experience, motivation for acquiring a specialty and CC, humanistic orientation of future teacher's personality, persona's orientation to "person to person" profession).

1 Introduction

Nowadays, the ideas of the New Ukrainian School are being implemented in Ukraine as well as the requirements for the quality of education are increasing and changing, including the requirements for teacher's education, both in the context of European integration and in the context of the Millennium Development Goals. According to the executive order No 776 of the Ministry of Education and Science of Ukraine dated July 16, 2018, "The Concept of pedagogical education development" was approved. It is stated in the concept that today there is a problem of imbalance "between the public demand for highly qualified teaching staff, the prospects for the development of society, global technological changes and the existing system of pedagogical education, as well as the level of readiness, ability of teaching staff to understand and implement the educational reforms in Ukraine"[1].

The quality of education is directly related to the teacher's willingness to work in the new environment. That is why the certification of teachers is offered today, and the opportunities for improving their qualification are being expanded. Nowadays, there is a rapid increase in the amount of information available to children. Teachers and books are not the only sources of knowledge today. The information that children get from the Internet is diverse, contradictory, and it can be threatening and provocative. It is easy to find the necessary material, samples of solving the problem, to gather the necessary material for training, and to make own solutions to the problem and form the necessary competencies on this basis. The teacher is a consultant, an assistant, a guide in acquisition of new material, a facilitator, even a mediator, but not a lecturer and not a translator of knowledge. This requires from a teacher to communicate with a student as a subject of education. This statement is also emphasized by in Law of Ukraine on "On General Secondary Education", according to which students, parents and teaching staff are equal participants in the educational process. Therefore all above mentioned requires teachers' high-level communicative competence (CC) in order to communicate and organize training and education in new roles as well as to organize learning process with students and their parents as equal participants. This brings communicative competence to the level of the educational process condition at the general secondary educational establishments, considering that communication permeates all kinds of human activity and at the same time is a separate kind of activity. It

* Corresponding author: trubavina@gmail.com

should be noted that competencies are the result of studying at a higher educational establishment; in the Law of Ukraine "On Higher Education" (in the context of competency approach), they are considered to be "a dynamic combination of knowledge, skills and practical know-how; ways of thinking; professional, world-view and civic qualities; moral and ethical values, which determines the ability of a person to perform successfully professional and further educational activities and it is the result of education at a certain level of higher education" [2].

V. Kremen distinguishes the following types of competences: key (general) competences (intersubject and super-subject). They are defined as "the ability of a person to perform complex multifunctional, poly-subject, cultural-purposeful activities, to solve effectively actual individual, social, and other problems" [3, pp. 9-17]. Communicative competences are general (key) competences, which are necessary for all future professionals due to the presence of it in all communication activities, but for the teacher they are also professional (special competencies), as the word is the main teacher's work tool. That means that the communication of a teacher with the students belongs to the intersubject and super-subject and at the same time to the professional competences, which proves the importance of formation of teacher's communicative competence in the aspect of improving the quality of general secondary education. This confirms that scientific and teaching staff in the educational process at higher pedagogical educational establishments should pay attention to the formation of future teachers' communicative competence.

The analysis of the recent researches has proved that the problem of introducing a new competency approach for higher education in Ukraine as well as the formation of teachers' communicative competence are the subject of modern pedagogical theory [4]. Thus, Yu. Tatur considers the competence to be an integral feature of a person and it characterizes his desire and ability (readiness) to realize his potential (knowledge, ability, experience, personal qualities, etc.) for successful activity in a certain field [5, p. 6-7].

O. Tsymbalyuk colligates competence with the ability of a person to perform professional functions. They include educational, upbringing and developmental functions, and the formation of pedagogical communication, communicative abilities and communicative skills is at the core of each [6, p. 8].

O. Kasatkina considers communicative competence to be the high-quality integral psychological quality, which is formed on the basis of a combination of communicative skills and empathy [7, p. 4-5]. D. Hodlevska adds communicative knowledge, speech skills, demonstration of reflection and creativity to the definition [8, p. 17-18].

I. Romanov notes that communicative competence is "the individual ability of a person to organize his speech activity adequately according to different situations of communication (by purpose, by role, by form, by content, etc.) in its productive and receptive types appropriate to each specific situation by linguistic means and methods" [9, 46-49].

I. Zymniaia defines that communicative abilities are person's individual and psychological features of a person, which ensure effective interaction and adequate mutual understanding between people in the process of communication or performing cooperation activities [10, p. 35-38]. The author emphasizes that communication skills allow a person to interact successfully with other people, to carry out communicative, organizational, pedagogical and other activities, to determine the qualitative and quantitative characteristics of the information exchange; to perceive and understand other people, to develop interaction strategies. S. Bucharskaya marks that communicative competence is a learning goal, "formed ability of a person to act as a subject of communicative activity, but communication in a foreign language can be taught. So, communicative competence as an ability only can be formed, developed and diagnosed" [11].

Therefore, nowadays, communicative competence of a teacher is his ability to communicate effectively with the students in subject-subject positions by creating a variety of conditions for them to learn new things. Its structure consists of knowledge in communication, pedagogical communication, communication skills, speech skills, communicative abilities, specific qualities and personality traits (empathy, reflection, creativity). That is why, the formation of future teacher's communicative competence is a long process, which is provided by different disciplines in the higher pedagogical educational establishments. So, the identification of the effectiveness of formation of future teacher's communicative competence is necessary, taking into account the results of teacher training for the analysis and correction in case of need.

Consequently, the purpose of the article is to determine the state of formation of future teachers' communicative competence at higher pedagogical educational establishments as well as to analyze the problems of formation of it. The tasks of the article are: to determine the state of formation of students' communication competence at higher pedagogical educational establishments and to identify the factors that influence the level of formation of students' communicative competence at higher pedagogical educational establishments.

2 Methodology

The theoretical analysis of the sources on the problem of formation of teacher's communicative competence, the analysis of documentation of higher pedagogical educational establishments, questioning (oral and written) of students and teaching staff, method of determining communicative competence by I. A. Martianova were used for the research [12]. It should be noted that today there are many methods for determining the communicative competence of future teachers; they are developed and substantiated by different scientists.

We have chosen the substantiated method of I. A. Martianova because it allows us to define several criteria for each component of communicative competence, which gives us a better understanding of the level of formation of communication competence through its components. It should be noted that this method is developed to determine the communicative preparedness of students of higher educational establishments. Other methods are dedicated to communicative competence of children of different ages, professionals from different fields, etc.

I. A. Martianova considers that the communicative competence is an effective tool for graduates` socialization and their adaptation in the modern multicultural society [12]. The communicative competence is considered in foreign researches as a student's ability to properly understand and use language to communicate in an authentic (rather than imitated) social and school environment [13]. At the same time, most scholars now consider linguistic competence as a part of communicative competence [4; 13]. Thus, Paulston believes that due to the communicative methodology of language initiation, the interest in learning language competence has expanded in order to include communicative competence [14]. That means, that the communicative competence in foreign studies is considered as a result of linguistic competence (while learning a foreign language) and as its component. We are interested in the teacher's communicative competence in the context of teacher's pedagogical communication and performing professional tasks that are mostly unrelated to foreign language learning.

In addition, the validity and reliability of diagnostics was verified by its author by an expert survey. This technique is a variant of self-diagnosis of students' communicative competence through a personal questionnaire based on closed questions, including a five-point rating scale of measurement [14]. These aspects led to the choice of this methodology for our study.

That is more important for the pedagogical activity results as well as for successful pedagogical communication among all the educational subjects. We are mostly impressed by the structure of the communicative competence, which I. A. Martianova offered to students: cognitive, behavioral communicative competence, communicative tolerance [12]. For pedagogical activity it is important for the teacher to know about effective communication, to be able to communicate with different children, their parents, colleagues, as well as to show tolerance to them, because all of them are different, moreover, it is important to work with all of them and qualitatively, in different directions.

Therefore, in our study, a simplified figure 1 illustrating the structure of the communicative competence can be used to identify the state of communicative competence formation of future teachers.

The basis for our research is: the legal framework and state documents for specialists' training at the current stage of development of national education, the concept of the New Ukrainian School, Laws of Ukraine "On Education", "On Higher Education", a competence approach to higher education, the regulations on communication as the main form of human activity, the theories of pedagogical communication and art of teaching.

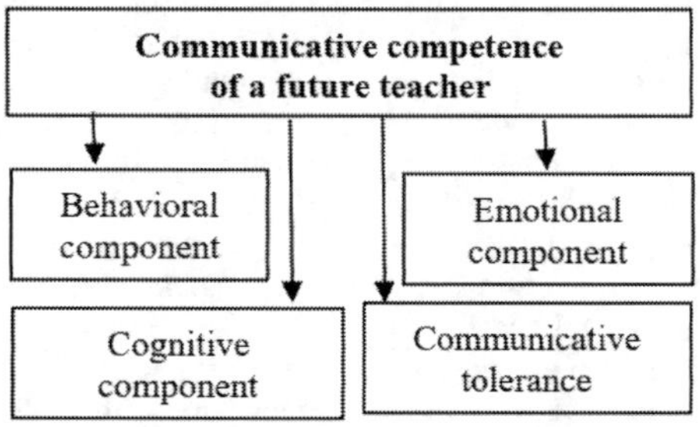

Fig. 1. Components of future teacher communicative competence

In order to conduct valid research dedicated to the determination of the level of communication competence, we chose students and took into account the current documents on the qualification levels of education. The National Qualifications System is developed in Ukraine in accordance with the European Qualifications System, it is a framework scheme that provides information on the generalized structure at all levels of qualifications of education. According to the law of Ukraine No776 "On Higher Education" dated July 16, 2018 qualification is considered to be "the official result of assessment and recognition obtained, when an authorized institution established that a person has attained competencies (education results) in accordance with higher education standards, which is certified by the relevant higher education document" [2]. We conducted our survey in third-year students of the first bachelor's level. According to the Law of Ukraine "On Higher Education", this level corresponds to the seventh level of the National Framework of Qualifications and identifies person's assimilation of theoretical knowledge and practical skills sufficient for successful fulfillment of professional duties in the chosen specialty [2]. We chose the third-year students, because the fourth-year students would have another year for correction the obtained results. After the third year of studying. many students work as teachers, that is why we consider that during the first three years of education teacher's communicative competence is formed at higher pedagogical educational establishments. Most of the theoretical subjects in general and professional training cycles (102 weeks from 132 weeks for pedagogical specialties), as well as a part of practical training (continuous propaedeutic pedagogical practice and pedagogical practice in general and specialized secondary educational establishments), which train future teachers for work in secondary educational establishments, are already taught, and only a part of theoretical preparation (30 weeks from 132 weeks of the budget time for pedagogical specialties) and practical training (pedagogical practice in general and special secondary educational establishments) remain unacquired by students.

The students of H. S. Skovoroda Kharkiv National Pedagogical University were interviewed in 2019 at the end of their third year, according to the method of I. A. Martianova [12]. The total amount of the interviewed students was 534. Among them there were 78 students of the Faculty of Foreign Philology, 76 students of the Faculty of Physical Education, 72 students of the Faculty of Natural Sciences, 74 students of the Faculty of Physics and Mathematics, 73 students of the Faculty of Arts and Graphics and the Faculty of Art, 80 students of the Primary Education Faculty, 81 students of the Psychology and Sociology Faculty. It should be mentioned that we conducted conversations with 20 teachers of each faculty. The total amount of the interviewed was 554, from the mentioned faculties of H. S. Skovoroda Kharkiv National Pedagogical University. The chosen specialties are related to natural-mathematical, artistic, humanitarian field, physical education and sports. The students were offered to answer 50 questions that consisted of four main scales: 1-10 – questions of the cognitive nature; 11-20 – questions which determine emotional parameters of communicative competence; 21-40 – questions which reflect the behavioral component of communicative competence; 41-50 – questions which reflect the level of communicative tolerance. The questionnaires fitted a five-point rank scale, and the questions were of closed type. The first half of the questions reflects the positive characteristics of communication, the second part – the negative characteristics.

According to the received answers, an individual score was calculated according to Table 1.

Table 1. Individual score calculations.

Question numbers	Options for ratings				
	almost always	in most cases	Some-times	rarely	almost never
From 1 to 5 11 – 15 21 – 30 41 – 45	5	4	3	2	1
From 6 to 10 16 – 20 31 – 40 46 – 50	1	2	3	4	5

The results had a direct correlance: the higher the score was, the higher the level of communicative competence was. So, 220 and above points meant the high level; points from 180 to 220 certified the medium level with a tendency to the high level; points from 140 to 180 meant the average level; points from 100 to 140 confirmed the medium level with a tendency to the low level; points under 100 meant the low level.

3 The results of research

We received the following data on the level of communication competence (Table 2):

Table 2. Generalized data of formation of the components of students' communicative competence (CC) according to faculties (in %).

CC compo-nents	Faculty	Al-most always	In most cases	So-meti-mes	Ra-rely	Almost never
Cognitive	Foreign Languages	*18.2*	*33.2*	*27.2*	*19.9*	*24*
	Psychology and Sociology	*5.6*	*46*	*16.7*	*21.7*	*20.7*
	Physical Education	*8.5*	*38.6*	*36.4*	*13.8*	*14.7*
	Primary Education	*0*	*42.2*	*23.3*	*37.8*	*16.7*
	Physics and Mathematics	*10.4*	*27.4*	*30*	*22.2*	*26.3*
	Nature	*27.3*	*28*	*28*	*20*	*23.3*
	Arts and Graphic, Arts	*17.9*	*25.8*	*33.7*	*20*	*20.8*
Emotional	Foreign Languages	*15.6*	*28.1*	*24*	*23.6*	*15.8*
	Psychology and Sociology	*11.3*	*32.7*	*24*	*32*	*14.9*
	Physical Education	*12.6*	*29*	*30*	*17.1*	*12.6*
	Primary Education	*5.56*	*48.9*	*23.3*	*20*	*34.4*
	Nature	*12.5*	*21.2*	*13.7*	*11.7*	*19.1*
	Physics and Mathematics	*21.1*	*27.4*	*18.9*	*22.2*	*26.3*
	Arts and Graphic, Arts	*20.4*	*38.3*	*18.7*	*21.6*	*25.8*
Behavioral	Foreign Languages	*32.5*	*32.9*	*21.6*	*16.5*	*14.8*
	Psychology and Sociology	*10.4*	*33.3*	*33.5*	*21.1*	*14.8*
	Physical Education	*19.7*	*33.3*	*35.7*	*20*	*13.1*
	Primary Education	*6.6*	*40*	*43.3*	*8.9*	*16.7*
	Physics and Mathematics	*27.8*	*29.6*	*27.8*	*8.1*	*23*
	Nature	*20*	*34*	*22*	*18.7*	*30.7*
	Arts and Graphic, Arts	*44.2*	*24.2*	*25*	*19.2*	*13.3*
Communicative tolerance	Foreign Languages	*22*	*30.7*	*28.3*	*23.02*	*16*
	Psychology and Sociology	*24*	*35.9*	*27*	*16.8*	*9*
	Physical Education	*31.9*	*36.7*	*21.4*	*12.3*	*0.48*
	Primary Education	*31.1*	*13.3*	*40*	*20*	*31.1*
	Physics and Mathematics	*27.8*	*21.5*	*21.1*	*14.8*	*17.4*
	Nature	*36.7*	*28*	*20*	*18.7*	*23.3*
	Arts and Graphic, Arts	*44.6*	*26.7*	*25*	*10.8*	*15*

The obtained data from the questionnaire on communicative competence are presented in Table 2 and indicate that the students tried to avoid critical and negative answers, in general. They had formed CC according to all criteria and indicators but in different ways in different faculties. The results indicate the influence of the teaching profession on formation of communicative competence, as well as the organization of the educational process at the faculties in different ways.

According to the obtained data students have mainly formed the cognitive and emotional components of communicative competence. This is caused by lack of professional pedagogical experience and activity, that is why the behavioral component is less formed.

The students had the high level of communication tolerance because of their love for children and desire to communicate with them all as well as humanistic orientation of future teaching profession and an orientation to the acmeological approach in education (the emphasis on the highest score) as a basis for the quality of education.

The further analysis of formation of students' communicative competence was conducted with a focus on students' answers "almost always" and "in most cases" as these indicators mark the highest level of their formation and orientation to the acmeological approach in education (resistance to the highest result).

Regarding to the state of formation of the cognitive component, we should note that the Faculty of Nature took the first place; it was followed by the Faculty of Foreign Philology, the Faculty of Psychology and Sociology, the Faculty of Physical Education, the Faculty Primary Education and the Faculty of Arts; the last place is occupied by the Faculty of Physics and Mathematics. This is caused by the fact that the students of the Faculty of Foreign Philology, the Faculty of Primary Education, the Faculty of Psychology and Sociology are more likely to be taught to communicate and work with children. It is the most pronounced focus on person's professional activity, while the future teachers of Physics and Mathematics are more focused on the professional "sign to sign" activity.

In order to explain why the students of the Faculty of Nature got higher score on the cognitive component of the communicative competence, than humanitarian students, we had to talk with students and teachers of these faculties and other faculties about how they achieved such results. They explained that the faculty pays attention to the harmonious development of students' personality, that is why the psycho-pedagogical disciplines are taken as much as possible in the curriculum, they are in all of its parts, unlike other specialties at the university, where these disciplines have been significantly reduced during the recent years, and are not present at all optional blocks.

According to the emotional component of the communicative competence, students of the Faculty of Arts and Graphics took the first place, followed by students of the Faculty of Primary Education, then by the Faculty of Foreign Philology, the Faculty of Psychology and Sociology, the Faculty of Physics and Mathematics, the Faculty of Physical Education, and the last one is the Faculty of Nature.

These results can be caused by the emotionality of artistic specialties that transfer emotions from art to communication, as well as by the fact that when working with children of younger age it is impossible to work successfully without emotions, understanding, because these children are just beginning to form logical thinking, they perceive everything emotionally and figuratively. Therefore, the future teaching profession occupation as well as the age of children affect the formation of the emotional component of communicative competence.

We got average results of emotional competence of students of other specialties, particularly, students of the Faculty of Psychology and Sociology, the Faculty of Foreign Philology, the Faculty of Physics and Mathematics, the Faculty of Physical Education.

The low results of students of the Faculty of Nature indicate a lack of understanding of the significance of emotions in the pedagogical profession, life, resistance to profession and rational thinking that is insufficient when communicating with children. Comparatively, students of the Faculty of Foreign Philology and the Faculty of Arts and Graphics and Arts received the best results in the behavioral component of communicative competence. This is caused by the experience of communication of students of both faculties in a foreign multicultural environment (most of the students of these faculties were abroad, worked there, communicated with foreigners to acquire language practice, gave concerts, were on internships).

The reason for the average results of students of the Faculty of Physics and Mathematics and Nature is in their focus on the profession of "man-nature", "man-sign", along with the teaching profession, which belongs to "person to person" orientation. We were surprised by low results on this component of the communicative competence in psychologists, sociologists and students of the Faculty of Primary Education.

The results of students of the Faculty of Primary Education can be caused by the fact that while communicating with young children of primary school age, students in practice do not have conflicts about the inequality of positions. The students perceived a teacher, and for this reason students do not have problems and experience in solving conflicts while communicating with children of this age, that is why their self-esteem is high. In addition, they do not have a certain language technique yet, which is formed by experienced primary school teachers: to speak briefly, little and clearly for everyone, to influence children non-verbally (because it takes into account the small amount of children`s attention, their fast fatigue, rapid refocusing of attention to more interesting things, unstable attention, its spontaneity, a small amount of arbitrary attention, which is just beginning to form in children).

All above mentioned creates communicative barriers for children, leads to problems with discipline in the classroom, and students understand their problems in the behavioral component of the communicative competence after having three types of practice. Regarding to the results of the questionnaires of the students of the Faculty of Psychology and Sociology, it turned out that they deal with quite different categories of children and adults, mainly with problematic clients, so they do not have a solid experience of communication and behavior yet.

https://doi.org/10.1051/shsconf/20207502006

The results of formation of the communicative tolerance were the best in students of the Faculty of Art and Graphic, Faculty of Arts, the Faculty of Physical Education, the Faculties of Nature and the Faculty of Psychology and Sociology. The students of the Faculty of Foreign Philology, the Faculty of Physics and Mathematics were separated from them. The best results are caused by the fact that students from the Faculty of Art have a figurative perception of people, all of them are beautiful, distinguished in some ways, and art requires understanding and respect.

The results of the athletes are interesting in the context of orientation towards winning in sports and tolerance in communication, which can be explained by special educational work at the faculty with future coaches and teachers of physical education (many different aged, short-term teams, team play and participation, where support and understanding are needed).

The results of psychologists and sociologists meet the requirements for the profession. The data on communicative competence of the students of the Faculty of Nature are explained by the fact that they are taught the most disciplines of the psychological-pedagogical cycle. Also we have counted the average score of communicative competence formation (Table 3).

Table 3. Distribution of average score of communicative competence formation by faculty (%).

Faculty	Average score
Foreign Languages	128
Psychology and Sociology	136
Physical Education	132
Primary Education	187
Physics and Mathematics	138
Nature	150
Arts and Graphic, Arts	147

We used the Pearson χ^2 criterion in order to evaluate the statistical reliability of the detected values regarding the communicative competence formation. The highest level of communication competence formation was obtained on the basis of the research conducted at the students of the Faculty of Primary Education – $\chi^2 = 36.96$) exceeds the critical value ($\chi^2_{0.05} = 7.815$ and $\chi^2_{0.01} = 11.345$). The lowest level of communication competence formation was obtained from students of the Faculty of Foreign Philology – 4.642.

The distribution of average grade point of communicative competence by faculties is represented in Figure 2.

Summarizing the results obtained by students (Table 3, Figure 1), we can say that the highest score of formation of the communicative competence was obtained by the students of the Faculty of Primary Education. They received 187 grade point, which corresponds to the average level with a tendency to the high level. At the same level, there are students of the Faculty of Nature but with fewer points (150 points) and the Faculty of Arts Graphic and the Faculty of Arts (147 points both). The received data can be explained by the fact that students of the Faculty of Primary Education have more psychological and pedagogical disciplines than students of other specialties. They study 8 methodics of teaching different subjects, which is one third of the studied subjects. More lessons are spent in practice (there are two lessons for each discipline at school, so there are 20 final lessons, while students of other faculties have only 10 final lessons). They are in the classroom during all days of practice with children and come only to their subjects. More disciplines of the humanities cycle according to the curriculum are studied by students of this faculty regarding language development (they are children's literature, the Ukrainian language, the Russian language, literature, etc.).

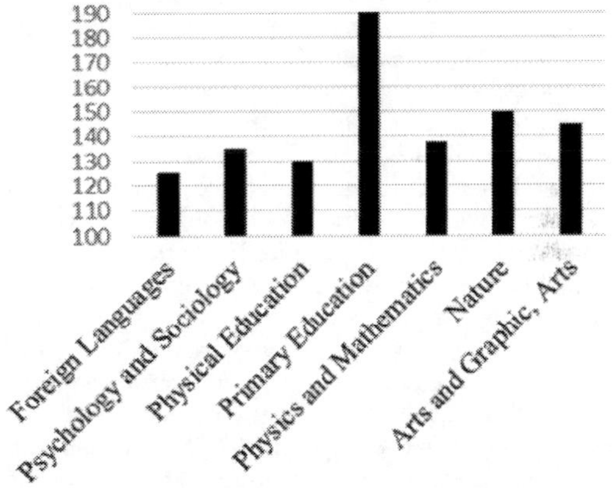

Fig. 2. Distribution of average grade point of communicative competence by faculties (%).

The students of the Faculty of Nature are also taught a lot of methodics of teaching (botany, zoology, biology, club activities).There are more lessons given by students during the practice. The high results of students of the Faculty of Arts and Graphics and the Faculty of Arts are caused by the emotional and tolerant components, the necessity to make children and adults interested in arts, a lot of practices and exercises, which are accustomed to choreographers, singers, musicians. Therefore, when doing exercises, they fulfill pedagogical disciplines in full, which is confirmed by our experience and work with students of these specialties.

The students of the Faculty of Foreign Philology (128 points), the Faculty of Sociology and Psychology (136 points) and the Faculty of Physical Education (132 points) are almost at the same medium level. These data in the aspect of athletes, for example, correlate with our data on the average level of formation of the general competence, including communicative competences, for students of the specialty "Physical education" [15, p. 338-348]. We should note that the average level of formation of the communicative competence of students of the Faculty of Foreign Philology indicates that the study of a foreign language, even at a high level, does not provide a high level of formation of communication competence yet.

The pedagogical experiment was conducted for three years at this faculty together with the British Council.

The issues of the experiment were the content of teaching, the training of translators and foreign language teachers. During the experiment the curriculum was reduced by 50% hours of the humanities as well as pedagogical disciplines, relatively to other faculties for studying. The study of the subject of pedagogical skills and the fundamentals of general pedagogy was replaced to pedagogical practice at school. As a result, students were at school more than students from the other faculties. However, they did not have enough theoretical knowledge of general pedagogy, pedagogical skills in sufficient quantity, which, in our opinion, was one of the reasons for the decreasing in the development of communicative competence.

These disciplines form the pedagogical skills, which include the communicative competence, give theoretical foundations for formation of it, provide exercises for development of it in the student environment. It should be noted that in general, the students of the 3-rd year of higher pedagogical educational establishment have an average level of development of communicative competence, which is sufficient level for work at school. At the same time, the reserves of formation of the communicative competence through the analysis of curricula and the obtained results were revealed: disciplines of the psychological and pedagogical cycle, a large number of methodics of teaching various subjects, a cycle of humanitarian preparation, educational work taking into account the students' specialty and their harmonious development, internship abroad, and an increased number of internships, attention to the independent professional exercise of students, who form the habit of self-working in all disciplines.

On the basis of the stated above, we distinguish the internal and external factors of formation of students' communicative competence at the higher pedagogical educational establishments. The internal factors include students' attitude to self-development, self-education, life and social experience of students, motivation to acquire a specialty and communicative competence, humanistic orientation of the future teacher', person's orientation to "person to person" profession.

The external factors include the formation of communicative competence: organization of subject leisure of students; professionally directed educational work with them; content of education in the educational institution: specialty of the chosen profession, increased cycle of humanities and psycho-pedagogical disciplines, active pedagogical practice of students; the age of the students to work with, the different categories of children and adults to communicate with. We have found out that disciplines of humanities, social and economical training, where students can spend more time discussing, proving their opinions, defend their positions, perform exercises on pedagogical communication have the greatest influence on the communicative competence development.

4 Conclusions

Solving the problem of imbalance between the public demand for highly qualified teaching staff and increasing the quality of general secondary education will contribute to the development of future teachers' communicative competencies. The communicative competence is considered to be a person's ability to carry out professional and further educational activities successfully as well as it is the result of training at a certain level of education, and at the same time, a teacher's professional competence. Therefore, the purpose of the article was to determine the state of formation of future teachers' communicative competence at higher pedagogical educational establishments. On the basis of determination of the level of formation of communicative competence according to the method by I. A. Martianova the following results were obtained: students of the Faculty of Primary education obtained the highest result, which corresponds to the average level with a tendency to the high level. The lowest average score was received by students of the Faculty of Foreign Philology. This proves that the presence of speech competence does not indicate the high development of communicative competence. Based on the study of formation of communicative competence through its components, it was found out that the students' behavioral component is formed at insufficient level, which can be explained by the lack of experience of professional pedagogical activity of the communicative competence. The high results of the formation showed cognitive and emotional components of communicative competence. The high level of the formation of the cognitive component of communicative competence is explained by the organization of the educational process, which is aimed more at mastering the methods of communication and teacher's work with children, students' professional orientation to people.

The emotional component of communicative competence is formed better in the students of the Faculty of Arts, which is explained by the emotionality of artistic specialties. The high level of communication tolerance of students' communicative competence is related to the humanistic orientation of the future teacher's personality. The behavioral component of the communicative competence is formed better in students of the Faculty of Foreign Philology and the Faculty of Arts. That is facilitated by the experience of communication of students of both faculties in foreign and multicultural environment. Students of the Faculty of Psychology and Sociology and the Faculty of Primary Education showed low scores on the behavioral component. The obtained results of students of the Faculty of Primary Education can be caused by the specificity of communication with young children of school age, which requires a special style of communication. Students of the Faculty of Psychology and Sociology showed such results due to the lack of practical experience with complex clients. The communicative tolerance of the communicative competence is formed better in students of the Faculty of Arts and Graphics and the Faculty of Physical Education. That is explained by the imaginative perception of students of the Faculty of Arts and Graphics and the specificity of educational work at the Faculty of Physical Education, which is focused on winning and team game.

The obtained data can be explained by internal and external factors influencing the formation of communicative competence. The internal factors of the formation of communicative competence include students' attitude to self-development, self-education, life and social experience of students, their motivation to acquire a profession, the humanistic orientation of the future teacher's personality, the orientation of students to "person-person" profession. The external factors of the formation of communicative competence include: organization of subject leisure of students; professionally directed educational work with them; content of education in the educational institution: specialty of the chosen profession, increased cycle of humanities and psycho-pedagogical disciplines, active pedagogical practice of students; the age of children to work with, various categories of children and adults to communicate, pedagogical skill and teaching staff.

The prospects for further research are to determine the degree of influence of various factors on the development of students' communicative competence according to their chosen profession and course of study.

References

1. Pro zatverdzhenij Koncepciї rozvitku pedagogichnoїi osvity (On approval of the Concept of teacher education development) (2018), https://mon.gov.ua/ua/npa/pro-zatverdzhennya-koncepciyi-rozvitku-pedagogichnoyi-osviti. Accessed 16 July 2019
2. *Pro vyschchu osvitu* (About higher education) (2019), https://zakon.rada.gov.ua/laws/show/1556-18. Accessed 09 August 2019
3. V.G. Kremen (ed.), *Rozroblennja osvitnih program. Metodichni rekomendaciї* (Development of educational programs. Guidelines), (Priorytety, Kyiv, 2014)
4. R. Nordquist, Communicative Competence Definition, Examples, and Glossary (2019), https://www.thoughtco.com/what-is-communicative-competence-1689768. Accessed 25 November 2019
5. Yu.G. Tatur, *Kompetentnosnyy podkhod v opisanii rezul'tatov i proyektirovanii standartov vysshego professional'nogo obrazovaniya* (Competent approach in describing the results and designing standards of higher vocational education). (Issledovatel'skiy tsentr problem kachestva podgotovki spetsialistov, Moscow, 2004), http://fgosvo.ru/uploadfiles/npo/20120325221547.pdf. Accessed 25 August 2019
6. O. M. Tsymbalyuk, Dissertation, Rivne State Humanitarian University, 2011
7. O. V. Kasatkina, Dissertation, Vasyl Stefanyk Precarpathian National University, 2007
8. D. Hodlevska, Dissertation, National Pedagogical Dragomanov University, 2007
9. I. Romanov, *Psykholohichni osoblyvosti yunats'koho viku ta yikh vplyv na formuvannya inshomovnoyi komunikatyvnoyi kompetentnosti na zanyattyakh z inozemnoyi movy* (Psychological peculiarities of juvenile age and their influence on the formation of foreign language communicative competence at English classes). Science And Education A New Dimension **VI** (164), 46–50 (2018). doi:0.31174/send-pp2018-164vi68-11
10. I. A. Zymniaia, *Klyuchevyye kompetentnosti – novaya paradigma rezul'tata obrazovaniya* (Key competencies – a new paradigm of educational outcome). (Rossiyskiy Gosudarstvennyy Gumanitarnyy Universitet, Moscow, 2006), http://aspirant.rggu.ru/article.html?id=50758. Accessed 25 August 2019
11. S. M. Buchachka, *Komunikatyvni zdibnosti studentiv u profesiynomu spilkuvanni inozemnymy movamy* (Communicative abilities of students in professional foreign languages communication). Problemi suchasnoy pedagogichnoy osviti. Ser.: Pedagogika I psihologija 1(34), 232–237 (2009), http://library.vspu.net/bitstream/handle/123456789/1628/komunik%20zdibn.pdf?sequence=1&isAllowed=y. Accessed 01 December 2019
12. I. A. Martianova, *Diagnostika urovnja sformirovannosti kommunikativnoj kompetentnosti studentov vuza v uslovijah polikul'turnogo obshhestva* (Diagnosis of the level of formation of the communicative competence of university students in a multicultural society). Pedagogicheskie nauki **46-2** (2016)
13. Communicative Competence (2019), https://www.learnalberta.ca/content/eslapb/about_communicative_competence.html. Accessed 11 December 2019
14. C.B. Paulstone, Linguistics and Communicative Competence: Topics in ESL (Multilingual Matters, Clevedon, 1992)
15. I.M. Trubavina, S.O. Vasylieva, N.O. Agarkova, in *Theory and practice of introduction of competence approach to higher education in Ukraine*, ed. by I.M. Trubavina, S.T. Zolotukhina (Premier Publishing, Vienna, 2019), pp. 338–347

The gender component in communication education in Ukraine: conflict communication problems

Yulia Kekalo*

Kryvyi Rih State Pedagogical University, Kryvyi Rih, Ukraine

Abstract. Communication is a universal phenomenon that is explored by many social sciences and humanities. In the structure of communication science has long been distinguished conflict communication. However, the gender aspect of the language conflict in Ukrainian studies is not solved, but highly relevant and in-demand today by scholars. In our work, we emphasize the need for a detailed study of the linguistic gender conflict within the framework of forming a communicative culture in Ukraine. The article analyzes the state standards of higher education in speciality 035 "Philology" for the first and second levels of higher education, where we pay attention to the formation of communicative competence. We argue that the gender identity of the speaker is not taken into account in the structure of communicative competence. This can be the result of conflicts between a man and a woman. Therefore, the author talks about the importance of gender competence within the communicative. The article also analyzes the program in the course of teaching conflict communication for postgraduate students, which allows characterizing gender conflict from different positions (national linguistic representation, space-time parameters, social characteristics, age, etc.).

Introduction

In today's world, conflict is an everyday reality. They are deployed in all spheres of life of individuals; actively go to the pages of newspapers, TV screens, the Internet. Because of this, people are often in a stressful state, feeling stress, which makes it impossible for a comfortable life. Therefore, world science is increasingly involved in finding ways to resolve conflicts.

One of the reasons for the disharmony of the social order can be considered hierarchical relations between a man (recognized in the society leading, the main, the first) and a woman (recognized in the society subordinate, the second). In the history of culture, there is a distribution of roles in society. The spheres of external activity, the sphere of formation of culture and society were assigned to the men. According to the woman – the sphere of reproduction of life, as if the sphere of nature. And this distribution has been maintained for many centuries.

However, gender issues become relevant when civilization encounters fundamental difficulties not resolved within norms and cultural norms. The complexities in the foundations of industrial civilization, which are manifested in different cultures, are forcing society to address a gender problem that has become socially aware recently, although the conflict between man and woman is perhaps the most ancient of the many contradictions that define the diversity of social space.

Despite the fact that in Europe the problems of gender identity of a person, gender communication have long been explored, in Ukraine the same issues are still relevant. The status and roles of men and women are constantly changing, accompanied by a clash of interests. Today in the Ukraine there is still a certain dissonance between men and women's perceptions of sex-role behaviour, unwillingness to identify and take into account gender stereotypes in behaviour, actions or speech. However, it is becoming increasingly apparent that gender norms are being transformed with the penetration of both women and men into opposite spheres of activity. Because of this, a new type of conflict is emerging – gender [1].

In Ukraine, there is an active fight for equality between people of different sexes. From the history of culture, we know that women submitted to men. But in the twentieth century, women began to fight for their rights, and this process continues today. Therefore, it is commonplace nowadays those women are increasingly not only paying attention to family affairs but also working hard to provide for their families. Men, for their part, do not always like it, especially when a woman is more affluent than a man. These circumstances often lead to domestic family conflicts. The reason for confrontation can be psychological, physiological features of a man and a woman, because they have a different emotional background, different interests, values, preferences, habits.

Unfortunately, Ukrainian society remains a hostage to gender stereotypes: women, unlike men, receive less pay; Ukraine is still in the group of countries with the lowest representation of women in power; almost 90% of Ukrainian victims of violence are women. The issue of

* Corresponding author: kekaloyulia@ukr.net

recognizing people with alternative gender identities, who are often subject to psychological abuse, remains very painful.

All of this can be the cause of gender conflicts, which has been little explored in Ukrainian studies. Therefore, we focus our research on the issue of recognition of gender conflict communication. After all, conflicts are mainly represented by means of language. Therefore, linguistic thought does not stand alone, but, on the contrary, studies them, looks for the most successful models for neutralization or harmonious communication.

The anthropocentricity of the scientific paradigm has led to an appeal to gender aspects of language, to the identification of discordance and differences in the linguistic representation of men and women in culture. The spread of gender studies, the development of new directions and methods of study in the field of linguistics made it possible to form *gender linguistics*. The interest of Ukrainian scholars in the study of gender aspects of language and speech is due, on the one hand, to the continuation of the feminist movement, and, on the other, to the general tendency of the linguistic sciences to study the social conditions under which communication, in particular, the human factor in language, takes place. Therefore, the purpose of this article is to share our views and our experiences in analyzing the features of gender-conflict communication in the context of shaping communicative culture in Ukraine. We would also appreciate any feedback from anyone interested in this matter.

1 Interest in gender manifestations in communication abroad and in Ukraine

Communication is a universal phenomenon that comes into view of many social sciences and humanities. The concept itself is quite broad and not fully defined, because communication is caused by: 1) a variety of ways of fixing information; 2) specifics of the purpose of information transfer; 3) the discretion / indiscretion of the generation and perception of information; 4) the way the information is addressed [2].

It can be argued that communication is an important factor in the existence of the individual and society in general. Compliance with the harmonious process of information transfer is the key to a successful state. Therefore, researchers from different countries are interested in the dynamic development of their own "information infrastructure" and the improvement of communication tools.

Traditionally, communication is seen as an interdisciplinary subject that is closely linked to language, social cognition and psychology. Since the 1960s, scholars have actively studied communication as one of the central aspects of human existence.

In the scientific literature differentiate communication. In particular, American sociologist C. H. Cooley said that communication is a certain mechanism by which human relations exist and develop, including all symbols of the mind with the means of transmitting them in space and preserving them in time [3]. Modern studies highlight communication as a mechanism that affects all spheres of society. In particular, B. D. Ruben and R. A. Gigliotti study communication as a strategic tool through which leadership qualities of the individual are manifested. They reveal this concept through the prisms of leadership, defining communication as the process by which people create and use information that expresses their attitude to the world and to others. The conceptualization of human communication leads to a view of leadership as a process of social influence, which is ascertained through both verbal and non-verbal communication [4].

H. Giles and T. Ogay outline the basic principles of communication theory, which include: 1) communication is influenced not only by the particularities of the situation and the orientation of the participants in the communication, but also by the socio-historical context; 2) communication is not only an exchange of information about facts, ideas and emotions, but also a clear identification of the social category of participants; 3) communicators expect optimal communication based on social stereotypes; 4) communication participants use specific communication strategies to express their relationship to one another or to a social group [5]. As we can see from the above, the researchers talk about communication as social relations between people.

In domestic science, they rely on the views of foreign colleagues, but they have their own view of the problem. H. Pocheptsov understands the processes of transcoding verbal into non-verbal and non-verbal into verbal spheres [6]. A. Pokhresnyk analyzes the modern information society within higher education and says that communication is a message that is interpreted through the political, cultural, economic, semiotic, hermeneutic and social aspects of their contexts. With the expansion of information devices and "communication" there is a growing need for the emergence of new science, which is not limited to social-humanitarian knowledge, but also takes into account the exact sciences [7].

O. Bohuslavska emphasizes on effective communication, noting that it is a complex multidimensional process of interaction between people, which is implemented by a specific model and consists in the exchange of information, as well as in the perception and understanding of each other's partners [8].

In the higher education system, communication is considered, on the one hand, as a type of social process, and on the other, as a type of communicative social influence and interaction of participants in the educational process [9].

In Ukrainian linguistics, communication is defined as social interaction by means of linguistic and paralingual (facial expressions, gestures, etc.) means to convey information [10].

Therefore, we argue that communication is a broad interdisciplinary concept that includes linguistic and non-linguistic means of information exchange, perception and understanding.

Within communication science, conflict communication is singled out as one form of communication. In general, there are several names in linguistics: verbalized conflict (I. Frolova), verbal conflict (V. Borysevych), language conflict (L. Bilokonenko, V. Tretiakova), conflict discourse (S. Formanova), and others. In particular, V. Borysevych defines: verbal conflict is the result of a violation of the process of human communication, the most important means of which, as we know, is language [11]. I. Frolova, based on the English language environment, it concludes that a verbalized conflict is a form of communication that has historical and cultural roots and is linked to natural features of a national character, but is recognized in the modern day as not conforming to the norms of communication [12]. V. Tretiakova notes that language conflict is a state of opposition between two parties (parties to the conflict), each of whom knowingly and actively acts to the detriment of the other party, explaining their actions by verbal and pragmatic means [13]. Although these definitions have differences, it can be said that language conflict is a violation of norms of communication between people.

However, the issue of gender in the language conflict has not been resolved, although such an analysis is in demand in Ukrainian studies, especially regarding the specific language behaviour of men and women.

In general, issues of gender specificity in conflict have long been explored abroad. In particular, the French researcher C. Dijkema stresses the importance of studying gender and conflict together, since ignoring the gender aspect of social reality complicates the resolution of key elements of the conflict. In general, conflict can be seen as an opportunity to change the gender roles of men and women [14].

Conflict scientist J. Birkhoff notes that gender influences and really permeate the dynamics of conflict at the social and individual levels. It has an impact on how individuals identify, perceive, and understand a conflict [15]. L. Handrahan draws attention to the impact of ethnicity in gender conflicts and says that gender analysis of ethnicity has problematized the complexities of social realities, community concepts and understanding of conflict and post-conflict dynamics [16].

H. Giles and T. Ogay also speak about the peculiarities of communication between men and women, emphasizing that both sexes use different communication strategies and tactics in communicating with each other. However, both men and women do not have the same communicative behaviour in same-sex and heterogamous groups. This means that they adjust their communicative style to the gender of the interlocutor [5].

As we can see, gender issues and conflict have been explored abroad for a long time. Therefore, in our opinion, there is a need to speak about the unification of these two scientific directions and the identification of a new direction in Ukrainian studies – *gender conflict communication* science, which opens the prospect of gaining new knowledge on specifying the functions of the Ukrainian language in gender conflict communication.

Our thoughts are based on domestic research. Researchers point out that the gender factor influences the use of different language means by men and women, and communicative failures occur in different societies. Gender-related verbal conflicts arise primarily from differences in cognitive information models contained in the psyche of participants in a communicative act and of a sociocultural nature [17]. When interacting in heterogamous groups, women and men use multi-vector communication strategies, resulting in many misunderstandings, sometimes leading to communicative deviations or communicative conflicts that illustrate a particular relationship between men and women in communication.

We treat *gender communicative conflict* as a representation of the language of the process of creating a communicative space, which is conditioned by the opposition of gender values, social roles, interests of persons of different sexes, which covers all spheres of life and aims at rethinking gender stereotypes in the minds of people.

2 Communicative educations in the state standards of Ukraine (with gender competence)

The development of a person and becoming him / her as a future specialist is through the prism of mastering and correct ability to use the set of competencies (general and special / professional). These competencies are enshrined in the updated provisions of the Law of Ukraine "On Higher Education" of 2014. It is this Law that is central to the formation and differentiation of state standards and competencies for particular specialities.

The Law treats competence as a dynamic combination of knowledge, skills and practical skills, ways of thinking, professional, ideological and civic qualities, moral and ethical values, which determines a person's ability to successfully pursue professional and further educational activity and is the result of teaching at a certain level of higher education [18].

According to this Law, Section III defines standards of educational activity and higher education. In particular, the values of the category "higher education standards" are defined, which are defined as the set of requirements for the content and results of educational activity of higher education institutions and scientific institutions for each higher education level within each speciality [18]. Each standard should have a number of components, including a list of competencies.

Having analyzed the state standards of higher education of Ukraine in speciality 035 "Philology" for the first (bachelor) and second (master) levels of higher education, we noticed that one of the key competencies is communicative [19–20].

In general, communicative competence is a complex multicomponent concept, a set of personal qualities of an individual (values and personal orientations, abilities, motives, knowledge, skills, experience), which ensure her ability to establish the necessary contacts with other people, to build social communication taking into account and with social norms and systems of cultural representations and values of language. However, in the structure of communicative competence, we did not notice gender identification through communication norms.

The issue of gender competence as one of the central components of personality development remains relevant in Ukrainian education. The importance of the gender competence of the future teacher is determined by the fact that it is the educator who is one of the important subjects of gender socialization of the growing generation and has a great influence on the formation of ideas about the quality of men and women and their role in society.

Unfortunately, in Ukraine, the gender issue in education remains relevant today. In general, gender competence is represented by a three-layer structure that reflects the logic of the scientific interpretation of the concept: knowledge – experience – skills. Knowledge is teachers' awareness of gender issues (its relevance to Ukrainian society, the essence of gender theory, understanding of the gender category). Experience is the degree of development of a teacher's gender culture (in particular, gender sensitivity) that influences their readiness to perceive these issues through their genetic connection to feminism. Skills are the level of mastery of a teacher by the method of gender education and training [21].

We believe that gender competence should also be included in the structure of communicative competence. The perception and understanding of the opposite articles by means of language is equally important. In linguistics, there are concepts of both masculine and feminine communication styles that differ. We believe that future philological teacher needs to acquire the skills and abilities of gender communication.

In the modern world, in the study of speech and in general communicative behaviour, gender is considered as one of the parameters by which the social identity of the speaker is constructed in communication. Today, a great deal of scholarly work is devoted to the interrelation of language and gender, which reveal the specifics of male and female discourses. Identifying the differences inherent in all aspects of communicative behaviour of persons of different sex is not only theoretical but also of practical importance. Knowledge of gender peculiarities of communication allows to avoid communicative failures, promotes mutual understanding of men and women, gives an opportunity to detect gender stereotypes and overcome negative or, conversely, to promote their reproduction if they are embedded in the system of social values.

The emergence of conflict between men and women is the result of "gender illiteracy". In order for both sexes to provide effective communication, it is necessary for men to know and take into account the peculiarities of female communicative behaviour, and for women – male.

Therefore, our study is aimed at regulating optimal communication between men and women. We try to take into account the different communicative behaviour of both sexes.

3 Theoretical parameters of the description of conflict communication: my conclusions

Within the framework of postgraduate studies in the speciality 035 Philology, we study the discipline "Ukrainian-language communicative conflictology", in the structure of which we try to resolve the issue of gender conflict and identify ways of its neutralization [22].

The curriculum of the discipline stipulates that as a result of studying the discipline, the graduate students should learn the basic concepts and provisions of communicative conflictology, presented by methodologically and theoretically diverse linguistic trends; have an understanding of the structure and mechanisms of functioning of the Ukrainian-language interpersonal conflict, as well as the principles and technologies of its management; learn to analyze the communicative laws of conflict and cooperative communication taking into account social, cultural, cognitive, psychological, ethnic, gender factors; be able to apply the acquired knowledge in practical professional activity.

As a result of studying the discipline, the graduate student should know the laws, mechanisms, linguistic means of interpersonal interaction of speakers in different spheres of interpersonal communication; understand the laws of the conflict process; to analyze the Ukrainian-language representation system of harmonious and conflict communication; be aware of the public inquiry to raise the level of knowledge of the society about effective communication.

Upon completion of the course, the graduate student should be able to: creatively use knowledge of the laws of language communication development in teaching, research and teaching activities; be aware of the nature and mechanisms of functioning of this phenomenon; to disclose its deep causal relationships, functional features of harmonious and conflicting statements; constructively analyze barriers to communication; to know the basic ways of language neutralization of the conflict risk situation and the end of the conflict itself; independently think, analyze and synthesize, evaluate the information obtained.

The program provides a general awareness of conflict communication as a phenomenon caused by various factors. We pay particular attention to the following topics: *Lexical conflict markers*, *Phraseological markers of conflict*, *Morphological and syntactic markers of conflict*, *Linguopragmatic aspects* of discrepancy between communicative actions of persons that enable us to characterize gender conflict from the point of view of national linguistic representation of gender behaviour of men and women. The gender factor influences the use of different language means by men and women. This is important because we can describe the most distinctive features of the use of linguistic means of opposing articles in a conflict, which allows for a greater understanding of each other and promotes harmonious communication between individuals.

The issues of space-time parameters of conflict gender communication cannot be overlooked. Here we have to take into account the scope (political, economic, administrative, public, family, friendly) of the origin and manifestation of the conflict, the conditions and factors of its occurrence, the means and actions to which the parties involved in the conflict, the duration, the frequency and recurrence of the conflict, the temporal characteristics of each stage of the conflict.

Another significant aspect is the influence of social characteristics, age, profession, education on the communicative behaviour of the initiator of the conflict, whether a man or a woman. This is due to the fact that in society there is a certain differentiation of the social statuses of the opposite sexes, which are influenced by gender stereotypes. Human social status is one of the basic parameters of communication, which manifests itself in the postulates of communication and speech acts, ways of influencing the opponent. The age of the communicators is another important factor in the social organization of communication in general and speech interaction in particular. For example, the age of the older generation determines the level of accumulated knowledge and experience and is a factor in the expansion of the individual cognitive space and cognitive base of the individual.

If we are talking about the course of gender conflict in the light of national specificity, it is obligatory to consider the topic of Ethical and cultural principles of cooperative communication. After all, Ukrainian ethical and cultural norms imply changes in worldview values. Although Ukrainian culture remains a hostage to gender stereotypes, for the most part, gender norms are increasingly being transformed today.

Summarizing the previous results of our work, we can say that the Ukrainian mental opinion is that the male style of communication is more active and substantive, competitive and conflicted, as well as emotionally restrained. Instead, women are more emotional in communication, able to empathize in conversation.

That is why the Ukrainian culture shows the peculiar types of male and female verbal behaviour. Among the most characteristic features are the following: for women the fact of communication is important, for men – the result of communication; women prefer private communication, men prefer public communication; the story of men is usually short, without detail; women express more exclamation and interrogative sentences; men are prone to irony at their opponent; in conflict situations, women are more likely to abandon their views, are prone to emotional resolution, men insist on their positions, use a rational way to end the conflict; women are characterized by a cooperation strategy, and men by competition.

In our work, we analyze interpersonal gender conflict, where language material indicates that most gender-specific communication is expressed at the lexical-semantic level. At this level, women are more likely to use vocabulary with hyperbolized connotation, and they are the initiators of conflicts, especially household ones, as opposed to men who, on the contrary, seek to resolve the situation and conflict peacefully. Male representatives, in turn, are more likely to resort to obscene, vulgar, and stylistically reduced vocabulary.

We note that men and women in conflicting communication are equally inclined towards the image toward their opponent. However, these images differ in their evaluation characteristics: men often indicate excessive talkativeness of women; mental disability; deceitfulness; appearance based on age; anger and meanness. Very often, men resort to the names of women with zoonomenes, as well as words that indicate immorality.

Although vulgar vocabulary is not characteristic of women, in conflict, in our observations, they are more active in the use of abusive words against men: most often, the offences refer to the tendency to act inappropriately; ignorance; social level; deceitfulness; audacity.

We believe that gender competency is one of the priorities of higher education. After all, this knowledge can help in communicating with the opposite sex.

Therefore, we believe that the above facts give us the right to speak about a new approach to gender in a language that integrates theories of gender linguistics and conflict communication. We call this approach *gender-conflict*. Its essence is that it makes it possible to distinguish:

1) the influence of gender on the verbal interaction between men and women in interpersonal conflicts of different types;
2) the circumstances in which certain communication strategies and tactics are chosen;
3) resources of the national language that determine the communicative actions of men and women in the area of conflict and / or conflict risk;
4) pragmatic markers that can cause interpersonal conflict between men and women;
5) models of linguistic behaviour of men and women at the pre-conflict, conflict stage and completion stage;
6) methods of effective communication between men and women [5].

Conclusion

Therefore, the study of language gender conflict is relevant to Ukrainian culture. Following on the research of foreign colleagues, Ukrainian scientific thought should distinguish its own system of linguistic signs used by men and women. This will promote harmonious communication in society.

That is why in our work, we emphasize the need for a detailed study of the gender conflict, which opens up perspectives for new knowledge of the functioning of the Ukrainian language.

We believe that the gender-conflict aspect should be introduced in the structure of communicative competence and gender-communicative competence should be formed. After all, knowledge of the gender features of communication will allow the future philological teacher to avoid communicative failures and to foster "gender literacy" in the young generation, which will ensure effective communication between a man and a woman. The new gender-conflict approach allows different perspectives to characterize the conflict between the persons of both sexes and to find new ways of avoiding (mitigating) gender conflicts.

References

1. Yu. Kekalo, in *Southern Archive* (Helvetyka, Kherson, 2018), pp. 34–38
2. O.O. Selivanova, *Modern linguistics: directions and problems* (Dovkillia-K, Poltava, 2008)
3. C.H. Cooley, *Social Organization: A Study of the Larger Mind* (Charles Scribner's Sons, New York, 1909)
4. B.D. Ruben, R.A. Gigliotti, J. Of Lead. & Org. Stud., **23**(4), 467–479 (2016). doi:10.1177/1548051816641876
5. H. Giles, T. Ogay, Exp. Com., 293–310 (2007)
6. H.H. Pocheptsov, *Communication theory* (Refl-buk, Kyiv, 2001),
7. A. Pokhresnyk, Hig. Educ. in Ukr. **4**, 41–47 (2017)
8. O.H. Bohuslavska, Hum. Bull. **66**, 45–54 (2016)
9. H.H. Sillaste, Not. of the Khar. Hum. Univ. **24**, 112–123 (2018)
10. F.S. Batsevych, *Fundamentals of Communicative Linguistics* (Akademiia, Kyiv, 2009), pp. 8–10.
11. V.V. Borysevych, Cult. of the Bl. Sea peop. **44**, 186–190 (2003)
12. I. Ye. Frolova, Bull. of Sumy St. Univ. **11**(95), 77–82 (2006)
13. V.S. Tretiakova, Dissertation, The Moscow State University, 2003
14. C. Dijkema, Why study Gender and Conflict Together? (2001), http://www.irenees.net/bdf_fiche-analyse-801_en.html. Accessed 11 Dec 2019.
15. J.E. Birkhoff, Gender, conflict and conflict resolution (1998), https://www.mediate.com/articles/birkhoff.cfm# Accessed 18 Dec 2019.
16. L. Handrahan, Sec. Dial. **35**(4), 429–445 (2004)
17. L.V. Chaika, Scient. Not. of the Nat. Univ. of Ostr. Acad. **23**, 192–194 (2012)
18. Law of Ukraine "On Higher Education" (2014), https://zakon.rada.gov.ua/laws/show/1556-18. Accessed 15 Dec 2019
19. Higher education standard of Ukraine (first bachelor's level, field of knowledge 03 Humanities, specialty 035 Philology (2019), https://mon.gov.ua/ua/osvita/visha-osvita/naukovo-metodichna-rada-ministerstva-osviti-i-nauki-ukrayini/zatverdzheni-standarti-vishoyi-osviti. Accessed 15 Dec 2019
20. Higher education standard of Ukraine (second master's level, field of knowledge 03 Humanities, specialty 035 Philology) (2019), https://mon.gov.ua/ua/osvita/visha-osvita/naukovo-metodichna-rada-ministerstva-osviti-i-nauki-ukrayini/zatverdzheni-standarti-vishoyi-osviti. Accessed 15 Dec 2019
21. T. Doronina, Scient. Not. of Tern. Nat. Ped. Univ. **4**, 163–170 (2008)
22. L.A. Bilokonenko, The course syllabus "Ukrainian communicative conflictology" (Kryvyi Rih, 2019)

Narrative tolerance: on the road to unbiased history, cultural tolerance and civil society

Olena Mishalova[1,2,*]

[1]Nanjing Normal University, Philosophy Department, Nanjing, 210000, China
[2]Kryvyi Rih State Pedagogical University, Philosophy Department, Kryvyi Rih, 50086, Ukraine

Abstract. The article examines a concept of narrative tolerance as the basic structure of cultural tolerance and modern civil society. It is proposed to consider narrative tolerance as an effective methodological tool to create a tolerant inclusive sociocultural world paradigm. The tolerant sociocultural approach in history studying and teaching is a crucially important condition that makes it possible to resolve protracted historical conflicts both within and beyond countries and, as a result, leads to reconciliation and mutual forgiveness.

1 The notion of narrative tolerance

The modern information society is an open society. This circumstance makes it extremely vulnerable for manipulation of public consciousness, especially when it concerns the problematic issues of the past. Every careless word about the past can cause public outcry inside the country, provoke an inter-state conflict or cause the renewal of an old conflict which does not exist any more in reality, that belongs to the past, but it still continues to exist in the historical space (so-called "wars of historical narratives"). A striking example of the latter is the uneasy history of Ukrainian-Polish relations in the middle of the last century ("Volyn massacre", the operation "Wisla"), which was full of mutual aggression, and which remains very problematic up to nowadays and requires attention of both historians and philosophers of history.

It should be emphasized that every country in the modern world has similar problematic historical pages in its history. As an example, we can take the history of the United States. Ukraine and the United States have much in common: both countries have a complicated history of struggle for their independence, both of which are multinational, multicultural and multireligious. This similarity is also reflected in the similarity of socio-political issues facing Ukraine and the United States nowadays. I mean a very controversial issue regarding the monuments to the Confederate Army generals, which causes serious social indignation and divides the American citizens into two groups: supporters who consider them as part of their history and opponents who emphasize that the monument to General Robert Lee and others are symbols of the race and social inequality, discrimination and, therefore, should be demolished. A similar situation occurs in Ukraine, it concerns the numerous monuments to the Soviet Union political leaders and military generals who were involved in the capture and enslavement of Ukraine, the repression of the Ukrainian people and the organization of Holodomor of 1932-1933.

These complicated situations clearly show us the close connection between history (its various versions) and the formation of our national, cultural and political identity. Thus, cultural and religious differences of various groups of the population and conflictual issues of the past – all this requires constant rethinking from the philosophy of history. This is extremely important for Ukraine as well as for the rest of the world.

I formulate the notion of narrative tolerance as the basic structure of cultural tolerance and civil society. Narrative tolerance as the basic structure of cultural tolerance has an extensive foundational importance for supporting and developing the concept of modern democratic civil society in the information-oriented globalizing world. Manipulating knowledge of the past and historical consciousness is a real threat to the future of the modern world, since it prevents the existence of a free peaceful society.

It is commonly known that fundamental principles of a democratic rule-of-law state are guaranteed by the force of law. In this context it is of crucial importance to analyze if the narrative theory of historical knowledge can also be a ground for the development of democracy and sociocultural (historical and religious) reconciliation. I am convinced that the modern analytic narrative philosophy of history possesses extraordinary methodological possibilities for successful resolution and prevention of protracted political and sociocultural conflicts, as it is capable of providing us with a whole system of effective conceptual tools for working with historical information in the conditions of modern hybrid wars.

The notion of narrative tolerance is one of such conceptual methodological tools. In my opinion,

* Corresponding author: elenmishalova@gmail.com

narrative tolerance is the basic structure of cultural tolerance and civil society. Considered as a principle, narrative tolerance is a strategy for conducting historical research which presupposes two imperative steps: first, the refusal to impose its own national historical narrative on other countries and, secondly, the gradual renunciation of the national narrative and the replacement of it with a regional one or transnational one or a narrative of cultural convergence (for example, interpretation of the Ukrainian history as part of Eastern Europe, the united Europe or the Western democratic world). Considered as a framework, narrative tolerance is a way of writing a historical narrative, which in principle supposes at least two points of view on a particular sequence of historical events. It also presupposes avoiding the following expressions: "old enemy", "endless enmity", "glorious victory," etc. in the practice of historical writing, as they act as triggers to restore past conflicts, adjusting to hostility and militancy [1, p. 52]. While the modern world is extremely in need of a willingness to understand "the other", openness to dialogue and compromise, the spirit of solidarity.

2 The role of narrative tolerance in history studying

The contemporary narrative theory of historical knowledge arose out of fruitful searches, heated debates and even disputes between leading American philosophers of history. In 1965 Arthur Danto published his important book *Analytical philosophy of history* which summarized a number of still relevant problem issues and contradictions of historical knowledge that can be called "paradoxes of historical knowledge". The most famous of them, for example, are the statement about the impossibility of an uncontroversial and complete historical description, because any description of the past is essentially incomplete, and the paradox of the Ideal Chronicle that can only be an invention or an abstraction. "It is not even an ideal for history to achieve, and that *in the nature of the case* historians are obliged to aim, not at reproduction but at a kind of organization of the past. … We cannot conceive of history without organizational schemes, nor of historically organizing schemes apart from specific human interests. … The difference between history and science is not that history does and science does not employ organizing schemes which go beyond what is given. Both do. The difference has to do with the *kind* of organizing schemes employed by each" [2, p. 111].

Danto emphasizes that historians use narratives. A narrative as itself is a way of organizing things which describes and explains at the same time. Therefore, a narrative "goes beyond" what is given and involves historians in something one might call "giving an interpretation". This shows us that according to Danto, the maximally detailed account – an ideal duplicate of history-as-actuality – would not be a narrative [3, p. 136, 138]. To tell a story means to exclude some events, a story must leave some things out. That is why any narrative is a structure imposed upon events, grouping some of them together with others and ruling some of them out as lacking relevance [3, p. 20, 132]. "Accounts … must by their nature leave things out, and in history as elsewhere it is the mark of someone capable of organizing a subject that he knows what to exclude, and is able to assert that some things are more important than others" [2, p. 131]

Danto maintains that our method of organizing past happenings changes, that is, the historian selects various events and constructs his narrative in the other way. This organization will be influenced by what is revealed in retrospect about what has happened. The same event may be a part of unlimited number of historical narratives [3, p. 162]. The reason is that we do not explain events as themselves, but the one that are given specific retrospective descriptions. The choice of description is important in equal measure in both history and science [3, p. 237].

Here we come across another Danto's paradox of historical knowledge: a historical fact is an event under a description. The key structure element for providing such a description is the construction of narrative sentences. Danto proposes to consider narrative sentences as the most distinctive feature of historical knowledge. A narrative sentence refers to at least two time-separated events but it only describes the earliest event to which it refers. He adds further that it logically requires, if it is to be true, the occurrence of *both* events [3, p. 139, 160].

Narrative sentences express our temporal perception of the world. In this context, we should elucidate one more feature of history that is usually underestimated. Historians have a unique privilege of seeing actions in temporal perspective. This means that historians have an advantage which the actor or his contemporaries could not have had in principle. According to Danto, it would be incorrect to complain that we, being at a temporal remove from the actions which concern us as historians, cannot know them in the way in which witnesses might have. "For the whole point of history is *not* to know actions as witnesses might, but as historians do, in connection with later events and as parts of temporal wholes. To wish away this singular advantage would be silly, and historically disastrous, as well as unfulfillable. It would, in analogy to Plato's image, be a wish to re-enter the cave where the future is still opaque. Men would give a great deal to be able to see their actions through the eyes of historians to come" [2, p. 183].

One of the most interesting Danto's conclusions is that we do not see our own actions, at the time we perform them, as having the significance we will later attach to them in the light of further events to which they are to be related. But this turns out to be a general rule for the historical organization of events - they are continually being re-described and their significance re-evaluated in the light of later information. And because historians have this information, historians can say things that witnesses and contemporaries could not justifiably have said [3, p. 20]. In this regard, Lidia Goehr remarks that although like present historians, future historians may elicit from a past event a significance of which past historians were not and could not have been aware, no claim by present historians should be made as if they were speaking on behalf of future historians as well. Taking into account Danto's idea of incompleteness of the past, she says about narratival openness which means that rather than being content with the material at hand, historians must wait to see how future events in the world will change their contemporary narratives of the past [4, p. 7, 28].

Danto also argues that the unobservability of the object of study, which until now is considered as the essential weak point in history, is a general feature of scientific research in all advanced sciences. The scientist, just like the historian, very often does not have access to a direct observation of the subject of his study, for example, it concerns electrons or genes. "It is perhaps a logical truth that we cannot observe the past, and merely a matter of contingent fact that we cannot observe electrons or genes. But that the grounds should be logical in the one case and factual in the other does not entail any difference in current *practice*. There may be branches of science in which the subject-matter is observable: certain parts of chemistry, of zoology, and geology, for instance, come to mind. But these are apt to be very elementary sciences in contrast with, say, atomic physics, and the fact that the most highly developed sciences are concern specifically with unobservable shows, I think, that the unobservability of subject-matter is not an overwhelming disadvantage for science, or that access to subject-matter is an overwhelming advantage" [2, p. 95]. So, dealing with unobservable objects, he is forced to develop theories and some conceptual tools. This circumstance does not pose any special problems for the history, moreover, precisely because we do not have a direct access to the past, there is a history as a science [2, p. 95].

This idea about the narrative as a sort of organizing theory was taken up by Hayden White. In his seminal work *Metahistory: The Historical Imagination in the Nineteenth-Century Europe* he stresses that the historical narrative is a verbal structure in the form of a narrative prose discourse which plays role of a sign (a model) of past events and processes in the interest of "explaining what they were by representing them" [5, p. 2]. Historical texts (narratives) are all what historians can study with the direct observation. This puts White close to the further deep analysis of the structure of the historical narrative. He proposes to view the historical narrative as a puzzle consisting of facts united through narrativizing strategies. These strategies White understands as having distinct parts. There is a formal argumentation, a plot construction and an explanation by means of ideological implication. Each of them presupposes four possible variations. As a result, we actually have twelve categories of historical consciousness [5, p. 7]. A specific combination of them forms one or another type of historical narrative. Consequently, historical knowledge is the second-order knowledge, it is based on hypothetical constructions of possible objects of research that require interpretation by means of narrative forms.

It should be stressed that White was the first one who clearly realized the incredible danger of such a situation and the consequences that can result from the manipulations with historical information and consciousness in general, especially in the era of the information revolution.

At the same time, White makes a clear division between the event (the state of affairs happened in the particular time and space) and the fact (the statement about this state of affairs). Events occur and more or less adequately reflected in historical texts, facts are conceptually constructed in the thought and/or figuratively in the imagination, that is, exist only in the thought, language and discourse. In *The Content of the Form: Narrative Discourse and Historical Representation* White adds that these events are real not because they occurred but because they were remembered and are capable of finding a place in a chronologically ordered sequence. In order to be qualified as historical, an event must be susceptible to at least two narrations of its occurrence. Unless at least two versions of the same set of events can be imagined, there is no reason for a historian to take upon himself the authority of giving the true account of what really happened. The authority of the historical narrative is the authority of reality itself. The history, then, sums up White, belongs to the category of what can be called "the discourse of the real" [6, p. 20].

These main ideas of Hayden White caused a lively discussion in which (directly or indirectly) many distinguished scholars from all over the world took part, and this discussion is still going on. White's idea of explanatory strategies aroused a general interest because of its deep connection to some fundamental philosophical concepts, above all the concepts of truth and objectivity of historical knowledge.

Following the constructivist position of White, Leon Goldstein in his famous *Historical Knowing* emphasizes that there are two realities of the past: the real past which exists outside the totality of historical texts and direct experience of historians and the historical past (the totality of historical texts or narratives) which is created by historians [7, p. 136]. He also insists that the historical fact is not found in the past because a fact is a statement about the past formulated by the historian on the basis of his research. "Historical facts are the products, or outcome, of historical research. … We have simply no approach to the human past other than by means of historical research" [7, p. 89]. As well as Thomas Kuhn, Hilary Putnam and Karl Popper, Goldstein underlines the role of prior theoretical beliefs. The historian's description is true because it is reasonable to believe that some part of the past had such characteristics, and not because it corresponds to an actual event as a witness may have observed it [7, p. 211], but it does not mean the subjectivity of historians as historians constitute the historical past (this is the core point of their work). Historical knowledge, Goldstein emphasizes, is relative to the discipline of history in the same way that any sort of knowledge is relative to the disciplined way in which it is produced. According to the methodological view, historical objectivity is an intersubjective agreement between historians [7, p. 213].

It is important to note that White, using Michael Oakeshott's distinction between "the historical past" and "the practical past", demonstrates that it is very useful for distinguishing between modern professional historians' approaches to the study of the past and the ways in which lay persons (not historians) and practitioners of other disciplines call on, recall or seek to use "the past" with a variety of purposes, for instance, as a basis for all kinds of judgments and decisions in daily life. The kinds of purposes is the key difference between the historical past and the practical past. The historical past is that past which can be studied scientifically, disinterestedly, as an end in itself and for "its own sake". It is a past constructed by historians, its authenticity (but not its reality) is guaranteed by other professional historians. The practical past is made up of all those memories, illusions, bits of vagrant information, attitudes and values which the individual or the group summons up as best they can do to justify, dignify or excuse for actions to be taken in the prosecution of a life project. The practical past is a version of the past that most of us keep in mind and draw on in the performing to our daily task where we compelled to judge situations, make decisions and solve problems [8, p. 16]. White stresses that these two kinds of past are rather more ideal types than descriptions of actual points of view or ideologies. It becomes clear especially if we recall the fact that professional historiography was set up in the universities (in the early nineteenth century) to serve the interests of the nation-state, to help in the work of creating national identities, and was used in the training of educators, politicians, imperial administrators and both political and religious ideologues in manifestly practical ways [8, p. 17]. As a result White concludes that the practical past needs to be the subject of consideration of the philosophy of history which could bring the order and reason more than the historical past [8, p. 18]. One cannot disagree with this White's position, but nevertheless, it should be remarked that the analytic narrative philosophy of history (more particularly, a notion of narrative tolerance) can successfully be involved in work on both the professional historiography and the ordinary historical information.

Based on works by Danto, White and Goldstein, Paul Roth develops an interesting and fruitful irrealist approach (an irrealist account of history). According to his *The Pasts*, we can speak about a plurality of pasts which exists because constituting a past always depends to some degree on socially mediated negotiations of what could or could not happened. Roth argues for pasts as constructed (not found) and stresses the priority of classification over perception in the order to understand and verify historical knowledge [9, p. 339]. He stresses that only descriptions create a past in which human actions have meaning. Of course, what has been done cannot be undone. But "that will turn out to depend on what one takes a "doing" to be. If what happens in the world is at least in part a function of human actions, and if what actions are Goodmanian kinds, that is, exemplifications of ways a given community descriptively collates behaviors in particular ways, then when new description, new ways of collating physical doings, become available, this changes what actions happened" [9, p. 333].

Roth notes that nothing *a priori* anchors practices of classification, therefore, no sense can be attached to claims that some single structure must or does determine what events take place in human history. Given alternative modes for structuring what happens, changes in description can alter relations among events imputed to a past, and so how a past thus structured impacts what becomes possible going forward. "A plurality of pasts results because constituting a past depends to some degree on socially mediated negotiations of a fit between descriptions and experience. Even what we take to mark what can change and what cannot itself depends on the descriptions deployed. Unless for reasons now unknown there ceases to be a possibility of descriptive change or reclassification, human histories will continue to reveal a multiplicity of pasts" [9, p. 339].

There is no principal distinction between empirical knowledge in general and historical knowledge at the most general epistemological and metaphysical level. This point suggests that the forms of inference (inductive, deductive and abductive) arise for all cases of knowledge. Such things as events, facts, actions, kinds exist and have explanation only in a theory. The temporal distance may accentuate some problems of making sense of others and what they did, but the problems posed turn out not to be unique at all [9, p. 336].

So, the narrative theory of historical knowledge can provide us with a whole system of methodological tools to study conflicting historical past in order to establish inclusive unbiased vision of history. Narrative tolerance has an extraordinary significance in constituting such an unbiased tolerant paradigm of history.

3 The role of narrative tolerance in history teaching

The investigation of narrative tolerance gains extra significance in view of developing modern information civil society in the present-day world. To build it we need tolerant historical thinking as well as critical thinking.

According to Gabrielle Spiegel, the greatest issue facing the practice of history today is to understand its relationship to ethical goals [10, p. 505]. Its means an ethical response to the catastrophes of the last century and, in a more general sense, a turn from epistemological to ethical commitments in the study of the past, creating a place and an opportunity for a new historical ethics. Spiegel underlines that this ethical approach to history cannot and should not mean the abandonment of the search for evidence, the responsibility to seek to "get it right" in our investigation of the past, or the insistence on a critical approach to knowledge in all its manifest forms as the fundamental practice of the historian [11, p. 177].

In the middle of the last century Karl Popper in his famous work *The Open Society and its Enemies* drew attention to our ordinary mode of history writing as mainly political history. From his point of view, there is no history of mankind, there are only many histories of all kinds of aspects of human life. And one of these is the history of political power. This is elevated into the history of the world. But this, as Popper stresses, is an offence against every decent conception of mankind. It is hardly better than to treat the history of embezzlement or of robbery or of poisoning as the history of mankind. "For the history of power politics is nothing but the history of international crime and mass murder (including, it is true, some of the attempts to suppress them)" [12, p. 257]. Popper rightly accentuates the fact that this history is taught in schools, and many of the greatest criminals are presented as heroes. The main reason of such a situation is that many historians as a rule wrote their texts under the supervision of the authority ("the generals and the dictators") [12, p. 257].

That is why Bertrand Russell especially defeats a feeling of intellectual independence of teachers in his essay *The function of a Teacher*. It is essential to the proper fulfillment of the teacher's functions, since it is his business to instill what he can of knowledge and reasonableness into the process of forming public opinion [13, p. 113].

Russell puts the point that the thing that a teacher should endeavor above all to produce in his students if democracy is to survive, is the kind of tolerance that springs from an endeavor to understand those who are different from us. Russell believes that it is a natural human impulse to view with horror and disgust all manners and customs different from those to which we are used. And those who have never travelled either physically or mentally find it difficult to tolerate the queer ways and outlandish beliefs of other nations and other times. This kind of ignorant intolerance is the antithesis of a civilized outlook, and is one of the gravest dangers to which our overcrowded world is exposed. The educational system ought to be designed to correct it, but much too little is done in this direction at present. In every country nationalistic feeling is encouraged, and school children are taught in such a manner that the inhabitants of other countries are morally and intellectually inferior to those of the country in which the school children happen to reside and they are absolutely ready to believe it [13, p. 121].

Russell stresses that in every country history is so taught as to magnify that country: children learn to believe that their own country has always been in the right and almost always victorious, that it has produced almost all the great men, and that it is in all respects superior to all other countries. He illustrates this point with a simple example: the facts about the battle of Waterloo which are known in great detail and with minute accuracy. But the facts as taught in schools will be widely different in England, France, and Germany. He notes in *Education* that "the ordinary English boy imagines that the Prussians played hardly any part; the ordinary German boy imagines that Wellington was practically defeated when the day was retrieved by Blucher's gallantry. If the facts were taught accurately in both countries, national pride would not be fostered to the same extent, neither nation would feel so certain of victory in the event of war, and the willingness to fight would be diminished. It is this result which has to be prevented. Every state wishes to promote national pride, and is conscious that this cannot be done by unbiased history. The defenseless children are taught by distortion and suppressions and suggestions" [14, p. 382].

Russell adds further that there is a widespread belief that nations are made strong by uniformity of opinion and by the suppression of liberty. But nations have been brought to ruin much more often by insistence upon a narrow-minded doctrinal uniformity than by free discussion and the toleration of divergent opinions. "It is obvious that organized party spirit is one of the greatest dangers of our time. In the form of nationalism it leads to wars between nations, and in other forms it leads to civil wars. It should be the business of teachers to stand outside the strife of parties and endeavor to instill into young the habit of impartial inquiry, leading them to judge issues on their merits. ... The function of the teacher, however, is not merely to mitigate the heat of current controversies. He has more positive task to perform. Teachers are more than any other class the guardians of civilization" [13, p. 117].

The false ideas as to the history of the world which are taught in various countries are of a kind which encourages strife and serves to keep a bigoted nationalism alive. If good relations between States were desired, one of the first steps ought to be to submit all teaching of history to an international commission, which should produce neutral textbooks free from the patriotic bias which is now demanded everywhere [14, p. 382].

This circumstance requires narrative tolerance which rests upon the idea of philosophical pluralism. Advancing the narrative tolerance as well as the narrative theory of history in philosophy of history and social philosophy provides us with effective methodological tools for working with historical information in the conditions of conflicting historical past, divided societies and modern hybrid wars.

Eva Domanska finds Ethics as the practice of care emphasizing that Ethics is of key importance for historiographical criticism and is not understood as a set of abstract rules but as the product of the student's own everyday experience. Historiographical criticism combines a close reading of texts with ethical reflection which reaches beyond the text [15, p. 200].

We should remember here an interesting thesis of Russell's: "I believe that the control of our acts by our intelligence is ultimately what is of most importance, and what alone will make social life remain possible as science increases the means at our disposal for injuring each other. Education, the press, politics, religion – in a world, all the great forces in the world – are at present on the side of irrationality; they are in the hands of men who flatter King Demos in order to lead him astray. The remedy does not lie in anything heroically cataclysmic, but in the efforts of individuals toward a more sane and balanced view of our relations to our neighbors and to the world. It is to intelligence, increasingly wide-spread, that we must for the solution of the ills from which our world is suffering" [16, p. 39].

Domanska emphasizes that we must address the question of the limits or the frameworks of interpretation, the question of the "ethics of analysis". The idea of multiple meanings of texts and of the impossibility of a definitive interpretation does not mean that every interpretation is equally acceptable. The student's decision to use a given theory, interpretative category, or way of thinking is related to his/her world-view, existential situation, and the choice of intellectual tradition that corresponds to his/her view of the world and the human being. On the deep level, therefore, the choice of method is an existential and ethical choice. At the same time, Domanska notes that the problem of the "ethics of analysis" and the limits of representation is an aporia, touching upon the dilemma of being situated between the freedom of choice and the abstract ideals or accepted values which impose constraints on that freedom [15, p. 201].

White develops a thesis that every historical narrative has as its latent or manifest purpose the desire to moralize the events of which it treats. He suggests that narrativity certainly is related (if not a function of) the impulse to moralize reality, that is, to identify it with the social system that is the source of any morality that we can imagine. White defines the social system as a system of human relationship governed by law, creates the possibility of conceiving the kinds of tension, conflicts, struggles, and their various kinds of resolutions that we are accustomed to find in any representation of reality presenting itself to us as a history [6, p. 14].

As Keith Jenkins shows, history is one of a series of discourse about the world. These discourses do not create the world (that physical stuff on which we apparently live) but they do appropriate it and all the meanings it has. More precisely, history is a discourse about the past but categorically different from the past. History and the past are not stitched into each other so that only one historical reading of the past is absolutely necessary. The past and history float free of each other, they are ages and miles apart. For the same object of enquiry can be read differently by different discursive practices [17, p. 7]. "No discourse – and therefore no contribution to, and/or comment on, aspects of an existing discourse – is of 'a natural kind'. You cannot find a historical or geographical or scientific or literary discourse just out there, just growing wild. Discourses are cultural, cultivated, fabricated and thus ultimately arbitrary, ways of caving up what comes to constitute their 'field" [18, p. 15].

That is why it is crucially important to develop and promote the principle of narrative tolerance in history teaching. Because there can be no history of the past as "it actually happened". There can only be historical interpretations, and none of them final; and every generation has a right to frame its own. But not only has it a right to frame its own interpretations, but it also has a kind of obligation to do so [12, p. 255].

I want to sum up this essay with White's vision about the role of History (as a science) and historians in the present world. History can provide a ground upon which we can seek that "impossible transparency" demanded for the distracted humanity of our time. Only a chaste historical consciousness can truly challenge the world anew every second, for only history mediates between what is and what men think ought to be with truly humanizing effect [19, p. 134]. "The burden of the historian in our time is to re-establish the dignity of historical studies on a basis that will make them consonant with the aims and purposes of the intellectual community at large, that is, transform historical studies in such a way as to allow the historian to participate positively in the liberation of the present from the burden of history" [19, p. 124].

References

1. O. Mishalova, Naratyvna tolerantnist: vykhid iz labiryntu natsionalnykh istorychnykh naratyviv (Narrative Tolerance: Overcoming the Maze of National Historical Narratives). Actual Problem of Mind. **20**, 46–71 (2019)
2. A. Danto, *Narration and Knowledge* (Columbia University Press, New York, 2007)
3. A. Danto, *Analiticheskaya filosofiya istorii* (Analytical Philosophy of History). (Ideya-Press, Moscow, 2002)
4. L. Goehr, Afterwords: an introduction to Arthur Danto's philosophies of history and art. History and Theory **46**, 1–28 (2007)
5. H. White, *Metahistory: The Historical Imagination in the Nineteenth-Century Europe* (The Johns Hopkins University Press, Baltimore and London, 1975)
6. H. White, *The Content of the Form: Narrative Discourse and Historical Representation* (The Johns Hopkins University Press, Baltimore and London, 1987)
7. L. Goldstein, *Historical Knowing* (University of Texas Press, Austin and London, 1976)
8. H. White, The practical past. Historein **10**, 10–19 (2010)
9. P.A. Roth, The Pasts. History and Theory **51**(3), 313–339 (2012)
10. G.M. Spiegel, Above, about and beyond the writing of history: a retrospective view of Hayden White's Metahistory on the 40th anniversary of its publication. Rethinking History **17**(4), 492–508 (2013)
11. G.M. Spiegel, The future of the past: History, memory and the ethical imperatives of writing history. Journal of the Philosophy of History **8**, 149–179 (2014)
12. K. Popper, *The Open Society and its Enemies*, Volume II. The High Tide of Prophecy: Hegel, Marx and the Aftermath (George Routledge and Sons, London, 1947)
13. B. Russell, *Unpopular Essays* (Simon and Schuster, Wolf H. Book Manufacturing Company, New York, 1950)

14. R.E. Egner, L.E. Denonn (ed.), *The Basic Writings of Bertrand Russell* (Routledge Classics, London and New York, 2009)
15. E. Domanska, in *Manifestos for History*, ed. by K. Jenkin, S. Morgan, A. Munslow (Taylor and Francis Group, London and New York, 2007), pp. 197–203
16. B. Russell, *Sceptical Essays* (Routledge Classics, London and New York, 2004)
17. K. Jenkins, *Re-thinking History* (Routledge: Taylor and Francis, London and New York, 2004)
18. K. Jenkins, *On "What is History?": From Carr and Elton to Rorty and White* (Routledge: Taylor and Francis, London and New York, 2005)
19. H. White, The Burden of History. History and Theory **5**(2), 111–134 (1966)

The learner-centred paradigm of education: its features and philosophical basis

Olena Olifer*

Kryvyi Rih State Pedagogical University, Department of Philosophy, Kryvyi Rih, 50086, Ukraine

Abstract. Nowadays, educators notice swift changes in education as a social institution. One possible way to describe and analyze its transformations is T. Kuhn's paradigmatic approach applied to education. The paper states that the contemporary educators work within the learner-centred paradigm. However, it may seem that education is still influenced by the previous paradigm. So, the paper has two main objectives. The first one is to consider the two educational paradigms: the teacher-centred and the learner-centred to show their differences. Secondly, it is essential to know what concepts are topical in contemporary philosophy, and if they coincide with the learner-centred paradigm. The author attempts to distinguish philosophical concepts that have a considerable impact on the worldview at present. They are the following: four-dimensional ontology, constructivist epistemology, the problem of agency, personal identity, and the narrative approach to the questions of identity and cognition. These concepts are explained and analyzed through the lenses of the learner-centred paradigm.

1 Introduction

At present, the humankind experiences dramatic changes caused by technological development. As a result, one faces transformation in all spheres and social institutions. Y.N. Harari states that rapidly developing technologies cause disorientation and anxiety [1]. The system of education is not an exception. The changes happening there are so fast that it is not able to adapt to them.

K. Robinson remarks that as a result of economic and technological development, we face the situation of academic deficiency, which has two important outcomes. Firstly, due to economic transformation, the amount of workplaces decreases. As a result, more graduating students apply for a job, for which they are overqualified. Accordingly, a degree is not a guarantee that a student will find a job. Moreover, it leads to an increase in educational standards and pupils and students' loss of motivation to learn. Secondly, employers are disappointed with the content of academic programmes. It is necessary for them to find an employee who is flexible, adaptable, and communicative. Academic programmes are not designed for developing those personality traits and competencies [2].

According to E. Blass, the system of education contradicts the employers' requirements. Behaviour, considered as cheating at school, is often treated as a good strategy at work [3, p. 130]. J.W. Cook says that the impact of technological development is considerable, and as a result, the purpose of education becomes unclear at present. That is why the school system should be redesigned [4, p. 3].

Thus, there is a necessity to reconsider the development of education. One of the possible approaches to achieve this task is the paradigmatic approach introduced by T. Kuhn in the book „The Structure of Scientific Revolutions". He defined the concept of a paradigm as "accepted examples of actual scientific practice – examples which include law, theory, application, and instrumentation together" as the paradigm provides "coherent traditions of research" [5, p. 10]. Though T. Kuhn applied the concept of paradigm exclusively to describe the development of science as a social institution, the researchers think that the concept can be used in other domains. For example, S. Crowell says that the paradigmatic approach can be applied to culture and education, as well as to science. Such educational paradigm is regarded as a dynamic unity [6, p. 61]. In "A Companion to the Philosophy of Education", the word "paradigm" is used to describe a philosophical concept of education in a certain period. Therefore, there can be distinguished the Aristotelian paradigm, Neohumanists' paradigm, the classical Marxian paradigm, the feminist paradigm, and others [7]. S. Loomis and J. Rodriguez use the notion of "paradigm" as a certain way to organize education as a social institution that is treated as a shape of human interaction that "evolves from prior philosophical commitments" [8, p. 53].

Concerning paradigms in education, R. Barr and J. Tagg prefer the metaphor of a game. They write the following: "A paradigm is like the rules of a game: one of the functions of the rules is to define the playing field and domain of possibilities on that field. But a new paradigm may specify a game played on a larger or

* Corresponding author: xafany@gmail.com

smaller field with a larger or smaller domain of legitimate possibilities" [9, pp. 14-15]. In their opinion, a paradigm sets the rules and basic principles of people's activity.

According to W. Huitt, a paradigm is a mental representation or a shared worldview [10]. It is also the representation in a mind of an agent of "how an entity is structured (the parts and their interrelationships) and how it functions (behaviour within a specific context or time dimension)" [11]. However, W. Huitt admits that these two concepts are not synonyms: a worldview gives an overall image of reality, while a paradigm shows how the worldview is understood and investigated [10, p. 74]. Thus, a paradigm is a framework with a more precise presentation of factors and their relationship. W. Huitt distinguishes four main paradigms in science: mechanistic, existential or phenomenological, organismic or systems, and process paradigm [10, p. 69]. He admits that the paradigms of education should be based on the following frameworks:
1. Context or socio-cultural background;
2. Input or subjects' characteristics before learning and teaching;
3. The process including factors that describe teachers and pupils thinking, feeling, planning and other activities and behavioural patterns in learning;
4. Output including assessments and results of learning, compared with the aim of education [11].

D.F. Labaree does not use the word "paradigm" in his research. However, he describes long exiting systems of education as a hierarchy of such levels as the rhetorical level, the system's formal structure, teaching practice, and student learning. The rhetorical level means the public opinion towards schooling and the primary goal of education. This level is the easiest to influence. The second level of the system's formal structure concerns all elements in the school organization: educational policies, curriculum and the content of education, the classroom management, textbooks and manuals, the teachers' professional development. The next level is regarded as the core of the educational reforms. It concerns the questions about how a teacher works and hat strategies and methods he or she uses. D.F. Labaree admits that if teachers do not put the demands of the reform into the classroom practice, it faces the problems and eventually stops. The fourth level is the students' behavioural strategies, attitudes toward the process of education, and ideas of academic success or efficiency of the education systems [12, pp. 110-111].

Thus, the concept of educational paradigm can be defined as educational traditions shared and accepted by all educators at a certain time. These traditions include peculiarities of education as a social institution, the purpose of education, the organization of the school, including types of classroom management, the methods to teach and learn, the process of assessment, and values transmitted through the process of training and upbringing.

However, the most important component of an educational paradigm is philosophical concepts, concerning metaphysical and epistemological views, key ideas about personhood and personal development. Philosophical concepts shape the worldview in society. They are the basis of the educational paradigm as well. Therefore, the primary task of the philosophy of education is to analyze such concepts and observe their impact on education.

There are different ways to distinguish educational paradigms in the philosophy of education. On the one hand, the scholars distinguish analytic, transcendental (Kantian) pedagogy, critical rationality, phenomenology, existentialism, hermeneutic, critical-emancipatory and other paradigms [13, p. 10]. On the other hand, Kuhn's idea of scientific paradigm states that two or more well-developed paradigms cannot exist simultaneously at a certain time. A paradigm exists within the stage of normal science followed by the scientific revolution and paradigm shift. Thus, phenomenology, existentialism, hermeneutic should be treated as approaches within one paradigm, but not as separate paradigms since they exist simultaneously. There is no significant change in the organization of education as a social institution. Moreover, education as a social institution cannot be separated from the socio-cultural and socio-economical context. Therefore, each paradigm should represent the socio-economic type of society. A. Collins and R. Halverson admit that the industrial revolution created universal schooling as a new education system different from the apprenticeship, and the digital revolution is transforming education now [14].

Taking these factors into account, the researchers distinguish two main paradigms of education: the learner-centred and the teacher-centred. It is necessary to admit that the educators may call them in different ways. For example, J. Dewey, the founder of the philosophy of education, distinguishes two paradigms of education: traditional and progressive [15, p. 17]. M. Lipman speaks about reflexive education, based on the development of critical, caring, and creative thinking and traditional (non-reflective) education [16]. P. Freire considers two ways to characterize the relationship between the teacher and the pupil: through banking education, leading to the oppression of the pupils, and through dialogic education, which is considered as "the practice of freedom" [17, p. 81]. W. Huitt distinguishes two paradigms: based on instructivism (teacher-centred) and based on constructivism (learner-centred) [11]. G. Jacobs speaks about a new person-centred paradigm and the previous one which remained unnamed [18]. S.L. Watson and C.M. Reigeluth call these two paradigms: sorting-based and learning-based [19]. R. Pring analyses subject-centred and child-centred educational paradigms [20]. R. Barr and J. Tagg outline the instruction paradigm and the learning paradigm [9].

We can say that the teacher-centred paradigm predominated before the twentieth century, and the learner-centred paradigm has been developing since that time. R. Pring states that the beginning of learner-centred paradigm is associated with the philosophy of pragmatism, developed by C.S. Peirce and W. James, and especially with J. Dewey's views on education, outlined in "Democracy and Education" and "Experience and Education" [20, p. 81]. Since that time, the learner-centred paradigm has been developing.

However, there is an opposite view. V. Terziev thinks that the teacher-centred paradigm continued throughout the twentieth century, whereas the learner-centred paradigm has formed in the twenty-first century [21, p. 83]. J.C. Bodinet also writes that the changes in education happen slower than it is expected, so we are still within the education system created for the needs of the eighteenth and nineteenth centuries [22]. M. Kochetkov admits that the new paradigm is still not formed, especially in Russian higher educational establishments, and there are views that the new paradigm will deteriorate the situation and even led to the decrease of graduates' competence level [23, p. 37].

In out opinion, the educators work within the learner-centred paradigm. However, this paradigm is still forming. The paper has two main objectives. The first one is to characterize the teacher-centred and the learner-centred paradigms to show their peculiarities and differences. The second objective is to outline the philosophical principles, on which the learner-centred paradigm is based.

2 The teacher-centred paradigm of education

The teacher-centred paradigm is well described in P. Freire's book "The Pedagogy of the Oppressed", in which he develops his "banking" concept of education. According to P. Freire, the main characteristic of the paradigm mentioned above is that the teacher is considered to be an active subject, who transmits information in narration. The teacher is often the only source of information, and the teacher's main task is to "fill" the pupils' consciousness with ready-made answers to the worldview questions [17, p. 71]. The pupils are regarded as passive objects. They listen to the teacher's instructions carefully and memorize the information. In the teacher-centred paradigm, the teacher does not demand the pupils to understand the information because it will be enough for them to learn it mechanically and reproduce it on tests. The main method of learning is to drill. The pupils' primary task is to obey, but not to think critically [17, pp. 73-74].

In the teacher-centred paradigm, the teacher uses narration as the main method to present information. The teacher tells different facts about reality that are collected and systematized by the pupils. As a result, the pupils treat reality as unchanged, static, and predictable. That is why the content of the provided information is detached from reality, and its meaning and significance are lost. Therefore, P. Freire states: "Words are emptied of their concreteness and become a hollow, alienated, and alienating verbosity" [17, p. 71].

In P. Freire's concept, education is viewed as "an act of depositing", where the teacher is a depositor and the pupils are depositories [17, p. 72]. The teacher transmits knowledge to the pupils, and this procedure is the same as to deposit money in a bank. The pupils collect knowledge like the bank stores the deposit. If the pupils are hard-working and collect knowledge well, they can get "interests" on deposits. It means that they cumulate transmitted knowledge and add to it some own conclusions. Thus, in the banking concept of education, the best teacher is someone, who "fills the receptacles" more completely, and the best pupil is someone, who "permits to be filled" [17, p. 72].

P. Freire gives the following characteristics of the society of oppression and the teacher-centred paradigm:

"(a) the teacher teaches and the students are taught;

(b) the teacher knows everything and the students know nothing;

(c) the teacher thinks and the students are thought about;

(d) the teacher talks and the students listen - meekly;

(e) the teacher disciplines and the students are disciplined;

(f) the teacher chooses and enforces his choice, and the students comply;

(g) the teacher acts and the students have the illusion of acting through the action of the teacher;

(h) the teacher chooses the program content, and the students (who were not consulted) adapt to it;

(i) the teacher confuses the authority of knowledge with his or her professional authority, which she and he sets in opposition to the freedom of the students;

(j) the teacher is the Subject of the learning process, while the pupils are mere Objects" [17, p. 73].

Concluding the discussion of the teacher-centred paradigm, P. Freire emphasizes that such educational process increases contradictions between educators and children, which results in the society of oppression, which is full of adults, devoid of creativity and critical thinking. Therefore, the teacher-centred paradigm is typical for non-democratic and revolutionary societies [17, p. 78].

3 The learner-centred paradigm of education

In the book "Experience and Education" J. Dewey outlined the principles of pragmatic education, which formed the basis of the teacher-centred paradigm. Firstly, the primary task of the school is to cultivate individuality and self-expression. Secondly, free activity should be instead of discipline imposed from above. Thirdly, learning is regarded as the acquisition of experience in activities, which prepare the pupil to life. Finally, J. Dewey emphasizes that the aim of education has not to be static. Educators are to take into consideration the changing world. Thus, the learner-centred paradigm is regarded as opposed to the previous one [15, pp. 19-20].

The first considerable change happened in the learning-centred paradigm concerns the purpose of education. According to J. Dewey, the main objective of education is to prepare schoolchildren "for future responsibilities and for success in life, by means of acquisition of the organized bodies of information and prepared forms of skill which comprehend the material of instruction" [15, p. 18]. In J. Dewey's opinion, education is regarded as preparation for work and executing the duties of a citizen. M. Lipman also treats the purpose of education as preparation for life since the pupils should become reasonable adults. However, he calls education "the great laboratory for rationality", so the purpose of education is not only to teach children to complete different tasks at work but also to teach them how to think rationally and prepare the schoolchildren for making decisions [16, p. 22].

According, K. Robinson for contemporary educational establishments are not only academic but also market-orientated, the purpose of education is developing a pupil's creativity, because being creative is a vital skill in the changeable world that helps to adapt to any situation, and it is always a demanded skill [2]. E Blass thinks that in the twenty-first century, the primary purpose of education remains the same. It means to prepare a pupil for life. However, it does not mean to survive. In his opinion, the purpose of education is to prepare pupils for the new changeable world. It is possible to achieve if the learners will have the habit of learning [3, p. 132]. Thus, the purpose of education is to teach pupils how strategies of learning.

J.W. Cook states that in the twenty-first century, education should be focused on sustainable wellbeing, which means having a successful and meaningful life [4, p. 14]. This idea can be realized in practice on different levels. On the individual level, sustainable wellbeing stands for physical and mental health, life-management skills, and inclusion into social groups. On the level of a community, this idea is put into actions in local communities where citizens share their vision of the future and cooperate. On the governmental level, sustainable wellbeing concerns the decentralization of authority and horizontal approaches to policy challenges. On the planetary level, sustainable wellbeing presupposes caring attitude towards nature and wise usage of the resources. The primary task of education is to transmit these values. Going to school, a child is not separated from real life, so the task of the teachers is to show how to communicate efficiently, how to be a responsible citizen, and how to solve real-life problems [4, p. 14].

As for the context of education, formal knowledge is replaced by practical activities in the learner-centred paradigm. P. Freire admits that education should be problem-posing because such education based on solving practical problems stimulates a person's creativity and desire to get better. According to the researcher, education and practice are inseparable as he states that "education is thus constantly remade in the praxis" [17, p. 84]. Moreover, it also highlights the importance of the principle of life-long learning. As J. Dewey stated: "The aim of education is to enable individuals to continue their education – or that the object and reward of learning is continued capacity for growth" [24, p. 63].

In the twenty-first century, life-long learning is one of the major trends in education. C.A. Chiţiba defines the life-long learning as "the continuous development and improvement of the knowledge and skills needed for employment and personal fulfilment through formal and informal learning opportunities" [25, p. 1944]. The researcher also distinguishes four principles of life-long learning: learning to know, learning to do, learning to live, and learning to be. The first principle means mastering learning tools but not getting structured knowledge. The second one stands for an adaptation of learning to future work. Learning to live is efficient and non-violent interaction with others. Finally, learning to be is based on the idea that learning should contribute to personal development [25, p. 1945].

The learner-centred paradigm proves that life-long learning is possible when learning is based on the pupils' activity. According to J. Dewey, learning in activity is the only way to get the necessary experience, which he defines as "a practical enterprise, made possible by knowledge of the recurrent and stable, of facts and laws" [26, p. iv]. The philosopher outlines two principles of experience: continuity and interaction [15, p. 44]. Without them, it is impossible to realize a person's activity. Even the first knowledge of how to walk and talk is an activity [15, p. 74]. The process of communication in society is also an activity. Therefore, learning by doing is a natural process for a human being. M. Lipman also believes that it is possible to educate successfully when education presupposes the practical application of getting knowledge in the problematical setting [16, p. 199].

However, the learning in activity is possible only when a pupil is motivated since coercion and pressure abrogate the significance of his or her actions. So the learning is based on the intrinsic factors: motivation, desires, and interests. That is why the traditional roles and behavioural patterns of the teacher and the pupil are changed. In the learner-centred paradigm, the teacher cannot be the only active participant in the process of education if the pupil remains passive. They are both active. That is why the teacher is viewed not as a mentor but as an assistant or an advisor. M. Lipman says that a teacher facilitator. Such a teacher is not authoritarian, but democratic [16, p. 18]. The teacher should guide pupils and encourage them to get new achievements, focuses on children's perspectives and helps to develop their inclinations. In the research conducted by A-M. Iversen, A.S. Pedersen, L. Krogh, and A.A. Jensen, it is emphasized that a teacher is a facilitator and a resource person, whereas pupils and students participate in collaborative teaching and learning. A teacher gives them control and responsibility over their education [27].

This shift in relations between the teacher and the pupil demands a change in methods of teaching and different organization of the classroom. In the learner-centred paradigm, learning is treated as dialogical, so the dialogue is one of the most important methods. P. Freire writes that dialogue is the only possible way to educate a human being. It is the essence of education and its "existential necessity" since action and reflection are united in the dialogue [17, p. 88]. With the help of dialogue, a person can learn how to think critically. M. Lipman compares dialogue to conversation. He states that dialogue is unstable, unbalanced, and collaborative, that is why the dialogue is thought provocative. M. Lipman defines dialogue as a mutual exploration or inquiry that leads to personal grows and the development of mental skills [16, pp. 87-88]. In the dialogue, the participants develop self-criticism and self-correction. Moreover, the pupils learn how to achieve practical results: how to make decisions and reach to the conclusion [16, p. 111]. A.-M. Iversen and A.S. Pedersen call this new method of teaching "co-creative generative dialogue" [28, p. 17]. It is interdisciplinary and focuses on innovation. The primary task of the method is creating new ideas together. The scholars emphasize that the co-creative generative dialogue turns into the team co-creative generative dialogue when it occurs in the classroom. This dialogue is based on non-violent communication, emphatic listening and conflict management [28, pp. 17-18]. J.C. Bodinet shares her experience of dialogical learning implementation in the classroom. She offers such methods as the visioning workshop, discussion and large circle discussions, concentric circles, small group work, Socratic inquiry, the world cafe method, and guest lectures [22].

In the learner-centred paradigm, one more important method and a way of classroom management is the project method implemented by J. Dewey and W. Kilpatrick. According to Dewey, when a teacher is aware of pupils' motivations, needs and past experiences, he or she can suggest a plan or project, which contributes and organizes students into the group. The plan or project is called "a co-operative enterprise", it is a suggestion, and it cannot be imposed from above [15, p. 72]. The teacher tries to manage the group in such a way that all pupils' experience should contribute to the development of the project. Then, the process of interaction or "the reciprocal give-and-take" occurs, and pupils learn through the development of social intelligence [15, p. 72].

Moreover, in the information age, the information increases greatly, so there is a necessity to develop a skill on how to differentiate reliable and unreliable informational sources. P. Freire writes that the present demands critical spirit and flexibility [29, p. 6]. According to M. Lipman, critical thinking is a trend in education [16, p. 44]. M. Black, J. McPeck, I. Scheffler, K. Popper, R. Nickerson, and many other prominent philosophers and educators published books, concerning the development of critical thinking. H. Siegel emphasizes that critical thinking is organized on rationality. He defined critical thinking as principled thinking, based on reasons. H. Siegel believes that to teach critical thinking means "to teach students how the reasons are assessed, what principles govern such reasons, and why (we think) such principles are to be adherent to" [30, p. 8]. Thus, critical thinking is vitally important not only to cognition but also to the ethics of education.

However, N. Kozachenko admits that the issue of critical thinking is so popular that there are many definitions and approaches to it [31, p. 165]. Critical thinking is often viewed as rational thinking based on logical laws. According to M. Lipman, critical thinking as formal logical thinking can be treated as "a disconnected, discontinuous fragment, shouldered with responsibility for upgrading the whole of education" [16, p. 6]. Such logical thinking is not enough for developing rationality. Therefore, M. Lipman believes that critical thinking should be reflective, which means not only critical but also creative and caring. It is developed through the dialogue between a teacher and a pupil and within the class [16, p. 84].

The development of critical thinking and the realization of the project method are not possible without fruitful and favourable school environment. In "Democracy and Education", J. Dewey states that every environment can play an educative role in upbringing. He writes: "We never educate directly, but indirectly by means of the environment" [24, p. 22]. Furthermore, the educator emphasizes that school is the environment of a special kind, as it has a constant and controllable educative influence on an immature person. School is a simplified environment because it has a clear purpose of its influence: to prepare a person for life and develop mental and moral faculties. School as a controllable environment can eliminate negative or "unworthy features" of society to "establish a purified medium of action" [24, p. 24].

Nevertheless, school as a special environment cannot achieve this task without the principle of democracy, lying on its basis. For school is a society in miniature, it should be based on democracy because democracy is "mode of associated living, of conjoint communicated experience" [24, p. 101]. Firstly, without democratic communication between educators and pupils vales and experience cannot be transmitted because dialogical education presupposes democracy. M. Lipman thinks that inquiry in the classroom is possible when "students listen to one another with respect, build on one another's ideas, challenge one another to supply reasons for otherwise unsupported opinions, assist each other in drawing inferences from what has been said, and seek to identify one another's assumptions" [16, p. 20]. So he admits that democracy and reasonableness are two principles or regulative ideas which lie in the basis of the contemporary paradigm of education. According to M. Lipman, democracy is crucial for the development of social structure [16, p. 204]. Secondly, for school resembles society, the organization of education based on democracy strengthens democratic institutions.

P. Freire states that increase of the general level of education and increase of the democratic level are interconnected: educated people cannot obey the authoritarian government, so the development of critical thinking causes the development of democracy [29, p. 32]. J. Dewey thinks that cultivating democracy at school brings up responsible citizens in all spheres and vice versa: the development of democratic institutions forms the request in society to educate a person on the principle of democracy [24].

In the twenty-first century, the principle of democracy should be in and out of the classroom. E. Blass mentions that nowadays the teacher makes an assessment at fixed time points, without paying attention to the pupil's readiness. That should be eliminated, because in the twenty-first century, "assessment is submitted when the learner feels ready" [3, p. 133]. J.W. Cook writes that today's school has to move towards complexity. It means shared responsibility between all school administration, teachers, pupils, their parents and members of a local community in governance [4, p. 17]. Obviously, shared responsibility is not possible without horizontal management based on democratic principles and mutual respect.

4 The philosophical ideas for the learner-centred paradigm

The learner-centred paradigm was influenced greatly by the philosophy of pragmatism. However, it is necessary to outline several important philosophical issues that are developing nowadays and shaping the worldview of contemporary generations. These philosophical ideas concern metaphysics, epistemology, and practical philosophy. They are viewed by many contemporary philosophers, especially in analytic philosophy.

Firstly, the learner-centred paradigm emphasizes that the society and the world are unstable and changeable, and they are in sustainable development, so the pupil should be ready for these changes. Therefore, it is important to understand what change is and how it occurs. To give an explanation, we should refer to metaphysics since this branch of philosophy describes abstract qualitative characteristics of being as they are the basis of a person's understanding of reality. In the twentieth century, analytic philosophy develops a new metaphysical theory, which tries to explain the existence of an object or a person both in space and over time. Thus, the new theory considers four dimensions of an object, so it is known as four-dimensional ontology. It was introduced by D. Lewis and it is developed by R. Nozick, H. Noonan, R. Chisholm, and other contemporary philosophers. Four-dimensional ontology describes a thing changes or remains the same over time.

D. Lewis defines four-dimensional ontology as perdurance of an object or a person over time. Perdurance is opposed to the ides of endurance at a certain time. Endurance theories, also known as three-dimensional ontology, regard an object as being wholly present at different periods. So it means that an object shows its features wholly at a time, there are no hidden qualities in an object. If a feature at one period of time is not the same as a feature at another time, it means that an object has changed. If an object perdures over time, it consists of temporal parts or stages, which D. Lewis defines in a similar way to spatial parts. Thus, the philosopher considers temporal parts as events and processes [32, pp. 202-203].

According to four-dimensional ontology, an object is not wholly present at a time, because at a certain time it exists only as a stage or an episode in its development. Therefore, wholly present objects exist as a sum of their temporal parts. In four-dimensional ontology, an object can have some properties at a certain period on time, but it cannot show them on another time, and it does not mean that the object has changed. It means that certain property belongs to one stage of an object, but it does not belong to the other one. Thus, changes over time are treated as more complicated. They are changes in the appearance of an object, but they are not changes in its nature [33, pp. 11-13]. K. Hawley shows the difference between three- and four-dimensional worldviews on the example of a banana. Let us imagine that a banana is green and then it becomes yellow the next day. According to three-dimensional ontology, a banana has changed its colour. However, in such a situation, four-dimensional ontology treats a banana as multicoloured, so it has not changed in its features, but it has changed some properties of how the features exist [33, p. 13].

For educators, such approach to understanding reality and changes means that a person is a complicated creature, who never wholly exists in all properties at a time, so at a time a person is considered as a person-stage – a stage in the development. Thus, it is impossible to make an opinion about a person once and for all. One should constantly reconsider their attitude to a person from time to time. On the other hand, when a person shows different properties at a time, it may not mean a change in a person, so more attention should be drawn to personal development. The idea of the world as complicated and developing over time highlights the importance of life-long learning. To realize an object as the sum of its temporal parts, a person needs to reconstruct it constantly in mind over time.

The second philosophical concept that is important for the learner-centred paradigm concerns epistemology. As A. Abdula notes, the contemporary philosophy shapes specific attitude towards the cognition. On the one hand, the researchers develop theories of cognition that define knowledge as the way to reveal the truth and to construct reality in a person's consciousness. On the other hand, there is a problem of revealing and justification of the degree of certainty that cognitive forms contain by constructing reality. The subjectivity loses its rational foundation and the reality is constructed on an irrational basis [34, p. 193]. So knowledge is regarded as something unique and subjective. It is constructed in the process of interaction in the community.

The modern epistemology highlights the importance of environment and interaction in the process of cognition. For example, in the book "The Social Construction of Reality" P. Berger and T. Luckmann emphasize that a person is world-open to the environment, and the environment constructs a person as an agent, that is an active self-conscious participant of the socio-cultural process. They admit: "Humanness is socio-culturally variable" [35, p. 67]. What we call "knowledge" is everything an agent learns in socialization and social interaction, so knowledge meditates the objective reality or social world and an agent's consciousness. Knowledge is realized in two senses: "in the sense of apprehending the objectivated social reality, and in the sense of ongoingly producing this reality" [35, p. 84]. For knowledge is defined as the socially and culturally determined process of reality production, it cannot be static or unchanged. Thus, the teacher in a learner-centred paradigm cannot keep using the "banking" concept of learning, describing the picture of the world in static facts.

For a person constructs the understanding of reality in the course of socialization, the question arises how personhood is constructed and how it is changed or stays the same over time. This question is about personal identity and agency. It lies between metaphysical and epistemological domains and unites the concept of four-dimensional ontology and socially determined knowledge acquisition in interaction. For educators, the problem of personal identity may be interesting as the construction of self and agency. In our opinion, these concepts are important because of two reasons. Firstly, a person has to have the same self over time to be capable of realizing the information and the self remains the same and unified when the person interacts with the social environment and develops cognitive faculties over time.

The concepts of personhood and an agent are interconnected. C. Korsgaard, M. Schechtman, C. Rovane state that a person is an agent. According to C. Korsgaard, a person is an active being. However, she admits that an action requires a doer or an agent to whom this action is ascribed and who is regarded as its author [36, p.18]. That is why a person is an agent. C. Rovane adds that a person wants to be an agent, and that is a universal goal of agency [37, p. 85]. She also admits that a person becomes an agent because he or she is involved in the agency-regarding relations, which are necessary and inevitable. In the agency-regarding relations, people mutually recognize themselves as agents and shape their attitude to others, taking into consideration their agency. Besides, being involved in agency-regarding relations a person has got motivation and a purpose to act [37, pp. 48-49].

C. Korsgaard emphasizes that a person as an agent must be unified. It means an agent should have one undivided self to be capable of ascribing actions. This unity of self should remain unchanged over time, so being a unified agent over time presupposes the concept of personal identity [36, p. 7]. E. Olson states that the problem of personal identity has several aspects. Firstly, there is the persistence question that means a person is the same over time and conditions of staying the same person. The second aspect is the problem of evidence, which means that deciding whether a person now is the one who was yesterday. The third issue is about the unity of consciousness that raises the question if it is possible to have a split personality and change the self over time. Finally, personal identity presupposes the question about personhood that is the conditions of being a person. In our opinion, the aspect that should be mentioned concerns the appearance of the unity of consciousness and how it is kept over time [38, pp. 352-355].

According to C. Korsgaard, personal identity is a result of actions and choices [36, p. 1]. While acting, a person can have different motivations. Sometimes, the motivations are mutually exclusive, so a person appears in the situation of inner conflict that threatens the unity of the self. For a person needs to eliminate the conflict, a person needs to choose, and a choice is done, according to the individual understanding of good. When a person makes a choice, he or she ascribes this intention to the self and becomes its author. The next action can lead to a new struggle of motivations. So a new choice should be made. Moreover, the new choice should correlate with the previous one not to deteriorate the struggle of motivations or not create a new one [39, p. 121]. C. Korsgaard also states that a person as an agent is involved in various social practices. The social roles a person plays in society are practical identity that can be defined as a principle, which influences choices in the inner conflict. Thus, a person constitutes the unity of self in actions [36, p. 20].

According to M. Schechtman, to be a person means to live the life of a person. Three elements make up a person's life: individual capacities, typical activities and interactions, and social infrastructure [40, p. 115]. The last component means belonging to various social institutions and the cultural determination of a person's lifestyle. All of these constitute different experiences united as self. M. Schechtman states that the unity of self remains the same over time as it is apprehended as a story or the narrative of a person's life [39, p. 112]. A. MacIntyre says: "The unity of human life is the unity of a narrative quest. Quests sometimes fail, are frustrated, abandoned or dissipated into distractions; and human lives may in all these ways also fail. But the only criteria for success or failure in human life as a whole are the criteria of success or failure in a narrated or to-be-narrated quest" [41, p. 219].

The narrative is considered as an autobiography, which unites different experience and makes life stages coherent. Considering life as narrative, a person can plan the future, because the future is a result of the present and past. So a person understands life as coherent, whole, and unique. For educators, the narrative approach of personal identity and agency is essential as it forms the basis of what is called "individual learning trajectory". Understanding life as a story, one can analyze strengths and weaknesses, inclinations and motivations to choose a path in education. When a person narrates a story of one's life, he or she understands different events better and is more responsible for choices. Such a person is more self-sufficient and prepared for the future. In the learner-centred paradigm, the teacher's task is to help the pupil become an efficient agent of social interaction. This task can be realized when a pupil learns to apprehend life as narrative.

Moreover, the concept of narrative is also applicable to the understanding of the nature of consciousness. K. Atkins says that thinking is narrative, and the process of cognition can be also described as the construction of a narration. The philosopher says that a person can ascribe the word "I" to herself it means to have the first-person perspective. When a person thinks what other people think about him or her, a person acquires the third-person perspective. The second-person perspective is an ability to interact with others. K. Atkins admits that the unity of consciousness is achieved when the first-, second- and the third-perspective unite in the whole self [42, p. 69]. This can be achieved in an inner dialogue while thinking [42, p. 141]. For the thinking is dialogical in nature, learning has to be dialogical too, then, it will become a natural activity for a person. Thus, the narrative approach to personhood justifies the idea of dialogical learning.

5 Conclusions

If one compares the educational principles introduced by J. Dewey with later ideas, one will see that they have not changed. However, when they were put into practice by the contemporary educators, they became more exact. Instead of preparation for life, educators speak about learning to learn. Such a slight change in wording does not stand for change in the essence of the educational process. The purpose of education formulated in the twenty-first century the only emphasizes J. Dewey's statement about the dynamic nature of education and the necessity for educators to take into consideration world changes and, perhaps, economic tendencies. In the twentieth century, many prominent educators emphasize that learning should ground on a pupil's activity and motivation. At present, it is not a piece of advice but a priority. J. Dewey, P. Freire and M. Lipman wrote about the importance of democracy at school. Nowadays, democracy is in teaching, learning and school administration. It is a key component of sustainable wellbeing, which is proclaimed to be the priority in education. Therefore, one can conclude that the learner-centred paradigm was introduced in the twentieth century. Nowadays, it is and improving actual teaching practices and in the process of school administration.

The paradigmatic approach suggested by T. Kuhn shows that it takes time to a paradigm to be formed. For changes in education take place quite slowly, it may take more time for the learner-centred paradigm to be settled down. Besides, the pace of its development is slower than the economic and technological changes. That is why education as a social institution faces a crisis at present. So, the learner-centred paradigm has to develop faster to catch-up with the world changes.

Any paradigm is shaped by worldview ideas, philosophical concepts, and principles. If they remain the same, a paradigm does not shift. The learner-centred paradigm was influenced by the philosophy of pragmatism. Nevertheless, the outlined discussions in contemporary philosophy fully coincide with the tendencies in education. So, the teachers influenced by philosophical concepts maintain the learner-centred paradigm. It means that while the described philosophical concepts remain unchanged, and while they make influence upon the educators' worldview, the paradigm will not change. Thus, the paradigm shift is not expected in the nearest future.

The further perspectives of our work are studying teachers' opinions towards the development of the learner-centred paradigm. It is also necessary to find out what philosophical concepts shaped the educators' worldview, especially in Ukraine, and how they influence actual teaching practices and the educators' attitudes towards the process of reformation and readiness to maintain the learner-centred paradigm.

References

1. Y.N. Harari, *21 Lessons for the 21st Century* (Spiegel & Grau, New York, 2018)
2. K. Robinson, *Out of Our Minds: Learning to be Creative* (Capstone Publishing, Oxford, 2001)
3. E. Blass, White paper: a 21st century education paradigm. JESP **5**(3), 128–133 (2018). doi:10.30845/jesp.v5n3p16
4. J.W. Cook, in *Sustainability, Human Well-Being, and the Future of Education*, ed. by J.W. Cook (Palgrave Macmillan, New York, 2019)
5. T. Kuhn, *The Structure of Scientific Revolution* (University of Chicago Press, Chicago, 1996)
6. S. Crowell, Educ. Leadership **47**, 60 (1989)
7. R. Curren (ed.), *A Companion to the Philosophy of Education* (Blackwell Publishing Ltd, Maiden Oxford Melbourne Berlin, 2003)
8. S.R. Loomis, J.P. Rodriguez, *C.S. Lewis: a Philosophy of Education* (Palgrave Macmillan, New York, 2009)
9. R.B. Barr, J. Tagg, From teaching to learning – a new paradigm for undergraduate education. Change Mag. High. Learn. **27**(6), 12–25 (1995). doi:10.1080/00091383.1995.10544672
10. W. Huitt (ed.), *Becoming a Brilliant Star: Twelve Core Ideas Supporting Holistic Education* (IngramSpark, La Vergne, 2018)
11. W. Huitt, Analyzing paradigms used in education and schooling (2019), http://www.edpsycinteractive.org/topics/intro/paradigm.pdf. Accessed 4 Mar 2020
12. D.F. Labaree, *Someone Has to Fail: the Zero-Sum Game of Public Schooling* (Harvard University Press, Massachusetts London, 2010)
13. N. Blake, P. Smeyers, R. Smith, P. Standish, *The Blackwell Guide to the Philosophy of Education* (Blackwell Publishing Ltd, Maiden Oxford Melbourne Berlin, 2003)
14. A. Collins, R. Halverson, Rethinking education in the age of technology: the digital revolution and the schools (2009), https://www.researchgate.net/publication/264869053_Rethinking_education_in_the_age_of_technology_the_digital_revolution_and_the_schools. Accessed 4 Mar 2020
15. J. Dewey, *Experience and Education* (Touchstone, New York, 1997)
16. M. Lipman, *Thinking in Education,* 2nd edn. (Cambridge, New York, Melbourne, Madrid, Cape Town, Singapore, São Paulo, 2003)
17. P. Freire, *Pedagogy of the Oppressed* (The Continuum International Publishing Group Inc., New York, 2005)
18. G. Jacobs, Cadmus **2(2)**, 116 (2014)
19. S.L. Watson, C.M. Reigeluth, Educ. Technol. **48(5)**, 42 (2008)
20. R. Pring, *Philosophy of Education: Aim, Theory, Common Sense, and Research* (Continuum, London New York, 2005)
21. V. Terziev, Lifelong learning: the new educational paradigm for sustainable development. IJASOS **5**, 82–98 (2019). doi: 10.18769/ijasos.531370

22. J.C. Bodinet, Pedagogies of the futures: Shifting the educational paradigms. Dig. Eur. J. Futures Res. (2016). doi: 10.1007/s40309-016-0106-0
23. M. Kochetkov, in *Proceedings of the 5th International Conference on Education, Language, Art and Inter-cultural Communication (ICELAIC 2018)*, ed. by R. Green, Y. Zhang, I. Rumbal, M. Zhang. 5th International Conference on Education, Language, Art and Inter-cultural Communication (ICELAIC 2018), Moscow, November 2018. Advances in Social Science, Education and Humanities Research, vol. 289 (Atlantis Press, Paris, 2018), p. 36
24. J. Dewey, *Democracy and Education: An Introduction to the Philosophy of Education* (The Macmillan Company, New York, 1920).
25. C.A. Chiţiba, Lifelong learning challenges and opportunities for traditional universities. Procedia Soc. Behav. Sci. **46**, 1943–1947 (2012). doi:10.1016/j.sbspro.2012.05.408
26. J. Dewey, *Experience and Nature* (George Allen & Unwin Ltd, London, 1929)
27. A.-M. Iversen, A.S. Pedersen, L. Krogh, A.-A. Jensen, Learning, leading, and letting go of control: learner-led approaches in education. SAGE Open **5**, 1–11 (2015). doi:10.1177/2158244015608423
28. A.-M. Iversen, A.S. Pedersen, in *Co-creation in Higher Education: Students and Educators Preparing Creatively and Collaboratively to the Challenge of the Future*, ed. by T. Chemi, L. Krogh (Sense Publishers, Rotterdam Boston Taipei, 2017), p. 15
29. P. Freire, *Education for Critical Consciousness* (Continuum, London New York, 2005)
30. H. Siegel, Critical thinking as an educational ideal. Educ. Forum **45(1)**, 7–23 (1980). doi:10.1080/00131728009336046
31. N. Kozachenko, Actual Probl. Mind. Philos. J. **18**, 165 (2017)
32. D. Lewis, *On the Plurality of Worlds* (Blackwell, Oxford, 1986)
33. K. Hawley, *How Things Persist* (Oxford University Press, Oxford New York, 2001)
34. A. Abdula, Actual Probl. Mind. Philos. J. **17**, 193 (2016)
35. P.L. Berger, T. Luckmann, *The Social Construction of Reality: A Treatise in the Sociology of Knowledge* (Penguin Books, London, 1991)
36. C.M. Korsgaard, *Self-Constitution: Agency, Identity, and Integrity* (Oxford University Press, Oxford New York, 2009)
37. C. Rovane, *The Bounds of Agency: An Essay in Revisionary Metaphysics* (Princeton University Press, Princeton, 1998)
38. E.T. Olson, in *The Blackwell Guide to Philosophy of Mind*, ed. by S.P. Stich, T.A. Warfield (Blackwell Publishing Ltd, Maiden Oxford Melbourne Berlin, 2003), p. 352
39. C.M. Korsgaard, Philos. Public Aff. **18(2)**, 101 (1989)
40. M. Schechtman, *Staying Alive: Personal Identity, Practical Concerns, and the Unity of a Life* (Oxford University Press, Oxford New York, 2014)
41. A. Macintyre, *After Virtue: A Study in Moral Theory*, 3rd edn. (University of Notre Dame Press, Notre Dame, 2007)
42. K. Atkins, *Narrative Identity and Moral Identity: A Practical Perspective* (Routledge, New York, 2008)

Practical philosophy of education: the ecological-discursive paradigm as the complex problem

Halyna Baluta*

Kryvyi Rih State Pedagogical University, Department of Philosophy, 50086, Kryvyi Rih, Ukraine

Abstract. In the 21st century, more attention is drawn to the development of ecological thinking. Instead of explicating treatment of nature, a human being should take care of it. Such a caring attitude should become an educational universal, so it is to be transmitted as value. However, the ecological-discursive paradigm is still problematic in education. Therefore, the objective of the paper is to analyze it as a complex approach in philosophy of education. The paper considers environmental anthropology as the methodological basis of the ecological-discursive paradigm. Ecological culture is analyzed as ethnics of conservation. The paper emphasizes that ecological thinking cannot be developed without changing an epistemological model. Thus, instead of facts and competences, the person's cognition should be based on values. The paper highlights the idea of diagnostic cognition, which is based on values. Its development in the process of education is the first step to the ecological-discursive paradigm.

1 Introduction

Philosophy of education is an area of contemporary practical philosophy that actively implements the "globalization parameter" because it takes into account the necessity to include ecological culture as a basic cultural form in the educational space. There is a question of what modern education should be, and it becomes an issue for consideration that takes into account the scenarios of global development. The progressive paradigms of education involved in the educational space are particularistic. Therefore, they are partially suitable for successfully overcoming the global crisis as "unpredictable obviousness". There are some reasons for that. Firstly, they remain essentially specialized and mostly based on science. Secondly, they are not intended for the formation of situational competencies, correlating with changeable parameters of the biosphere and anthropogenic consequences. Education remains beyond important educational values, which include readiness for life and understanding of the vital value of nature as "the home of being". The specialization, which is required for the development of civilized thinking, had a negative impact on the educated people's philosophical thinking and, as a result, on the development of institutions created to promote education. A. Whitehead, a classic of philosophy of education and one of its founders, admitted that the progress of science, which has caused a considerable separation of universities and to their autonomy, contributed to narrowing the breadth of thinking.

The thinker paid great attention to education as an indicator of intellectual progress. The philosopher writes: "As science grew, mind shrank in width of comprehension. The nineteenth century was a period of great achievement, suggestive of an anthill. It failed to produce men of learning with a sensitive appreciation of varieties of interest, of varieties of potentiality. It criticized and exploded, where it should have striven to understand" [1, p. 44]. Unfortunately, the well-described educational situation has not changed in the 21st century. What does A. Whitehead mean when he speaks about understanding as a way of thinking and as an important criterion for the quality of education, educational value? He means the awareness of the meaning of education, which is the nature of our existence, the essential content of its depth [1, p. 44]. A. Whitehead motivates to draw more attention to understanding "if civilization is to survive" [1, p. 45].

A. Whitehead's philosophical methodology was defined by the principles of the unity of nature, the understanding of life as the ordered process, so a person must be inscribed in the general order of nature. This prompts the actualization of ethical, environmental, and communicative aspects of education, which reproduces the ethics of conservation, "ethics of intelligent transformations", as opposed to the unmanaged "ethics of pragmatic transformations". Therefore, the discursive-environmental approach can be suggested as an educational universal that highlights the ethics of the subjects' shared responsibility to save the environment and transform it wisely.

In the monograph "Social Ethics and Ecology. The dignity of human – respecting of nature", A. Yermolenko writes the following: "The Kantian question "what should I do?" is up-to-date, complemented by the question of what we should do together, how to substantiate our shared responsibility, in

* Corresponding author: moment71164@gmail.com

which individual responsibility would be identified [2, p. 25]. The researcher emphasizes: "There has never been a question of responsibility for one's existence as a species because until now humankind has not been able to destroy itself as part of nature" [2, p. 182]. A. Yermolenko draws attention to the imperative of the present, formulated by Hans Jonas' ethics. He incorporated a metaphysical element into the dimension of ethical discourse. The ecological issues of "taking care of oneself" are the complex that catalyzes the change in the vector of social action, the value-based intentions of education on the basis of understanding morality as an important life-saving principle, and the intentions of what it means to be a human being or a person in the today's civilization.

2 The principle of understanding

The critique of the disadvantages of traditional educational culture, observed today, is reasonable since the traditional educational culture does not take into account the current social context and its challenges. Among the educational problems, the most topical issue is the gap between education and social dynamics, since the researchers do not consider the civilization sense of education. Neither personal success nor the social success of a certain country can be considered as real on the background of global challenges and constant environmental threats that are extremely unpredictable. The serious disadvantages are the fragmentation of knowledge, the lack of methodological bases that provide a picture of the unity of the world, and the "family affinity" of its phenomena. As mentioned above, the reproduction of the general picture is possible by introducing into the cognitive horizon the ethical "understanding" discourse, offered by Paulo Freire, Yuval Noah Harari, Dario Salas Sommer, Vladimir Lefebvre, Philip Zimbardo. They discussed the informational and methodological insufficiency of traditional education, which excludes the value of being, and minimizes reflective practices. Simultaneously, traditional education keeps the spirit of monologue in teaching, intellectual competition, and individual pragmatics. Moreover, traditional education does not take into account objective reality as a complex entity. It does not take into account the latest achievements of psychological, ethical and environmental studies.

For example, Dario Salas Sommer suggests the concept of "physics of morality", which is based on the idea of the universal order of being. He proves that nowadays "physics of morality" is a reliable tool for the practice of meaningful learning that takes into account the global perspectives: importance and inevitability of the global problems. Dario Salas Sommer greatly broadens the horizons of the stereotyped images of morality. He also discusses progressive features of modern education as practical philosophy through the lenses of anthropological pedagogy, focused on global thinking. He thinks that the study of "physics of morality" is very different from the usual intellectual and psychological techniques, based, as a rule, on mechanical repetition. D. Sommer thinks that we need to start with understanding ourselves and understanding our relationship with nature, for which we will proceed from several prerequisites. Some of them are based on quantum physics and allow proving in practice that every human action, depending on the moral qualities, in one way or another, influences the entire universe [3]. It is necessary to emphasize that the researcher brings up the theme of ontological or "noospheric" understanding of morality, in contrast to the academic interpretation of morality that regards it as the historically formed norm of social interaction, rooted in education, and as a sphere of autonomy, detached from nature. The researcher emphasizes that the notion of morality in our culture is limited and unrelated to the laws of nature [3]. True morality keeps witching the cosmic laws, expressed in the intention of understanding.

As we know, there are different interpretations of the concept of understanding. In the hermeneutic tradition, understanding is defined as the discovery of the meaning of a text, carried out in the process of interpretation. It is also viewed as the ontological principle, underlying the worldview. We are situated in being without the ability to embrace it as a whole. Therefore, knowledge of the world is preceded by cognitive instruction – pre-understanding, which sets the horizon of the process of cognition as interpretation. The change in epistemological perspective-tradition leads to a change in understanding/interpretation. The hermeneutical model of knowledge, passed through the lenses of tradition, has been taken into account by representatives of radical constructivism, justifying cognition as a tool of adaptation to the changeable living environment. The radical constructivism was developed by Humberto Maturana and Francisco Varela. Modern hermeneutics does not oppose understanding and interpretation since they tend to be regarded as complementary correlates.

In linguistic philosophy, the fundamental difference between interpretative schemes has been substantiated for different cultural types. Despite such a difference, "understanding" is treated as a universal platform of shared socio-cultural practices. This approach to understanding was developed by Carl Jung, Ludwig Wittgenstein, and Jürgen Habermas. In social science, Max Weber's approach is also taken into account. According to the scholar, understanding determines the meaning of social action and its value orientation. Nowadays, the idea of the social world as a construct of its agents acquires methodological value. It is considered by A. Schütz, P. Bourdieu, and A. Giddens. The social constructivism implies the real possibility of socio-cultural transformations, caused by global challenges. Such challenges are viewed as the stable civilizational basis, independent of cultural forms and traditions. We define the value of life and its continuation as a common basis that removes linguistic and cultural barriers. It also unites personal interests and values with universal meaning. The diagnostics is considered to be a new meaning concerning the theme of understanding. It is treated as special knowledge, aimed at the timely detection of natural and social deformities and anomalies.

A. Whitehead asks: "What is understanding?" It is a self-evident thing, which has different parameters both in terms of magnitude and its structure. He writes: "In mathematics, as understood, the ideal fact stands out as self-evident" [1, p. 46]. The philosopher introduces the concept of "composition" to characterize the understanding as the basis of the methodological construct. The composition is defined in two ways: a) as the study of the system-forming factors that constitute a thing, its integrity, its obviousness; b) as the analysis of things as the unity of components, its effect on the environment. He emphasizes: "The first mode may be called the internal understanding, and the second mode is the external understanding" [1, p. 46]. In the first aspect, a thing is viewed as a result. In the second aspect, it is regarded as a causal factor. The sequence of fragments as self-evident details forms an argument that is consistent with understanding. A. Whitehead says that the sense of reasonableness and expediency is correlated with the growth of understanding. Using the achievements of science and philosophy, in particular, the anthropic principle, we can think of the natural state as the probable variety of natural laws, based on the idea of their unity and expediency, correlating the existing state with the value-intensive and transformative activities of a person, who is included in the natural horizon.

3 Environmental pedagogical anthropology

The formation of the environmental paradigm is associated with the names of A. Leopold, A. Schweitzer, G. Picht, E. Odum, H. Jonas, K. Lorenz, etc. The active mission of the leaders of the Club of Rome contributed greatly to the development of the environmental approach in philosophy. Such researchers as O. Höffe, D. Böhler, J. Passmore, M. Rock, L. Tribe, R. Spemann, A. Neugeisler, W. Frankena, K.-M. Meyer-Abich are representatives of environmental ethics in its variety of theories.

Despite the diversity of environmental approaches, communicative ethics is also an important approach, as it is the ethics of shared responsibility that actualizes the intersubjective philosophical paradigm. The conventional basis of environmental ethics was represented by J. Habermas, K.-O. Apel, M. Riedel, etc. J. Habermas considers that the discourse participants need to understand each other. Social action is possible through communicative action – a common understanding of the situation, its mutual adequate interpretation [4].

G. Picht was at the beginning of environmental anthropology as the doctrine of human essence. He considered a human being through the lenses of relation to nature. Moreover, the philosopher substantiated the principle of "human ecology" (Humanökologie), and predicted the connection between a human being and social institutions, believing that identity is contained not within a person, but outside personhood [2, p. 181].

It is necessary to mention that the anthropological turn in philosophy was determined by the contribution of M. Scheler, H. Plessner, A. Gehlen. It was caused by the attempt to eliminate the one-sided understanding of a human being as a purely spiritual being or a purely biological being. They promoted the principle of the holistic human being, who is a carrier of biological and cultural programmes. Furthermore, the holistic human being possesses a particular super-biological ability to enter the transcendental realm. Therefore, culture emerges as a "second" inalienable, humanized nature. Philosophical anthropology also played a great role in the formation of humanistic-oriented education, as philosophical anthropology contributed to the formation of the basic educational principles. M. Scheler and A. Gehlen defined and outlined the concept of "environment" (Umwelt). They understood it as the whole world. This definition was as opposed to the functional concept of "environment". According to the later, "every living creature has its environment related to its constitution" [2, p. 206]. The emphasis on reasonable spiritual parameters has somewhat replaced the natural foundation, so it can be regarded as secondary or self-evident nowadays. At the same time, the Christian and Renaissance understanding of nature was lost. That understanding treaded nature as spiritual creation – the perfect culture of the Creator himself when "unnatural" was understood as sinful, anti-good.

In Ukrainian tradition, the environmental theme in education is presented in the works of A. Yermolenko, I. Popova, S. Klepko, S. Kutsepal, T. Andryushchenko, and others. In this context, special attention should be paid to the research "Educational paradigm of ecological development of transformational society", conducted by I. Popova, in which introduction ecological ideas in educational space is considered as the main task of modernization of education in the 21st century [5]. It should be noted that the researcher raised an important problem in educational culture. This fact can be considered as a good start on the way to the improvement of national education. In the research, we suggest enriching the educational content of the paradigm of environmentally friendly development with additional philosophical argumentation and methodology.

Therefore, to be a civilized person in the modern sense is to combine two ethical principles: concerning a human being and towards nature, guided by the ethics of concern for the world, its preservation. Environmental anthropology forms a new ethical image: human morality cannot be the result of mechanical subordination to social norms, but be based on autonomous reasoning and volition to pursue the highest transcendental values [3]. It should be noted that a considerable number of publications do not practically concern the educational practice in the aspect of specific steps towards the implementation of the unified type of educational standard, which would provide for a wide range of relevant competences. In particular, ecological competence would appear not as secondary among others, but as an educational fundamental value. Thus, environmental pedagogical anthropology must be treated as a theory and it should be concretized by a conceptual and methodological platform that implements the successful development of ecological culture and ecological thinking as higher-order thinking.

https://doi.org/10.1051/shsconf/20207503003

4 Ecological culture as the ethics of conservation

At first glance, the elimination of the extremes of particular education seems possible due to the affirmation of the original relation "a human being – nature – the universe" and due to the replacement of the value of education, which is in moral respect of nature of a person. However, it should be noted that in this case, the principles of environmental ethics do not appear as categorical, but only as hypothetical. They are implemented to some extent in the modern educational process, but they are not as effective as they should be. The general scientific scheme is preserved. It explains the universe from the lower to the higher, from nature to a human being as the intelligent but arbitrary master of the natural creation, who destroys the biological foundation. The scheme of explanation of the "higher through the lower" absolutizes the natural principles or the instincts that take sublimated or rationalized forms. Therefore, this scheme is also problematic, as the anthropological basis and the value sphere disappear. A human being with his system of values that humanize the world disappears from the spiritual horizon, remaining in the space of facts and instincts.

The unity, the synergy of a human being and nature can be understood through the lenses of "systems and synergy". Classical philosophical postulates lead to the following conclusion: everything that is in a person and consciousness, such as rationality or expediency is also inherent in nature, but in an implicit form. These postulates include Plato's, Socrates' and Newton's ideas. Plato wrote that the essence of being is to be implicated in causal action on other beings [6, p. 153]. Socrates believed that "a man is the microcosm in the macrocosm". According to Newton, "a book of nature written in the language of mathematics". Therefore, as A. Whitehead highlighted: "Physics abstracts from the element of soul, but due to that fact structural and mathematical relations (eternal objects) are part of the worldview" [7, p. 50].

The constant appeal to sensory perception as a source of knowledge has time delimited thinking from nature for a long. That idea has been expressed in Cartesian dualism. I. Kant considered the united world of phenomenal and noumenal. As A. Whitehead notes: "Mentality is an agent of simplification; and for this reason appearance is an incredibly simplified edition of reality" [6, p. 273]. The person's rationality as the permanent attribute is complemented by the moral criterion, which is powerful, but problematic from the standpoint of scientific proof. For a person as a natural being is the bearer of morality, can nature be moral if morality is regarded as the highest measure of expediency? Are there any correlations between a person's morality and nature's morality determined? Could it happen that the existence of a person and its consequences may appear to be inappropriate for the existence of nature?

Such hypothetical, unfinished model of worldview is important for the ethics of saving in its various modalities. In the process of forming an ecological culture, one can use "an appeal to the spirit". It means that a person does not create nature, but creates and saves it by appearing as the image of God. Moreover, there is also "an appeal to life". It means that a person is the bearer of life. Therefore, people save and continue it, as they are guided by the survival instinct. I. Mechnikov wrote about the survival instinct the following: "The survival instinct is undoubtedly developed in human beings. It is little expressed when they are born, but it is already sharply manifested in young children" [8, p. 94]. A. Schopenhauer, S. Kierkegaard, S. Freud discussed the unconscious force of "the fear of death". Finally, there is "an appeal to understanding". It is the presence of a fragile boundary, separating a person and nature, the real world and its appearance.

Therefore, there is a question of what directions and activity results should be included in the logic of scenarios development of a person and nature. The scenarios can be both ethical as well as anti-ethical, concerning arbitrary interference of a person in natural ties and relations. Paradoxically, the person's intrusion in natural processes is more threatening to a person than to nature. Nature with its hidden powerful potentialities has less impact on humankind.

5 The epistemological pyramid and cognitive technologies

Education is an important strategic resource that not only describes reality but also conserves, reproduces and transforms its contexts. Any problem, idea or phenomenon correlates with the system of knowledge, caused them, so the successful solution of a problem situation requires refinement, improvement or replacement of the previous system. Knowledge and its outcomes become socially and ethically oriented. The polyphony of philosophical discussions focused on the question of what knowledge and understanding are in its alternative forms, which establishes the relevance of hermeneutical discourse in pedagogical epistemology. In educational theory, there are noticeable transformations associated with the revision of educational priorities, focused on understanding cognition as an adaptive activity and knowledge itself as an "instrumental value", as opposed to knowledge as the sum of justified facts. In the history of philosophy, there were such epistemological models as contemplative, representative, hermeneutic, projective-constructive, etc. They generally characterized cognition as a kind of mental activity, rejecting its biological side. The representation of reality as the system of connections and relationships involving thinking is the product of evolutionary epistemology, which views knowledge as a way of adaption to the environment in the form of intellectual production. Therefore, cognition is regarded as a starting point, a function of universal evolution, and, what is more important, as an effective action that allows a living being to continue its existence in the environment.

The incorporation of thinking into the picture of reality creates the idea of thinking as an adaptation process – autopoiesis. H. Maturana and F. Varela treat the process of autopoiesis as the cognitive specificity of a person immersed in the environment of interaction with the environment as an expression of his or her ability to be an autonomous living system. However, this problem remains unsolved. The researchers admit that the biological mechanism tells us that operational stabilization in the dynamics of the body does not embody the way it is generated. Life leaves no imprint on its beginnings [9]. The shift of interaction to another mode pushes us out of the ordinary situation, and then we begin to think of the event as an observer, explaining it to those who did not notice, did not know or believed and so understood. The researchers say that the framework of such non-knowledge is the tradition within which knowledge functions. The tradition is a way not only to see and act but also a way to hide. Options for possible developments remain hidden until tradition and its scenarios become unfit for use, and that fact actualizes reflection. People and cultures are different, but their biological heritage is the same as a biological basis. Therefore, cognition may really belong to the biological realm, but in the realm of the spirit, it is always different, as it resides in one or another cultural tradition as a reflection – "following a general rule" on the principles of an awareness of the commonality of the living environment, taking into account the cognitive cultural diversity that is not inherent in contradict the "general rule" to conserve environment.

Thus, the research of cognition in evolutionary epistemology has formed a new understanding of human activity as determined by the tradition. Can the "diversification" tradition in education be based on moral and environmental bases? The epistemological pyramid of education based on facts, competences with values as its top can be rebuilt in the following way: values should be in the base of the pyramid, as they join facts, giving them an eco-conservation force, defining competencies; the meaning of the facts, and giving them the direction for development. I. Chernikova admits that incising risk is accompanied by adequate use of knowledge, which is the ability to think well [10]. R. Shenk's report at the 4th International Conference on Cognitive Science outlined the suggestions for organizing the education system based on the achievements of cognitive science in understanding the nature of thinking. I. Chernikova outlines the following: "His program identifies 12 fundamental cognitive processes: modelling, experiment, prognosis, assessment, diagnosing, planning, causality, judgment, negotiation or negotiation skills, influence, teamwork, description. These principles should be taught to everyone as the basis for education, from early childhood, then at school and even at university [10, p. 107]. Among the basic skills description, planning and diagnosing are identified as the most important [10, p. 107]. Knowledge in its essence is increasingly becoming "diagnostic", as there is no field in which there is no need for timely diagnosis.

6 Diagnostic cognition

For the sake of environmental protection, health, self-improvement, the ethics of conservation as a leading imperative of the global-discursive paradigm includes in the spectrum of educational procedures the idea of "caring for life", which is in line with the thinking of the diagnostic type. According to E. Krotkov and T. Nosova, diagnostics is not a doctrine, but a kind of cognitive activity that covers biological and social systems [11, p. 37]. The specificity of diagnostic is a combined intellectual skill associated with the formation of its cognitive images. Unlike other types of cognition, mostly aimed at evident facts, the diagnostic type of cognition encompasses the horizon of both evident and non-evident, which, in a methodological sense, corresponds to the substantiated picture of the world as the unity of reality and appearance. Diagnosing as a process of recognition always involves some obscurity or non-obviousness of the object, which may change the parameters or remain hidden in different situations or modes. An important component of diagnosing is also the process of identification, which means ascending from single facts to the general. The specificity of diagnosing also implies the presence of an a priori knowledge base, a database of the object of diagnosing, the presence of justified criteria, the construction of a cognitive image of the studying phenomenon by the mental transfer of noumenal features to the object of diagnosing [11, p. 48].

Therefore, diagnostic cognition is based on an understanding of the complex structure of reality – the presence in it of noumenal and phenomenal levels that explicitly show symptoms of the invisible. Furthermore, it requires complex thinking: logical culture, the culture of critical and philosophical thinking; consideration of contextual values, or, conversely, the ability to reduce context.

7 The conclusion

The diversity and unity of the world are not limited to the empirical sphere – the sphere of the visible. That was emphasized by I. Kant, A. Whitehead, R. Penrose, R. Grossman, and others. In particular, in the book "The Existence of the World: An Introduction to Ontology", R. Grossman emphasizes that the world is not limited to the physical universe. Philosophical ontology illuminates the structure "not so much of our real (actual) world, but any possible world whose existence is potentially permissible" [12, p. 101].

Timely diagnosis of environmental phenomena, analysis of their course and changes, comprehensive understanding of the possible consequences of intervention in the biosphere and designing strategies for their timely correction are valuable social skills and the result of effective civilized education. The above arguments outline the following:

– the need to change the epistemological tradition, which lies in the universal understanding of the systemic nature of reality, its ties and relationships, as opposed to fragmentary knowledge of the world and a person;
– the prospect of incorporating the educational paradigm of eco-safety development into universal contexts of ecological-discursive educational theory and practice;
– directions of formation of ecological culture and ecological thinking as thinking of the highest order;
– the actualization of eco-friendly ethical systems in educational and training practices;

– reviewing the pyramidal structure of pedagogical epistemology as a unity of values, facts and competences, as opposed to facts, competences, and values;
– the use of cognitive technologies in educational processes as key intellectual procedures;
– the orientation of educational technologies to the ethics of conservation and diagnostic knowledge.

As H. Jonas wrote: "Any living being is an end in itself, which needs no further justification, and from this point of view man is no different from other living beings except that he may also have responsibility for them, that is, to care on the preservation of their purpose" [13, p. 152].

References

1. A. Whitehead, *Modes of Thought* (The Free Press, New York, 1968)
2. A. Yermolenko, *Sotsialna etyka ta ekolohiia. Hidnist liudyny – shanuvannia pryrody* (Social ethics and ecology. The dignity of human – respecting of nature). (Libra, Kyiv, 2010)
3. D.S. Sommer, *Internal Ecology – Morals for XXI Century* (Codex, Moscow, 2013)
4. J. Habermas, *From the Pictures of the World to Life World* (Idea-Press, Moscow, 2011)
5. I.V. Popova, *Osvitnia paradyhma ekobezpechnoho rozvytku transformatsiinoho suspilstva* (Educational paradigm of ecosafety development of transformational society). (The National Pedagogical Dragomanov University, Kyiv, 2016)
6. A. Whitehead, *Adventures of Ideas*, the 19th edn. (Cambridge University Press, London New York Toronto Bombay Calcutta, 1939)
7. A. Whitehead, *Izbrannye raboty po filosofii* (The selected works on philosophy). (Progress, Moscow, 1990)
8. I. Mechnikov, *Etyudy o prirode cheloveka* (Essays on human nature). (Medkniga, Kyiv, 2015)
9. H.R. Maturana, F.J. Varela, *The Tree of Knowledge: The Biological Roots of Human Understanding* (Shambhala Publications, Boston, 1992)
10. V.I. Chernikova, Epistemol. & Philos. Sci. **27**(1), 101 (2011)
11. Ye.A. Krotkov, T.V. Nosova, Epistemol. & Philos. Sci. **10**(4), 36 (2006)
12. Y.V. Shramko, Actual Probl. Mind. Philos. J. **18**, 101 (2017)
13. H. Jonas, *Pryntsyp vidpovidalnosti. U poshukakh etyky dlia tekhnolohichnoi tsyvilizatsii* (The principle of responsibility. In search of ethics for technological civilization). (Libra, Kyiv, 2001)

The relationship between education, income, economic freedom and happiness

Olena Stryzhak*

Simon Kuznets Kharkiv National University of Economics, Tourism Department, 9-A Nauky Ave., Kharkiv, 61166, Ukraine

Abstract. Education is a factor for economic prosperity and social development in modern society. As well education allows its owner to receive a higher income and gives the opportunity for self-expression, creative fulfilment, as well as moral satisfaction from current activities. The life of educated people is not only longer, but also more interesting and informative. Moreover people with a higher level of education are happier also. In this aspect the purpose of the article is to determine the relationship between the quality of education, the degree of economic freedom, the level of income and the feeling of happiness. This paper presents results of the correlation analysis between indicators of education, income, happiness and economic freedom for 145 countries for 2018. The author of the work calculated Pearson (product-moment) correlation, the Spearman rank correlation, and Kendall's Tau correlation in the Statistica. The analysis showed that Education Index has a strong relationship with Happiness Index, Economic Freedom of the World Index, Index of Economic Freedom and GDP. The analysis showed that education is closely related to the level of income and self-awareness of happiness. Education is also closely related to economic freedom. Also, the results of the study suggest that education not only contributes to an increase in income, but also makes persons happier.

1 Introduction

Contrary to numerous predictions about global job cuts in an era of rapid technological development, this did not happen. This is evidenced by the statistics of the World Bank. According to annual reports, the technology does not reduce the number of jobs, and the impact on the change in the proportion of jobs in the labor force. Development of information technologies stimulates the emergence of entire sectors of the economy. Because of this, there is the release of employment in certain sectors by increasing demand for labor in the others.

As noted in the World Bank report technology is reshaping the skills needed for work. The demand for less advanced skills that can be replaced by technology is declining. At the same time, the demand for advanced cognitive skills, sociobehavioral skills, and skill combinations associated with greater adaptability is rising. Already evident in developed countries, this pattern is starting to emerge in some developing countries as well [1].

Education is becoming a major factor in ensuring the competitive advantages of countries and companies in the global community. Education not only raises the level of income, it expands the boundaries of consciousness, opening up new opportunities and creating prospects. That is why the study of the relationship of education, income, economic freedom and the feeling of happiness is relevant and timely.

Many researchers consider the questions to determine the relationship between indicators of well-being, income, freedom, happiness, etc.

Veenhoven revealed a positive correlation between freedom and happiness for 46 nations. However, he found that freedom and happiness do not always concur. Freedom is positively related to happiness among rich nations, but not among poor nations. On the other hand, economic freedom is positively related to happiness in poor nations, but not in rich nations. The results of the study show that freedom does not always produce happiness. The analysis included available data in the early 1990's [2].

Gropper, Lawson, and Thorne explored the empirical relationship between liberty, as measured by economic freedom, and happiness across more than 100 countries. Scientists found a positive relationship between national levels of happiness and economic freedom. GDP per capita also has a strong positive effect on happiness. Around the world, freer people, as a rule, are richer, live longer and happier [3].

Mehrara and Musai investigate the causal relationship between education and GDP in developing countries for the period 1970-2010. The study revealed a strong causality from investment and economic growth to education in the analyzed countries. However, education does not have a significant impact on GDP or investment in the short and long term. This means that it is the formation of capital and GDP that determines education in these countries, and not vice versa [4].

Webster researched direction of causation pertaining to economic expansion and increased measures of freedom by examining the economic growth-economic freedom relationship. A study of data from 141 countries

* Corresponding author: ssssellennnn@gmail.com

showed that economic freedom did precede economic growth. However, reciprocal analysis for models building has shown that economic growth does not cause economic freedom [5].

Nikolaev analyzed the relationship between economic freedom and indicators of quality of life. Data cover the period from 1972 to 2010 for developed and developing countries. He found that economic freedom is strongly and positively correlated with most of areas of well-being such as community, safety, and life satisfaction. The analysis showed that changes in economic freedom foster human development in both the short run and the long run. All studies showed a positive relationship between the analyzed indicators [6].

Gregorio and Lee investigated the relationship between education and income distribution. Panel sample covers a large number of countries for the period from 1960 to 1990. They found that education factors play a significant role in making income distribution more equal. The results of the study are mixed. On the one hand, a high level of education and its even distribution in society leads to a more even distribution of income. On the other hand, the level of education does not explain the differences in income between countries. The study examined both interethnic and intertemporal relations between education and income [7].

Yilmaz and Tag explored the relationship between economic freedom and happiness in their work. The results of evaluations have shown that economic freedom has a strong impact on subjective well-being in society. In the same place, scientists found a strong negative relationship between limited government and subjective well-being [8].

Esmail and Shili investigated relationship between the level of happiness and economic growth. The work is devoted to the analysis of how happiness influences GDP and economic development. The results of the study show that social factors are the most important in determining happiness. At the same time economic and political factors are also important [9].

2 Research hypothesis

In this work we assume that the relationship between educational level and awareness of being happy exist. Education not only allows its owner to receive a higher income, but also gives the opportunity for self-expression, creative fulfilment, as well as moral satisfaction from current activities.

The life of educated people is not only longer, but also more interesting and informative. Therefore, we can assume that people with a higher level of education are happier also. In addition, happiness in a subjective understanding is impossible without satisfying basic needs. In this case, the ability to meet these needs is determined by the level of income and economic freedom. Therefore, if there is a relationship between happiness and education, then there should be a relationship between education and income, since a high level of education contributes to the mobility of labor resources, as well as to more interesting, promising and highly paid job. Consequently, the relationship between education indicators and indicators of income and economic freedom should be clearly expressed. Thus, the study assumes that the level of education is closely related with all the analyzed indicators.

The purpose of the study is to determine the relationship between the quality of education, the degree of economic freedom, the level of income and the feeling of happiness.

3 Methods

Studies were conducted on panel data for 145 countries for 2018. The data sources for the study were World Development Indicators, The Education Index, The World Happiness Report, The Index of Economic Freedom and Economic Freedom of the World Index [10-14].

We did not divide panel sample data into groups for achievement of statistically significant research results and objectivity of analysis. For example, research [2] is severely limited in that it covers a small number of countries. As noted by the author of this study, as a result, many of the differences observed do not reach statistical significance.

The Pearson (product-moment) correlation, the Spearman rank correlation, and Kendall's Tau correlation were used in the research process. The calculations are implemented in the Statistica. This approach to data analysis was successfully tested in the study [15].

Pearson correlation coefficient is calculated by the formula:

$$r = \frac{\sum_{i=1}^{n}(X_i-\overline{X})\times(Y_i-\overline{Y})}{n\times\sigma_x\times\sigma_y}, \quad (1)$$

where:

X_i, Y_i – compared quantitative traits;

$\overline{X}$, $\overline{Y}$ – selective arithmetic averages;

n – a number of compared observations;

σ_x, σ_y – standard deviations in the databases compared;

$(X_i-\overline{X})\times(Y_i-\overline{Y})$ – product of moments.

The Spearman`s rank correlation coefficient is calculated by the formula:

$$r_S = 1-\frac{6\times\sum d^2}{n(n^2-1)} \text{ or } r_S = 1-\frac{6\times\sum(x'-y')^2}{n(n^2-1)}; \quad (2)$$

where:

d – the difference between the ranks of corresponding variables X and Y;

x' – values of ranks, replacing the actual variants or qualitative features of the argument x;

y' – values of ranks, replacing the actual variants or qualitative features of the function y;

n – number of observations.

The Kendall correlation coefficient is calculated by the formula:

$$\tau = 1-\frac{4R}{n(n-1)}, \quad (3)$$

where $R=\sum_{i=1}^{n-1}\sum_{j=1+1}^{n}\left[[x_i<x_j]\neq[y_i<y_j]\right]$ – the amount of "mess" (inversions) formed by the values y_i and

placed in ascending order of the corresponding x_i; n – the number of set objects.

The founders of the theory of human capital the Nobel laureates Theodore William Schultz and Gary Stanley Becker have long proved the effectiveness of investments in education. It is confirmed that education contributes to increasing earnings in the long term. Therefore, investment in education is beneficial both to the individual and to the state.

It is widely known that education is an important factor in the formation of income, but it is advisable to determine how closely related these two indicators. The existence of a relationship between educational level and economic freedom is also interesting. In addition, it is advisable to determine the relationship between education and the level of happiness, and as far as education allows a person to be free and independent in an economic aspect. If educated people as a whole receive more income compared to uneducated, it is necessary to determine whether they are happier if they perceive themselves as such.

According to human capital theory, education is not the only factor in earnings growth, but also allows a person to self-actualization, self-assertion to raise their social status, etc. That is, a high level of education should have a strong correlation with economic freedom and self-awareness of happiness, because additional knowledge opens up new opportunities for experience and self-realization not only in the professional sphere.

It should be noted that the correlation relationship as opposed to functional, shows only the tendency of change of one value under the action of another. Therefore, based on the correlation one can argue only about the degree of relationship between variables, but not about the existence of a causal relationship between them.

Education is an important component of well-being and it is used to measure quality of life and economic development, which is a determining factor whether a country is a developed, developing or underdeveloped country.

The Education Index is calculated on the basis of two indices: Mean Years of Schooling Index and Expected Years of Schooling Index [11]. Expected years of schooling is number of years of schooling that a child of school entrance age can expect to receive if prevailing patterns of age-specific enrolment rates persist throughout the child's life. Mean years of schooling is average number of years of education received by people ages 25 and older, converted from education attainment levels using official durations of each level [15].

Happiness is understood as a subjective assessment of satisfaction with one's life. Since the level of happiness is not an absolute indicator, only its subjective assessment is possible.

The level of happiness is measured by the Happiness Index, which is calculated on the basis of three main happiness measures: life evaluations, positive affect, and negative affect. Happiness assessment is carried out using a variety of subjective well-being measures, which in subsequent reports are referred to as "happiness".

Happiness is rated on a scale ranging from 0 to 10, where 0 means the worst possible life and 10 means the best possible life. A positive influence includes the average frequency of happiness, laughter and pleasure on the previous day, and a negative affect comprises the average frequency of anxiety, sadness, and anger on the previous day [12].

Freedom in the general sense is understood as the possibility of making a choice in all spheres of life – economic, political, social, personal, etc.

Economic freedom is the fundamental right of every person to control his labor, property, produce, consume, and invest. The Economic Freedom Index measures the impact of freedom and free markets around the world. The index includes 12 freedoms – from property rights to financial freedom.

As an indicator of the quality of education, we take Education Index, one of the three components of the Human Development Index calculated by the UNDP, per capita income measured in GDP US dollars и GDP US dollars PPP for the whole group of countries, Happiness Index Happiness Index, Economic Freedom of the World Index, Index of Economic Freedom. We calculate the correlation between these indicators for the entire sample.

4 Results

The analyzed sample includes 145 countries for 2018, while the sample is balanced, that is, the sample contains data for all indicators for each country. The results of the correlation analysis are presented in Table. 1.

Table 1. The Relationship between Education Index, Happiness Index, Economic Freedom of the World Index, Index of Economic Freedom and GDP.

Variable	Marked correlations are significant at p < ,05000; N=145 (Casewise deletion of missing data)		
	Pearson Correlations	**Spearman Rank Order Correlations**	**Kendall Tau Correlations**
HI	0,771259	0,769307	0,578069
FIW	0,578534	0,642119	0,466908
EF	0,514523	0,631251	0,458684
GDP US$	0,653197	0,867830	0,684567
GDP PPP $	0,668527	0,862748	0,676320

Source: own calculation based on data from [10-14]

The results of the analysis are presented in table. 1, show a strong and a very strong direct relationship between the analyzed indicators.

Moreover, this dependence is revealed when calculating the correlation coefficients of The Pearson correlation, the Spearman rank correlation, and Kendall's Tau correlation, which gives grounds to speak about the objectivity of the results obtained.

Our study for a panel sample from 145 countries of the world showed quite a strong direct relationship between the Education Index and Happiness Index (Fig. 1).

More educated people generally feel happier than less educated people. The results of the analysis confirm the conclusion that education makes an individual's life happier, expanding the possibilities of self-realization, development of personal qualities, choice of place and working conditions. In addition, education gives a person the chance to be freer in the economic aspect.

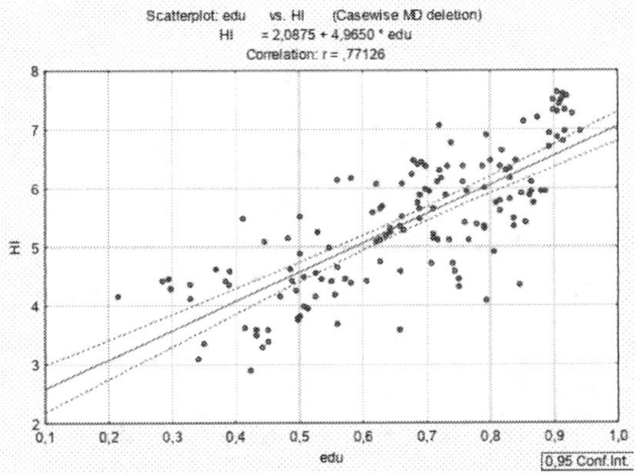

Fig. 1. The relationship between the Education Index and Happiness Index, 2018

In addition, there is a higher average dependence between education and economic freedom, represented by Education Index and Economic Freedom of the World Index (Fig. 2-3).

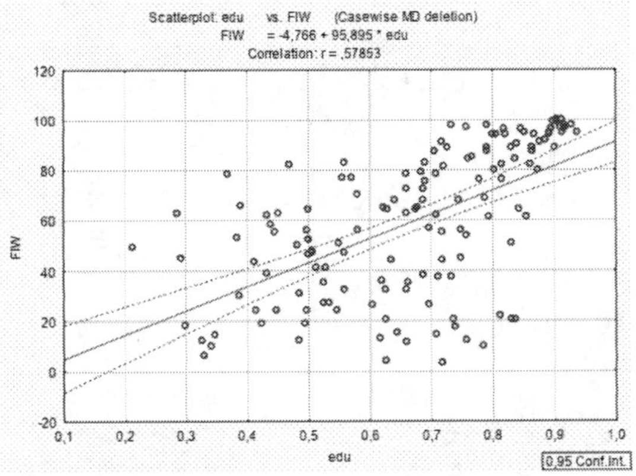

Fig. 2. The relationship between the Education Index and Economic Freedom of the World Index, 2018

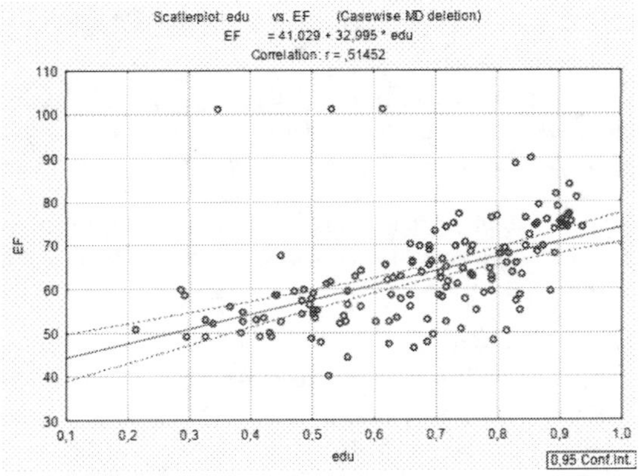

Fig. 3. The relationship between the Education Index and the Index of Economic Freedom, 2018

It should also be noted that there is a stronger variability of the analyzed indicators presented in Fig. 2 and Fig. 3. This can be explained by the fact that the Economic Freedom of the World Index and Index of Economic Freedom calculation methodology focuses on various aspects of economic freedom. Economic Freedom of the World Index has 25 components. Many of the components are themselves made up of several sub-components. In total, the index incorporates 43 distinct variables. Index of Economic Freedom based on 12 quantitative and qualitative factors, grouped into four broad categories.

You can also find a strong relationship between the Education Index and the level of economic development, determined by GDP per capita. For the analysis, we used the indicators of GDP per capita in US dollars and GDP in US dollars at the purchasing power parity of currencies (Fig. 4-5).

The relationship on Fig. 4-5 shows a pronounced direct relationship between level of education and income.

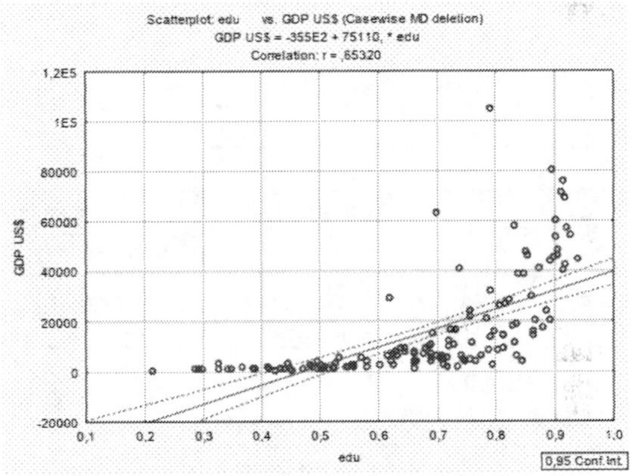

Fig. 4. The relationship between the Education Index and the level of economic development (in GDP, US dollars), 2018

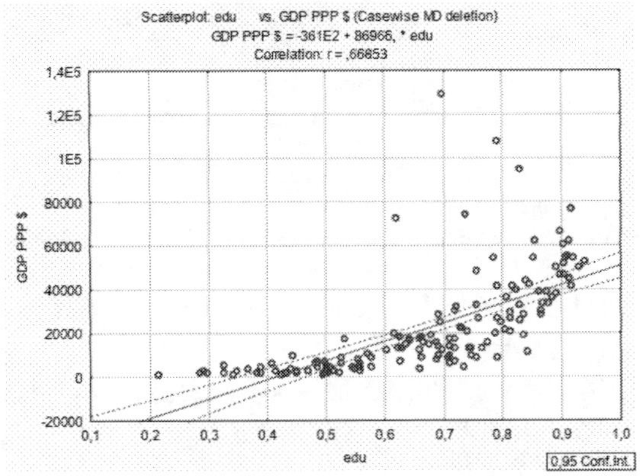

Fig. 5. The relationship between the Education Index and the level of economic development (in GDP, US dollars PPP), 2018

It is noteworthy that the income level represented by GDP, US dollars PPP has a stronger correlation with the education index than GDP, US dollars. This can be explained by the fact that the GDP, US dollars PPP indicator reflects the real income level of the country's population, adjusted for inflation.

The analysis showed that education is closely related to the level of income and self-awareness of happiness. This can be explained by the fact that a higher income received through education provides more opportunities for implementation and opens up new perspectives. Education is also closely related to economic freedom, but the graphs show a stronger variation between these indicators, which can be explained by the fact that not free countries can provide a fairly high level of education of the population, mobilizing state financial resources in this direction, as was the case in the former USSR.

The top of 20 countries by rating for the analyzed indicators are presented in Table 2. Those countries that are in the top 20 in all the analyzed indicators are highlighted in red those countries that are in the top 20 by 5 indicators are highlighted in green, yellow – by 4, blue – by 3, gray – by 2, those countries that are in the top 20 once out of six are not highlighted.

Table 2. Top 20 countries by indicators of Education Index, Happiness Index, Economic Freedom of the World Index, Index of Economic Freedom and GDP [10-14].

	Happiness Index	Education Index	Economic Freedom of the World Index	Index of Economic Freedom	GDP US$	GDP PPP $
1	Finland	Germany	Finland	Hong Kong	Luxembourg	Qatar
2	Norway	Australia	Norway	Singapore	Switzerland	Luxembourg
3	Denmark	Denmark	Sweden	New Zealand	Norway	Singapore
4	Iceland	Ireland	Canada	Switzerland	Iceland	Ireland
5	Switzerland	New Zealand	Netherlands	Australia	Ireland	United Arab Emirates
6	Netherlands	Norway	Australia	Estonia	Qatar	Kuwait
7	Canada	United Kingdom	Luxembourg	Canada	United States	Switzerland
8	New Zealand	Iceland	New Zealand	United Arab Emirates	Singapore	Norway
9	Sweden	Netherlands	Uruguay	Ireland	Denmark	Hong Kong
10	Australia	Finland	Denmark	Chile	Australia	United States
11	Israel	Sweden	Iceland	United Kingdom	Sweden	Iceland
12	Austria	United States	Portugal	Georgia	Netherlands	Netherlands
13	Costa Rica	Canada	Ireland	Luxembourg	Austria	Denmark
14	Ireland	Switzerland	Japan	Lithuania	Hong Kong	Saudi Arabia
15	Germany	Czech Republic	Malta	Netherlands	Finland	Austria
16	Belgium	Belgium	Switzerland	Denmark	Canada	Germany
17	Luxembourg	Slovenia	Austria	United States	Germany	Sweden
18	United States	Lithuania	Belgium	Sweden	Belgium	Australia
19	United Kingdom	Israel	Germany	Latvia	New Zealand	Belgium
20	United Arab Emirates	Estonia	United Kingdom	Mauritius	United Arab Emirates	Bahrain

As seen from Table 2, most of the countries are represented in almost all of the analyzed ratings. This fact confirms the results of the correlation analysis conducted in this work, and also indicates the presence of a relationship between the all analyzed indicators.

5 Discussion

Despite the identified relationship between indicators of education, happiness, income and freedom, this study does not explain the direction of the identified relationship. The issue of causality relationship of the analyzed indicators remains unresolved. The solution to this problem in further studies can be implemented based on the use of Granger causality tests. In other words, the answer to the question is of certain interest, is education a source of income and happiness? Or people feel happier and have more educational opportunities in a higher income society.

Also, in our opinion, the issues of identifying differences in the interdependencies between the analyzed indicators in countries with different levels of economic development need to be addressed. In particular, is the level of happiness and economic freedom determined by the size of income in a particular group of countries? Or income does not affect the level of happiness.

6 Conclusion

The importance of education should not be underestimated, since it is interconnected with all spheres of the economy and society. However, the value of education depends not only on its level, but also on the institutional conditions that are necessary for its implementation.

For illustration, the poor quality of the institutional environment and weak formal institutions lead to the emergence of corruption. Accordingly, corruption is an indirect indicator of the quality of the institutional environment. Corruption has a negative impact on existing institutions, including the institute of education.

As an example, the functioning of the institute of education can be mentioned. In the conditions of normal market activity, the knowledge and qualification, the level of which is certified by a diploma, is of paramount importance. On the contrary, because of the high level of corruption and the use of pseudo-contests for occupation of vacant positions, the level of knowledge is not taken into account; only the formal availability of a diploma is required. In such conditions, the demand for diplomas without the corresponding level of knowledge increases, and the demand for corrupt services is also increasing. When the knowledge gained is not a competitive advantage for obtaining a vacant position, incentives for their receipt are reduced. Accordingly, the value of education in society decreases. Therefore, the state should not only finance the education system, but also create favorable conditions for the implementation of skills acquired in the learning process.

Summing up, we can conclude that when the development strategy of the economic system is determined, it is necessary to take into account not only economic indicators. Happiness, freedom, education are also important. This study confirms the conclusion about the importance of education and therefore need to improve the level of education in society. Also, the results of the study suggest that education not only contributes to an increase in income, but also makes persons happier.

References

1. The World Development Report 2019. The Changing Nature of Work (2019), http://documents.worldbank.org/curated/en/816281518818814423/pdf/2019-WDR-Report.pdf. Accessed 28 Nov 2019
2. R. Veenhoven, in *Culture and Subjective Wellbeing*, ed. by E. Diener, E. Suh. (MIT Press, Cambridge, 2000)
3. D.M. Gropper, R.A. Lawson, J.T. Thorne, Cato Journal **31**(2), 237–255 (2011)
4. M. Mehrara, M. Musai, International Letters of Social and Humanistic Sciences **5**, 55–62 (2013)
5. A.L. Webster, Applied Econometrics and International Development **13-1**, 11–24 (2013)
6. B. Nikolaev, The Journal of Private Enterprise, **29**(3), x–xx (2014)
7. J. De Gregorio, J. Lee, Review of Income and Wealth **48**, 395–416 (2002)
8. I. Yilmaz, M. Nasih Tag, Happiness and economic freedom: Are they related? *SHS Web of Conferences*, **28** (2016). doi:10.1051/shsconf/20162801109
9. H.A.H. Esmail, N.N.J. Shili, The Relationship between Happiness and Economic Development in KSA: Study of Jazan Region. Asian Social Science, **14**(3), 78–87 (2018)
10. World Development Indicators (2020), http://databank.worldbank.org/data/reports.aspx?source=world-development-indicators. Accessed 21 Mar 2020
11. The Education Index (2018), http://hdr.undp.org/en/data. Accessed 25 Oct 2019
12. The World Happiness Report (2018), https://worldhappiness.report. Accessed 25 Oct 2019
13. The Index of Economic Freedom (2018), https://www.heritage.org/index/ranking. Accessed 25 Oct 2019
14. The Economic Freedom of the World Index (2018), https://www.fraserinstitute.org/studies/economic-freedom. Accessed 25 Oct 2019
15. O. Stryzhak, Assessment of the relationship between the tourism sector development and other sectors of economy. Economics of Development **2**, 10–18 (2019)
16. Human Development Report 2019. Beyond income, beyond averages, beyond today: Inequalities in human development in the 21st century (2019), http://hdr.undp.org/sites/default/files/hdr2019.pdf (2019). Accessed 28 Nov 2019

The subject of education in the information society: cognitive identity of a media culture carrier

Iryna Pokulyta, and *Mariana* Kolotylo*

[1]National Technical University of Ukraine "Igor Sikorsky Kyiv Politechnic Institute", 37 Peremohy Ave., Kyiv, 03056, Ukraine

Abstract. The article substantiates that the temporality of being of the subject as synchronization of communicative, cognitive demands on a person with dynamic changes of information reality, is the basis for the development of media education. It is justified that the formation of competencies of the carrier of modern culture is a difficult task, since such traits as critical thinking, free possession of judgments about content quality, preventive measures in guarantees of dangers of communicative activity and others, fall into the scope of the main threats – leveling of important conditions of cognitive activity, multiplicity of models of communicative interaction, complex of semiotic systems of media culture, etc. Identified symptomatic cognitive features of the modern subject of information activity indicate the need for an urgent educational response, which means the formation of requirements and criteria for media literacy, a balanced approach to the consideration of temporality, procedural thinking in terms of knowledge, memory, self-identification. It is stated that mastering the semantics, topology, and navigation of the basics of audiovisual communication is an important vector of media education. The article emphasizes the importance of a comprehensive response from the education system to the civilizational challenges of the transition to the information society.

1 Introduction

Informatization, as enrichment, saturation of the world around us with new and convenient technologies, cannot but please everyone. Informatization of education is no exception. The processes of learning, upbringing are dynamic; there are unprecedented forms, niches and adaptation mechanisms in the organization of the educational segment of culture.

At the same time, the changes of the subject to whom all these efforts are directed are evident. Externally and superficially, the situation looks very positive (it should be noted immediately that we do not include pathological cases, such as Internet addiction, gambling, etc.): modern youth are growing, socialized in the conditions of active use of digital technologies. There can be no other: this is their world, they live in it and carry on the relay of history. Therefore, it is quite logical that young people are adaptable and responsive (unlike the older generation) in obtaining information: applying a minimum of effort, gaining maximum variability, and so on. In addition, generational dissimilarity is a prerequisite for social development, since posterity is the next step in the civilization movement. We also need to capture the fact that the advanced capabilities of the digital technology age are not conferred by institutionalized education. Youth, young people, children underwent their primary socialization as early as the transition of society to the information society. That is, they are shaped by this culture, so further growth of a person through education is primarily characterized by media literacy, a set of competencies, including professional ones. But, thinking about the "further growth of man", it is necessary to understand the present state of things: how fundamental is the dissimilarity of different generations in the aspect of the culture of cognitive activity.

It is important to note that the amount of knowledge in terms of equivalence of the content of modern science is not commensurate with the range of what the educational institutions could have given to their graduates in the recent past. Even these characteristics today are enough to formulate the purpose of the study – to elucidate the historical uniqueness of the individuals, who receive educational services. Or else: who is the modern Homo cogitas? What are our perspectives on the relay of history, cultural heritage?

Education as a breeding platform of culture has the same nature as humanity. The very possibility of anthropo-sociogenesis involves, along with other factors, two components of human and social development. It is the education in its heuristic-consolidating project and the image of a person in the demand of the future generation. In other words, evolution in the social context envisaged (today is no exception) the self-reproduction of humanity through the tools of education. The continuity of being is based on the fact that community, and in the case of man, the social community, can continue in the future, embedding the

* Corresponding author: kolotylo.mariana@gmail.com

information about the world of culture into the knowledge of the next generation.

The peculiarity of this process is that all factors, including the semantic orientations of the above, are variable and extremely poorly predicted. These are circumstances such as natural conditions (e.g. climatic or other aspects of biogenesis), historical (e.g. external conquests and, as a consequence, assimilation, acculturation, etc.) and many other contexts. Of course, the very content of knowledge is constantly updated. Due to such a fragile, unstable resource for the self-preservation of mankind, the possibility of evolution (progress) of culture, the appearance of civilizations was and still remains today highly dependent on the whole spectrum of circumstances.

The modern world community has acquired in the course of history a number of significant components of difficult-to-predict development. It is a techno-dependency, and now a digital extrapolation of culture (both as a convenient prospect and as a new platform for information accumulation). The only one that confronts these challenges of social being is the human cognitive resource. It feeds on the knowledge acquired in the process of education; it is able to convert, modify information into another educational package for future generations.

2 Literature review, methodology

The current stage of history – entering the information age, has several features. Outline their basic parameters, stating that information becomes the dominant of different social processes, structures, systems, etc., is insufficient. Researchers such as E. Toffler [1], D. Bell [2], M. Castells [3], etc. addressed the question of characterizing the principles of "postindustrial", "informational" society. Their views had a warning context about the transition to a new type of organization of society, some of which (e.g. the Internet galaxy (M. Castells), the development of virtual communication, the formation of virtual reality) became our reality. Among the various aspects of this without exaggerating global theme, we consider it necessary to emphasize the factor of a new type of information dependency in this kind of public relations. Independence, the insecurity of the subject from the opportunities that reveal information technologies in the external immersion in the personal, even confidential space of the person, have a direct bearing on the lack of media literacy, hence the requirements that are faced with the organization of the media education system.

Increasing the capacity of media literacy is directly related to understanding the specifics of the media communication process. It is, first of all, a study of a new type of communication, drawing on the powerful intelligence of M. McLuhan [4] and L. Manovich [5]. Particular attention needs to be given to the subject's cognitive activity, which includes the need to understand thinking in terms of the laws of logic, starting with Aristotle, the contexts of memory, intuition, subconscious – H. Bergson [6], C. Jung [7], and thinking as creativity – W. Benjamin [8]. It is also necessary to highlight the features of semiotic systems of information interaction. In this article, we focus on the semiotic concept of U. Eco [10].

3 Problem statement and the aim of the study

Education in the given civilizational past and present is an institutionalized form for most cultures of the world. Therefore, we can state the social strategies of different societies in their forecasting of the future. In this aspect, it is important to note that, despite the differences in current education models in many countries, the challenges associated with forming a new type of communication, a culture of perception and information processing are common. Exploring new factors in communicative, cognitive, and semiotic directions is the aim of the study. The various challenges need to be addressed separately.

4 Communicative concepts of media education

Firstly, it is advisable to consider the creation of a new type of communication. Replaced by the "Gutenberg Galaxy" (M. McLuhan) came electronic media, today – "new media" (L. Manovich). Again, at first glance, this is just a change in the way it is transferred – the dissemination of information that should not affect its content. Moreover, broad parameters of the choice of the medium of information and the method (channel) of communication are formed. And, if the New Age is the era of the printed text, then the present is a variability of both visual semantics and audio semantic codes. Information is actively, often very aggressively, through various sensory systems crashing into our consciousness. In the Gutenberg era, being educated, being able to read books was a blessing, a value, a social privilege. At present, this norm of the path to knowledge is being retained, but it is increasingly reminiscent of the nominal one: what is the difference between reading a text or listening to an audiobook, or perceiving a combined digest of information with entertainment elements? In our opinion, there was a communicative atmosphere not only of variability, but also of unstructured semantic flow.

The information made by the wave of history on top of the axiosphere of a globalized society, under the press of its own power, mass, significance, destroys this pyramid of meanings. Pluralism of thoughts and knowledge turns to the absence of a strategy for obtaining them. Along with the depreciation of information, due to its excess at the level of media everyday life, the way to master information is not appreciated. It is illusory to know, not necessarily to remember, even harmful, because tomorrow there will be new information. And if it doesn't (in the case of historical literary works, fundamental discoveries of the past, etc.), there is always a media device, a gadget that will fill the gaps of ignorance. Thus, the modern multimedia type of communication, providing an incredible archive of information, reaching the various possibilities of its implementation, eliminated the procedure of comprehension of information.

The latter, in turn, determines the second factor of challenge – the culture of perception and processing of information. In our opinion, its main parameters are the "mosaic" of the communicative environment itself; clickability (not even clipping) as the temporality of the subject's attitude to content and the effect of implantation, the "germination" of the media information space into consciousness and vice versa.

Concerning the third parameter, it is necessary to explain separately what exactly we mean: the deconstruction of the procedure of comprehending information literally consists in the fact that it is melting, as snow, the time to receive it, but this is compensated by the very space of its existence. Thus, dependence is formed in terms of the cognitive function of consciousness in the continuity of nourishment, information saturation. Gradually, the subject loses identification of what he/she knows or does not know, and most importantly, somebody knows, but not him/her. That is, "where?" ends knowledge, principles, beliefs, etc. and "how?" to distinguish, concentrate their own competence.

The practice of educational processes often reveals the following symptom of such a problem: the student as a person, attuned to intellectual activity, occasionally takes his eyes off the smartphone and talks about something. If he/she is stopped and asked to repeat what has been said without the help of the gadget, at best, the last phrases will be reproduced. To present the content of his/her own message becomes an unfulfilled task. Analyzing this symptom, we can make only two assumptions: the first – problem with memory, the second is that the content of the information is irrelevant to the subject, and the information process creates only acoustic noise.

5 Cognitive dimensions of media literacy

The topic of "memory" as a philosophical problem is directly related (in the formulation of the question and in the methodology of research) to the cognitive potential of the individual. It is worth pointing out H. Bergson's views on the understanding of the nature of intuition, and the "subconscious" in the theory of psychoanalysis and analytical psychology of C. Jung. Memory, its irreplaceable role, involvement in various manifestations of the entire front of thinking, today falls into the risk zone.

The dependence of the professional, cognitive activity of the person on the amount of knowledge possessed by the person gradually becomes smaller (in other words, "what?" is remembered by the subject). Possession through accessibility, ease of information retrieval, simulates the learned, meaningful knowledge, the value of which does not allow us to relax our cognitive efforts in revitalizing memory structures and mechanisms. Remembering as an activity is offset by knowledge of the roadmap for finding information. Memory does not become a function of consciousness to hold, accumulate, transform the essential parameters of information, but a reservoir for its navigation. That is, the person remembers not the content, but the skills of its search. And what is left to human intelligence: how to create, dream, on what to fantasize, where are the images that nourish our imagination?

The question is rhetorical - in the media space, as an incubator of meanings. It may not be all that sad, but there is clearly a tendency to do so. In the context of this problem, the course of modern education, namely, to teach to learn, does not seem unambiguously successful. On the one hand, this is a necessity because it is extremely difficult to predict future educational preferences. Therefore, the subject must be flexible and adapted to form his own life project of acquiring more and more competences through education, self-education and so on. But today, at least for the above symptoms, loss along the way is noticeable.

The effectiveness of the technological progress of civilization is determined by how much a person is "liberated" from uncreative, monotonous, physically difficult actions through various devices. But along with ease of movement, everyday comfort, etc., modern technology "unloads" the thinking of man, freeing from memory, atrophy the imagination. As for the latter, almost a century ago, W. Benjamin ("A work of art in the age of technical transformability") warned about it. Creativity is not perceived in close connection with creative activity, but only as the ability to achieve what is desired in an unexpected way. Due to the volume, the diverse specification of knowledge, one does not even need to speak about the universal type of personality. It can be a question of integrating knowledge into the modern world, which seems to correlate with the scientific picture, significantly diluted by probabilities, scenarios of reality with virtual character. However, the generation formed one step before the advent of information technology, is forced to combine in their consciousness different civilization platforms. They need to keep in touch basic knowledge with modern requirements of narrow specialization of activity. This means that the mechanisms of critical perception of information still remain (although this does not apply to all older generations and we believe to be directly dependent on education).

For young people, the decision to use this or that information is more often driven by the factors of effort minimization, convenience, the accessibility of the channel to receive it, and the accompanying novelty and entertainment. It is important for students who have a clear purpose in the future profession to apply the knowledge in practice. Although thinking about what theoretical intelligence is behind these empirical techniques is also unnecessary. For the most part, even the motivation to become a unique specialist and to choose the practical options in their professional activity, understanding the causal bases, basic knowledge, does not work. The main features are those that fit into the "Procrustean bed" of competencies.

Critical thinking is nowadays at the forefront of requirements for future professionals. Based on the properties of critical thinking, it is possible to apply them to the critical understanding of the concept of "critical thinking". Reflecting on the anthropologically determining property of consciousness – the ability to think in the integrity of the active (including practical) attitude of mankind, it is important to note that thinking is the cause and effect; the purpose and means of human existence. The cognitive activity of consciousness in the bowels of the process of anthropogenesis is formed as a holistic mechanism of development. More than a two-thousand-year history of studying thinking (beginning with the laws, principles of Aristotle's formal logic), from the point of view of organizing this process for the comprehension of truth, is a difficult way of organizing the most incomprehensible horizons of human desire for ontological meanings.

Scientific discoveries, the creation of essential social development platforms – all this is directly related to the self-reflection of the subject of such activity. It is quite obvious that the order of judgments, the level of understanding – the scale of understanding reality as a sign of this creativity does not imply mass embodiment. Therefore, there are different qualitative levels of thinking, among which one of the highest stages of its development is critical thinking. However, it should be remembered that at its core it is about the integrity of the cognitive system, to climb to the top of which, it is possible only from the plain plateau. Various predicates, such as projective, pre-emptive, transgressive, creative, critical, etc., can only describe an element of a single system. However perfect it may be, this element does not function in isolation from the holistic structure. Therefore, it is necessary to put firstly the task of formation, development of thinking, which in itself cannot be reduced to the invention of the "simulator" of the mind. Or, by choosing a more delicate, imaginative comparison: we are not faced with the choice of Mahayana or Hinayana, the short or long path to truth, and we cannot replace the path by a convenient short way of "transferring" the mind to the "destination".

We make mistakes in the above formal and logical considerations, because we identify the thinking subject with the carrier of consciousness – man. However, what is the independence of judgment, controllability, when critical thinking is considered today in relation to the need to analyze the masses of information materials, a kind of filtering device for various messages, judgments, and ideas. So it is another cognitive process that has the ability to transform reality. Accordingly, its structure is also implanted in the information flows of such a society - the collective consciousness of a globalized network of knowledge, meanings, emotions, values, etc. In this formulation, it is necessary to consider a person as a subject, who produces critical judgments, as they become a means of preventing obstacles to the formation of favorable conditions for the development of a real subject of thought: a global subject of the information society. However, it is also worth noting that this is only a hypothesis that we have a basis for in the scientific research.

Critical thinking, as a cornerstone of contemporary educational requirements, is complemented by media literacy. In parallel with the formation of a new type of communication, there are changes in the way of being in this audiovisual environment. In addition to the advanced communication capabilities, the subject also receives new requirements for his own competencies. Particular attention is paid to the technical side of the issue: the main task is to master modern convenient devices that ensure the status of active users of information networks. Today, the subject has the opportunity, and representatives of the younger generation are realizing it in every possible way, to stay almost without restrictions (time, space) in the system of information contacts through a different kind of media communication platforms. For the older generation, the technical component of the issue sometimes becomes a barrier to this kind of user status in the network communication. However, because of its cognitive nature, this problem is not limited to technical measurement. As users of media communication, we are simultaneously different in terms of tempo-rhythm and saturation of semantic dimensions, which are connected solely by the cognitive structures of our consciousness. In such circumstances, the following question arises: how does it affect the thinking process in the mental, psychological, and identification aspects.

Consider a specific example. When we read a fascinating book, and in the process, someone interferes with the outside, asks about something, then we need some time to "return" to the surrounding reality. We do not even immediately understand what is being said, we do not respond immediately, and so on, because all our attention, cognitive effort, "ontic" (M. Heidegger) immersion was in another semantic reality [9]. Nowadays, the condition of multimedia culture is decentralization and deconstruction of attention. Modern living space is a woven mosaic panel of blocks of different nature, so the very germination of this culture into a personal sense space is at the cost of the loss of a mental, cognitive order. However, it is a current requirement, even a challenge. The educational task in forming media literacy is the answer.

The main trajectory of acquiring media literacy is to train the subject to recognize, classify by quality, and independently create potentially competitive content. Therefore, it is necessary to master the ability to "read" and "write" the whole semantic series contained in the media message, located on the information site, etc. Literacy, as a mastery of the audiovisual semantics of the information space, is one component of success in this activity.

6 Semiotic context of communication

Semiosis of media culture is a process that has been going on for more than a century and is conditioned by the emergence of a new type of audiovisual communication, and in recent decades, due to the emergence of "new media" (L. Manovich), its specificity has been modified. In terms of grammar, the semiotic system can be regarded as a sphere of signs, symbols, each of which captures a certain value. For example, consider the linguistic system as one of the varieties of the semiotic. Culture is a mechanism of social interaction, having as its component a dialogical basis embodied through a communicative function. Meaningful orientations arise in sociality when they are made out through signal –message –information.

The dialectic of "meaning – thinking" is awakened in the beginnings of human history by the need to stand in solidarity. The information contained in the message, which minimally distorts meaning as an intention, and is accessible to the recipient's perception, already initiates the foundations of the signaling semiotic system. For example, an audio message, such as a signal of danger, because of its naturalness, intonation, acoustic immersion in imitation of sounds of the surrounding world, gradually becomes convenient and accessible by virtue of the anatomical features of the person through communication.

The question of the appearance of language or what type of communication was the first: acoustic, tactile, "audio-tactile" (M. McLuhan) etc., for academic science for obvious reasons (lack of evidence base, impossibility of its reconstruction) has no solution. But the very fact of language creation, relatively speaking the emergence of a "social contract" on the legitimacy of its existence, lays the foundation for development and enables the mode of growth, permanent "replenishment" of the semiotic sphere of culture. Subsequently, there is a need to duplicate a particular message, or rather to capture its meaning in another long-term design: rock drawing - pictogram - visual symbol. In the absence of the subject, the message will still reach the addressee.

It is clear that the information potential of these signs was minimal, since knowledge of the decryption code was limited by the number of carriers of this information. The so-called "mimetic semiotic code" (U. Eco) appears. Tens of thousands of years will remain until the "iconic" (U. Eco) code (e.g. writing) appears, but the pictogram is already a second-order sign system. Thus, language and writing, as semiotic varieties of the linguistic communication system, provide an example that communication and information exchange can be carried out in different ways, using separate semiotic systems. In this case, they have different sensory nature.

In parallel with the improvement of communicative "expansion" – the media (M. McLuhan), the semiosphere of culture diversifies and new semiotic systems, such as phonetic writing, emerge. The diversity of cultures and the multidimensionality within one culture affect the process of semiosis in such a way that even a detailed study of them will not cover the entire spectrum of the semiosphere. It can only be stated that the emergence of new social needs, forms of activity, communication methods are dialectically related to the development, functioning of certain semiotic systems. It is especially important to dwell on the difference between mimetic and iconic codes. It should be noted that the advantages of the first is the existing visual path – the link of visual imitation. It allows you to map the subject, thing, action, etc. with their image. Even the imperfect, distant drawing allows us to use this sign (system of signs) without additional grammatical structure – a superstructure of rules. Therefore, literacy as a requirement to use mimetic code is absent (or almost non-existent). In other words, in terms of user requirements, this is the easiest way to convey information.

But there is a downside to the problem: the number of characters. It should be unlimited. Any limitation will mean either at least minimal rules or restrictions on the subject of the image. Regarding the rule-setting option, it is a way of developing a written culture with multiple modifications. In our view, the variant of limitation of the subject of the image is a component of the mimetic code of modern media culture. For example, we all know "emoticons", without which we can no longer imagine the daily conversations with people. Each emoticon represents an emotion, if our feelings are not reduced to one image, we can use several. But, if the situation is not typical, we cannot do without words.

Also in the context of examples of mimetic code, it is important to mention another example – frames (E. Goffman) [11]. The reality itself is framed as a kind of storyboard. On the one hand, it narrows the reality, but it gives the opportunity to cross, balance different communication planes not digital and virtual reality. Thus, mimetic code gains new horizons in contemporary culture for use in media communication. As for the iconic code – the signs of the second order, which have no analogy, resemblance to things, actions, sound, meaning, which encode, then their semiotic capabilities are several orders of magnitude greater. The minimal number of characters combined with the grammar system opens up a huge range of usage. But the condition of their functioning is knowledge of the rules of this grammar. This is also a limitation that can only be overcome by learning the grammatical basis of such a system of signs. An example of a modern icon code is a binary system in programming: zero and one. It is versatile, but proficiency in programming is tied to a deep professional knowledge of applied mathematics. Therefore, we cannot claim that the binary system is the basis of social communication and, accordingly, an element of media literacy.

In the case of semiotic media coverage requirements, it is a matter of knowing, first and foremost, the visual icon code, which is linked to the specifics of various programs, information platforms, which we can use quickly and optimally. This is a visual learning of combined perception, "reading" of the information provided by the frame or interface in combination with the target visual acuity: to see the necessary, decipher the sign and get information among different things.

Another feature of media literacy in the context of mosaic audiovisual culture is the maximum combination of different semiotic codes. Deep, perfect knowledge of only one nowadays limits the subject of activity, both in the communicative everyday life and in the professional spheres of being. For example, at first glance, it is as far from the media profession as a composer – the author of academic music. Classical instruments, live performance, classical acoustic nature of sounds, but now more and more composers (and composers of the young generation without exception) create music, using not the usual, many centuries-proven musical notation, but modern computer programs where sound is born in digital system of timbres, bands, musical instruments, etc. This ability requires a special skill: the music that the author's inner hearing hears and touches the piano keys is now being transcribed into the digital version of the composer. With all respect to amateur musicians and amateur versions of this art, however, when it comes to professional authors, performers, it is necessary to point out their training and the requirements of proper possession of this relatively new semiotic system of music.

This version of transcoding the classic profession of artist, architect, is increasingly in need of skills in computer graphics, design, etc. It is not just a tool replacement: it is a computer keyboard and a mouse instead of a paintbrush and paint. It is a change in the semiotic system that is complicated and simplified at the same time. New grammar is emerging - a second-order sign, so the movement goes from mimetic to iconic code. And as discussed above, it broadens the range of possibilities and at the same time narrows the range of people who possess this knowledge. In architecture, it is an opportunity to travel the space that previously existed only in the imagination of the author. People had to wait for the building to be created and only then could the architect's design be imagined. Current technologies, virtual projects make it possible to carry out a kind of testing of the building before this building will either be a city decoration or change its appearance for many years.

We have given examples of the arts professions, in which the special requirement of media literacy is mastering a new communication system. However, there are examples of industries in which the profession is receiving a different semiotic, communication basis, and their number is increasing in different areas of activity. Such sciences as theoretical physics, microbiology, etc. do not need optical devices of their own development and discoveries. With the help of media - computer programs, virtual reality is simulated a world that anyone can look into. It enormously enriches our ideas about the universe, creates a media picture of the world, which acts as a kind of compensation for those knowledge and ideas that each of us cannot get in full because of their incredible scale and industry specificity.

The organizing specificity of mimetic and iconic codes in the formation of modern media communication is also considered by another author – Ch. Gere [12], but in the aspect of screen culture. The screen itself today acts as a certain topos – the place of our worldview, cognitive, leisure existence. And despite the physical size of a particular screen – it actually unfolds the panorama of the media world, which prevails informally in relation to the surrounding non-screen reality. This in turn means that the frame, the boundary of this reality, we are gradually learning not to notice. The behind-the-scenes communicative space embraces us wholeheartedly, and we are sensory focused on the media image.

The world around is dimming not only in spectacular but also in temporal and spatial features. But screen magic has, according to the author, its nature – the genealogy of visual communication. Rectangular, square, oval – various forms of the image, which are locked inside a geometric figure, have passed their approbation in the history of fine arts. Thus, the visual culture of modern civilization incorporates historically formed "screen" models, focusing on two main types: "portrait" and "landscape", each of which corresponds to a corresponding semiotic code, semantic load and communicative environment.

The semiosis of the contemporary audiovisual media culture is based on the factors of information society development discussed above. First of all, it is an easy-to-identify and interactive sign system. There is also a semiotic series of characters that carry an emotional and psychological color and have an anthropomorphic imprint. The verbal segment of information remains important, but for the most part it does not play a central role. A common feature of the semiotic ordering of the message is the compositional scheme of information blocks scattered in the screen plane. The visual topology corresponds to a meaningful navigator of the information space. Today, a large number of specific audiovisual codes have been developed. Awareness of this issue, in our opinion, is a significant part of formation of media literacy.

7 Conclusions

All the issues considered point to the tendencies of cultural transformations in connection with the new communication strategies as the civilizational value of the formation of the information society. This transition makes considerable demands on the subject in terms of mastering the competencies of the carrier of the modern way of human existence. The main attention was paid to the dangers that accompany changes in the ontic characteristics of media culture: cognitive, mental etc. The process of thinking: new tempo-rhythmics, speed and quality of processing, evaluation of information, creation of competitive content; sensory experience: audiovisual defocus, at the same time targeting perception; semiotic polyphony: depth of ownership and, at the same time, variability in the use of different in nature and functions of the screen codes of media communication – all of them are the main guidelines of the named cognitive, mental and identification factors of formation of the subject of the modern educational process.

However, the material discussed above concludes that the learning process today cannot be guided by these characteristics, such as skills, knowledge, competencies, that are amenable to training or isolated development in the context of different educational tasks. If these media literacy requirements are the subject of training, then we are encroaching, even destroying the subjectivity itself. Therefore, it is necessary to create a semantic vertical – a symphonic score of fostering a mentally whole self-sufficient personality. Therefore, in our opinion, the content of modern education, in particular media education, should be unprecedented approaches in which information resources, social goals and means of their implementation are based on preserving the identity of the subject. The person is tasked not only to fit into the information reality of the present, but also to acquire knowledge, the ability to actively build their own semantic space.

Therefore, the development of media education, the depth of awareness of the problems and the development of methods, approaches to ensure the media literacy of the subject – the carrier of the culture of information society, is the basis of self-preservation of human civilization.

References

1. A. Toffler, *The Third Wave* (Bantam Books, New York, 1980)
2. D. Bell, The coming of the post-industrial society: a venture in a social forecasting (2011), https://www.os3.nl/_media/2011-2012/daniel_bell_-_the _coming_of_post-industrial_society.pdf. Accessed 15 November 2019
3. M. Castells, *The Internet galaxy: reflections on the Internet, business and society* (Oxford University Press, Oxford, 2003)
4. M. McLuhan, *The Gutenberg Galaxy: The Making of Typographic Man* (University of Toronto Press, Toronto, Buffalo, London, 2011)
5. L. Manovich, The language of new media (2001), https://dss-edit.com/plu/Manovich-Lev_The_Language_of_the_New_Media.pdf. Accessed 28 December 2019
6. A. Bergson, *Tvorcheskaya evolyutsiya* (Creative Evolution). (Harvest, Minsk, 1996)
7. K. Yung, *Arkhetip i simvol* (The archetypes and the collective unconscious). (Renessans, Moscow, 1991)
8. W. Benjamin, The Work of Art in the Age of Mechanical Reproduction, in *Illuminations*, ed. by H. Arendt (Schocken New York Books, 1969)
9. M. Khaydegger, *Bytiye i vremya* (Being and Time). (Folio, Kharkov, 2003)
10. U. Eko, *Otsutstvuyushchaya struktura. Vvedeniye v semiologiyu* (The absent structure: Introduction to Semiotics). (Petropolis, Saint Petersburg, 1998)
11. E. Goffman, Frame Analysis: An Essay on the Organization of Experience (Northeastern University Press, Boston, 1986)
12. Ch. Gere, Visual Communications **5**(2), 141–152 (2006)

Academic writing as an academic discipline: problematic issues

Zhanna Koloiz[1], and *Yuliia* Yelovska[2,*]

[1]Kryvyi Rih State Pedagogical University, The Ukrainian Language Department, Kryvyi Rih, 50086, Ukraine
[2]Kryvyi Rih State Pedagogical University, The Department the English Language and Methods of Teaching, Kryvyi Rih, 50086, Ukraine

Abstract. The paper studies peculiarities of introducing the course of "Fundamentals of Academic Writing," into the Ukrainian education and scientific process. The integration of the Ukrainian science into the European space caused the necessity of including "Academic Writing" into the Ukrainian curricula. Fundamentals of a new academic discipline are represented through the prism of world and national achievements, which determine such concepts as "academic communication", "academic etiquette", "academic integrity", "the quality of academic text". This research focuses on the following major issues: What is the essence of "Academic Writing"? Is it a brand new course? How does it affect the ranking of higher education institutions? How should intertextual tools be used to keep the academic integrity? How does the level of English proficiency affect the quality of the Ukrainian scientific products? Is the course of "Academic Writing" a tribute to fashion or a vital necessity? It also investigates the peculiarities of academic fundamentals required for development of certain knowledge, skills and abilities of the Ukrainian language academic writing. The latter aims not only to promote academic integrity by following ethical principles and statutory rules that should guide the participants of the education process while studying, training and conducting research activities, but also to represent the concepts, strategies and tactics of organizing the processes of academic text creation, analysis and editing.

1 Introduction

Structural reforming in the national system of higher education started after signing the Bologna Declaration and integrating Ukraine to the Bologna Process, and to the European scientific and educational space. Nowadays this reforming is at an active stage, as it involves all the participants of scientific and education activities. The process of state certifying higher education institutions directly depends on the level of qualification of the scientific and education process participants, lecturers in particular, whose original research achievements are recognized and highly appreciated by the international scientific community due to the high citation index. The presence of eminent researchers is the "stock-in-trade" of any education institution, and simultaneously it confirms favorable conditions necessary for preparing competitive graduates.

Unfortunately, recently the vast number of national higher education institutions makes attempts "to shop" themselves not by means of quality but quantity indicators. This is explained both by subjective and objective factors. Among the variety of reasons the following should be distinguished: the lack of knowledge, skills and abilities to create an academic text and an insufficient level of English.

In this regard, the ideas of introducing certain academic disciplines ("The Principles of Academic Writing", "Academic Writing", "Academic Writing in English" or "English Academic Writing", in particular) into the national education process gain their popularity [1-3].

2 Presentation of the main research material

The implementation of the above mentioned courses and consolidation and systematization of world and national research achievements have led to the emergence of modified theoretical ideas and new (or updated) terms such as *scientific / academic genre*, *scientific / academic information*, *scientific / academic communication*. These concepts are considered in the ratio of common / partial. Cf.: 1) *scientific information* includes reports on certain known and studied issues of objective reality in any genre of academic (learning) communication; *academic information* is related to reports on certain known and studied issues of objective reality in the academic genres of scientific (academic) communication; 2) *scientific communication* is the process of exchanging scientific, scientific and education, scientific and popular information between the representatives of certain fields of study, industry, education, using both verbal and non-verbal means, well as oral or written forms; *academic communication* is the process of exchanging particularly academic (scientific) information between the representatives of certain fields of study, industry, education, using both verbal and non-verbal means, oral or written forms [4]. Some papers represent synonymic

* Corresponding author: julia.yelovska@gmail.com

identification of attributive components *scientific / academic* within these terms. This inconsistency of scientific approaches causes some confusion in the scientific and academic environment. Sample tasks with the following instructions "*Write a Scientific (Academic, Scientific-Academic) Essay*" loosens both current theoretical principles of the *scientific language style* and at the same time complicates understanding of the term "*Academic Writing*".

2.1 Analysis of recent research and publications

Over the last decade *academic writing* has become the subject of interest among the Ukrainian scientists (O. Asadykh, L. Kozolup, L. Matsko, I. Mishchynska, M. Rogozha, L. Sazanovych, O. Semenog, etc.), who focus on the formation of academic literacy, observance of the academic integrity, constructing both English and native language academic texts, etc. It should be noted that such research is usually carried out through the prism of different foreign practices [5-10], based on modern approaches to the interpretation of *the academic style* primarily ("a functional sub-genre of literary language that provides academic communication, mechanisms and processes of exchanging scientific information within state scientific and education institutions, focused on the development and dissemination of scientific ideas") and *academic writing* ("the result of an independent research activity embodied in written form using a graphic symbol system of a particular language that manifests new ways of problem studying and solving" [4]). Cf.: *the scientific language style* is "a functional type of literary language that serves different fields of science, industry, education, represents the relevant results of research activities of individual scientists and scientific teams, manifests and promotes their scientific ideas in specialized texts of different genres" [4]. In any case, it does not raise any objection that the result of academic communication, its structural and linguistic component, and at the same time its representation should be *an academic text*, which is considered as a structurally integral, logically complete communication block.

Despite certain scientific contribution in the relevant field, there are many problematic issues which are worth being the matter of further scientific consideration.

At present the striking fact is that the basis of theoretical determination and interpretation of *academic writing* derives from not the primary sources of certain Western European approaches, but the results of research activities already tested in the Russian-speaking scientific space (Ye. Bakin, Ye. Barbashina, I. Korotkina, G. Makovich, F. Nasybullina, I. Nuzha, A. Syrazeyeva, N. Smirnov, A. Suslov, I. Shchemeleva, L. Valyeyeva, etc.). These studies outline the following statements: the essence of the term *academic writing*, the aspects of developing skills of English and native language academic writing, the formation of certain methodological approaches in this field, the development of the academic writing competences as a tool for the implementation of the state program for improving the competitiveness of national science and education [11], the ways of achieving successful results [7, 8].

It is obvious that world universities are focused on raising their final rating, which is based on the most main activity indicators. Among them there are the following: academic reputation; positive feedback; percentage ratio. The latter includes such aspects as: a) academic teaching staff; b) the number of foreign lecturers and the total number of higher education teaching staff; c) the number of foreign students and the total number of students; d) citations of teaching staff scientific publications, etc. [12]). These criteria are to ensure the investment attractiveness of higher education institutions. These observations draw to the idea that the citation, which is directly related to the academic text and the academic writing, gains high importance.

According to I. Korotkina, applying the new academic discipline is one of possible positive steps. The other way is creating university centers of academic writing, since it is necessary primarily to train those who will teach further. The relevant project has been successfully maintained in Russian Federation [13, 14].

Over last two or three years, the Ukrainian academic community investigates the problems of academic communication and academic integrity within numerous forums held by both national and international higher education institutions. Roundtables, workshops, scientific conferences such as "Academic Integrity: Compliance Issues and Spreading Priorities for Young Scientists" [15], "Researcher Academic Culture in Education: European and National Experience" [16], "Academic Honesty as the Basis of Sustainable University Development" [17], "The School of Academic Writing: From Theory to Practice", etc. encourage their participants to search effective mechanisms to ensure the implementation of the ethical aspects of academic communication.

Recently, the Ministry of Education and Science of Ukraine has approved the Integration Guide of Ukraine into the European Research Area (ERA-UA). It updates six key priorities for developing effective national research systems and efficient international cooperation and competition, opening labor market, providing research gender equality and valuable exchange of scientific knowledge [18]. It is stated that "integration will give access to new markets and new knowledge, giving additional opportunities for the development of the national research system" [18]. It is clear that successful joining the ERA (European Research Area) requires the demonstration of our scientific and creative potential, involvement into the discussion of current world problems, offering probable solutions, etc. This integration is usually fulfilled due to the implementation of various scientific activities such as participation in conferences, seminars, workshops etc., or in the way of scientific publications. Therefore, it brings to the necessity of "the development of competencies required for text writing in accordance with international rhetorical conventions and training independent researchers who may become active participants of global academic discourse" [8].

Theoretically everything seems to be clear, but in practically there are some problems that we aim to study within the proposed research.

2.2 Representation of disputable points

2.2.1 The first problem

The first problem is connected with the issues whether "Academic Writing" is a completely new course, or simply a modified term of an officially accepted course, such as "Scientific Prose Style" or "Culture of Scientific Prose Style", which is approved not only in the Ukrainian education system but also abroad.

Current studies show that the term of "academic writing" has received several definitions: on the one hand, it is interpreted as the result of independent scientific or scientific and education activity, embodied in writing. It uses the system of graphic signs of a certain language which reveals new cognition and problem solving ways [4]. On the other hand, it is a "scientific and methodological branch of knowledge, the field of scientific and pedagogical research". Some researchers determine this term as an academic discipline, "studying the methods and tools of constructing an academic (scholastic) and scientific text" [8]. However, it should be mentioned that "the methods and technologies of constructing an academic (scholastic) and scientific text" have recently been represented within other courses [19]. Moreover, certain studies demonstrate simply the replacement of an "old-fashioned" term with the "modern" one. As a result, we observe so-called term updating: *fundamentals of research studies → the fundamentals of academic writing* [20], *constructing an academic (scientific) text → constructing an academic text* [6, 21] etc.

In addition, the list of knowledge, skills and competences that students should acquire while studying these old and new courses (for instance, to distinguish scientific prose style from journalistic or literary one; logically arrange the text and organize its elements; construct the text starting from hypothesis suggestion, abstract formation to conclusion drawing; critically evaluate, select, summarize and use information from various sources, etc. [7]) may be enriched by adding the peculiarities of "international norms and requirements of the academic text creation".

It should be noted that many scientists use the terms "a scientific text" and "an academic text" as synonyms (sometimes even within the same paper), or consider them in total / partial interrelation (cf.: "basic principles of nonlinear constructing a scientific (academic) text as a coherent system" [7]). In terms of an academic discipline, I. Korotkina defines the term *academic* through the word *scholastic* and counters it as a scientific one. Such interpretation is probably related to the traditional differentiation of the *academic (scientific)* as the subtype of the scientific language style (partial → general). In this case, it is not appropriate to equate academic style and education (learning) one, because the education and scientific subtype is another functional sub-genre that serves the field of education, science, industry and technology [19]. On the contrary, the word "*academic*" in the national usage has consolidated with the meaning "*school*" (cf.: "an academic year"). However, it should be emphasized that the corresponding meaning is the derivative, and the original definition is interpreted in the relation to the academy ("the name of state higher education institutions aimed at the development of sciences or arts"). Hence, *academic* is "related to education and scientific activities in higher education institutions". Taking into account these ideas the issue of the academic literacy, based on academic writing of university students, and even comprehensive school students seems to be ambiguous.

Another issue that should be pointed at is that the word "*writing*" in the Ukrainian language has clearly fixed meanings, among which is out-of-date "written or printed work". Moreover, none of these connotations actualizes the meaning "oral". For that reason it is probably not entirely correct to use it in terms of academic writing, since an academic text as a unit of academic writing functions in both written and oral forms. Furthermore, by means of the latter it becomes possible to train a person to generate his / her own ideas, prove or negate certain statements, logically arrange the scientific facts and data, etc. Therefore, the course of "Academic Writing" is sometimes represented as "Academic Writing and Rhetoric". Consequently, it is obvious that the corresponding name is an attempt of tracing the English language term and adapting it to our conditions in order to affirm the integration of Ukraine into the European educational space. As a result, we face to another problem.

2.2.2 The second problem

The second problem reveals whether "Academic Writing" influences on the promotion processes of higher education institutions.

This statement sounds rather peremptory. In such a case, it is necessary to differentiate the comprehension of the native language "Academic writing" and the English language "Academic writing". Unfortunately the Ukrainian language as many other languages, does not take the leading place in the global information space, since it has been totally conquered by English.

Millions of students graduate from national universities (especially today, when higher education is no longer considered to be elitist), but only a small number of them dedicates themselves to professional research (academic) activities. A student, who has poor knowledge of his /her mother tongue and, perhaps obtains even worse English language skills, can hardly show outstanding results. The assessment of their mother tongue "Academic writing" is usually marked with "satisfactory" level and the review of a "created" English language academic text receives the following comments of "Poor English". It is fair to say that to get a perfect English text you need to create a good native language text. It is well known that a bad national product, even if it is wrapped in the attractive English language cover, is not suitable for export to the world market.

To change the situation for the better means to improve the quality of Ukrainian academic products. It becomes the priority requirement for those who plan to study at higher education institutions. Therefore, they are obliged to join the English-speaking scientific community, foremost to ensure the citation index, which is supposed to allow the objective evaluation of scientist qualification. However, neither good knowledge of the native language, nor a high level of English language skills and the best training results in studying "Academic Writing" can guarantee unconditional success in academic text creation. The latter depends on the scientific field and the level of problem significance. It is obvious that world humanity field faces certain difficulties (the Ukrainian language in particular). For instance, it is quite complicated for Humanities such as the Ukrainian language Philology, to compete on the world market. The point is not only that some of the Ukrainians consider this language "for home use only", but also in the peculiarities of its formal structure and the specificity of linguistic units.

In addition, the presence of the English language works, their publishing in editions registered in the scientometric databases, etc., does not mean worldwide recognition of their authors due to various reasons. Among them there are the following: it is much easier to publish the paper in the national scientometric publications than in the foreign ones. On the one hand, it is good as it uncovers the rating of both the scientist and the higher education institution. On the other hand, this raises the issues of scientific significance of the national English language academic products in solving global scientific problems. Furthermore, a high citation index can hardly indicate the worldwide recognition of a scientist.

2.2.3 The third problem

In fact, the third problem is related to the citation and references that influence on scientist ratings. To paraphrase a well-known saying, we refer, and we are referred. It is especially important to choose among the variety of scientific information which is the most useful for creating your own academic product. Moreover, it is necessary to use this information in such a way so that the author is not accused of violating academic integrity, quote verbatim or plagiarism.

Considering above mentioned facts the course of "Academic Writing" is a rescuing solution. Nowadays the problems of applying other people's scientific ideas, peculiarities of their citation and reference methods, the interaction of the newly created texts and source texts [22], are widely discussed by the researchers all over the world. At first, it seems to be fairly simple: you should not give out other people's ideas as your own ones, and practice an excessive citation. While putting certain quotations you should fill in the proper reference marks, paraphrase the texts, etc. Though following all these requirements does not guarantee 100% originality of this newly created academic product. Consequently, the more quotations you have, the lower the indicators of originality. Intertextuality is also one of the category features of any academic text, and, (cf. intertextuality is a category feature of an academic text based on its interaction with semiotic cultural environment, which brings the traces of one academic text in the other [4]). If penetrating deep into the essence of this term, we are able to verify the statement that "all words have already belonged to someone" (L. Kostenko). Framework and situational language clichés (pattern phrases) reflect the specifics of a large academic text, and therefore they often reduplicated in different samples. Corresponding clichés are required when creating both native language and English academic texts.

In addition, the analyzed course and its "prototypes" successfully use concepts of reference (the fragment of an academic text where the author appeals to another scientific source to dispute, add or illustrate his / her own position with facts from other academic works, confirm or negate someone's opinion, compare different scientific approaches to problem analyzing); quotations (the fragment of an academic text, which is appropriately designed and literally transmits the ideas of other people on a certain problem; it is usually used for confirmation of author's opinion or disputing with the cited author); recitation (the fragment of an academic text in which the author appeals to another scientific source indirectly, through mentioning the third one); invocation (using other people's statements in the structure of one's own academic text with a slightly modified form and content); rendering (retelling the results of one's research activities scientific ideas, scientific concepts, scientific concepts, etc. in other words), paraphrasing (using one's statements in the structure of one's own academic text with a slightly modified form or content) [4]. It is impossible to "build your own house" without the use of "additional" building materials. The same situation is observed in academic writing. One can not build his / her own text without applying other people's thoughts, ideas, or so called quoting, paraphrasing, etc. Creating a high quality academic text encourages researchers to analyze the requirements of using foreign texts established by foreign higher education institutions and scientific publications, which are guided primarily by "logics of scientific communication", "respect for the reader and other authors" [22].

2.2.4 The fourth problem

The fourth problem is connected with the fact that the native scientific community often applies English usage as the medium language to join the global information space.

Consequently, the level of the English language skills becomes of high importance. The Ukrainian researcher, who attempts to integrate into the European scientific space, should obtain major foreign language knowledge and skills, which can be used in terms of academic communication or academic text creation. The situation is considerably complicated under the conditions of poor language skills which do not allow the researcher to demonstrate his / her significant academic potential. Moreover, when constructing an academic text, alternative ways of transmitting information are unacceptable. The level of the English language skills used for the academic text creation should correspond to high English proficiency. Beginner, Elementary (A1), Pre-Intermediate (A2) or even Intermediate (B1-B2) English are considered unsatisfactory. Unfortunately, not many Ukrainian researchers and senior students demonstrate Advanced (C1) or Proficiency (C2) English skills, necessary for perception, comprehension, and thought verbalization in English, conveying the most significant shades of meaning, using peculiar grammar structures while creating a proper academic English text. Otherwise, it turns out to be a forgery, which, in fact, we still have.

The basis of intellectual development, the cognition of the definite objective reality concepts, their comprehension, awareness and understanding, of course, is accomplished by means of the mother tongue. In turn, this draws the following conclusion: an English language academic text is a secondary result, built on the primary Ukrainian one. In other words, English language academic writing is basically a formal English translation of the Ukrainian language academic text. Obviously, one should consider the specific structural and grammatical organization of each language, to transmit the accurate original content and avoid the distortion of scientific facts.

In this case, the native language and the English language "Academic Writing" should not be contrasted and studied as two separate courses. They can function as different content blocks of one training course, where the English language academic writing, specifies the knowledge acquired during the study of the native language academic text construction (for instance, for PhD students). Alternatively they can be manifested as separate cycles illustrating the following hierarchical system (from the simple to the more complex): "Essentials of Academic Writing" (for Bachelor training) → "Academic Writing" (for Master training) → "English Academic Writing" (for PhD training). This approach also presents new requirements for lecturers, who should be fully aware of their purpose, obtain good knowledge of both languages in order to teach properly, and show their own samples of high quality native language and English language texts. Furthermore, it does not exclude the possibility of making the co-work of two different lecturers.

2.2.5 The fifth problem

The fifth problem encourages reconsidering the statement, whether the course of "Academic Writing" is a tribute to fashion or a vital necessity.

Nowadays the idea of "Academic Writing", as well as academic integrity, becomes essential, gradually affirmed according to the present-day reality and highly demanded while carrying out education and scientific activities.

Higher education institutions should concern their reputation as "one of the mechanisms of internal self-regulation in the professional community", since the sphere of higher education serves to form "future social and professional elites", and is responsible for the interpretation, dissemination and use of new knowledge [23]. Academic etiquette, academic ethical priorities, academic values are widely disputed nowadays in terms of societal demands. It is worth noting that all these concepts are not new, as the relationships in each academic community have long traditions based on certain principles and rules and they are hardly related to academic corruption. This phenomenon depends equally on each participant of academic research activity and the object of academic communication. Academic dishonesty, including academic corruption, is a negative phenomenon and needs to be excluded. However, today the question is how quickly the elimination of such "immoral standards" will affect the quality of academic writing and an academic text. It is also not clear whether all current results of "global academic culture" should be considered as a model.

Analyzing present day situation we observe the following: counteracting the "excessive commercialization of national higher education institutions", their "intense bureaucracy", we commercialize scientometric publications of foreign countries by sponsoring their economies. No one can be completely be sure about the validity of these publications, as the cases of so-called scientific fraud are quite frequent.

Still, there are some risks associated with the overload of the academic market with low-quality scientific products, so-called "Chinese fakes". The emergence of a new subject of discussion, like anything new, arouses the interest, promotes the creation of certain products and increases their demand, consequently determining their supply. A large number of academic publications, which study modern issues of academic communication, academic integrity, academic literacy, academic writing, etc., is partially considered as a tribute to fashion. Unfortunately the quality of some of academic achievements is unconvincing [2].

The study does not enumerate all the problematic issues as the national theory of academic writing is now at its initial stage and requires applying effective mechanisms for its development not only at the level of higher education institutions.

2.3 The Ukrainian language academic writing

Within the proposed research we plan to dwell on the peculiarities of the course which is aimed at build primary basis for the development of knowledge, skills and abilities of the Ukrainian language academic writing.

Firstly, this course focuses on the promotion of academic integrity appealing to ethical principles and rules defined by law, which should be followed by participants of the education process while studying, teaching and conducting scientific (creative) activities to ensure the validity of learning and scientific (creative) outcomes. Secondly, it represents the concepts, strategies and tactics of academic writing, the technology of organizing the process of academic text creation, its analyzing and editing.

Therefore, the concepts of *academic etiquette*, *academic literacy*, *academic text originality*, *plagiarism*, *compilation of other people's thoughts* are naturally updated. Proficient understanding of the laws, principles, ethical rules of academic communication in the scientific and education community, observance of academic integrity, support of academic standards, prevention of fraud, fabrication plagiarism is a duty of each participant, who should generate his / her own scientific (creative) ideas, carry out new competitive research projects that would meet contemporary scientific requirements bringing theoretical and practical significance.

This course provides a number of specific tasks that contribute to the above mentioned goal, such as: 1) to provide basic information about academic writing as one of the scientific (academic) communication sub-genres, its theoretical and methodological fundamentals; 2) to develop practical skills of communicatively justified language usage in creating appropriate samples of academic writing; 3) to train to create different kinds of academic texts depending on the form of communication and the way of presenting information; 4) to update information in terms of the academic text interpretation (content, intertextual connections, composition and architectonics, etc.), and psychological, logical and linguistic principles of its editing; 5) to represent peculiarities of academic essay writing as one of the genres of academic writing.

At the end of the course, students should acquire the following knowledge: the traditions of academic communication and international experience; the interpretation of academic writing, its category features, content and methods of information presentation; genres of academic writing; rules of academic integrity and ways of plagiarism detecting and preventing; types of academic texts according to the forms of communication and methods of material presentation; peculiarities of academic text content and composition; typical errors common to the process of creating an academic text; psychological, logical and linguistic principles of academic text editing.

The knowledge, obtained through practical experience, allows gaining such results as: to operate the concepts "scientific style", "scientific text", "academic communication", "academic writing" and identify their integral and differential characteristics; to generate scientific ideas, model scientific knowledge and organize the process of academic writing; to use presentation and argumentative forms of academic communication; to create fragments of texts-stories, texts-descriptions, texts-arguments, texts-proofs, texts-definitions, and take into account the peculiarities of their construction; to conduct scientific debate, validate or negate updated statements; compress foreign texts, group sources, systematize information, refer to reputable scholars, cite their positions; to analyze an academic text through the prism of content and structuring by means of pretextual and textual units; to provide relevant arguments on the topic, to represent the state of the problem, to formulate the object, subject, purpose and tasks of made research; to accomplish the semantic analysis of composite academic text fragments; to apply different types of academic text correction; to follow the psychological, logical, linguistic and stylistic principles of academic text editing; to write an academic essay with its subsequent analysis and linguistic editing.

3 Conclusions

This course aims to form the responsibility to follow academic integrity and academic literacy among the national educational and scientific researchers while creating different samples of academic texts. It should gain the compulsory status of in all education levels: from Bachelor to Ph.D.

Despite its recent and selective functioning in the modern education and scientific circulation, we conclude that it does not emerge as an entirely new course, but absorbs the most advanced and future-oriented achievements of the theory of scientific language (scientific language style), providing not only written but also oral forms of verbalization various scientific ideas, knowledge keeping in mind "there is never a new fashion but it is old" (Geoffrey Chaucer).

Theoretical information on organizing academic writing, mechanisms of academic text creating, etc. should be gradually represented (from simple to more complex; from the creation of a high quality academic language text to the publication of the same quality English language version). It is obvious that the knowledge, skills and qualifications of higher education applicants are different and depend on both their level and system of values.

Academic integrity and related issues should be permanently integrated into the theory and training practice in higher education institutions and the course "Academic Writing" is focused on understanding the basic concepts of academic communication, improving the latter and developing personal unique way of academic writing in order not to assent to such negative phenomena as academic copying or imitation, since even high quality copies are fakes, and usually not in demand.

References

1. T. Babenko, Youth and the market **3**(134), 112 (2016)
2. T. Tymoshenko, N. Hrechykhina, Naukovi zapysky **9**, 214–219 (2009)
3. N. Shlikhta, I. Shlikhta, *Osnovy akademichnoho pysma* (Academic Writing Basics) (2016), https://drive.google.com/file/d/0ByePGdGpHh6WbnRzT2pGNUhoY2c/view. Accessed 26 Sept 2019
4. Zh. Koloiz, *Osnovy akademichnoho pysma* (Academic Writing Basics). (FOP Marynchenko S. V., Kryvyi Rih, 2019)
5. S.V. Bogolepova, Higher Education in Russia 1, 87–94 (2016)
6. A. Ibraeva, T. Ippolitova, *Akademicheskoye pismo: printsipy strukturirovaniya i napisaniya nauchnogo teksta* (Academic Writing: Principles for Structuring and Writing a Scientific Text). (SKGU im. M.Kozybaeva, Petropavlovsk, 2015)
7. I. Korotkina, *Akademicheskoye pismo: protsess, produkt i praktika* (Academic Writing: Process, Product and Practice). (Iurait, Moscow, 2017)
8. I. Korotkina, Dissertation, Institute for Education Development Strategy of the Russian Academy of Education, 2018
9. N.V. Smirnova, Vestnik Voronezhskogo gosudarstvennogo universiteta. Seriia: Lingvistika i mezhkulturnaia kommunikatciia 1, 140–147 (2017)
10. N.V. Smirnova, I.Yu. Shchemeleva, Vestnik Sankt-Peterburgskogo universiteta. Jazyk i literatura **9**(2), 185–197 (2015)
11. G.V. Makovich, Development of the academic letters 'competencies as a tool for implementation of the state program for increasing the competitiveness of

science and education. Voprosy upravlenija 1(50), (2018). doi:10.22394/2304-3369-2018-1-12-18,

12. Times Higher Education, World University Rankings, https://www.timeshighereducation.com/world-university-rankings (2019). Accessed 28 Nov 2019
13. E.V. Bakin, Higher education in Russia 8–9, 112–116 (2013)
14. I.B. Korotkina, Higher education in Russia 1(197), 75–86 (2016)
15. N.H. Sorokina, A.Ye. Artiukhov, I.O. Dehtiarovoi (eds.)., *Akademichna dobrochesnist: problemy dotrymannya ta priorytety poshyrennya sered molodykh vchenykh* (Academic Integrity: Compliance Issues and Dissemination Priorities for Young Scientists). (DRIDU NADU, Dnipro, 2017)
16. O. Semenog, *Akademichna kultura doslidnyka v osvitnomu prostori* (Academic culture of the researcher in the educational space). (SumDPU, Sumy, 2018)
17. T.V. Finikov, A.Ye. Artiukhov (eds.), *Akademichna chesnist yak osnova staloho rozvytku universytetu* (Academic honesty as a basis for the sustainable development of the university). (Takson, Kyiv, 2016)
18. Vyznacheno 6 priorytetiv intehratsii ukrainskoi nauky v Yevropu – Kolehiia MON zatverdyla proekt Dorozhnoi karty (Six priorities for the integration of Ukrainian science into Europe identified – Collegium of MES approved draft Roadmap) (2018), https://mon.gov.ua/ua/news/viznacheno-6-prioritetiv-integraciyi-ukrayinskoyi-nauki-v-yevropu-kolegiya-mon-zatverdila-proekt-dorozhnoyi-karti. Accessed 28 Nov 2019
19. O. Semenog, *Kultura naukovoii ukraiinskoii movy* (Culture of the Scientific Ukrainian Language). (Akademiia, Kyiv, 2010)
20. V.K. Novikov, *Osnovy akademicheskogo pisma* (Academic Writing Basics). (MGAVT, Moscow, 2016)
21. E. Yarskaya-Smirnova, *Sozdaniye akademicheskogo teksta* (Creation of an academic text). (Variant, Moscow, 2013)
22. I.B. Korotkina, Higher education in Russia 2, 142–150 (2015)
23. A.Ye. Artiukhov, T.V. Finikov, Akademichna dobrochesnist: vyklyky, dii, uspishni istorii (Academic Integrity: Challenges, Actions, Success Stories) (2019), https://examenarium.sumdu.edu.ua/info/courseinfo/100. Accessed 28 Nov 2019

Fostering students' autonomy in learning English in nonlinguistic university

Oksana Pershukova[1,*], *Nina* Nikolska[2], and *Oksana* Vasiukovych[1]

[1]National Aviation University, 1 Lubomyra Guzara Ave., Kyiv, Ukraine
[2]National University of life and environmental sciences of Ukraine, 15 Heroiv Oborony Str., Kyiv, Ukraine

Abstract. The study aims to find out whether it is possible to foster students' learner autonomy in the context of ESP language learning in non-linguistic universities by using a special approach. The experiment was carried out at National Aviation University in Ukraine with two groups of first-year students of electronics (experimental and control) in 2018-2019. Testings to determine students' level of communicative competence in English and surveys to identify students' level of learner autonomy development were conducted in September and May of the same year of education. The control group did not receive any special training, while in the experimental group were created special learning conditions. With the purpose to prepare students to accept responsibility for their learning, they were given the opportunities to choose educational materials; to set goals of their learning; to reflect the process and evaluate the results of learning, etc. Modern technologies were widely used as well as scaffolding (if necessary). According to the results of the experiment, it was stated that only a part of the most active students used the created conditions and gained experience in autonomous learning. The conclusion was made about creating such an environment. It is a challenge that is appropriate to realize to give an autonomous learning experience to aspiring students.

1 Introduction

1.1 The problem statement

Education in the age of globalization and integration is closely linked to the level of competitiveness of university graduates, which today largely depends on their ability to work with information, self-study and acquire knowledge independently. At the same time, the new Ukrainian educational standards introduction has resulted in the transfer of a large part of the work aimed at the mastering of educational material to outside the classroom, leaving the volume and content of the programs unchanged. The discrepancy between the amount of knowledge a student has to absorb and the time allotted for this makes him or she look for effective ways to organize students' independent work. The problem of students' willingness to learn for a lifetime depends on the development of self-directed learning activities. In this regard, students' independent learning activity has a special role and involves the transition to a 'subject-subject' interaction between a teacher and a student, shifting a focus on active methods of mastering knowledge. Therefore, the whole learning process should be re-oriented towards promoting students' self-development, motivation raising, and learners' autonomy fostering.

In the current conditions of globalization and integration, the willingness to communicate freely with representatives of different peoples and cultures is a demand of time, which is why the whole world is now engrossed in mastering foreign languages. Modern professional education in Ukraine in this context aims at training a specialist ready to work in the flow of foreign-language scientific information and communication with foreign colleagues able to work with professional scientific literature and to learn new technologies because the level of competitiveness of a modern specialist depends largely on using a foreign language. English is considered now as a lingua franca, a means of international and intercultural communication. At the same time, the modern realities in Ukraine are that: a) students of non-linguistic universities have a rather heterogeneous level of competence in English (because sometimes they studied German or French as a foreign language at school); b) students are not always interested in a foreign language, because they poorly conceive the importance of this discipline in his/her future professional activity; c) the amount of time allocated for classwork at the university is constantly decreasing, extending the amount of students' independent work.

In these conditions, there is a contradiction between the objective necessity of a high level of foreign language competence of a modern specialist and the real state of affairs in non-linguistic universities in Ukraine, since the level of such training does not meet the requirements of the present and needs improvement. This problem can be considered in many aspects, and in our opinion, the aspect of creating conditions for

* Corresponding author: pershoks@gmail.com

students' autonomous learning in the current circumstances is one of the requiring solutions. Ways of fostering the development of students' autonomous learning English should be considered as one of the urgent tasks of the learning process at a non-linguistic university. Learners' autonomy has a high priority in the new education system since its formation through the mechanical transfer of knowledge from teacher to student is not possible, it can be developed. Knowing how students' autonomy can be fostered, how to make every student an autonomous learner and successful user of English would help us to make our educational system more efficient.

1.2 Literature review

Students' autonomy in the foreign language learning process. As far as researches for a long time failed to reach a consensus on what the notion of students' autonomy exactly is until recently some scientists have characterized students' autonomy as a controversial and problematic term. Little indicates that: 'it is often confused with 'self-instruction' [1]. Karastateva states that 'it is loosely used along with 'autonomous learning' [2]. Benson points out the number of related but not fully synonymic to autonomy terms (self-access, self-study, self-education, out of class learning, distance learning), and explained the difference in such way: 'they describe the way and degrees of learning by yourself, whereas autonomy refers to abilities and attitudes… as well as the capacity to control own learning'. The researcher also mentions that among other terms used as synonyms in discussions on the issues of students' autonomy is self-directed learning or independent learning. That's why he stresses the necessity to check what the author exactly meant by using them [3].

Learner autonomy has become a topic of interest and discussions over the last two decades. As Benson indicated, it was caused by some learner-centred approaches to language education, which included different aspects of independence of learning. The early history of the notion of *autonomy in language education* begins with Holec's seminal report to the Council of Europe's Modern Languages Project [4]. According to still influential Holec's [5, p.3] definition of learner autonomy in it is 'the activity that a learner has developed to take charge of his or her learning'. The researcher underlines that this activity includes: a) determining the learning objectives; b) defining its contents; c) selecting methods and techniques; d) monitoring the procedure, as well as evaluating the results. He stressed that self-assessment is an important and integral element of learner autonomy. This point of view is supported by many pieces of research, among them is Little [6, p.175], who asserted that 'the basis of learner autonomy is that the learner accepts responsibility for his/her learning'. According to Benson, the wide notion of autonomy can be classified as technical autonomy: the act of learning language outside the framework of an educational institution and without the intervention of a teacher; psychological autonomy: a capacity which allows learners to take more responsibility for their learning; political autonomy: control over the processes and content of learning [cited in Finch, 7]. Benson & Voller made a significant contribution to the field by clarifying the difference between autonomy and independence in language learning.

They stated five ways the term autonomy can be used:

1) for situations in which learners study entirely on their own;
2) for a set of skills which can be learned and applied in self-directed learning;
3) for an inborn capacity which is suppressed by institutional education;
4) for the exercise of learners' responsibility for their learning;
5) for the right of learners to determine the direction of their learning [8, p. 2].

Ommagio determined some features of the ideal autonomous learners. They are considered to have insights into their language styles and strategies, to take an active approach to the learning task, to be willing to take risks, to be good guessers, to attend to form as well as content and to have a tolerant as well as the outgoing approach to the target language…[cited in Thanasoulas, 2000, 9]. According to Littlewood, responsibility and ability are the two main dimensions of learner autonomy. Students must take responsibility for their learning, because, they are the ones who do the learning. Besides that, the students are responsible for some process which traditionally belonged to their teacher. These processes include deciding on learning objectives, selecting learning methods, evaluating the results of learning. Ability means the students' capability to complete the processes or tasks connected to responsibility [10]. Some later researches supported these ideas, as Bajramy who indicated that 'learners should be able to build their criteria for the quality of their work…be independent… and be able to make judgments for their strong and weak points of learning' [11, p.425]. So, to become an autonomous learner it is expected that one will be able to 'set personal learning goals, to identify learning strategies and develop study plans to achieve these goals, select relevant resources and support, assess and reflect on one's progress' [12].

Holec pointed out that autonomy is an attribute of the learner. He stressed that 'it's not inborn ability so it must be acquired by the learner either by natural means or by formal learning, i.e. in a systematic, deliberate way" [5, p. 3]. Dickinson indicated that 'learners do not automatically accept responsibility' for their learning and 'do not necessarily find it easy to reflect on the learning process. Teachers must… provide them with appropriate tools and opportunities'. Dickinson also stressed that the liberalization of the classroom will help the students to take responsibility for their learning. Thus, the learner has the responsibility to make decisions and take charge of his/her learning [13]. Little supported this idea considerably with words 'learners who take responsibility for their learning are more likely to achieve their learning targets'. But without suitable guidance, the student will not be able to become aware of the possibilities [6, p. 176]. So, the teachers' role is very important in students' autonomous learning and cannot be ignored. The relationship between teachers' autonomy and learners' autonomy has been in the centre of many researchers' attention. Among them is Little, who called attention to the importance of having autonomous teachers to promote students' autonomy as well as highlighting the necessity to analyze and research this concept thoroughly. He indicated the idea that 'learner autonomy develops through pedagogical dialogues in which teachers exercise their own autonomy' [6]. Littlewood reviewed teachers' autonomy from two different perspectives: as the capacity for independent decision making, which involves having abilities and skills for action: and as willingness, which includes motivation and confidence to make a choice [14]. Benson named the concept of teacher' autonomy as 'one of the most significant and problematic' because 'in the language teaching literature, there is a much greater emphasis on teacher autonomy as a professional attribute and the link between teacher autonomy and learner autonomy' [4]. Thus, to be able to foster students' autonomy in the classroom the teacher has to be autonomous him/herself.

There is a strong need for support and supervision of language learners' as they are moving toward their autonomy developing. The process to make the students become masters of their autonomous learning is not easy, because the teacher and the students 'must shift their roles so it can be created a positive learning atmosphere' [11]. Dickinson described autonomy as 'the learning situation in which the learner is responsible for all the decisions made and implemented concerning his learning and the teacher becomes a skilled manager of human beings' [13, p. 11]. To reach this goal the researcher pointed out to the teachers six ways for promotion learners' independence in his paper [15, p. 2]. They are:

1) legitimizing independence in learning by showing that teachers are approving and encouraging the students to be more independent;
2) giving learners successful experiences of independent learning and convincing them that they are capable of greater independence;
3) giving learners opportunities to exercise their independence;
4) helping learners to develop learning techniques (learning strategies) so they can exercise their independence;
5) helping learners to become more aware of language as a system so that they can understand many of the learning techniques available and learn sufficient grammar to understand simple reference books;
6) sharing with learners something of what we know about language learning so they have a greater awareness of what to expect from the language learning task and how they should react to problems that erect barriers to learning [15].

Thus, learners do not automatically accept responsibility for their learning in formal contexts and do not find it easy to reflect on the learning process. Teachers must provide them with appropriate tools and opportunities to practice using them [6, p.176]. The significance and importance of learner autonomy functioning at the university level pointed out Bajrami 'Learner autonomy ...undertakes the outcomes at the university level such as flexibility, adaptation, self-initiative and self-direction'... promotes democratic education societies, prepares individuals for a lifelong learning process,... makes the best opportunity for learners to use their creative ideas in and out of the classroom' [11].

So, according to scholars' papers mentioned above, we can conclude that *students' autonomy* is a complex notion. It means the activity that a learner has to develop for the most successive learning and it anticipates that every student has to take charge of his/her learning, which includes: determining the learning objectives, defining its contents and selecting methods and techniques, monitoring the procedure, as well as evaluating the results. Students' autonomy has three dimensions: responsibility, ability and capability. The basis of learner autonomy is that the learner's acceptance of responsibility for his/her learning, while the ability is the language learning aptitude, as well as his/her capability means readiness to complete the processes or tasks connected to responsibility. Student's autonomy is an attribute of the learner, it's not inborn ability Teachers have to provide students with appropriate tools and opportunities to develop it through the liberalization of the classroom and creation a positive learning environment as well as pedagogical dialogues in which teachers exercise their own autonomy.

The concept of *teachers' autonomy* in our interpretation means teachers' willingness and capacity to take control of teaching and learning to make the learners active and independent in their learning.

We understand *learners' autonomy in ESP language learning* as educational work performed by students under the methodological and scientific guidance of the teacher in the context of gaining knowledge and skills to accept responsibility for their learning in meeting the specific needs of the future professional English communication requirements and in this meaning we use it in the paper.

2 The study

Based on the above considerations, the current study attempted:
- *to explore the concept of students' autonomy in the ESP language learning process;*
- *to create a learning environment* that is favourable for students' autonomy fostering;
- *to examine* how much students' activity in autonomous work associates with their progress.

The following *research questions* were posed:
- How is it possible to create a favourable learning environment for students' autonomous learning in the context of ESP?
- Whether factors of learning environment for students' autonomy developing make a real difference in students' educational success?

It was anticipated that forming students' autonomy in learning will make a positive influence on their motivational intensity, they will be more satisfied with learning English and their results in education will be better. The study is based on the analysis of scientific papers and materials on the topic of research and exploring the effectiveness of the created learning environment, in which the students of the experimental group used special strategies for autonomous learning, while the students of the control group studied in traditional conditions.

2.1 Participants

The participants of our study were 38 first-year students of the National Aviation University specialty: 151 'Automation and Computer-Integrated Technologies'. Out of 38 participants, 9 were females and 26 males who ranged from 17 to 20 years. At the time of research, English for Specific Purposes (ESP) lessons were part of their official university curriculum. Although the students were registered in two different groups (control and experimental, matched by English language proficiency level) the research was conducted by the same teacher who followed the same procedure and used the same materials for questioning and forming the ability to independent learning activity in both groups. According to the curriculum, two-hour classes were held once a week for two semesters. Testing was conducted in the control and experimental groups at the beginning and the end of the academic year.

Students' language competence was assessed in the control and experimental groups by the course teacher and her two colleagues at the beginning and the end of the academic year and grades were considered as an indication of their achievement. To determine students' characteristics in terms of autonomy they were suggested to answer (anonymously) the questionnaire to investigate their views on their responsibilities in learning and decision-making abilities. As some other researchers [16] we used Joshi's (2011) Learner Autonomy Questionnaire in our work, which was conducted in English and was supplemented by a translation into Ukrainian (students' native language). To answer the questionnaire there was suggested the Likert scale with 5 ready answers (strongly agree – agree – neutral – disagree – strongly disagree). We suggested students of both (experimental and control) groups answer the questions twice at the end of September (at the beginning of the year of education) and in May (at the end of the same year of education).

It is worth noting that students' level of communicative competence in both groups was miscellaneous. As it was found that 7 students of the experimental group and 5 students from the control group had a Pre-intermediate level of language competence, which is not enough to master the course of ESP (Electronics) and absence skills of independent work. They were suggested special consultations on grammar and language practice improvements. For such students of both groups, the amount and content of their independent work were defined. Besides, all the students were taught how to navigate background materials and use different sources of information. At the language practice classes, students of the control group were taught using traditional teaching methods. In the experimental group, we used a special learning approach which is based on a special environment creation and using strategies for students' autonomy fostering. The purpose of the approach was to prepare students to accept responsibility for their learning:

- the analysis of the situation and determination of the goal;
- development of the plan of the desired achievement;
- determination of available means and resources;
- implementation of the plan and development of the corrective actions plan (if necessary);
- real-life incorporation into the process of students learning.

To prepare students for the gradual transfer them functions of the organizer, during the learning process they were suggested:

- to discuss problems about the future profession (individually or in groups) in the form of short talks (to make them feel the importance of learning English for the future profession);
- to pick the topics (for projects, guidelines, and rubrics), and present the results of work in the class (to make them feel responsible for what happens during the lesson);
- to keep personal blogs or written journals reflecting their learning experiences (positive and negative);
- to obtain self-evaluation constantly (students are given the possibility to be in charge of their learning);
- to create resources for learning, that relate to the content of the lesson (make them feel involved in the process of learning).

2.2 Methods

During the course preparation, we used the ideas expressed by the scientists whose works we analysed in the chapter Literature review in our interpretation. Active teaching methods were used – including simple tasks, which are based on a students' creativity, such as projects, presentations, topics related to current problems of the future students' speciality. Students were asked to keep personal blogs or written journals to help them to identify their strengths and weaknesses, to make it possible to find ways to improve the learning results, to inform the teacher about likes and dislikes in the learning process. Such tasks were of particular interest to students, they contributed to the development of their critical thinking, analysis, and synthesis of information. At the end of the year, preference was given to assignments that motivate students to self-improvement, learning activities with a taste of competitiveness, including participation in competitions, discussions, and debates. In such circumstances, the teacher acted as a consultant, and the level of students' independence in the tasks was increasing, as was the quality of the educational activity. We consider that several types of tasks can be used at different stages for the development of students' independent work, namely: *educational* (at the very beginning the teacher instructs the way of doing, later the students use scaffolding if necessary); *training* (model implementation is provided); *searching* (to be performed independently by each student). They are united by a common requirement — the clarity and controllability of results and accounting in the appraisal. Educational and training tasks are certainly necessary, especially at the initial stage as preparatory to demonstrating possible actions within the complex independent educational activity; search tasks are relevant at an advanced stage.

It is expected that an autonomous learner is able:
a) to set personal learning goals;
b) to identify learning strategies and develop study plans to achieve these goals;
c) to select relevant resources and support;
d) to assess and reflect on one's progress;
e) to accept responsibility for learning.

That is why we have set the task of identifying changes in the level of student autonomy during the academic year (at the beginning and at the end of the experiment). For this we need:
1. To diagnose students' awareness in language learning which includes students' perceiving of own abilities, possibilities and capabilities.
2. To find out students' level of willingness to development and their ability to set personal learning goals.
3. To discover students' readiness to accept responsibility for learning, assess and reflect on progress.
4. To identify students' learning strategies and their readiness to develop study plans and achieve stated objectives.
5. To come to know about students' ability to select relevant resources and support.
6. To distinguish students' attitude to the role of the teacher.

2.3 Results

The results of the study are presented in this section. The data, which were mainly elicited from the survey questions, are reported in Tables 1-9. According to Joshi's questionnaire numbers of questions are used for demonstrating subheadings:
Items 1-3 – Learner awareness in language learning.
Items 4-8 – Students' self-effort in learning English.
Items 9-13 – Learners' broader autonomous activities beyond the class.
Item 14 – Learners self-esteem.
Items 15-16 – Learners' use of references materials.
Item 17 – Learners self-motivation in learning.
Item 18 – Learners' use of computers and the internet for English learning.
Items 19-23 – The role of a learner;
Items 24-29 – Learners' of perceptions of teachers' role.

In commenting them we can affirm:

Table 1, items 1-3. The data of the experimental group in September indicates that only 40% of respondents believe they can learn, 35 % know how to study languages well in free time and only 15% were able to make their own decisions and set goals. The data received in May demonstrates the changes in the situation: 70% believe they can learn, 65% can make their own decisions and set goals, and 75% know how to study languages well in free time. The data of the control group in September and May has some differences, but they are not so significant (44, 4 %→ 38, 9%; 27,8%→ 33,3%; 22,3%→ 33,3%) as in the experimental group.

Table 1. Learner awareness in language learning.

No	Items	experimental group (%)		control group (%)	
		Sept	May	Sep	May
1.	I think I can learn English well.	40	70	44,4	38,9
2.	I make decisions and set goals for my learning.	15	65	27,8	33,3
3.	I make good use of my free time studying English.	35	75	22,2	33,3

Legend: No – Item Serial Number, % – Responses in percentage

Table 2. Students' self-effort in learning English.

No	Items	experimental group (%)		control group (%)	
		Sept	May	Sep	May
4.	I preview before the class.	0	35	5,6	16,7
5.	I try to use every opportunity to take part in activities where I can speak in English.	25	70	22,2	22,2
6.	I speak confidently in front of the people.	10	60	5,6	27,8
7.	I make notes and summaries of my lessons.	5	30	5,6	16,7
8.	I talk to the teachers and friends outside the class in English.	0	35	0	11,1

Table 2, items 4-8. The data of experimental group demonstrated that at the beginning of the experiment nobody (0%) systematically previewed their tasks before the class and only 25%tried to use all possible opportunities to take part in the English speaking activities, while later data show that 70 % of them made plenty of efforts by using every opportunity of participation in the English speaking activities during the lessons. Item 6 demonstrates that respondents were almost not ready to speak confidently in front of the people or to talk English to the teachers and friends outside the class, and didn't find it necessary to make notes and talk English outside the class. But at the end of the year, they agreed that they became more confident in English speaking, recognized the benefits of making notes and summaries of lessons and some of them became ready for English communicating outside the classroom. The data of the control group shows that 1 student (5, 6%) previewed in September and his example was followed by one more student. It can be observed that 22,2% of respondents tried to use every opportunity to take part in English speaking activities and this data remained unchanged. The ability to speak confidently in front of the people demonstrates changes: from 5,6 % in September to 27,8 % in May. Analyzing the data on making notes and summaries we can admit the growth from 5,6% to 16,7% of respondents. The ability to communicate out of class grew from 0% to 11,1%.

Table 3. Learners' broader autonomous activities beyond the class.

No	Items	experimental group (%)		control group (%)	
		Sept	May	Sept	May
9.	I practice English outside the class also such as record my voice; speak to other people in English.	0	45	0	22,2
10.	I use the library to improve my English.	0	0	0	0
11.	I use audio-visual materials to develop my speech (listen to the BBC, watch movies, read newspapers, etc.)	35	75	33,3	61,1
12.	I attend seminars, training courses, conferences to improve my English.	15	55	22,2	22,2
13.	I take a risk in learning the English language.	0	30	0	27,8

Table 3, items 9-13. The data demonstrated significant growth in practising English in the period from September till May in both groups (experimental: 0%→45% and control: 0% → 22,2%) It was rather strange to reveal that no one student from both groups used a library during the period of research. The results of responses showed that students actively used a wide range of audio-visual materials to develop their language ability (experimental: 35%→75% and control: 33,3%→61,1%). The results of attending training courses, seminars and conferences are rather different. They grew in the experimental group but remained unchanged in control (experimental: 15%→55% and control: 22,2% → 22,2%). It can be seen that students of both groups demonstrated readiness to take some risk in learning English, especially significant growth was in experimental group 0% →30%, while Control demonstrated 0% → 22,2.%, which proves students' autonomy development.

Table 4. Learner's self-esteem.

No	Item	experimental group (%)		control group (%)	
		Sept	May	Sept	May
14.	I note my strengths and weaknesses in learning English and improve them.	0	85	11,1	33,3

Table 4, item 14. This item demonstrated striking growth of positive answers in the experimental group (0% → 85%), which means students reflecting their learning experiences, attempts to improve quality of learning, as well as significant changes in students' self-evaluation in the control group 11,1% → 33,3.

Table 5, items 15-16. The data of experimental group demonstrated that only half of the total respondents' number revised lessons, and read extra materials in advance at the beginning of the experiment, but this number significantly changed by more in May (50% →90%; and 45% →75%), which proves students' self-initiative development. The data of the control group remained almost without changes (66,7% →66,7% and 50,0% →55,6%).

Table 5. Learners' use of references materials.

No	Items	experimental group (%)		control group (%)	
		Sept	May	Sept	May
15.	I revise lessons and seek the reference books	50	90	66,7	66.7
16.	Besides the contents prescribed in the course, I read extra materials in advance	45	75	50,0	55,6

Table 6. Learners self-motivation in learning.

No	Items	experimental group (%)		control group (%)	
		Sept	May	Sept	May
17.	When I make progress in learning, I reward myself such as: buy new things, celebrate parties, etc.	35	85	16,7	50,0

The data in Table 6, item 17 demonstrated the respondents' attitudes toward self-motivation. Both groups showed the intensive growth (35% → 85% and 16,7 → 50,0%), which proved that this activity is popular among some students.

Table 7. Learners' use of computers and the internet for English learning.

No	Items	experimental group (%)		control group (%)	
		Sept	May	Sept	May
18.	I use the internet and computers to study and improve English.	60	95	55,6	100,0

The data in Table 7, item 18 indicated that modern technologies are very popular among language learners. The data of the experimental group demonstrated 60% →100%, and the control group 55,6 % 72,2%. It looks like this item explains the students' refusal to visit the libraries.

Table 8. The role of a learner.

No	Items	experimental group (%)		control group (%)	
		Sept	May	Sept	May
19.	Students have to be responsible for finding their ways of practicing English.	30	95	27,8	38,9
20.	Students should use many self-study materials to learn English.	15	55	16,7	33,3
21.	Students have to evaluate themselves to learn better.	20	70	11,1	11,1
22.	Students should mostly study that has been mentioned under the course for exam purposes.	60	60	66,7	72,2
23.	Students should build a clear vision of their learning English.	65	75	44,4	61,1

Table 8, items 19-23. The data of the experimental group demonstrated significant growth in students' understanding responsibility for finding their ways of practising English in as Item 19 (30% →95%), while control groups' data grew more moderate: 27,8 % → 38, 9%. The same situation can be observed in item 20, demonstrating students' realizing the necessity to use self-study materials (experimental: 15% → 55%, control: 16,7 → 33,3%). Item 21 is demonstrating students' awareness to evaluate themselves to analyze the strengths and weaknesses to learn better. The data of the experimental group (20% →70 %) and control group (11,1% → 11,1%) is evidence of a favourable environment for students' autonomy fostering. Item 22 demonstrated students' believing whether they mostly have to study information mentioned under the course for exam purpose: the experimental remained unchanged 60%, control grew 66,7 %→ 72,2 %). Item 23 showed the respondents' ideas on the importance to build a clear vision of their learning before learning English. The data of the experimental group demonstrated confidence and full support (65% → 75%), which proves students' autonomy development. While the control group's data grew more moderate (44,4% → 61,1%).

Table 9. Learners' perceptions of teachers' role.

No	Items	experimental group (%)		control group (%)	
		Sept	May	Sept	May
24.	A lot of learning can be done without a teacher.	40	50	27,8	33,3
25.	Teachers have to be responsible for making students understand English.	70	55	83,3	88,9
26.	Teachers should point out the students' errors.	85	45	83,3	77,8
27.	Teachers not only have to teach 'what' but should also teach 'how' of English.	55	40	61,1	61,1
28.	Teachers have to provide exam-oriented notes and materials.	100	60	100,0	77.8
29.	The failure of the students is directly related to the teachers' employment	65	30	77,8	61,1

Table 9, items 24-29. The data of experimental and control groups demonstrated respondents believing that a lot of learning can be done for themselves as is shown in item 24 (experimental 40% →50% (which proves some students' autonomy development), while the control group remained the same 27,8% →27,8%). The data of Item 25 demonstrated students' difference in perception of the idea about teachers' responsibility for making students understand English (experimental: 70%→55%, control: 83,3% → 88,9%). This result is evidence of students' autonomy development in the experimental group because it proves some changes in students' understanding of teachers' role as a partner or a facilitator, but not a person only responsible for the result of students' progress. Item 26 showed developments in students' ideas about error correction by a teacher (experimental: 85% → 45%, more than half of a total number of respondents were able to understand the need to correct the mistakes for their own; control 83,3 % → 77,8%. Item 27 demonstrated the changes in respondents' understanding of what and how teachers have to teach. The data of the experimental group is 55% → 40 %, which means that students decide for themselves how they should learn information, and the data of the control group remained unchanged:61,1% → 61,1%. Item 28 showed students' attitudes to teachers' provision exam-oriented notes and materials (experimental group: 100% → 60% and control group 100% → 94,4%). Item 29 demonstrated the connection between the students' failure with the teachers' classroom employment. The data of the experimental group showed significant changes in students' understanding: 65% → 30%, and the control group 77,8% → 61,1%).

The changes in students' autonomy level at the beginning and the end of the experiment demonstrate some changes between experimental and control groups (E→E vs C→C):

1. To diagnose students' awareness in language learning which includes students' perceiving of own abilities, possibilities and capabilities. Items:
- 1 (40%→70 vs 44,4%→ 38,9%);
- 6 (10%→60% vs 5,6% →27,8%);
- 14 (0%→85% vs 11,1 → 33,3).

The experimental group demonstrated significant changes in comparison with the control group.

2. To find out students' level of willingness to development and their ability to set personal learning goals. Items:.
- 2 (15%→65% vs 27,8%→33,3%);
- 3 35%→75% vs 22,2%→33,3%);
- 4 (0%→35% vs 5,6→16.7%);
- 5 (25% →70 %vs 22,2% →22,2%);
- 8 (0%→35% vs 0%→11,1%);
- 14 (0%→85% vs 11,1% →33,3%);
- 17 (35%→85% vs 16.7%→ 50.0%);
- 21 (20%→70% vs 11,1→11,1%);
- 23 (65% →75% vs 44,4%→61.1%).

The experimental group demonstrated considerable positive changes in comparison with the control group.

3. To discover students' readiness to accept responsibility for learning, assess and reflect on own progress. Items:
- 9 (0% →45% vs 0% →22,2%);
- 13 (0%→30% vs 0% →27,8*);
- 14 (0% →85% vs 11,1%→33,3).

Positive changes in the experimental group compared to the control group are more noticeable.

4. To identify students' learning strategies and their readiness to develop study plans to achieve these goals. Items:
- 7 (5%→30% vs 5,6%→16,7%);
- 9 (0% →45% vs 0% →22,2%);
- 11 (35%→75% vs 33,3% →61,1%).

The experimental group demonstrated positive changes in comparison with the control group.

5. To come to know about students' ability to select relevant resources and support. Items:
- 9 (0% → 45% vs 0% → 22,2%);

- 10 (0%→0% vs 0% →0% --students don't use the traditional libraries);
- 11 (35%→75% vs 33,3%→61,1%);
- 12 (15%→55% vs 22,2,3% →22,2%);
- 13 (0% →30% vs 0% →27,8%);
- 15 (50% → 90% vs 66,7 →66.7%);
- 16 (45%→75% vs 50,0% →55,6%);
- 18 (60% → 95% vs 55,6 →100%).

The experimental group demonstrated considerable positive changes in comparison with the control group.

6. To distinguish students' attitude to the role of the teacher. Items:

- 19 (30% →95% vs 27,8% →38,9%);
- 24 (40% →50% vs 27,8% →33,3%);
- 27 (55%→70% vs 61,1%→61,1%).

Positive changes in the experimental group compared to the control group are more noticeable.

7. Items which do not demonstrate the students' autonomy development are:

- 22 (60%→60% vs 66.7%→72,2%);
- 25 (70%→55% vs 83,3% →88,9%);
- 26 (85%→45 vs 83,3→77,8);
- 28 (100% →60% vs 100%→77,8%);
- 29 (65%→30% vs 77,8% →61,1%).

Table 10. Results of students' language competency testing before and after the experiment.

Marks	Experimental group					Control group				
	Before the experiment		After the experiment		Differences	Before the experiment		After the experiment		Differences
	Students	%	Students	%	%	Students	%	Students	%	%
A	6	30	11	55	+25	6	33,3	8	44,4	+11,1
B	7	35	4	20	-15	7	38,8	6	33,3	-5,5
C	7	35	5	25	-10	5	27,7	4	22,2	-5,5
Total	20	100	20	100		18	100	18	100	

Table 10. The results of the testing students' in the control and experimental groups at the beginning of the academic year demonstrated nearly the same level of competence in experimental and control groups. The results of the tests were received using the ECTS grading scale (A – 100-90; B – 89-82; C – 81- 75; D – 74-67; E – 66-60). According to the results of tests conducted at the beginning of the year,30% of students of the experimental group had A-marks. After the experiment, the percentage of students with A-marks increased to 55%. The difference reached +25%. The percentage of students with B marks was 35% before and 20% after the experiment, the decrease constituting -15 %. Similarly, the percentage of students in the experimental group with C marks decreased from 35 % to 25 %with a difference of -10 %. In the control group, the differences in the level of learning progress before and after the experiment were not so significant. Level A after the experiment had 44,4% (vs.33,3% before the experiment), B level before the experiment had 38,8 % and after 33,3%. There was also a less significant (in comparison with the experimental group - 5,5% vs.-10%) decrease in the number of the students with C level after the experiment. The results of the testing are summarized in Table 10.

3 Conclusion

Learner autonomy is a deeply-rooted phenomenon in the educational systems of Western countries, whereas in Ukrainian educational traditions it is rather new and unnatural. In our country, although official educational policies, which encourage the implementation of learner autonomy, many teachers at secondary schools and university levels tend to use methods they are familiar with. The special approach to ESP learning used in the experimental group was rather unusual and uncomfortable at the beginning for the students as well as their teacher. But soon all the participants began to fulfil the conditions of the experiment. The main task of the teacher was to teach the students how to learn the language, which means, from our point of view, to outline own plans of mastering content, to gain experience in applying structures, to analyze own mistakes made in the text and many actions more. But through the experimental work students realized their capabilities. The results of the questionnaire prove the fact of the special learning environment in the experimental group was created (at least its initial stage). This environment is characterized by favourable conditions for students' autonomy fostering. They are neutral attitude to mistakes and positive for their analysis; a comprehensive friendship with students creativity in learning; gentleness in planning changes, but a clear requirement to capture all aspects of learning content.

According to the results of the experiment, we can state that only a part of the most active students used the created conditions and gained experience in autonomous learning. They participated in a selection of study materials, willingly maintained diaries (blogs) in which analyzed own work and the work of peers. They were able to find opportunities to practice particular English vocabulary in and out of the classroom, tried to set own goals and honestly evaluated the work. As a result of continuous work, they became able to identify own strengths and determine own pace in learning. Being persistent they attended lessons regularly and worked hard acquiring necessary skills during the year of the project work. In addition to the above, we can mention that these students are intrinsically motivated and have a high level of academic performance. No wonder such students have succeeded. Their number in the experimental group is 11+4= 15 (55% (A) and 20% (B)), which makes up 65%.

The limited teaching hours at the non-linguistic university, which devoted to ESP, were not enough to change completely the learning habits and styles of all the students and make them active autonomous learners.

Some students from the experimental group (35%) were not a success. We see the main problem in their general unpreparedness to change, absence skills of information acquiring, inability to choose the most important among the minor, unwillingness to take responsibility for own training. We consider it appropriate to note that many students from the control group became interested in the work of the project and expressed a desire to participate in the work of the next project, but in the experimental group.

As an important way to promote the development of autonomy, we consider:
1) the favourable atmosphere in the classroom, mutual assistance, and scaffolding provision;
2) the maintenance of students' blogs or educational diaries.

We can conclude that during the work of our experiment we used the special approach which involved:
- active teaching methods (such as doing projects and presentations based on students' creativity);
- developing students' independent work at different stages of the experiment (e.g. educational, training, and searching tasks united by common requirements (the clarity and controllability of results and accounting in the appraisal));
- paying due attention to the use of students' feedback and self-reflection (students were asked to keep written journals or blogs to help them to identify their strength and weaknesses of learning);
- maintaining a high level of motivation (learning topics were strictly related to problems of the future speciality);
- stimulating the development of critical thinking (using tasks for analyzing, synthesizing, and, evaluating information, providing different perspectives upon the same material).

In such circumstances, the teacher was informed about students' likes and dislikes in the learning process, and was acting as a consultant.

As a result of the experimental work, new skills have been formed among the students of the experimental group, which include abilities to determine the learning objectives, to select methods and techniques of content assimilation, to monitor the procedure of learning, and to find reference materials as well as evaluate the results. That is what we mean under the notion of student's autonomy.

This approach was useful for the development of writing skills and the ability to express feelings and attitudes to the studied material. Such work enabled students to analyze their studies and see further prospects. This type of work significantly expanded students' professional horizons, made them feel involved in the process of learning. We believe it can be applied to all learning activities.

We also understand that our research is based on the student questionnaires result, which can be called rather subjective. At the same time, the indicators of students learning success are becoming evidence of the usefulness of the ESP approach mentioned above. The approach made a positive influence on students' motivational intensity, they became more satisfied with learning English and their results in education became better. This allows us to state that our experience contains a rational grain, is useful and deserves further development.

Creating an environment fostering learners' autonomy at non-linguistic university is a challenge that is appropriate to realize to give an autonomous learning experience to aspiring students. In our opinion, the development of students' autonomy is an indispensable condition of study at a modern university. Such an environment for learning is an important factor in students' learning activities intensification.

As we see the problem future researches should be dedicated to a need for learners' autonomy levels determination, because of lack of certainty in this realm. Another problem worthy to be investigated is the dependence of students' personal qualities and their autonomy development.

Besides, we consider the necessity to start the formation of students' autonomy in learning foreign languages at a young age as soon as possible. Because this process has many psychological factors worthy, to begin with at school and foster at the university level.

References

1. D. Little, Learner autonomy and second/foreign language learning, in *The Guide to Good Practice for Learning and Teaching in Languages, Linguistics and Area Studies* (CIEL Language Support network, 2003), https://www.researchgate.net/publication/259874624_Learner_autonomy_and_second_foreign_language_learning. Accessed 01 Dec 2019
2. V. Karastateva, M. Krasteva, Fostering Autonomy in ESP. Proceedings of the University of Ruse **49**(6.3) (2010)
3. Ph. Benson, *Teaching and researching autonomy in language learning* (Longman, London, 2001)
4. Ph. Benson, Autonomy in language teaching and learning. Language teaching **40**, 21-40 (2006). doi:10.1017/S0261444806003958
5. H. Holec, *Autonomy and Foreign Language Learning* (Pergamon, Oxford, 1981)
6. D. Little, Learning as dialogue: The dependence of learner autonomy on teacher autonomy. System **23**(2), 175–181 (1995). doi:10.1016/0346-251X(95)00006-6
7. A. Finch, Autonomy: Where are we? Where are we going? https://www.academia.edu/6022949/Autonomy_Where_are_we_Where_are_we_going. Accessed 12 Dec 2019
8. Ph. Benson, P. Voller, *Autonomy and Independence in language learning* (Longman, London, 1997)
9. D. Thanasoulas, Learner Autonomy in ELT. Newsletter **32.9**(2) (2000)
10. W. Littlewood, Defining and developing autonomy in East Asian contexts. J. App. Ling **20**, 71–94 (1999). doi:10.1093/applin/20.1.71

11. L. Bajrami, Teacher's new role in language learning and in promoting learner autonomy. Procedia-Social and Behavioral Sciences **199**, 423–427 (2015). doi:10.1016/j.sbspro.2015.07.528
12. G. Gamble, M. Wilkins, Learner autonomy dimensions: What motivated and unmotivated EFL students think. Lingua Posnaniensis (2018). doi:10.2478/linpo-2018-0003
13. L.Dickinson, *Self-Instruction in Language Learning* (Cambridge University Press, Cambridge, 1987)
14. W. Littlewood, "Autonomy": anatomy and a framework. System **24**, 427–435 (1996)
15. L. Dickinson, *Learner Autonomy 2: Learner Training for Language learning* (Authentik, Dublin, 1992)
16. N. Okumus-Ceylan, Thoughts on learner autonomy in a computer-assisted language learning program. IOJET **6**(2), 339–354 (2019)
17. K. Joshi, Learner perceptions and Teacher Beliefs about Learner Autonomy in Language learning. NELTA **16**(1–2), 13–29 (2011). doi:10.3126/nelta.v16i1-2.6126

Development of junior students' emotional intelligence in terms of New Ukrainian School: realities and prospects

Svitlana Prakhova[*], *Natalya* Makarenko, and *Iryna* Shumilina

Kryvyi Rih State Pedagogical University, Kryvyi Rih, 50086, Ukraine

Abstract. The article presents a theoretical overview of the phenomenon "emotional intelligence" (EQ). The perspectives of different foreign and domestic psychologists on the concept of EQ, its origin and structure are analised. The importance of studying emotional intelligence and the possible ways of its development as a priority of the New Ukrainian School (NUS) are emphasized. The basis of EQ study is determined as the interrelation of the concepts of "emotional – intellectual". Certain positive and negative characteristics of advanced emotional intelligence are highlighted. A methodological toolkit for EQ study is developed and presented. Some prospective ways of its development in ontogeny under appropriate conditions are substantiated. There is an algorithm elaborated for the complex consistent work on EQ development.

1 Introduction

Due to certain education reforms and the transition of classical domestic education onto the rails of the "New Ukrainian School", the issues related to diagnostics and developing strategies for the formation of "emotional intelligence" (EQ). EQ are gaining a new level of relevance. At the present stage of the educational process development, which attempts to function "in a new way", empirical studies are of particular importance, especially those devoted to comparing different aspects of the study of students' emotional intelligence in terms of "classical" and "New Ukrainian School" (NUS) and the search for the system of EQ development in the context of NUS.

It should be noted that among the psychic phenomena identified at the end of the twentieth century, it is difficult to find such a phenomenon that would attract more attention of scientists and representatives in practical (applied) fields of research than EQ. Almost every national scientific studies collection presents the results of research on this topic.

This topic is especially relevant in terms of NUS, one of the tasks of which is the development of communicative knowledge, abilities and skills of the child. It is also significant to investigate the problem of emotional intelligence and look for ways to develop it in a turbulent flow of information, in the presence of clipping type of thinking along with an excessive immersion into the virtual world.

The key formulas of NUS formulated in the concept of new content of education are: information and communication competence; civic and social competences. All the competencies in NUS possess a common feature: the ability to manage emotions; apply emotional intelligence. In this regard, it is necessary to define the peculiarities of the child's personal emotional states clearly; to account for new elements in the emotional sphere at different stages of ontogeny and to develop a step-by-step plan for the effective development of emotional intelligence.

2 Objectives of the study

Considering the relevance of the subject under study, the objectives of this article are to investigate the role of emotional intelligence of the individual, opportunities and factors of their development on the example of junior students studying in the realities of new Ukrainian school system and to compare their statistics data with the sample group of students who were junior students before 2018.

3 Methods

The experimental study of junior students' emotional intelligence level was organized in two stages. The first one covers the period before the reform implementation (academic year 2017/2018), and the second stage was realised in the course of the changes related to the functioning of the NUS (academic year 2019/2020). The diagnostic sections were performed with the junior students in Year 2 in the middle of September of each academic year. In total, 110 students of Kryvyi Rih Scientific and Technical Metallurgical Lyceum № 16 participated in the study. According to the time terms of the diagnostics, all the students were divided into two groups. The first group included primary school students who became first graders by 2018 (the year of the beginning of NUS introduction), and the second group consisted of students who started

[*] Corresponding author: svetlankaprakhova@gmail.com

their studies in September 2018, under the new curriculum.

Traditionally, there are two main approaches to measuring emotional intelligence: the questionnaire (aimed at using questionnaires that the researcher presents as self-report) and the test. Each of these approaches corresponds to a specific type of emotional intelligence model: ability/capability models as well as mixed (eclectic) models.

The first type of the models interprets emotional intelligence as a purely cognitive construct associated with the processing of emotional information. The second type - explains emotional intelligence as a set of cognitive, personal, motivational traits providing an effective understanding and management of emotions.

A complete description of the methodological toolkit is presented by S. Derevianko [8]. V. Zarytska provides a set of methods for diagnosing the level of self-control and self-regulation development [9].

In addition, the diagnostics with the help of the mentioned T-methods are supplemented by communication with the child (Q-methods), observations and expert assessments (L-methods). That is, there is currently present a really valid and reliable toolkit for determining the level of emotional intelligence development [11].

In the diagnostic part of our empirical study, we aimed to determine the general modality and frequency of manifestation of students' dominant emotions, with the latter determined by using the common psychodiagnostic method of K. Isard's Differential Emotions Scale (DES). The general level of junior students' emotional intelligence was determined by the results obtained through applying a modified version of the classic one of the complex method by M. Nguyen, which consists of two projective techniques: "Domestication: the world of things – the world of people – the world of emotions", "Three desires" and the methods "What – Why – How". Although in the classic version this diagnostic tool is aimed at studying the characteristics of preschool children, we decided to use its modified version with primary school students, most of whom are no more than six or seven years old.

4 Results and discussions

Realising that the concept of intellect by J. Piaget was of an extremely narrow character, G. Gardner was able to hypothesize about the multiplicity of manifestations of intelligence, among which are interpersonal and intrapersonal ones. G. Gardner believed that these types of intelligence were the key to self-cognition (cognitive insight), to understanding one's emotions and feelings, to managing one's behavior, to understand other people and interact with them effectively. Such characteristics are also inherent in emotional intelligence [6].

In the early twentieth century, E. Thorndike assumed the existence of social intelligence as the ability of a person to understand others in communication, to establish interpersonal relationships as well as to maintain friendly relations.

At the end of the last century, social intelligence was viewed as a component of emotional intelligence. The term "emotional intelligence" was coined by J. Mayer and P. Salovey [13]. The idea was then further developed in D. Goleman's book, the journalist and psychologist [5]. According to these scientists, emotional intelligence is a collection of cognitive abilities to identify, understand and manage emotions.

A global study of this phenomenon is presented in the writings of D. Goleman, who considers emotional intelligence as an ability to interpret one's own emotions and the emotions of others in order to use the information received and fulfill one's own goals. The significance of the issue is already covered in a subheading to his edition of "Emotional Intelligence: Why It Can Mean More Than IQ". According to the author, contrasting these concepts (EQ and IQ) in scientific research increasingly leads us to believe that "people with high intellectual abilities turn out to be staggeringly bad pilots of their own lives" [5].

The Israeli researcher-theorist on emotional intelligence R. Bar-On [3] developed his own concept of emotional coefficient to measure emotional and social competence, identifying five subcomponents characterizing the structure of EQ (self-understanding, communicative potential, anti-stress potential, etc.).

Summarizing the findings of the key founders of the study on the problem, we can identify five major components of emotional intelligence: a person's awareness of their own emotions; regulation of one's emotions based on self-awareness; the ability of a person to direct their emotions to achieve a specific goal of activity; empathy as the ability to understand other people's emotions; the ability to have a friendly relationship with others.

Quite a thorough study of the EQ phenomenon was also conducted by domestic scientists. Thus, a generalization of the theoretical foundations of the study of emotional intelligence (stages of forming the understanding of the EQ concept; the periods of formation and study of emotional intelligence; analysis of the features of studies of emotional intelligence concept content in the post-Soviet area) were presented by M. Savchuk [17].

E. Nosenko [14] sees EQ in the context of the unity of external and internal regarding the determination of the mental and considers that emotional intelligence as an aspect of revealing the inner world of the personality reflects a measure of reasonableness of the person's attitude to the world, to others and to themselves as the subject of life activities.

The Russian scientist D. Lusin proposed to interpret the concept of emotional intelligence as "the ability to understand their own and others' emotions and control them, i.e. cognitive ability" [12].

According to M. Shpak [19], emotional intelligence is an integrative personality trait, which is conditioned by the dynamic unity of affect and intelligence through the interaction of emotional, cognitive, connotative and motivational features and is aimed at understanding one's own emotions along with providing emotional experience management of emotional state, subordination of emotions to mind; it also promotes self-knowledge and self-realization through enrichment of emotional and social experience.

The influence of the level of emotional self-control and self-regulation development on the students' emotional intelligence development was investigated by V. Zaritskaya [9]. Presenting the results of the diagnostics, the author noted that the level of ability to self-control and self-regulation depends on the level of overall emotional excitability.

The influence of emotional intelligence peculiarities (emotional self-regulation, self-reflection, authenticity) of group members on group dynamics and individual dynamics of personality in the group is covered in V. Romanova's work [16]. It is proved that the general emotional participants positively influence the organization, responsibility and orientation of the group.

Emotional intelligence as the basis of communicative competence and the possibility of its development (with reference to the research of the French psychologist M. Kets de Vries) is presented by V. Fedorchuk [18].

It should be noted specifically that in the scientific literature the following synonymous with the concept of EQ terminology are presented as: "emotional reasoning" (E. Nosenko), "emotional self-awareness" (O. Vlasov), "emotional competence" (K. Saarny), "emotional thinking" (O. Tikhomirov), "emotional literacy" (K. Steiner), "emotional giftedness", "social intellect" (R. Ronald) to provide evidence of the emotional intelligence concept breadth and, in a certain way, define the ways of its development.

The priority area of modern research is the study of the problem of anxiety and frustration, their relationship with the level of students' emotional intelligence, the impact of models of learning in the NUS [15].

It is important, from our point of view, to carry out evaluation of the positive and negative properties of emotional intelligence. In particular, S. Boyko [4] explored emotional intelligence as a factor in resisting information stress. And according to the journalist-psychologist from *Scientific America*, Agata Blaszczak-Box, emotional intelligence has not only positive but also certain negative components. It contributes to increasing stress, makes us defenseless, it also makes us feel responsible for what we are unrelated to, and tempts us to manipulate others for personal gain. It was experimentally researched and featured in the September 2016 issue of *Emotion*. German psychologists Miriam Bechtoldt and Vanessa Schneider showed that students with high EQ level significantly increased their stress during the experiment. Therefore, it took more time to stabilize the emotional state of such subjects. The results obtained indicate the extreme vulnerability, perceptiveness of individuals with high levels of EQ. And in the context of the information society, this takes on new content and needs further and deeper study [2].

In the process of qualitative and quantitative analysis of the empirical data, the main focus was put on the comparative analysis of the peculiarities of junior students' emotional response to different-vector emotional situations.

As the results of K. Izard's Differential Emotions Scale (Fig. 1) show, the students of the first group (those who went to school before September 2018) turned out to bring such emotions as interest, surprise, fear, joy and shame at the top position of the rank distribution chart.

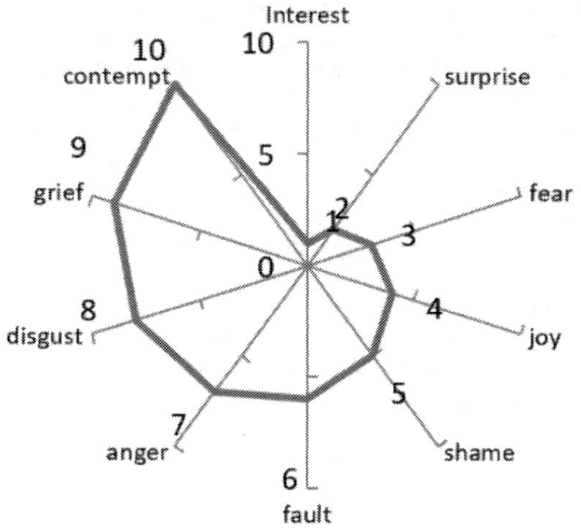

Fig. 1. The scale of differential emotions according to K. Izard's method ("classical" school students' diagnostic results)

At the same time, the results of the diagnostics of the junior students studying in the NUS system show that the positive emotions of interest and joy persist in the top five positions, with the difference that the students who went to school in September 2018, the joy moves from the fourth position in the rank of significance onto the first (Fig. 2).

The results of individual conversations with students in both groups revealed that the emotional experiences of students during this age period are primarily related to the learning process, interpersonal relationships with classmates and participation in various games and entertainment. Analyzing the curriculum of the new Ukrainian school and the variants of the programs that were developed for it, it can be stated that in previous years too little time was devoted to the students' emotional unloading in the course of their educational activity, which, in its turn, could not but affect the level of their emotional intelligence.

The statistics presented above are fundamentally important, since it is the junior school age that is sensually colored and characterized by a considerable emotional sensitivity that has an impact on the development of emotional intelligence. So, as the results of the first part of our empirical study show, the students of the NUS system are characterized by a more positive emotional coloring of the educational process. While negative emotional experiences almost did not change their rank positions, with the exception of the parameter "fear", which moved from the third to fifth position, which is significant in the context of the "fearless" attitude of the modern student to the learning process.

Fig. 2. The scale of differential emotions according to K. Izard's method (NUS school students' diagnostic results)

The quantitative and qualitative assessment of N. Nguyen's methods results showed the presence of differences in the levels of students' emotional intelligence of the first and second groups (Fig. 3). As the results of projective diagnostics showed, NUS students have significantly higher EQ parameters: high level – 20%, medium level – 60%, low level – 20%.

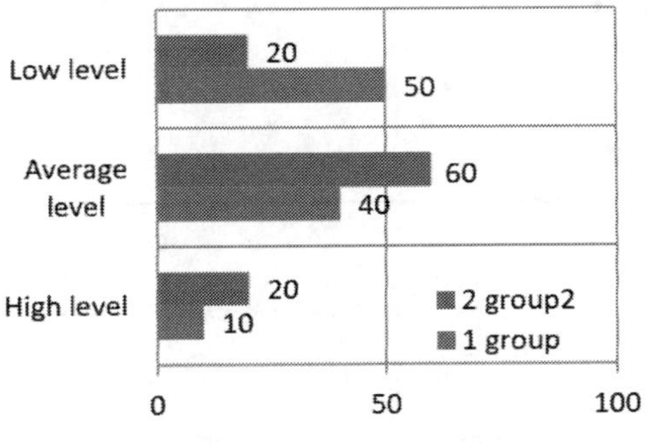

Fig. 3. Levels of junior students' emotional intelligence according to M. Nguyen's methods results

For comparison, students in the "classical" school had a parameter of 50% low level of emotional intelligence. This statistical correlation is the evidence that the new system of educational process in Ukraine contributes to the emotional unloading of students in a great way, which in its turn, creates the preconditions for the development of empathy related to interpersonal relationships among children.

In the new educational environment, the process of emotional intelligence and socialization is accelerating. Modern children are more likely to develop the ability to differentiate and classify emotional experiences and build their interpersonal relationships on this basis.

However, all this is possible only with the proper approaches to the system of students' emotional intelligence development at different stages of their maturation in the educational process.

There are several points of view regarding the possible ways of developing emotional intelligence. The founders of the concept J. Mayer and P. Salovey believe that it is impossible to raise the level of emotional intelligence, because it is a relatively stable characteristics of a man. D. Goleman is convinced that the development of emotional intelligence is possible at different stages of ontogeny, from childhood, with subsequent improvement in adulthood. R. Bar-On states that EQ develops over time, it can be improved with the help of trainings, exercises and therapy [7].

I. Andreeva [1] notes that the level of emotional intelligence of a child is related to a similar quality of the mother, but there is no connection with the parental characteristics. This fact refutes the "genetic" theory and works in favor of the "educational" one, since genes are inherited from their parents equally. But, on the other hand, there is a clear link between EQ and the dominance of the right or left brain hemisphere, which is a congenital feature. According to some studies, right-hemisphered people have much better emotional intelligence.

The researchers found another pattern: the higher the parental education level and financial position of the family is, the higher emotional intelligence of adolescents appears. However, in this situation, it is difficult to understand the cause and effect relation.

Thus, having analyzed the scientific literature on the problem, we are inclined to think about the possibilities of emotional intelligence development as a reality (and the experimental study of EQ in the context of NUS is therefore a testimony). This is possible through the introduction into the field of education of psychological and pedagogical technologies for the formation of emotional competence by K. Saarney (awareness of their own emotional states, the ability to recognize the emotions of others; the ability to overcome their negative experiences, etc.) and the scheme of emotions recognition, suggested by D. Lusin and N. Bylkin, according to which the eventual product of emotional intelligence is an ability to make decisions based on the reflection and comprehension of emotions, which represents a differential assessment of events with personal content.

The school activities should be systematic and cover not only students of NUS, but also parents and teachers. According to some scientists' survey [10] it is difficult for a teacher to determine what emotions a child has. Most often, teachers talk about personal characteristics (confidence-insecurity, kindness, modesty, mobility, etc.). Only 9% of those polled identified the emotional poles (positive-negative), but, unfortunately, could not identify their modality.

The emotional intelligence development realization will only be implemented by the algorithm:

✓ to familiarize participants of the educational process with theoretical knowledge

about the emotional sphere of personality (psychoeducation);
✓ the development of methodological recommendations for systematic formation of emotional competence, emotional thinking, emotional memory, etc. with children, parents, teachers;
✓ the diagnostics of participants' EQ level in terms of educational process;
✓ the study of existing disturbances in the emotional sphere (asynchronous disharmonious development, authentic reactive states, psychopathic behaviors, etc.);
✓ the implementation of exercises, trainings, workshops on the development of emotional intelligence in the educational activities;
✓ the results evaluation and correction.

Consequently, this is completely in line with the concept of NUS. Prior to this period, modern school was focused mainly on the development of students' academic abilities, on the cognitive sphere of personality (formation of knowledge, skills and abilities), whereas the emotional development of students was not given due attention. There is an indisputable fact: one of the elements of the educational content is the formation of emotional-value attitude of a person to themselves, to other people, to the surrounding reality (I. Lerner). The consequences of emotional "illiteracy" are manifested in a significant increase of interpersonal conflicts, an increase in cases of depression, cruelty or violence. Therefore, considering I. Andreeva's claim, the development of emotional intelligence can be viewed as a factor enhancing the psychological culture of society as a whole.

It is undoubtful, when creating a consistent process on the emotional intelligence development, it should be born in mind that only interactive methods of activating knowledge about the problem, innovative technologies will not only enable the participants of the educational process, but also help to actually develop their EQ. Unfortunately, the school still has a traditional monologic (mentoring) form of influence on the child, which can only cause negation and rejection. However, the introduction of cognitive innovators' findings into school practice will make this process interesting, understandable as well as consciously necessary. For example, acquainting participants of educational process with modern researches in the style of recommendations like «How ... enhance … develop … improve …" may finally lead to the Achiever Experiment. That is, after reading the book of recommendations how to regulate their own emotions (empathically listening, "reading" non-verbal signals, influencing another person, etc.), the student tries to put them into practice, evaluate the result and correct their mistakes. It is also necessary to use case-methods (with a detailed analysis of life situations) in the development of EQ. In our view, effective methods may be mental provocations, group work, volunteer activities, etc.

5 Findings

Thus, the junior students' emotional intelligence development is a part of the process of their socialization under the conditions of the NUS integrating. The empirical study results of the modality and levels of emotional intelligence of "classical" students and students enrolled in the NUS system reveals significant differences in the construction of cognitive-interpretative patterns of students' recognition of emotions and states. Apparently, the students of the NUS are more adaptable to the changing conditions of the educational process, they have a clear understanding of the objects and phenomena of the surrounding reality. Moreover, modern students quite easily interpret the complex emotions that arise in their familiar life circumstances.

However, despite the positive results of the experimental study of the level of students' emotional intelligence level in Year 2, it is worth noting that these are only the first steps in the educational realities of the New Ukrainian School. The issue of the current techniques and methods of students' EQ development at different stages of their adulthood still remains uncovered. This is where we see the prospects for further research in the field of emotional intelligence.

References

1. I. Andreeva, Emotsionalnyiy intellekt: issledovaniya fenomena. Voprosyi psihologii 3, 78-86 (2006)
2. A. Blaszczak-Box, Too Much Emotional Intelligence is a Bad Thing (2017), https://www.scientificamerican.com/article/too-much-emotional-intelligence-is-a-bad-thing/. Accessed 25 Oct 2019
3. R. Bar-On, *Emotional Intelligence in Men and Women: Emotional Quotient Inventory:* Technical Manual (Multi-Health Systems, Toronto, 1997)
4. S. Boiko, in *Naukova dyskusiia: pytannia pedahohiky ta psykholohii: zbirnyk materialiv Mizhnarodnoi naukovo-praktychnoi konferentsii* (Kyivska naukova orhanizatsiia pedahohiky ta psykholohii, Kyiv, 2017), pp. 125–128
5. D. Goleman, *Emotional intelligence* (New York: Bantam Books, 1995)
6. H. Gardner, *Multiple intelligences: the theory in practice* (Basic Books, New York, 1993)
7. D. Goleman, *Emotsionalnyy intellekt. Pochemu on mozhet znachit bolshe chem IQ* (Mani. Ivanov. Ferber, Moskva, 2013)
8. S.P. Derevianko, Metodychni zasoby diahnostyky emotsiinoho intelektu. Visnyk Chernihivskoho natsionalnoho pedahohichnoho universytetu. Seriia: Psykholohichni nauky 128, 95–99 (2015)
9. V. Zarytska, Vyvchennia rivnia rozvytku samokontroliu i samorehuliatsii emotsii yak elementiv emotsiinoho intelektu studentiv u protsesi fakhovoi pidhotovky. Aktualni problemy psykholohii. Zbirnyk naukovykh prats Instytutu

psykholohii im. H.S. Kostiuka **10**(5), 141–146 (2010)

10. E.P. Ilin, *Emotsii i chuvstva* (Piter, Saikt Petersburg, 2008),
11. A. Kozlova, *Emotsiinyi intelekt* (Shkilnyi svit, Kyiv, 2018)
12. D. Lyusin, *Organizatsiya znaniy ob emotsiyakh: vnutrennyaya struktura kategorii "emotsii"* (Razvitiye, Moskva, 1996)
13. J.D. Mayer, P. Salovey, The intelligence of emotional intelligence. Intelligence **17**(4), 433–442 (1993). doi:10.1016/0160-2896(93)90010-3
14. E. Nosenko, N. Kovryha, *Emotsiinyi intelekt: kontseptualizatsiia fenomenu, osnovni funktsii* (Vyshcha shkola, Kyiv, 2003)
15. S. Prahova, N. Makarenko, Napryamki prevenciyi shkilnoyi trivozhnosti molodshih shkolyariv: realiyi sogodennya ta perspektivi "novoyi ukrayinskoyi shkoli" (na prikladi longityudnogo doslidzhennya). Problemi suchasnoyi psihologiyi 44, 234–254 (2019)
16. V. Romanova, Doslidzhennia vplyvu emotsiinoho intelektu chleniv hrupy ta yikh osobystisnykh proiaviv na yavyshcha hrupovoi dynamiky. Aktualni problemy psykholohii. Zbirnyk naukovykh prats Instytutu psykholohii im. H.S. Kostiuka, **10**(19), 468–477 (2015)
17. M.R. Savchuk, Teoretychni osnovy doslidzhennia emotsiinoho intelektu (Theoretical basis of the research of emotional intelligence). Efektyvna ekonomika 5 (2017), http://www.economy.nayka.com.ua/?op=1&z=5608. Accessed 28 Nov 2019
18. V. Fedorchuk, in *Aktualni problemy psykholohii osobystosti ta mizhosobystisnykh vzaiemyn: materialy VII Mizhnarodnoi naukovo-praktychnoi konferentsii*, Kamianets-Podilskyi-Kyiv-Zheshuv-Brest, 2015, pp. 276–282
19. M. Shpak, Emotsiinyi intelekt v konteksti suchasnykh psykholohichnykh doslidzhen. Psykholohiia osobystosti 1, 282–288 (2011)

Systematicity of students' independent work in the course of operating systems

Oleksandr Kolgatin[1,*], *Dmytro* Holubnychyi[1], and *Larysa* Kolgatina[2]

[1]Simon Kuznets Kharkiv National University of Economics, 9a Science Ave., Kharkiv, 61166, Ukraine
[2]H. S. Skovoroda Kharkiv National Pedagogical University, 29 Alchevskykh Str., Kharkiv, 61002, Ukraine

Abstract. The paper is devoted to the study of systematicity of students' learning activity as a parameter of student's model, and influence of systematicity at learning results in the course "Operating Systems". The necessity to equip the student himself as the subject of the educational process with the skills and appropriate pedagogical forecasting tools for independent choice of the appropriate variant of educational activity is shown as theoretical framework. Parameters of models in such pedagogical diagnostics system are suggested and discussed. Empirical work has been realised on the base of learning management system *Moodle* and give possibility to analyse correlation between timeliness of completing the learning tasks by students and their educational achievements as well as to analyse the structure of students' time planning at homework. Recommendations to improve the educational process have been suggested

1 Introduction

1.1 Statement of the problem

Nowadays, effective educational process is not possible without active use of information and communication technologies. New educational environment puts forward advanced requirements to management of students' learning activity that become more independent. Such management should be grounded on comprehensive models. Theoretical basis of modelling of the open education organizational systems, theory of designing such systems have been expounded from systemic positions in monograph of V. Yu. Bykov [1]. A. E. Kiv, V. N. Soloviev, S. O. Semerikov [2] underline that Information technologies, especially, cloud technologies transform education, and have analysed according to results of the "Cloud Technology in Education" scientific conference modern approaches to managing students' learning activity in university educational environment. O. O. Triakina, O. O. Pavlenko, N. P. Volkova, and D. A. Kassim [3] have described the existing E-learning instruments that was designed by the international organizations for self-education and have suggested the ways of this tools implementation into professional training. K. Vlasenko, O. Chumak, I. Sitak, I. Lovianova, O. Kondratyeva [4] on the base of survey, conducted for teachers, suggested to develop an educational platform – an online environment for collaboration of the experienced professionals, whose joint activities should help in greatly enhancing their professional skills.

Independent work of students become one of the most significant part of modern educational systems. It is the demand of the curriculums and necessity to provide of the dual learning. Therefore, elements of distance learning are widely used in educational process as a form of education and as a form of management of students' independent work. Students' work is realised in specialised learning management environments without teacher's personal presence, and the teacher has no possibility to use traditional forms of pedagogical observation. The teacher needs in special system for management of students' independent learning activity instead of traditional intuitive management of learning process.

Learning management systems, for example *Moodle*, give us various new highly informative tools for pedagogical diagnostics. Management of students' learning activity in information and communication environment should be based on individual pedagogical prognosis for each student with use of an idealised student's achievements model, a model of student's real state and a model of available variants of learning methods [5]. The relationship between parameters of this models and indicators that can be directly measured in a learning management system should be studied experimentally and theoretically.

The field of interest in this paper is systematicity of student learning activity as a characteristic of student's leaning style and a parameter of student model in learning management system. Indicators of systematicity and its influence on student learning achievements are in the centre of our attention.

1.2 Analysis of previous researches

There are many scientific work devoted to students' independent work, its systematicity. But we want to pay attention to experimental data according to this problem.

* Corresponding author: kolgatin@ukr.net

In focus of interest are time planning and influence of systematicity of students' learning activity on educational achievements. So, A. B. Valynuk, S. O. Konovalenko [6] pointed on the basis of survey that only 14% of students prepare to classes systematically, 8% – occasionally, but 78% of students work at home only before practical and seminar classes. L. V. Klimenko [7] believes that systematic learning work of students promotes accumulation of knowledge, mastering in skills. But large amount of short structured task leads to obliviousness of educational material, so special work for systematisation should be suggested to students periodically. O. O. Lavrentieva, L. M. Rybalko, O. O. Tsys, and A. D. Uchitel [8] draw attention to the necessity of planning independent work with accounting the complexity of its various types and analysing new methods of organization of students' independent study activities together with the use of ICT and tools. O. H. Kolgatin, L. S. Kolgatina, N. S. Ponomareva, and E. O. Shmeltser [9] have analysed the results of survey and concluded that the most common problems in systematicity of learning activity during the independent work of the students are the lack of instructions, the lack of cognitive interest, students' mistakes in self-management of own learning activity, teachers' mistakes in time planning for the systems of learning tasks for students' independent work. There were suggested some requirements for management of students' independent work for fixing these problems.

1.3 Objectives

Despite a number of theoretical and empirical studies in the field of modelling the learning process, there is no integrated model yet. Available models are based on teacher intuition and personal pedagogical observation that complicates using such models for pedagogical prognosis at managing the learning activity in Internet-oriented environments. Great work for specification of models parameters and its indicators, detection of pedagogical criteria influence at efficiency of educational process is very actual. In this way the **aim** of this paper is to study systematicity of students' learning activity as a parameter of student's model, to study influence of systematicity at learning result in course "Operating systems".

2 Theoretical framework

In the context of mass education, the teacher cannot pay enough attention to each student to make pedagogical forecast for each student on the basis of own intuition, experience, theoretical and methodological knowledge. It is necessary to equip the student himself as the subject of the educational process with the skills and appropriate pedagogical forecasting tools for an independent choice of the appropriate variant of educational activity. The teacher should manage this pedagogical diagnostics system and provide the student with necessary help. Design of the computer-based pedagogical diagnostic system requires the development of a learning objectives model, a student psychological and pedagogical model (SPPM) and learning technologies model that would form the basis of this system. These models should be specially structured and should contain a limited number of parameters, which can be directly measured in the educational process.

Let fix our attention at one of this three major models in the system of pedagogical diagnostics – the psychological and pedagogical model of the student (SPPM) [5]. SPPM is built on the basis learning objectives model so that the parameters of the student model reflect the forthcoming to the intended learning goal. The SPPM is to allow comparison of successive academic achievements, reflecting the dynamics of learning process. That is, this model should be dynamic. Based on the analysis of pedagogical science data in the field of educational achievement modelling [10-14], a system of criteria has been proposed [5] according to such components: motivation and target, educational content mastering, self-management and activity, reflexion and prognosis (see Table 1). Further comprehensive developing of this model needs in a lot of experimental data on correspondence between criteria indicators that can be directly measured in educational process and real results of students' educational work. But we have not enough such data in modern publications as it was shown in 1.2.

Table 1. Structure of the student psychological and pedagogical model.

Component	Criteria
Motivation and target	Significance of the result of educational activity for the student
	Student's interest in the educational process, cognitive interest
	Conscious adherence to the educational discipline
Educational content mastering	Completeness of knowledge
	Promptness of knowledge
	Depth of knowledge
	Flexibility of knowledge
	Systematic of knowledge
	Lasting of knowledge
	Automation of activity
Self-management and activity	Stability of pace of educational work
	Ability of the student to mobilize energy, persistence and will
Reflexion and prognosis	Student's reflection on the result of activity
	Student's reflection on the process of activity

Estimation of the parameters that characterise the educational content criteria is carried out by means of pedagogical testing based on the concept of the level of educational achievements in accordance with the works of V. P. Bespalko [11] and I. Ya. Lerner [12] as well the Ukrainian educational standards. These works are not modern, but classic. The ideas of V. P. Bespalko correlate with B. Bloom's taxonomy [15], but V. P. Bespalko's approach is more simple and useful for practical automated pedagogical measurements. I. Ya. Lerner's ideas give us possibility to classify criteria according to indicators than can be measured directly as it was shown in [5].

The parameter of the lasting of knowledge has not included to composition of database for model parameters [5]. According to definition, lasting of knowledge is the permanent fixation in the student's memory of the system of essential knowledge and methods of their application or the willingness to derive the necessary knowledge from other based knowledge [12, p. 22]. A natural measure of lasting of knowledge is the ratio of the appropriate mastering coefficients according to the preliminary and current testing. If the mathematical model used in the automated system of diagnostics considers the parameters of student's academic achievements in dynamics (as a function of time), then a separate parameter "lasting of knowledge" is not needed. It is replaced by the functional dependence of all other parameters on the time that, definitely, carries more information.

The parameters of the student's psychological and pedagogical characteristics are determined by the teacher on the basis of pedagogical observation and analysis of the products of the student's educational activity. The student also takes active part in determining these parameters by introspection.

A high level of reflection on the result of the activity indicate the student's ability to objectively evaluate own results of the learning activity and his desire to complete the task qualitatively, to bring the work to a logical conclusion. The presence of an appropriate parameter in the student psychological and pedagogical model (SPPM) gives a reason to offer students, who have the developed reflection to the result of own activity, educational tasks of a creative nature. Otherwise, such tasks as projects, creative works etc. can be ineffective without student's own reflection, because it is difficult to build an objective and unambiguous algorithm for its checking.

High importance of the result of learning activity for the student is expressed in the desire to master given knowledge and skills as soon as possible, to get the result of the activity in the form of a fully completed task or project, a solved problem, etc. Of great importance is the student's sense of satisfaction from the successful completion of similar tasks in the past [13]. The organization of education of such students should provide for certain stop points at which the student can feel the completion of the stage of work. It is advisable to prevent the unexpected additional tasks and complications.

High interest in the process of learning is often native for students with research abilities, who can unlimited improve a computer program or laboratory equipment, collect some data from the Internet and so on. Modern multimedia tools and intelligent learning systems help to increase delight of the learning process itself. But the interest in certain activities in the absence of significance of the learning result leads to a shift in the focus on minor things and reduce the effectiveness of learning. Such students need in regular diagnose of the structure of academic achievement and control the implementation of the curriculum. They need in systematicity of learning activity according to curriculum. Such systematicity can be achieved by direct management of student's independent work or by training the student in skills of self-management. Cognitive interest as a separate parameter of the student model provides an opportunity to distinguish features of the student's motivation for learning activities. An important element of the emotional setting for learning activities is the conscious adherence to the educational discipline [14], which is expressed in the self-control of the correspondence of the learning activity to the work plan and culture of interaction with other participants of the educational process (timely completion of tasks, conscious fulfilment of requirements, accuracy in visiting classes and appointed consultations).

The strength and stability of the student's concentration on learning activities in a particular discipline largely depends on the peculiarities of the mental processes and physiological properties of the student and determines the style of educational activity. Therefore, it is important to add to the student model (SPPM) a parameter that characterizes a student's ability to mobilize persistence and will [13], and a parameter that characterizes the stability of the pace of student's academic work [14].

Activity of the student on introspection, observation of student's educational work, analysis of the style of educational achievements tests passing, analysis of the order of performance and presentation of educational products, analysis of the content of products of educational activity - are the sources of the information for the SPPM. It is advisable to measure parameters of reflection, emotional setting and volitional qualities on a scale of order (low, medium, high). The application of the equal-interval scale is problematic, because these parameters are complex and may include various indicators with significantly non-linear effect. Such measuring becomes a problem in case of lack of personal interconnection between student and teacher.

Summarising the above, it should be pointed that systematicity of learning activity is a complex indicator, which is connected with student's motivation, competency in self-managing, ability to stable work. But this indicator can be directly analysed in Internet-oriented learning management systems, such as Moodle. In this paper the systematicity of student's activity will be understood in context of activity by the plan given by teacher or designed by student in accordance to curriculum.

3 Methodology of empirical research

Study of features of students learning activity, connection between its systematicity and students' learning achievements was conducted in course "Operating Systems" with use of learning management system *Moodle*. Methods of learning the course "Operating Systems" are not a matter of this paper, but we should to describe the ground of our empirical work. This course combines theoretical and practical issues of operating systems concepts, models of its interconnection with hardware, applied software and users [16-18]. The first content module of this course is devoted to history and diversity of operating systems according to peculiarities its application. Students should understand the basic principles of computer hardware building, in particular, John von Neumann principles, shared bus architecture, address space, function of the registers, interrupts etc. One of the main fundamental issues of this module is to show the deep connection between the hardware and operating systems architecture. The simple operations in operating systems with use of command interpreter and graphical user interface were also the object of students' educational activity. Second module is devoted to detailed study of main abstractions in the theory of the operating systems: virtual memory, processes and threads. Students used built in and third party software as well as the authors' models to investigate the peculiarities of internal mechanisms of multiprogramming realisation, especially of scheduling CPU time and access to slow devices as well as RAM memory access. The third module covers wide spectrum of practical issues of booting the operating systems and logical organisation of disk drives, file systems, the structure of executable files, mechanism of management of the Windows operating system, security in operating systems.

Practical component of students' educational activity was dominant. Students of second year, future bachelors of computer science and software engineering completed the practical tasks on analysing structure, functionality, principles of design of some operating systems with use of virtual machines. Special software for virtualisation was used for supporting educational activity on installing different operating systems and third party software. Methods of study operating systems with using of virtualisation are enough developed in modern pedagogical works. As an example we can suggest the research of O. M. Spirin, O. S. Holovnia [19]. The tasks that were suggested for students assumed a part of work to be done in classes and other part was homework. Each student work according to unique variant of the tasks, but some steps were very similar for all students. There were suggested 11 tasks for every student for the semester according to the topics of the curricular (see Table. 2). Each of these task contained both reproductive and creative steps with problem solving.

As the result of this work, students prepared and submitted reports using *Assignment* activity in the university personal learning system based on *Moodle*. So, we have possibility to monitor the time of completing the task by the student. The grades for reports with late submission were less. The reports that were prepared later than 2 weeks after deadline were not accepted by personal learning system, and students presented such reports to teacher in printed form personally with oral discussion. These reports with very late submissions have not analysed in this paper. In total, 54 students took part in experiment. 274 reports were analysed. The final test in written form has been suggested to students for evaluating their educational achievements. The results of this test were the base for study the connection between systematicity of student's learning activity and his/her educational achievements.

4 Results and discussion

Specific values of final test results were used for analysing correlation between systematicity of students' learning activity and their educational achievements (Figure 1). This values were calculated as ratio of test result of each student to maximal test result. The indicator of systematicity was evaluated as a part of reports, submitted in time by a student, that is as a ratio of the number of reports, submitted by a student in time, to the number of reports according to plan (11 reports for 11 tasks). Pearson correlation between these two variables is 0.28. This correlation is statistically significant at the significance level 5% for samples size of 54 that is enough in pedagogical researches.

Table 2. Topics of the practical tasks.

Content module	Topic
Operating system concepts	1. Analysing the *Reactos* operating system (installing and customising the operating system, doing some work in it)
	2. Analysing the *Kolibrios* operating system (installing and customising the operating system, doing some work in it)
	3. Analysing the *Ubuntu* operating system (installing and customising the operating system, doing some work in it)
Memory, treads and processes	4. Analysing active processes and threads in the *Windows* operating system (operating with processes and threads, obtaining the information about the active processes and threads using built-in and third-party software)
	5. Analysing CPU and memory managing procedure in the *Windows* operating system (simulating of the operating system scheduling with use of the special designed model *WinMOS*)
	6. Analysing the *Windows* virtual memory (getting the information and optimising RAM memory with use of built-in and third-party software)
Operating system management and resources	7. Analysing the structure of the *Windows* executable files (getting the information about files and its structure with use of the fields map and third-party software)
	8. Analysing the *Registry* in the *Windows* operating system (using and changing the registry information for managing the operating system)
	9. Analysing system services and drivers in the *Windows* operating system
	10. Analysing data security in the *Windows* operating system (working with accounts, encryption algorithms, digital signature)
	11. Analysing and optimising the Windows operating system booting

So, we can conclude that the part of reports, submitted by student in time, positively connected with educational achievements. What is the kind of this correlation? Three variants are possible: 1) systematic work according to the plan, given by the teacher, contributes for increasing educational achievements; 2) students with high initial educational achievements easily execute the tasks and submit their reports in time; 3) students with high competence in self-management of their independent work have high educational achievements at all and, in particular, use their skills to complete the tasks in time for higher grading. Both the first and the third variants correspond the positive influence of systematicity on educational achievements.

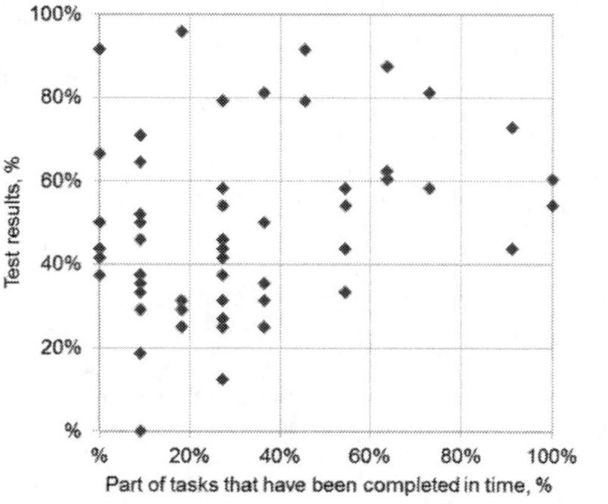

Fig. 1. Correlation between systematicity of students' learning activity and their final assessment results.

Analysing the diagram at Figure 1, we can see that the second variant was not realised: students with high test results (above 60 %) had systematicity indicator from 0 to 100 % and there was not any trend. Moreover, there was not any student with very high (above 80 %) systematicity indicator and test results simultaneously. So, we can see that the student with highest educational results did not work according to common plan even losing some grades. They, may be, worked systematically, but according to their own plans, so methodology of our experiment did not give us possibility to measure peculiarities of this work. Otherwise, they may be characterised by low level of importance of the learning activity results and high interest in the process of learning, and high cognitive interest. It should be appropriate to use for such students not the direct management of their independent work, but co-management or self-management.

Analysing the lowest boundary of points allocation at Figure 1, we can see that high value of the systematicity indicator (above 60%) guaranteed sufficient educational results (above 40 %). But students with highest systematicity indicator did not show excellent results in testing. This analysis gives grounds for hypothesis that the kind of management of student's independent work should be timely turned from direct management through co-management and subsidiary management to self-management according to the level of student's educational achievements and skills in self-managing for increasing the efficiency of educational process.

Choosing the day for completing the report, students taken into account many tasks in various sides of their life and study. But the fact that the number of reports, submitted in the last day, exceeds in near four times the number of reports, submitted in any other day (see Figure 2), show us the lack of students' competence in time planning and managing own work.

Did students work enough hardly in the educational process? Let us analyse diagram at Figure 3. We can see that students work at any time of day and night accept of period from 4 AM to 7 AM. In our opinion, such time scheme does not promote the learning of deeper questions of educational material, does not support productive and creative learning activity. This is not a problem of one course or one university, but a complex goal of development the methodology of education in direction to turning from reproductive methods of learning to more efficient students' activity with active use the information and communication technologies in education. We should take into account dual educational process, which is coordinated with professional-oriented practice work.

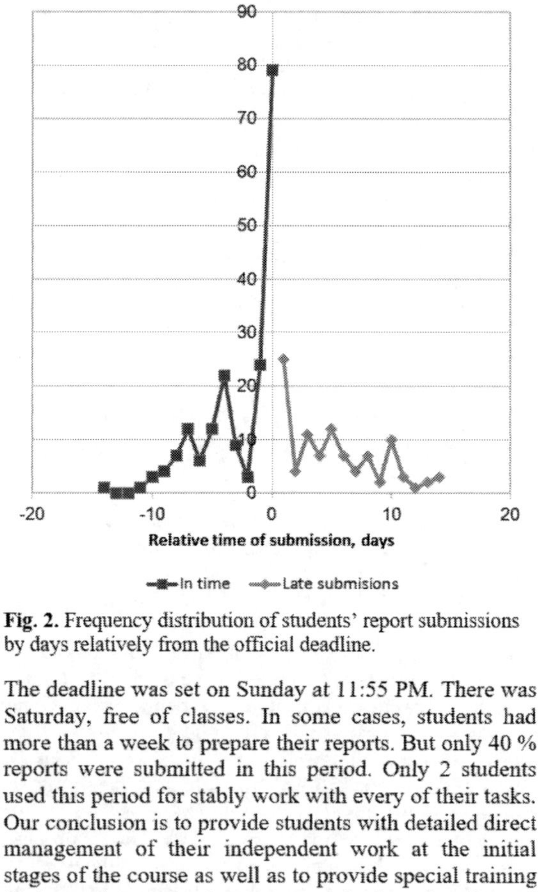

Fig. 2. Frequency distribution of students' report submissions by days relatively from the official deadline.

The deadline was set on Sunday at 11:55 PM. There was Saturday, free of classes. In some cases, students had more than a week to prepare their reports. But only 40 % reports were submitted in this period. Only 2 students used this period for stably work with every of their tasks. Our conclusion is to provide students with detailed direct management of their independent work at the initial stages of the course as well as to provide special training for increasing student's competence in time planning and self-management.

5 Conclusions

Analysis of obtained experimental data in context of our theoretical framework has given the base for such conclusions:

- systematicity of students' learning activity provide positive influence at students' educational achievements at low and sufficient levels;
- considering low students' competency in self-management and time planning it would be appropriate to provide students with detailed direct management of their independent work at the initial stages of the course

as well as to provide special training for increasing student's competence in time planning and self-management;

• kind of management of student's independent work should be timely turned from direct management through co-management and subsidiary management to self-management according to the level of student's educational achievements and skills in self-managing for providing the efficiency of educational process at highest level of educational achievements;

• students, who systematically complete educational tasks, have essential load that does not support productive and creative learning activity – turning from reproductive methods of learning to more efficient students' activity with active use the information and communication technologies in education is a complex goal of development of the educational methodology.

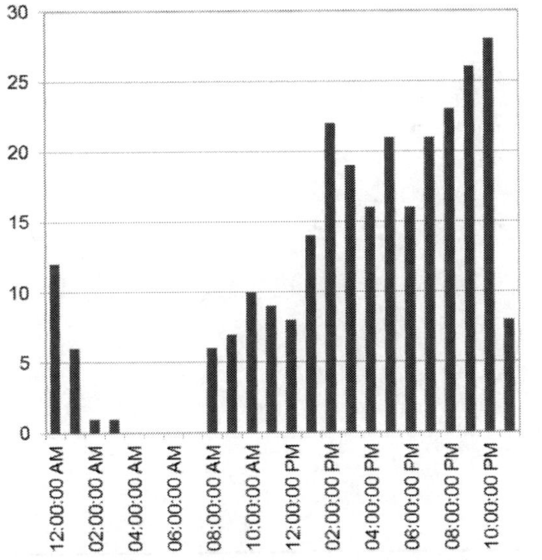

Fig. 3. Frequency distribution of students' report submissions by time, summarising all days (Deadline was 11:55 PM).

This study does not exhaust all the aspects in the field of creating of comprehensive student model for the systems of pedagogical diagnostics and prognosis. The main task in this direction is in obtaining a lot of experimental data about corresponding of some educational process indicators and efficiency of one or another methods of learning.

Concerning further development of the course "Operating Systems" we plan to introduce the management of student's independent work with more or less elements of self-management according to the features and educational achievements of a student. A good kind of developing learning management tools will be using such an interactive instrument as *Workshop* in the *Moodle* environment.

References

1. V.Yu. Bykov, *Modeli orhanizatsiinykh system vidkrytoi osvity* (Models of the open education organizational systems). (Atika, Kyiv, 2008)
2. A.E. Kiv, V.N. Soloviev, S.O. Semerikov, CTE 2018 – How cloud technologies continues to transform education. CEUR Workshop Proceedings **2433**, 1–19 (2018), http://ceur-ws.org/Vol-2433/paper00.pdf. Accessed 30 Dec 2019
3. O.O. Triakina, O.O. Pavlenko, N.P. Volkova, D.A. Kassim, Usage of E-learning Tools in Self-education of Government Officers Involved in Global Trade Activities. CEUR Workshop Proceedings **2257**, 173–181 (2018), http://ceur-ws.org/Vol-2257/paper16.pdf. Accessed 30 Dec 2019
4. K. Vlasenko, O. Chumak, I. Sitak, I. Lovianova, O. Kondratyeva, Training of Mathematical Disciplines Teachers for Higher Educational Institutions as a Contemporary Problem. Universal Journal of Educational Research **7**(9), 1892–1900 (2019). doi:10.13189/ujer.2019.0 70907
5. O.G. Kolgatin, Bazovi modeli v kompiuterno oriientovanii systemi pedahohichnoi diahnostyky (Base models in computer-based pedagogical diagnostics system). Information Technologies in Education **12**, 14–20 (2012). doi:10.14308/ite000309
6. A.B. Valynuk, S.O. Konovalenko, Eksperymentalne obgruntuvannia ta perevirka zaprovadzhenoi metodyky samostiinoi roboty studentiv (Experimental confirmation and verification of introduced methods of students' individual work). MEDSESTRYNSTVO **2**, 20–22 (2017). doi:10.11603/2411-1597.2017.2.8502
7. L.V. Klimenko, Samostiina robota studentiv yak obiekt systemnoho doslidzhennia (Independent work of students as the object of systematic research). Pedahohika formuvannya tvorchoyi osobystosti u vyshchiy i zahal'noosvitniy shkolakh **49**, 320–328 (2016), http://nbuv.gov.ua/UJRN/Pfto_2016_49_46. Accessed 30 Dec 2019
8. O.O. Lavrentieva, L.M. Rybalko, O.O. Tsys, A.D. Uchitel, Theoretical and methodical aspects of the organization of students' independent study activities together with the use of ICT and tools. CEUR Workshop Proceedings **2433**, 102–125 (2019), http://ceur-ws.org/Vol-2433/paper06.pdf. Accessed 30 Dec 2019
9. O.H. Kolgatin, L.S. Kolgatina, N.S. Ponomareva, E.O. Shmeltser, Systematicity of students' independent work in cloud learning environment. CEUR Workshop Proceedings **2433**, 184–196 (2018), http://ceur-ws.org/Vol-2433/paper11.pdf. Accessed 30 Dec 2019
10. V. P. Pustobaev, M. Yu. Saiapyn, Formalizatsiia elementov diagnostiki znanii uchashchegosia (Formalization of elements of student knowledge

diagnostics). Informatika i obrazovanie **7**, 120–123 (2005)

11. V.P. Bespalko, *Obrazovanie i obuchenie s uchastiem kompiuterov* (Education and training with the participation of computers). (MPSI, Moskva – MODEK, Voronezh, 2002)
12. I.Ya. Lerner, *Kachestva znanii uchashchikhsia. Kakimi oni dolzhny byt?* (The quality of students' knowledge. What should they be like?). (Znaniye, Moskva, 1978)
13. J. Raven, *The tragic illusion: educational testing* (Trillium Press, New York, 1991)
14. Yu.K. Babansky, *Izbrannye pedagogicheskie trudy* (Selected pedagogical works). (Pedagogika, Moskva, 1989)
15. B.S. Bloom, M.D. Englehart, E.J. Furst, W.H. Hill, D.R. Krathwohl, *Taxonomy of educational objectives: The classification of educational goals, handbook I: Cognitive domain* (Longmans, Green, New York, 1956)
16. A.S. Tanenbaum, H. Bos, *Modern Operating Systems* (Pearson Education Inc, Amsterdam, 2015)
17. M. Russinovich, D.A. Solomon, A. Ionescu, *Windows Internals. Part 1, 2* (Microsoft Press, Washington, 2012)
18. J. Bacon, T. Harris, *Operating Systems. Concurrent and Distributed Software Design* (Addison Wesley, 2003)
19. O.M. Spirin, O.S. Holovnia, Zastosuvannia tekhnolohii virtualizatsii Unix-podibnykh operatsiinykh system u pidhotovtsi bakalavriv informatyky (Using Unix-like operating systems virtualization technologies in training the bachelors of computer science). Information Technologies and Learning Tools **65**(3), 201–222 (2018). doi:10.33407/itlt.v65i3.2055

Design of educational environment for teachers' professional training

Nadiia Balyk, *Galina* Shmyger, *Yaroslav* Vasylenko, and *Vasyl* Oleksiuk*

Ternopil Volodymyr Hnatiuk National Pedagogical University, Department of computer sciences and methodology of its teaching, Ternopil, Ukraine

Abstract. The article analyzes the concept of the educational environment, its components and features of the structure. The concept of educational environment design was introduced. The components of educational environment design for the teachers' professional training of the Ternopil Volodymyr Hnatiuk National Pedagogical University (TNPU) are described: technological, didactic and social. The features of this environment such as: information saturation and openness, digital transformation, social practices and cooperation are considered. The study of the effectiveness of educational environment for teachers' professional development was carried out on the basis of TNPU. In total, 432 masters of all specialties of the University participated in this study. The study used the method of expert assessments for statistical processing of results. The study was conducted to determine the level of importance of all indicators of each component of the educational environment design. The results of the study illustrate the significant changes in the technological and social components of the university's educational environment design, which have a significant impact on the teachers' professional training.

1 Introduction

The key problems of the higher pedagogical school of today are the lack of manifestation of the cultural and historical context for the higher school, which sets the framework for higher education. The rapid development of educational management leads to process-oriented management of an educational institution and the emergence of terms such as "educational space", "educational landscape", "educational field", "educational environment".

Today, there are no universal indicators to quantify and qualify what an effective educational environment is. The new educational perspective on the development of the contemporary educational environment requires the reorganization of many aspects of future teacher training.

The transition from traditional educational models to modern ones envisages a change in the organizational, cultural, institutional dimensions, management models and educational environment design for the teachers' professional training.

2 Research review

The analysis of the sources shows that the issues of design and formation of the educational environment are an important component of the training of modern specialists both in Ukraine and abroad.

The educational environment is traditionally defined as learning, which depends on various environmental factors, a set of objective external conditions, factors, social objects. It is a system of influences and conditions of personality formation, as well as opportunities for its development, which are contained in the social and spatial-subject surroundings [1]. The educational environment is a contemporary temporal, spatial and social situation of learning, which consists of many different educational spaces of different levels, which have educational potential and interact in one way or another. In this environment, the interaction of different levels of the education system and personality happens and the corresponding cultural context is also included [2]. As a result of a detailed historical study, scientists of Kherson State University have determined that the most promising model for building an educational environment is a hybrid model [3]. Olena Glazunova and Mariya Shyshkina have been confirmed these findings for the case of university cloud-based learning and research environment [4].

In her monograph Liubov Panchenko determined that modern specialists should be able not only to use, but also to model and create an educational environment [5].

Today, the following structural components of the educational environment are distinguished:

1) physical environment – the room, its design, size and the spatial structure of the training classrooms;

2) human factor – the university contingent structure, its influence on the social behaviour of students, the quality of lecturer training, etc.;

3) training program – the nature of training programs content, technologies of training, style and methods of training, forms of educational activities, the nature of

* Corresponding author: oleksyuk@fizmat.tnpu.edu.ua

control [6]. The components of the educational space are united by certain ideas and values.

The main features of the educational environments that characterize the new millennium have been determined by many researchers [7-9]. They point out that the university's educational environment should include such components as the information and communication environment, the research environment, the organizational and management environment in accordance with the principles of intensity, psychological comfort and democratic possibilities of individualization of learning, openness and accessibility of information resources.

In most foreign studies, the educational environment is described in terms of "educational institution efficiency" as a social system – emotional climate, personal well-being, features of the microculture, quality of the educational process. The educational environment has a significant impact on students' learning and behaviour. There is a strong link between the learning environment and value components such as students' satisfaction and success. The educational environment defines physical and mental self-feeling and motivation and promotes emotional and behavioural responses. The study [10-11] analysed the physical environment using three aspects: the planning and size of study rooms, ergonomics and technology, the informal environment and comfort. Describing the ideal auditorium, students noted the importance of technology and comfort role. A broader understanding of the educational environment supposes the inclusion of various communications (press, radio, television, internet resources) created by young people in their own cultural microenvironment.

The work [12] focuses on the importance of the professional environment of teachers and not only on their professional training. This point should be emphasized, because over the past few years, academic research has forced many experts to assess not only the need to increase teacher effectiveness (for example, through qualification increasing), but also to change the educational environment by improving educational institutions policies, amending laws, and supporting by communities, improving decision-making process, digitizing education that can contribute to quality change in the education sphere.

Modern digitalization means the need to create a new educational environment [13, 14]. As digital technology becomes a central part of everyday work, teachers are forced to rethink and transform previous educational traditions through technology. These problems create insurmountable requirements for universities to develop teachers' professional training strategies in the context of mastering digital pedagogy and the digital educational environment [15-18].

3 Results and discussion

3.1 The structure of educational environment

During the research the following methods were used: analysis of scientific and methodological and technical literature in the field of educational environments design, state standards of higher education. In the course of the experimental research, the methods of observation, questioning and expert assessments were applied. The questioning of the respondents was conducted according to the methodology of expert assessments, with further processing of its results using the methods of mathematical statistics.

Analysing the views of various scientists about the particularities of the educational environment [19-21], we introduce the concept of educational environment design as a way of integrating and adopting many of its dimensions. In our study, as the concept of "design of the university's educational environment" we will consider systemic formation, which is a sociocultural surroundings of the subject of learning, which includes technological, didactic, social components that are able to provide quality professional training for teachers.

Such subjects (involved in the process of creating educational values) as lecturers, students, undergraduates, graduate students, educational institutions, organizations, scientific centres are important in the design of the university's educational environment for teachers' professional training.

Let us consider the components of the educational environment which were forming at the TNPU in recent years in the context of teachers' professional training.

The **technological component** of the educational environment design for teachers' professional training was provided through the creation of a digital environment for the university. The University's digital environment infrastructure is a system of software, computing and telecommunications tools that implements the providing of information, computing, telecommunication resources and services to all participants in the educational process. Various tools have been integrated into the university's digital environment, which enrich the educational process. In terms of infrastructure this environment is based on the use of university's LMS, cloud-based learning environment (CBLE), university's digital repository, Web 2.0 services. The researchers Olena Kuzminska, Mariia Mazorchuk, Nataliia Morze, Oleg Kobylin found 4 main components that group all the factors of the digital educational environment into such areas of focus as IT infrastructure and resources' provision, students' and teachers' digital competencies, scientific and educational communication between the students, teachers, and stakeholders, and educational process organization [22].

The effectiveness of CBLE in teaching and research has been investigated and tested by Ukrainian scientists under the guidance of Oleg Spirin [23], [24].

We consider that indicators of **technological component** development are:

1) **University network and Internet access**. TNPU provides access for students and lecturers from anywhere on campus to the resources of educational environment and the Internet. Local wired and wireless technologies have been used for this purpose. All resources are accessed using a single authentication data.

2) **Learning Management System and courses**. An advanced learning management system is functioning at the university. All subjects that are studying by students

have relevant e-courses in this system. In total, more than 600 courses have been developed by lecturers. Practically all kind of students' activities are recorded in this system.

3) **Cloud services and laboratories**. Since 2012, the lecturers of Computer Sciences Department and Methodology of Its Teaching have been working on the deployment of a cloud-oriented learning environment. It now operates according to a hybrid model and integrates many services of public and private platforms. CBLE provides unified, ubiquitous and secure access to file and computing resources (repositories, virtual computers, and networks). Cloud infrastructure provides management of educational resources, aggregation of computing resources, knowledge sharing services, increasing the flexibility of their use by participants in the educational environment.

4) **Hardware for 3D design and printing**. Within the frame of work of STEM-centre [25] promising technologies of 3D-modeling and 3D-printing, technologies of virtual and augmented reality, technologies of the Internet of things, robotics are being implemented at the University. These technologies ensure the execution of innovative projects through the formation of tool environments, the use of project management services. Work on educational projects (for example, a project on 3D-reconstruction and 3D-printing of the destroyed historical castles of Ternopil region) takes place inside a technologically equipped modern educational environment.

5) **Open environment**. An open, non-formal learning environment with lecturers and students has created at the University. The traditional academic hierarchy is gradually being replaced by an approach where students are respected as junior colleagues, and their opinions are appreciated and encouraged by more experienced colleagues. Such teaching is based on modern didactic approaches such as personality-oriented and synergistic. The technological basis of open education at TNPU is modern digital technologies, in particular cloud. This approach encourages dialogue and collaboration between students and lecturers, and creates new opportunities for the development of up-to-date professional training for future teachers.

6) **University archives and repositories**. The University has implemented a system of digital archives. The TNPU Institutional Repository contains materials published by lecturers, such as: monographs, books, manuals, articles, abstracts. Some faculties have digital archives for educational purposes. In addition to the materials of lecturers, they contain the results of students' learning – materials of practices, articles of students, master's works, etc.

Among the important components of the educational environment design of the university should be distinguished didactic, which includes the structure of students' activities, teaching style, nature of control, forms of study, the content of study programs. For example, the professorial and teaching staff of TNPU pays special attention to the modernization of educational programs in the context of the tasks of the New Ukrainian School through:

- implementation of a competency, personality-oriented approach in pedagogical education;
- formation of managerial skills for effective activity in the conditions of real autonomy of educational institutions;
- providing practical training through continuous pedagogical practice of students at different educational institutions.

In TNPU, the didactic component of the educational environment design for future teachers' professional training is characterized by digital transformation, student-centred education; using:

- thematic project studies;
- critical thinking;
- group work;
- social practices.

Let's take a closer look at these efficiency indicators of the **didactical component** for educational environment design:

1) **Digital transformation**. The digital transformation of the university's educational environment is a series of coordinated steps and changes in the information infrastructure, in the digital culture of lecturers and students. This makes it possible to embody new educational models and transform the activities of the university, aiming at value propositions and strategic directions for the development of modern society.

2) **Group work**. Group work is characteristic of many university disciplines. Its purpose is for students to practice teamwork in small groups, as well as to develop problem-solving and leadership skills. Group work is an important aspect of future teacher training with aim of real professional situations modelling.

3) **Critical thinking**. Critical thinking is encouraged in all activities at the university. At seminars, workshops, laboratory work the students analyse and present solutions to problems and tasks. Theoretical concepts are tested in practical situations, and practical experience is used to develop and enrich the theory.

4) **Student-centred education**. Studying at TNPU is student-centred. There is great support from educators, lecturers play the role of facilitators, helping students understand the content of the course. The focus is on giving students the opportunity to develop their critical and analytical thinking skills, self-study, group work, problem-solving and leadership skills to prepare them for careers.

5) **Thematic project studies**. The teaching methods used in university study focus on critical analysis of course content using real cases where possible. Invited teachers and speakers from schools, local authorities, and public organizations participate in the educational program to further link research with the professional environment.

6) **Pedagogical practices**. Much of the learning process takes place outside the classroom when students apply acquired professional competencies in real-life situations while undergoing pedagogical practices. Learning technologies are partly beyond the bound of university classrooms.

Let us characterize the indicators of the effectiveness of the **social component** of design of the modern educational environment of teacher training of the TNPU in the context of exploring ways of improving their professional development.

It is traditionally considered that university education is constructed based on the context of the surrounding reality, the cultural space and the environment in which the education takes place. Therefore, at TNPU the main indicators of the effectiveness of design of the modern educational environment of teacher training in the social aspect are: social innovation, leading development, corporate culture, leadership, social partnership, and social communication:

1) **Social innovation**. n our opinion, the departure from the traditional functions of TNPU and the implementation of innovative ones became important for the professional development of teachers:

- creation of conditions for the system of qualitative training and professional development of teachers through overcoming the fragmented responsibility of different educational institutions for different stages of becoming and professional development of the educator;
- transition from "translational" education to "active" based on the implementation of digital technologies, project and competency learning technologies.

2) **Leading development**. The essence of leading-edge development lies:

- in building curricula and learning programs around cross-cutting topics relevant to a particular public community, a united territorial community;
- in preparing graduates to organize the life of their local community in accordance with the principles of sustainable and successful development.

3) **Corporate culture**. We consider that not only structural components are the social component achievement of the university's educational environment design, but first of all – corporate culture. The key factor to the success of university education transformation projects has been the formation of a collective subject for change. The corporate culture of the university is based on a system of values that determine the philosophy of its activity, the attractiveness of the university brand in the scientific, educational and contemporary socio-cultural environment.

4) **Social leadership**. Social leadership means:

- engraftment of innovation as a way of thinking and a key leadership tool;
- distributed leadership in the development of new educational decisions and educational reforms,
- formation of teachers, as educators of leaders of the new generation, integral personalities.

The University promotes the growth of students as individuals through quality professional training of highly qualified professionals and personal growth.

5) **Social partnership**. TNPU's educational environment design serves as a catalyst for a new social reality in the region. The University is an active social partner and an element of the social system. The collaboration and partnership of the university's educational environment with various actors of the educational field and the public is developing. Lecturers share knowledge and experience in the educational environment, give the products of their professional and innovative activities in the public usage, participate in volunteer activities, assessments and expertise, and more.

6) **Social communication**. Communication has become a key prerequisite for the creation of new meanings, ideas and projects of the University, organization of applied research at the request of regional companies, authorities and the local community. It is important that the university is open to industry, government and other stakeholders. We believe that the greater the degree of openness of a university, the better it develops. The University successfully builds all necessary for its own existence and development of communications with other entities - authorities, manufacturing companies, civil society institutions.

3.2 The study on the effectiveness of educational environment design

In order to determine the effectiveness of the created educational environment design for the teachers' professional development in 2017/2019, a study was conducted in the form of a survey among future teachers. 432 masters of all pedagogical specialties of the University participated in the survey. We viewed undergraduates as internal stakeholders.

The questionnaire suggested to assess the importance of development each component of the university's educational environment design. In each component we have identified indicators of its development (see Table 1).

Table 1. List of indicators for assessment of the components of the university's educational environment design.

Component of the educational environment	Cipher of indicator	The name (description) of the indicator
Technological	T1	University network and Internet access
	T2	Learning Management System and courses
	T3	Cloud services and laboratories
	T4	Open environment
	T5	Hardware for 3D design and printing
	T6	University archives and repositories
Didactic	D1	Digital transformation
	D2	Group work
	D3	Critical thinking
	D4	Student-centred education
	D5	Thematic project studies
	D6	Pedagogical practices
Social	S1	Social innovation
	S2	Leading development
	S3	Corporate culture
	S4	Social leadership
	S5	Social partnership
	S6	Social communication

In each questionnaire, we explained to the experts the value of each indicator. To determine the most significant indicators of educational environment development, we used the ranking method. It was to determine the relative importance of the objects under study based on their ordering. A scoring system for assessment was proposed for each component. In each component of the educational environment development, the experts gave points. One point was awarded to the least significant indicator and six points to the highest significant one. The results of the survey are summarized in a table, the columns of which correspond to the codes of indicators, and in rows – sequence numbers of experts (see Table 2). The table data can be viewed in its entirety by the hyperlink: https://drive.google.com/file/d/1YHaqVE0NSVktz9GlwzqGVGy2HAK7CDWy/view?usp=sharing

Table 2. The final results of the study data processing.

Expert	Technological						Didactic						Social					
	Indicators																	
	T1	T2	T3	T4	T5	T6	D1	D2	D3	D4	D5	D6	S1	S2	S3	S4	S5	S6
1	5	6	3	4	1	2	6	5	1	4	2	3	6	5	1	4	2	3
2	6	5	4	3	1	2	6	4	5	1	3	2	6	3	4	1	5	2
…																		
432	6	5	4	3	1	2	4	5	6	1	2	3	6	4	1	5	3	2
S_j	2394	1946	1573	1405	770	984	2368	2153	1300	1119	969	1164	2369	1682	750	1736	1593	934
d_j	882	434	61	-107	-742	-528	856	641	-212	-393	-543	-348	857	170	-762	224	81	-578
$S(d^2)$	1810798						1758963						1734814					
W	0.55445265						0.539						0.53118692					

In order to prevent psychological clues that could influence the expert's choice of a certain ranking order, indicators of a certain criterion in the card were placed in alphabetical order.

An expert assessment method was chosen to work out the results of the survey, which was applied to each component of the university's educational environment individually due to the independent ranking of indicators within each component.

The most obvious value of assessment an indicator is its total rank, which is determined by all experts ($S_j = \sum_{i=1}^{m} R_{ij}$, where R_{ij} is the j-th indicator exhibited by the i-th expert, m is the number of experts).

However, such aggregate rankings will be objective if there is a certain level of agreement between the experts. The degree of such agreement is described by Kendall's coefficient of concordance W [26], which is defined as follows:

1. For each indicator, we find the difference between the totals and their average:

$$d_j = \sum_{i=1}^{m} R_{ij} - 0.5 \cdot m \cdot (n+1) \quad (1)$$

2. Find the sum of squares of values obtained from relation (1) $S(d^2)$

$$S(d^2) = \sum_{i=1}^{n} d_j^2 = \sum_{i=1}^{n} \left[\sum_{j=1}^{m} R_{ij} - 0.5 \cdot m \cdot (n+1) \right]^2 \quad (2)$$

3. The maximum value of $S(d^2)$

$$S_{\max}(d^2) = \frac{1}{12} \cdot m^2 (n^3 - n)$$

is achieved if all experts rank the criteria (indicators) equally.

4. The coefficient of concordance is equal:

$$W = \frac{S(d^2)}{S_{\max}(d^2)} = \frac{12 \cdot S(d^2)}{m^2 (n^3 - n)} \quad (3)$$

According to formulas (1) - (3) we find the values of the total ranks S_j, the values d_j, $S(d^2)$ and calculate the coefficient of concordance W for each component of the educational environment. The results of the calculations are presented in Table 2.

This value is always between zero and one. If $W = 0$, then there is no correlation between expert rankings, if $W = 1$, then the rankings are completely the same. We get the coefficient $W = 0.55$; 0.54; 0.53 is substantially different from zero, so it can be argued that there is objective agreement between experts.

However, such a value of W is not a criterion for objectivity, since it could be obtained by accidentally setting of ranks one or the other indicators.

The value $m \cdot (n-1) \cdot W$ is distributed by the law χ^2 with n-1 degree of freedom. Using the ratio

$$\chi_W^2 = \frac{12 \cdot S(d^2)}{m \cdot n(n+1)}$$

we find the value of χ_W^2 = 1197.62; 1187.77; 1147.36 for the relevant components of the educational environment. Comparing them with the table value for $\vartheta = n - 1 = 5$ degrees of freedom and for the significance level of $\alpha = 0.01$, we obtain $\chi_W^2 > \chi_t^2 = 15.1$. Hence, we conclude that there is consistency between experts' findings.

Consider the results of the survey regarding the importance of technological, didactic and social components of the university's educational environment design for teachers' professional development of teachers of the pedagogical university (see Figures 1-3).

From the conducted study it follows that:

- of the technological component, the most important for the teachers' professional development are University network and Internet access, Learning Management System and courses, Cloud services and laboratories;
- of the didactic component most important for the teachers' professional development are Digital transformation, Group work, Critical thinking;
- of the social component the most important for the teachers' professional development teachers are Social innovation, Social leadership, Leading development.

https://doi.org/10.1051/shsconf/20207503010

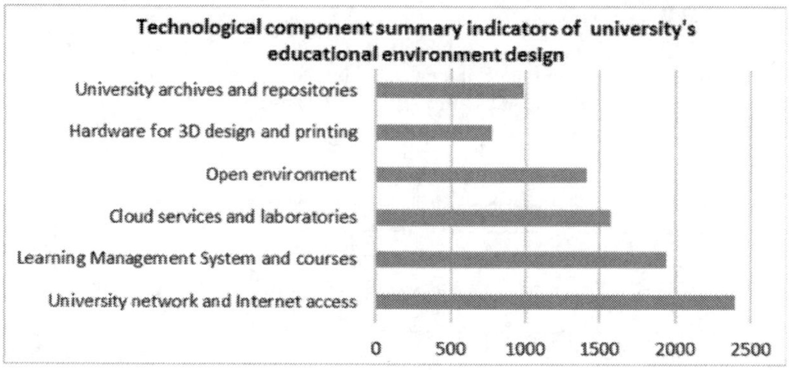

Fig. 1. Study's results of technological component importance of university's educational environment design of in the context of teachers' professional development.

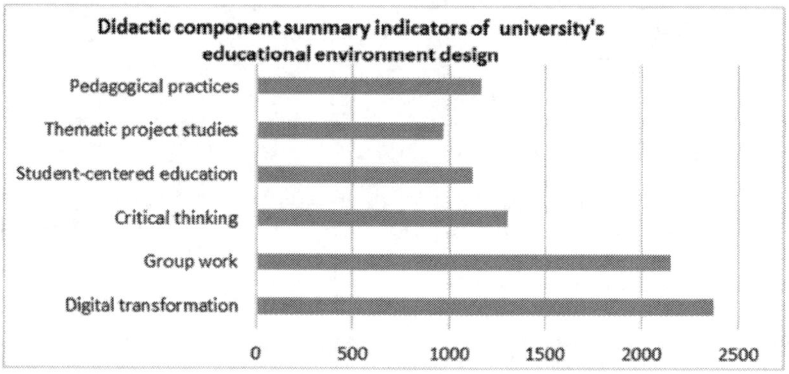

Fig. 2. Study's results of didactic component importance of university's educational environment design of in the context of teachers' professional development.

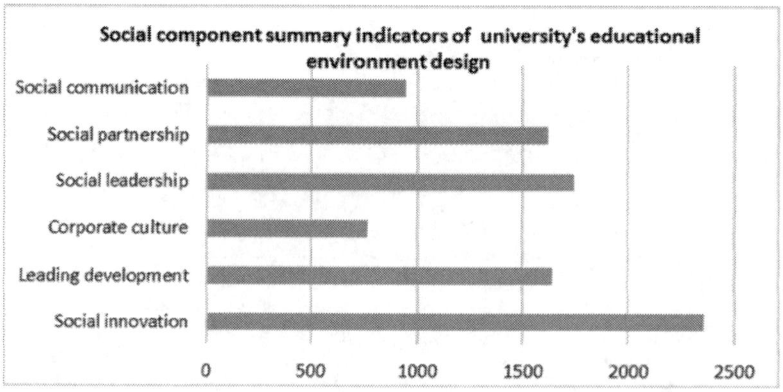

Fig. 3. Study's results of social component importance of university's educational environment design of in the context of teachers' professional development.

To determine the significance degree of each component of the educational environment, we calculated the arithmetic mean of the scores for each indicator (see Table 3). The indicator was considered positive if the arithmetic mean of expert estimates was at least 3.0.

The significance degree of each component was determined as follows:

- not significant enough – more than 50% of the criteria are negative;
- critically significant – 50% - 55% of the criteria are positive;
- significant enough – 56% - 75% of the criteria are positive;
- highly significant – 76% - 100% of the criteria are positive.

Table 3. Significance degrees of the university's educational environment.

	Technological component						**Didactic component**						**Social component**					
	Indicators																	
	T1	T2	T3	T4	T5	T6	D1	D2	D3	D4	D5	D6	S1	S2	S3	S4	S5	S6
Average value	5.54	4.50	3.64	3.25	1.78	2.28	5.48	4.98	3.01	2.59	2.24	2.69	5.48	3.89	1.74	4.02	3.69	2.16
%	66,7%						50,0%						66,7%					
	Degree of significance																	
	significant enough						critically significant						significant enough					

From the conducted study it follows that at the TNPU over the past three years, according to the view of undergraduates, technological and social components of the educational environment design have become crucial for teachers' professional development.

4 Conclusions

An analysis of the literature indicates that the term "educational environment" has no unambiguous interpretation. The study proposes to define the design of the educational environment as a systemic formation, which includes technological, didactic, social components that are able to provide quality professional training for teachers.

It should be noted that the design features of the modern educational environment of the TNPU are: openness and information saturation, student-centred education, thematic project studies, social practices, a harmonious blend of pedagogy and digital technology and, as a result, the digital transformation of educational environment design.

To identify the effectiveness of the created design of the university's educational environment for the teachers' professional development the components of their formation and their corresponding indicators were determined. In the process of research, the undergraduates noted that the greatest influence on their professional development has social (Social innovation, Social leadership and Leading development) and technological component of the educational environment design (University network and Internet access, Learning Management System and courses, Cloud services and laboratories).

We consider that in the development of educational environment design of pedagogical university promising directions are such as:

- developing educational strategies and monitoring their implementation and effectiveness;
- realization by the university of its socially transformative role - social and humanitarian innovations, humanitarian paradigm of education;
- organizing effective interaction between the university and external players in order to attract investments to create quality conditions for learning and nurturing successful and competitive human capital.

The perspectives of further research are in experimental testing the designed environment by other internal and external stakeholders like as lecturers, teachers, developers, IT-managers etc.

References

1. R. Hiemstra, Aspects of effective learning environments. New Directions for Adult and Continuing Education **50**, 5–12. (1991). doi:10.1002/ace.36719915003
2. H. Mandl, G. Reinmann-Rothmeier, in *International encyclopedia of the social and behavioral sciences*, ed. by N.J. Smelser, P.B. Baltes (Elsevier, 2001), pp. 4697–4701
3. A. Spivakovsky, L. Petukhova, V. Kotkova, Yu. Yurchuk, Historical Approach to Modern Learning Environment. CEUR Workshop Proceedings **2393** (2019), http://ceur-ws.org/Vol-2393/paper_420.pdf Accessed 22 Dec 2019
4. O. Glazunova, M. Shyshkina, The Concept, Principles of Design and Implementation of the University Cloud-based Learning and Research Environment. CEUR Workshop Proceedings **2104** (2018), http://ceur-ws.org/Vol-2104/paper_158.pdf Accessed 22 Dec 2019
5. L.F. Panchenko, *Information and educational environment of the modern university* ("LNU named after Taras Shevchenko", Luhansk, 2010)
6. Analitychnyi zvit pro virtualnyi kruhlyi stil «Yak zdiisniuvaty upravlinnia seredovyshchem vzaiemodii usikh uchasnykiv osvitnoho protsesu?» (Virtual Roundtable Analytical Report "How to manage the interaction environment of all participants in the educational process?") (2017), http://education-ua.org/ua/draft-regulations/966-analitichnij-zvit-pro-virtualnij-kruglij-stil-yak-zdijsnyuvati-upravlinnya-seredovishchem-vzaemodiji-usikh-uchasnikiv-osvitnogo-protsesu. Accessed 25 Dec 2019
7. OECD, Creating Effective Teaching and Learning Environments: First Results from TALIS. Executive summary (2009), https://www.oecd.org/berlin/43024880.pdf. Accessed 27 Dec 2019
8. A. M. Kalimullin, Z. I. Islamova, Formation of Information-Educational Environment in the Partner Universities of University of Shanghai Cooperation Organization. IEJME **11**(6), 1879–1890 (2016)
9. A. Potter, Powerful Learning Environments: Unravelling Basic Components and Dimensions. Internet and Higher Education **7**, 154–157 (2004). doi:10.1016/S1096-7516(04)00020-X

10. L. Licite, L. Janmere, in *Engineering for rural development*, Jelgava, 23-25.05.2018, pp. 1198–1203. doi:10.22616/ERDev2018.17.N361
11. C. Anderson, K. Day, Purposive environments: engaging students in the values and practices of history. IJHEEP **49**(3), 319–343 (2005)
12. Standards of Professional Competencies Required of Teachers (2011), http://cfieleon.centros.educa.jcyl.es/sitio/upload/modelo_competencias_profesionales_profesorado%20_traducido.pdf. Accessed 20 Dec 2019
13. O. E. Hatlevik, K. A. Christophersen, Digital competence at the beginning of upper secondary school: Identifying factors explaining digital inclusion. CE **63**(1), 240–247 (2013)
14. A.E. Kiv, V.N. Soloviev, S.O. Semerikov, How cloud technologies continues to transform education. CEUR Workshop Proceedings **2433**, 1–19 (2019), http://ceur-ws.org/Vol-2433/paper00.pdf. Accessed 20 Dec 2019
15. Digitalisation strategy for the higher education sector 2017-2021 (2017), https://www.regjeringen.no/contentassets/779c0783ffee461b88451b9ab71d5f51/en-gb/pdfs/digitaliseringsstrategi-for-uh-sektoren-engelsk-ve.pdf. Accessed 20 Dec 2019
16. J. Howell, *Teaching With ICT: Digital Pedagogies For Collaboration And Creativity* (Oxford University Press, Melbourne, 2012)
17. C. Kivunja, Embedding Digital Pedagogy in Pre-Service Higher Education To Better Prepare Teachers for the Digital Generation. IJSHE **2**(4), 131–142 (2013). doi:10.5430/ijhe.v2n4p131
18. J. Stommel, Critical digital pedagogy: A definition. Hybrid Pedagogy (2014). http://www.digitalpedagogylab.com/hybridped/critical-digital-pedagogy-definition. Accessed 8 Dec 2019
19. A. Lund, T. E. Hauge, Designs for Teaching and Learning in Technology Rich Learning Environments. NJDL **4**, 258–271 (2011)
20. W. Roth, Learning environments research, lifeworld analysis, and solidarity in practice. LER **2**(3), 225–247 (1999). doi:10.1023/A:1009953920993
21. K. Day, Creating and Sustaining Effective Learning Environments. AISHE-J **1**(1), 1–13 (2009)
22. O. Kuzminska, M. Mazorchuk, N. Morze, O. Kobylin, Attitude to the Digital Learning Environment in Ukrainian Universities. CEUR Workshop Proceedings **2393** (2019), http://ceur-ws.org/Vol-2393/paper_245.pdf. Accessed 22 Dec 2019
23. O. Spirin, Yu. Nosenko, A. Iatsyshyn, Current requirements and contents of training of qualified scientists on information and communication technologies in education. Information technology and learning tools **56**(6), 219–239 (2016)
24. O. Spirin, V. Oleksiuk, O. Oleksiuk, S. Sydorenko, The Group Methodology of Using Cloud Technologies in the Training of Future Computer Science Teachers. CEUR Workshop Proceedings **2104** (2018), http://ceur-ws.org/Vol-2104/paper_154.pdf. Accessed 20 Dec 2019
25. N. Balyk, G. Shmyger, Y. Vasylenko, V. Oleksiuk, A. Skaskiv, in *E-learning and STEM Education*, ed. by E. Smyrnova-Trybulska (Studio NOA, Katowice, 2019), pp. 109-125
26. P. Legendre, in *Encyclopedia of Research Design*, vol. 1, ed. by N.J. Salkind (SAGE Publications, Los Angeles, 2010), pp. 164-169

The problems of implementation of inclusive education in Ukraine: generalization the experience of empirical sociological researches experience

Iryna Hrabovets[1,*], *Liudmyla* Kalashnikova[2], and *Liudmyla* Chernous[1]

[1]Kryvyi Rih State Pedagogical University, Department of Sociology and Economics, 54 Gagarin Ave., Kryvyi Rih, 50086, Ukraine
[2]Petro Mohyla Black Sea National University, Department of Sociology, 68 Marines Str., Mykolaiv, 54003, Ukraine

Abstract. The article is devoted to current problems of the strategy of inclusion in the education system. The authors attempt to analyze documents regulating legal relations in the field of inclusive education, the accessibility of both educational and medical institutions, the readiness of the teaching community to work with children with special educational needs. The empirical basis formed by an array of empirical data collected by conducting a traditional analysis of documents, interviewing parents and teachers. The analysis result of conducted research allows us to conclude that informants are fully aware difficulties of inclusive education today. They are associated with the lack of architectural accessibility of the majority of school buildings, proper professional training of teachers, experience in communicating with children with various types and severity of diseases, and lack of readiness of society adequately to perceive the idea of integration.

1 Introduction

The objective of the study is to summarize the experience of random empirical sociological surveys (analysis of regulatory support and the results of teachers' and parents' surveys conducted in certain regions of Ukraine with the typical structure of the educational system) regarding the implementation of inclusive education, and identifying promising areas of sociological consulting in the context of the problem under study.

According to WHO data, published in the World Report on Disability in 2017, the total number of people with disabilities in different countries of the world is more than 1 billion people, with almost 200 million among them having serious problems with normal functioning in society [1]. According to the study of the United Nations Children's Fund (UNICEF), the number of children with disabilities in the post-Soviet countries increased almost three times on average from 1990 to 2015 [2]. The tendencies of increasing primary disability with various nosologies [3], in particular in Ukraine, determined primarily by the demographic crisis, the presence of a wide range of congenital anomalies, the diversity and extent of which are directly dependent on the social, spiritual, physical and economic well-being of certain countries. Thus, in Ukraine in 2016, there were 153547 children with disabilities, which is 2.02% of the total population of the country under the age of 18 years, and 12 267 123 diseases were registered, an average of 100 diseases per 11.6 people with disabilities [4]. According to the statistics from the Ministry of Health of Ukraine, throughout 2005-2015 the rating of causes of primary childhood disability remained relatively unchanged: in the first place – congenital anomalies, deformities and chromosomal abnormalities – 31.3% in 2015 (31.0% in 2014, 30.4% in 2013, 29.13% in 2011, 22.9% in 2005); the second place includes diseases of the nervous system – 17.0% in 2015 (17.1% in 2014; 17.6% in 2013; 18.9% in 2011; 20.4% in 2005); on the third place – mental and behavioural disorders – 14.0% in 2015 (13.9% in 2014, 13.9% in 2013, 15.5% in 2011, 16.6% in 2005) [5].

According to the Ministry of Education and Science of Ukraine in the academic year of 2016-2017, 4180 pupils were trained in inclusive classes, which is 53.6% more than in the previous year (2720 pupils in the studying period of 2015-2016). To provide qualified assistance in mastering educational material for children with special educational needs, 1,825 teacher assistant positions were introduced in the staffing table of general educational institutions. The question of the organization of accessible educational space remains topical. So, in the 2016/2017 survey data, 12652 educational institutions are available only up to the first floor (77.2% of their total number), 69% (0.4%) up to the second, 22 (0.13%) up to the third, and up to the fourth and next floors – only 11 structures [6]. As we can see, positive developments can be traced, but still many more questions remain unresolved.

The priority of solving problems related to the spread of childhood disabilities is actualized by the need to ensure the social protection of disabled children during a long period of time, since almost a quarter of cases of

* Corresponding author: 15srps2016@kdpu.edu.ua

disability in people under 50 years old are caused by diseases suffered in childhood. In addition, we should not forget that the effectiveness of solving problems of child disability depends entirely on the life quality of the population, integrating all social institutions, including state authorities, public organizations whose activities are aimed at creating the right conditions for full life activity of people with special needs in all its aspects. In this sense, the strategy of inclusion in the education system, in particular, can be considered as the direction of social investment in human development.

Benefiting from the successful implementation of inclusion is impossible without dialogical planning for implementation, long-term funding to create the right infrastructure, and creating a regulatory framework for settling legal relations in this area. The basis of the adoption and implementation of management decisions should based on the priority interests of stakeholders of inclusive education.

2 Literature review

A number of scientific studies are devoted to studying the problems of implementing inclusive education. So, T. Booth, M. Ainscow, M. Vaughan presented a selection of practical recommendations in their work 'Index for Inclusion' [7] for creating an inclusive educational environment for all members of the school community. In particular, T. Booth, studying the social practices of inclusive education in 17 countries of the world, concluded that the list of countries supporting the inclusion policy mainly included those that are characterized by decentralized systems where budgets for special needs are delegated to municipalities, districts and / or individual schools. Exploring the features of the implementation of the process of teaching children with developmental problems in different countries of the world, L. Shipitsyna noted that the principle of financing inclusive education depends on understanding the essence of the processes of inclusion and integration. Thus, in her opinion, Ukraine and Russia do not distinguish between these processes and, as a result, in these countries the majority of children with disabilities are neither included nor integrated into general education schools and instead receive education in special institutions. At the same time, resource centres are being established in Armenia, Belarus and Uzbekistan to support inclusive education in secondary schools [8].

According to UNICEF terminology, the term 'invalid children' (or children with health disabilities) is used to define a group of children who are diagnosed with impaired bodily functions of moderate or high severity, whereas special educational needs are inherent to children with milder functional disorders, who often do not have the social status of a disabled person, and are regarded as impaired learning abilities. According to the WHO, indicators of child disability in developed countries do not exceed 2.5% of the total population of children (1% of them are children with serious pathologies), but about 8% of children with special educational needs [9]. The indicator of children's disability in this case is considered as a basic criterion for the public to recognize the need to develop effective mechanisms for the social protection of children with disabilities in the field of education. According to the results of the research by A. Kolupayeva, in most countries of the post-Soviet space there are problems in harmonizing the methodology for registering children with special educational needs, as organizations are subordinate to different social departments, in particular to the Ministry of Health, the Ministry of Social Policy, the Ministry of Education and Science [10]. Actually, the first of these social institutions records the number of children with various groups of diseases, the second - the number of children with disabilities claiming monetary social assistance from the state, the third - the actual number of children with special educational needs who applied to general education and / or specialized schools at the place of residence for the provision of social protection in the field of education. The real statistics regarding the number of children with special educational needs, who have various kinds of disorders in the functioning of the body, can be several times higher than the official data of children's disability. Thus, the lack of interdepartmental interaction due to the methodological inconsistency of registering children with special educational needs is one of the problems of implementing inclusive education.

Studying the effectiveness of the functioning of the state social security system for people with disabilities, P. Romanov and E. Yarskaya-Smirnova argued that the social policy towards people with disabilities, implemented when the Soviet Union was in, was of a compensatory nature [11]. A sufficiently long period in a society where the attitude towards a person was utilitarian and the economy was economical, the life of representatives of this social group took place on the periphery of social relations. The system of special education has become the legacy of the post-Soviet countries, increasing the segregation of children with disabilities, depriving them of the possibility of being surrounded by their peers who have no health problems.

A. Kolupayeva [12], referring to UNICEF experts, argues that the main reasons why in post-Soviet countries parents raising children with disabilities still support the functioning of specialized international schools are a low level of material well-being of the family, lack of alternative provision of special assistance and support from both the state and society as a whole. In addition, she notes that before the transition, in the CIS and Baltic countries, only the main signs of disability (hearing, vision, intellectual development, speech impairment) classified children who had the opportunity to receive education in specialized institutions under corrective programs. This raises a second problem - the possibility of providing inclusive education for children with primary disabilities with other nosologies, since the spectrum of recorded diseases is enormous (12,267,123 diseases were registered in Ukraine in 2016) [5].

Systematizing the results of empirical studies of the problems of social and educational inclusion, T. Yudin and S. Alekhina concluded that the full implementation of the principles of inclusive education is impossible without overcoming psychological hostility from healthy citizens towards people with disabilities, the desire of people themselves with a disability to socialize and actively participate in society, the awareness of the differences between the integration and segregation model of education [13]. Thus, a third problem emerges – the formation of the readiness of parents raising children with special educational needs to uphold the right of educational inclusion for their children and educators to fulfil this right.

That is why the state educational policy should at the same time satisfy the educational needs of children with disabilities, promote the development of educational systems that would guarantee equal access to preschool, school, vocational and higher education for all children and young people, and also become an integral part of economic development. Of course, achieving positive changes in the implementation of children's inclusion with disabilities is impossible without analyzing the legal acts that ensure the existence of an inclusive component of the education system, an empirical study of the problems of families raising children with special educational needs, expert evaluation of the mechanisms for ensuring social protection of children with disabilities and children with special educational needs in education.

Certain attempts to identify promising areas of sociological consulting in the field of inclusive politics were taken by O. Panchenko [14], who focused on the need to develop an empirical sociological model of a stakeholder activity. In her opinion, a comprehensive analysis of this model, including objective-subjective factors of inclusion, inclusive environment, inclusive behaviour and practices, together will not only talk about the effectiveness of the implementation of inclusive education policies, but also carry out its correction for further optimization.

3 Materials and methods

The simultaneous use of structural-functional and activity approaches to the consideration of inclusive education allowed us to characterize the stakeholder agents that form an idea of the effectiveness of this process. Among them are the following categories of people: representatives of the authorities, developing regulatory support for the implementation of inclusive education; parents raising children with disabilities and children with special educational needs, acting as consumers of inclusive educational services; teachers as persons providing services in the field of inclusive education.

The source materials for the analysis were obtained as a result of processing an array of empirical data collected by conducting traditional analysis of documents (decrees, regulations, laws regulating legal relations in the field of inclusive education), as well as interviewing parents and teachers.

The first stage of the study was initiating by representatives of the Strong Together, public association, by the Mykolaiv Centre for Sociological Research of the Black Sea National Petro Mohyla University, with financial support from the Department of Labour and Social Protection of the Population of the Mykolaiv City Council in January-May 2018. In particular, 385 parents took part in the survey who raise children with special educational needs living in the city of Mykolaiv and the Mykolaiv region. In the course of forming the sampling, the available sampling method was used (the survey was conducted at the location of the rehabilitation centres), the snowball method (the participants of the social group on Facebook 'Special children of the Nikolaev region' were interviewed), the quoting method was used to select respondents from the territorial departments of social protection population in the city of Mykolaiv and Mykolaiv region [15, 16]. Considering the fact that there is no statistical record of families raising children with special educational needs, in order to verify the representativeness of the research results, an attempt was made to compare the data obtained in terms of nosology groups with the statistics of the Ministry of Health of Ukraine regarding the causes of primary childhood disability.

The second phase of the study, implemented by the staff of the Department of Sociology and Economics of Kryvyy Rih State Pedagogical University, suggested systematization and analysis of normative acts regulating the inclusion policy in Ukraine, as well as conducting an expert survey among teachers. The survey was conducted in October-November 2018. The study involved 30 teachers of secondary educational institutions and 10 employees of specialized boarding schools for children with disabilities in the city of Kryvyy Rih and Dnipropetrovsk region. The selection of the participants of the expert survey was conducted by the method of self-assessment of their competence in this problem.

In the course of interviewing parents and an expert survey of teachers, they were asked to answer a number of questions regarding the most appropriate types and forms of education for special children in the context of various nosologies, as well as the range of problems associated with the organization of inclusion, and the need to involve a number of related specialists in this process, including psychologists, social teachers, speech therapists, rehabilitation therapists, teaching assistants and others.

4 Research results

The effectiveness of inclusive education depends primarily on the willingness of the state to adopt appropriate legislation, clearly define policy directions, develop an action plan, build infrastructure as well as implementation capacity and benefit from long-term funding. According to A. Kolupayeva, I. Lutsenko [10], who analyzed the legislative base of Ukraine on issues of inclusive education, the main problem is the lack of the mechanisms description for its implementation at all levels of the education system, as well as the regulation of activities related to them in this area. After ratification of the UN Convention on the Rights of Persons with Disabilities by Ukraine, the first document in this context is the decree of the Cabinet of Ministers of Ukraine 'On approval of action plans for the implementation of inclusive and integrated education in

 https://doi.org/10.1051/shsconf/20207503011

general educational institutions until 2012' (No. 1482 of 12/03/2009). The first attempts to legitimize inclusive education were made in the decree of the Ministry of Education and Science No. 855 dated 11.09.2009 'On approval of the plan of measures for the introduction of inclusive education in general educational institutions for 2009-2012', which for the first time deals with the need for scientific substantiation to create programs, methodological support, training and retraining of teaching staff to work with special children and their parents in an inclusive education.

O. Krasyukova-Enns, S. Crocker, T. Lutsenko, Y. Naida [17], implementing the work on the Canadian-Ukrainian project 'Inclusive Education for Children with Special Needs in Ukraine', carried out an analysis of the compliance of Ukrainian legislation with the UN Convention on the Rights of Persons with Disabilities with the purpose of its subsequent harmonization. Among other remarks, voiced by the experts, our attention was drawn to the points dealing directly with issues of inclusive education. In particular, they noted the fact that in the Ukrainian legislation regulating the information and educational policy regarding people with disabilities, there are no norms on the obligatory conduct of an appropriate educational and educational function in society realised by the state. In addition, there are no laws regulating the creation of a unified system or program for collecting and summarizing statistical data on the number of people with disabilities and their needs, which are essential for the development and implementation of effective social programs, etc.

It took the Cabinet of Ministers 8 years to adopt the Order On Approving the Organization of Inclusive Education in General Education Schools (No. 588 dated 09.08.2017) [18] after ratification of the UN Convention on the Rights of Persons with Disabilities, but today there are still unresolved problems of accessibility of both educational and medical institutions, the readiness of the teaching community to work with children with special educational needs, etc. And how long will it take to develop specific mechanisms and their practical implementation, since the question remains open about the possibilities of financing inclusive education? In addition, the question about cultivating a tolerant attitude towards people with disabilities in society arises, as well as approving the reform of the education system in order to promote inclusion. The survey of both parents raising children with special educational needs and teachers allowed researchers to specify the range of problems that are present today in the field of inclusive education.

Moreover, parents,, answering the question about the acceptable form of education for their children, preferred full-time education at an educational institution (58.1% of the total number of respondents). Almost every fifth respondent (18.9%) approves of individual form, which provides for a combination of homeschooling and attending individual classes and events in an educational institution; 7.6% of parents consider it appropriate to use part-time education, which implies individual distance learning training without visiting an educational institution.

In the context of children's diseases, the data obtained are ambiguous (see Tables 1, 2), because the parents' opinion about acceptable types and forms of education is determined, above all, by the actual state of health of the child and the individual dynamics of the disease. On the other hand, from a purely mathematical point of view, the law of small and large numbers works – trends can be traced only for those cases that repeat more often. And since individual diseases in children are quite rare, the opinion of their parents in this matter is not indicative.

Table 1. Distribution of respondents' answers to the question 'What kind of training, in your opinion, is the most acceptable for your child?' according to the type of child's illness, *% of the total number of respondents*

Childhood Disease / Acceptable Type of Education	Education in an ordinary school class or a group of pre-school educational institution	Inclusive form of education in the ordinary school class (kindergarten group)	Training in a separate / special class (group) for special children in the framework of a general education institution according to an adapted program	Special education for special children	Individual tuition at home or at school	Difficult to answer
ICP (infantile cerebral paralysis)	16.1	30.4	21.4	16.1	12.5	3.6
Autism Spectrum Disorders	18.8	21.9	35.9	9.4	12.5	1.6
Down syndrome	12.5	31.2	6.2	37.5	6.2	6.2
Cardiovascular diseases	40.6	3.1	18.8	21.9	9.4	6.2
Retardation	28.6	25.0	17.9	25.0	0	3.6
Kidney disease	80.0	20.0	0	0	0	0
Eye diseases	41.7	16.7	16.7	16.7	8.3	0
Epilepsy	23.1	15.4	15.4	30.8	7.7	7.7
Organic brain damage	21.7	13.0	17.4	26.1	17.4	4.3
Diseases of the musculoskeletal system	56.8	18.9	0	5.4	16.2	2.7
Hearing disorders	40.0	6.7	6.7	33.3	6.7	6.7
Other congenital malformations	35.9	28.2	7.7	17.9	5.1	5.1
Immunodeficiency	33.3	33.3	0	33.3	0	0
Diabetes	40.0	10.0	30.0	10.0	10.0	0
Burns	0	0	0	0	100	0
Oncological diseases	100.0	0	0	0	0	0

According to parents' estimates, the number of specialists who are engaged to work with their children in school or kindergarten include: psychologists (35.0%), social teachers (18.9%), doctors, physical therapy instructors (13.2%), correctional teachers (12.0%), speech therapists (11.8%), rehabilitation therapists (4.8%), psychotherapists (2.8%), and pedagogues for deaf-and-dumb (1.2%). In the case of non-attendance of general education institutions or the lack of the required specialist, parents turn to other institutions where they receive almost equal services from psychologists (19.3%), social pedagogues (12.8%), doctors, physical therapy instructors (16, 0%), correctional teachers (12.1%), speech therapists (19.0%), rehabilitologists (13.7%). The difference in percentages regarding the availability of rehabilitation therapists requires special attention, which is explained by the fact that only 58.8% of parents surveyed attend rehabilitation centres with their children, while others, citing the lack of rehabilitation services due to the type of illness of the child, are forced to seek qualified medical assistance 'beyond the framework' Which, in its turn, is again the evidence of the lack of highly qualified professionals, and government support does not cover the cost of necessary treatment.

Table 2. Distribution of respondents' answers to the question 'In what form, in your opinion, is it best to train your child?' according to the type of child's illness, *% of the total number of respondents*

Childhood Disease / Acceptable Type of Education	Collective or group (all services are provided by teachers and psychologists of an educational institution directly in an educational institution) including an inclusive form	Individual (training at home with teachers of an educational institution in combination with attendance of classes, events in an educational institution)	Individual (on the recommendation of a doctor at home with teachers of an educational institution without attending classes and events in an educational institution)	Difficult to answer
ICP (infantile cerebral paralysis)	50.0	20.7	13.8	15.5
Autism Spectrum Disorders	58.5	23.1	7.7	10.8
Down syndrome	43.8	6.2	18.8	31.2
Cardiovascular diseases	62.5	15.6	3.1	18.8
Retardation	63.3	10.0	6.7	20.0
Kidney disease	80.0	20.0	0	0
Eye diseases	66.7	25.0	0	8.3
Epilepsy	53.8	30.8	7.7	7.7
Organic brain damage	47.8	26.1	4.3	21.7
Diseases of the musculoskeletal system	62.2	18.9	10.8	8.1
Hearing disorders	86.7	13.3	0	0
Other congenital malformations	56.4	12.8	2.6	28.2
Immunodeficiency	100.0	0	0	2
Diabetes	60.0	30.0	0	10.0
Burns	0	0	100.0	0
Oncological diseases	50.0	50.0	0	0

The analysis of the research data in the context of the types of children's diseases, has allowed us to make assumptions about the fact that in most cases the work on inclusion is more successfully organized mainly for children with diseases of the supporting-motor apparatus, musculoskeletal system. To solve the issue of assisting children with psychosomatic diseases, with developmental delay, who sometimes do not even have disability status, is more difficult.

The main argument of parents who oppose the education of their children in ordinary classes, groups of educational institutions, is that these institutions are not adapted to the needs of their children. According to parents, in general education institutions there are no opportunities to receive assistance from a teacher's assistant, a teacher (22.2%), support from peers and their parents (32.6%), to move freely within the institution (38.6%), to attend groups for free, sections/clubs (42.9%), attend the room of psychological relief (49.1%).

In addition, the interviewed parents believe that the most unacceptable educational programs (13.6%), the lack of appropriate training or experience in working with special children among teachers and educators (12.8%) are not introduced (or created only partially) nor the relevant conditions of accessibility (10.0%), lack of adequate funding for the school (7.9%), a large number of children in school classes, kindergarten groups (7.4%).

The majority of parents, having to struggle difficult life circumstances, are shortsighted in their plans regarding the possibility of continuing their education for their children. Only a third of the respondents (31.8%) answered the question about the prospects for their children to receive vocational education in the future. Among them, 4.0% said that their children have the ability and opportunity to continue their education in higher education, 12.0% - in vocational schools, college, and 4.0% - in labour workshops. The remaining 80% only suggested that such a possibility exists, but they have no specific plans, therefore, they are not definitively determined in their answers.

An important type of assistance to families with children with special educational needs is socialization, aimed at expanding communication and organizing leisure activities. Only 54.3% of respondents answered the question about the possibility of their children visiting out-of-school sports sections and clubs, almost a quarter of them do not know about the possibility of visiting such institutions, the other half know, but do not do that. In general, only every tenth adult tries to organize leisure of his child, however, the range of reasons for such passivity is quite diverse - from financial insolvency, physical condition of a child to psychological unwillingness of parents to adapt a child in society.

Only 70.6% of respondents are concerned about a similar picture of the informants' answers about attending cultural and festive events, theatres, cinemas, concerts, master classes, excursions, etc. However, we note that the majority of parents were interviewed precisely in rehabilitation centres, where they are involved, in particular, in the sphere of leisure activities. Those who attend leisure facilities on their own turned out to be only 1.6% of the respondents. Most parents

https://doi.org/10.1051/shsconf/20207503011

raising children with disabilities organize leisure activities with (or at) rehabilitation centre(s) or by an invitation of public organizations, charitable foundations, etc., however, this is done mainly at their own expense.

As we can see, the actual implementation of certain guarantees by law is associated with a number of difficulties, one of the most significant of which is the public passivity of the parents themselves raising special children. And this is not their fault directly, the problem is the existence of social distance in any post-Soviet society between healthy people and people with disabilities as a result of the implementation of compensatory state social policy until 1990 in order to develop inclusion and the actual encouragement of parents 'to fight' for the rights of their special children, are directly related to the possibility of bilateral interagency-governmental communication and media activities.

For inclusion to be successful, parents raising children with disabilities and children with disorders should provide complete information about the condition of their children, their basic needs, and public services, in turn, should not ignore their obligations to inform about their rights, opportunities, and meet, if not all, most of the needs of special children.

It is alarming that, according to the estimates of informants whose children mostly visit territorial rehabilitation centres, the main source of information about the types of state social assistance are social workers, doctors (23.7%), friends, acquaintances (23.6%), social networks, forums, chat rooms (17.3%), articles, messages on the Internet (16.3%). What can be said about the parents whose children are not able to attend rehabilitation centres for various reasons? In fact, they remain cut off from the world, remaining alone with their problems. Representatives of public organizations are ready to take on some of the concerns related to informing, and this could be a way out, since 33.0% of the parents surveyed are eager to learn more about their activities, 10.9% would like to take part in their events; 5.7% even expressed a desire to become their members, thereby expressing their active social position. The mass media, in turn, must step up work on cultivating a tolerant attitude towards people with disabilities, approving the reform of the education system with a view to developing inclusion, through the implementation of explanatory work on the need to improve an accessible and comfortable life environment in all respects for all citizens without any exception.

These and other questions remain open and require further deeper study, conducting a full-scale representative study that would justify recommendations regarding the potentially possible ways to improve social policy in relation to both children with disabilities and their parents.

The representatives of the expert group, answering the question about the acceptable types and forms of education for children with disabilities, preferred to study in specialized institutions for special children (60% of respondents), only 10% consider inclusion the most acceptable form of education in this case, another 20% agree that children with disabilities should be trained in specialized classes (groups) established in general education schools. At the same time, the majority of experts (60%) approve of an individual form of education, which provides for a combination of home schooling and attending individual classes and events in an educational institution. 30% of respondents consider it appropriate to use part-time education, which implies individual distance learning without visiting an educational institution. Only 10% of respondents advocated collective education for children with disabilities, which implies that all services are provided by teachers and psychologists in an educational institution (including inclusive education).

The comparison of answers about the most appropriate form of training and the sphere of employment of experts represents the greatest interest. Thus, 75% of those who support the education of children with disabilities or special needs in educational institutions (inclusive education or training in a specialized group in a general education institution) are employees of specialized boarding schools. Whereas the majority of supporters of isolated education for such children are teachers of secondary schools. This state of affairs is due, in our opinion, to the fact that for a long time there was a social distance between healthy people and people with disabilities, created by the policy of isolating children with disabilities in specialized educational, medical, sanatorium and preventive institutions. Teachers of secondary schools simply fear that they will be incompetent in solving certain tasks and difficulties they may face.

The main argument of teachers, who oppose the education of children with disabilities in ordinary classes, groups of educational institutions, as well as the parents of such children, is that these institutions are not adapted to the needs of children with health problems. According to the teachers, in general education institutions there is no opportunity to receive assistance from a teacher's assistant, a tutor (90%), support from peers and their parents (80%), to take part in holidays, competitions, contests (70%), to move freely within the institution (70%), schools lack the necessary equipment for training (50%).

In addition, the interviewed experts believe that the most difficult obstacles to obtain high quality education for children with disabilities are the missing (or only partially created) relevant conditions of access to education (70%), lack of adequate funding for an educational institution (50%), a large number of children in school classes (50 %).

According to teachers of general educational institutions, the specialists staff engaged in working with children with disabilities in their school include social teachers (90%), psychologists (80%), and speech therapists (50%). Only some of the experts noted (2.5%) that their school has the opportunity to invite a rehabilitation specialist or a pathologist from other institutions. The employees of specialized educational institutions responded that children with disabilities in their institutions have the opportunity to receive qualified assistance from all the necessary specialists,

https://doi.org/10.1051/shsconf/20207503011

regardless of whether they are in this institution or they need to be invited.

The carried out analysis of the possibilities for positive socialization of children with disabilities and their adaptation to communication with ordinary children showed that, in general education institutions, apart from occasional participating in sports competitions (from time to time – 30% of the surveyed experts) and visiting a theatre or class excursions on their own expense (50%), such opportunities are absent. The opportunity for children with disabilities to attend out-of-school sports clubs and circles is entirely the responsibility of their parents; the school, as a rule, does not deal with such issues (one might say, is not interested). Considering the fact that parents of children with disabilities, as can be seen from the survey data presented above, are not particularly engaged in organizing socially oriented leisure of their children for various reasons, such indifference on the part of general education employees becomes understandable. First of all, parents should insist on the usefulness of their special children and the protection of their rights.

Among the primary problems associated with children with disabilities, experts named the lack of teacher assistants (40%) and the need to create appropriate conditions in schools (ramps, toilet rooms, the ability to move around the floors and provide the necessary educational materials) – 30%.

5 Discussion and conclusions

An analysis of the results of the conducted research suggests that the opinions of teachers and parents regarding the possibility of integrating children with special educational needs in secondary schools are somewhat different. The fact of the architectural inaccessibility of most school buildings is important. Furthermore, almost a third of the interviewed parents, who repeatedly encounter the bureaucratic system of social relations, acutely experience the cases of neglecting their children and therefore more negatively perceive the idea of integration.

Drawing analogies with the research data conducted by the Central Institute for Disability Expertise and Labour Organization of Persons with Disabilities in 1990-1991 [19, 20], members of the Department of Social and Economic Systems and Social Policy of the Higher School of Economics Research Institute in 2004 and 2008 [13], we can conclude that, of course, attitudes towards people with disabilities in the post-Soviet countries is changing for the better. However, the social distance between healthy children and children with health problems is still quite noticeable today.

The first steps that need to be taken to solve the existing problems should be aimed at: regulating the interaction of state institutions regarding the collection and synthesis of data on the number of children with disabilities and their needs; the creation of an effective system of early intervention for the timely detection of deviations in the development of the child, coordinating work with their parents; development of effective control mechanisms to reduce doctoral (medical) errors; optimization of the work of institutions providing medical and social rehabilitation for special children, their parents and other family members; training teachers, their assistants, parents in order to teach them to work with special children in an inclusive education. The most important thing in this regard is the organization of information and educational work of the media, aimed at forming public opinion with targeted at destroying the usual stereotypes about people with disabilities, in particular, children with disabilities. In the context of interagency cooperation, increased transparency of control by public organizations, by parents raising special children, and by society as a whole, inclusive education will definitely develop, following international experience in line with the struggle for the civil rights of people with disabilities. The success of the further implementation of the social integration strategy fully and totally depends on the development of appropriate mechanisms for the implementation of legislative initiatives.

References

1. World health organization, The World Bank, World report on disability (2011), http://www.who.int/disabilities/world_report/2011/summary.pdf. Accessed 12 Oct 2018
2. INICEF, Progress for children beyond averages: learning from the MDGs (2015), https://www.unicef.org/publications/files/Progress_for_Children_No._11_22June15.pdf. Accessed 12 Oct 2018
3. The Commonwealth of Independent States (CIS interstate statistical Committee), Population, employment and living conditions in the countries of the Commonwealth of Independent States in 2017 (2018), http://www.cisstat.com. Accessed 15 Oct 2018
4. G.O.Slabkiy, V.V.Shafranskiy, O. O.Dudina, Disability of children as a problem of public health: prevention and effective rehabilitation. Bulletin of social hygiene and organization of health care of Ukraine 3(69), 4–9 (2016)
5. Ukrainska Baza Medyko-Statystychnoi Informatsii (Ukrainian base of medical statistical information) (2018), http://medstat.gov.ua/ukr/news.html?id=203. Accessed 17 Oct 2018
6. Statystychni dani (Statistics) (2018), https://mon.gov.ua/storage/app/media/inkluzyvne-navchannya/statistika-inklyuziya.pdf. Accessed 17 Oct 2018
7. T. Booth, M. Ainscow, *Indikatory inkliuzii: Prakticheskoe posobie* (Indicators of inclusion: Practical Guide), ed. by M. Wogan (Perspektiva, Moscow, 2007)
8. L. Shipicyna (ed.), *Obuchenie detei s problemami v razvitii v raznyh stranah mira* (Training of children with development problems in different countries of the world) (Rech, Saint Petersburg, 1997)
9. European Health for All Database (HFA-DB) (WHO/Europe, 2018), http://www.euro.who.int/en/data-and-evidence/databases/european-health-for-all-family-of-databases-hfa-db/national-health-indicator-

databases. Accessed 17 Oct 2018

10. A.A. Kolupaeva, I.V. Lutsenko, Normatyvnyi biuleten (Regulatory Bulletin). Osvita osib z osoblyvymy potrebamy: shliakhy rozbudovy 11, 242–260 (2016)
11. P.V. Romanov, E.R. Jarskaja-Smirnova, The disabled and society: twenty years later. Sociological research 9, 50–58 (2010)
12. A.A. Kolupaeva, *Inkluzivna osvita: realii ta perspektyvy* (Inclusive education: realities and prospects) (Sammit-Kniga, Kiev, 2009)
13. *A sociological study of disability and rehabilitation in the Russian Federation. Analysis of the main results of the study* (Moscow, 2009)
14. O.L. Panchenko, Sociological model of the implementation process of inclusive education. Vestnik of Tatar Institute of assistance to business 1, 96–102 (2015)
15. U merii obhovoryly aktualni problemy liudei z invalidnistiu ta shliakhy yikh vyrishennia (The city hall discussed the current problems of people with disabilities and ways to solve them) (2018), https://mkrada.gov.ua/news/8823.html. Accessed 17 Oct 2018
16. Shanovni batky (opikuny), yaki vykhovuiut ditei z invalidnistiu! (Dear Parents (Guardians) of Parenting Children with Disabilities!) (2018), https://mkrada.gov.ua/news/6876.html. Accessed 12 Oct 2018
17. Inclusive education for children with special needs in Ukraine: training modules, Module. Policy Advisory. (Canadian-Ukrainian project, Kiev, 2011)
18. On approval of procedure for organization of inclusive education in secondary schools (2017), http://zakon.rada.gov.ua/laws/show/872-2011-p. Accessed 19 Oct 2018
19. T.A. Dobrovolskaya, N.B. Shabalina, A persons with Disabilities: discriminated minority? Sociological researches 5, 3–8 (1991)
20. T.A. Dobrovolskaya, N.B. Shabalina, Socio-psychological features of relationships between disabled and healthy. Sociological researches 1, 62–67 (1993)

Supervision as a model of inclusive education retraining and professional advancement of the school community

Kateryna Bondar*, and *Olena* Shestopalova

Kryvyi Rih State Pedagogical University, Department of Practical Psychology, 79a Heroiv ATO Str., Kryvyi Rih, 50086, Ukraine

Abstract. In this article, we describe the model of supervision in support of the Education Reform in Ukraine and the retraining of the school's key stakeholders in an inclusion: the case of Kryvyi Rih. Empirical study has introduced a model of retraining through a socio-political and educational description of the current situation in Kryvyi Rih. That is why 124 key stakeholders from 20 schools with inclusive education were tested to plan the implementation model of supervision (teachers, tutors, staff, psychologists and parents from 7 districts of Kryvyi Rih were invited to participate). In this article, we present a different approach to supervisory interagency groups and understanding, with the focus on the model supervision retraining of inclusive education professionals. The supervision consists of four stages: case report, exchange, conceptualization, summary. For successful work, we propose in the article an adaptation of an unstructured interview by a specialist who describes the situation of training supervision. The analysis of the organizational, methodological and information retraining system, the implementation of the pilot project of supervisory groups for the psycho-educational support team in inclusive education allowed to offer the supervision program: I. Training of the team members to work with the goals of the individual child development plan: working with tools for preparation of the case; II. The main features of observing the lesson; III. Analysis of environmental factors; IV. Oversee team support.

1 Introduction

The concept of supervision in inclusive education in Ukraine based on national education reform and the new law. The Education Reform (Education act of Ukraine – Document 2145-VIII, effective, current version – adopted on 05.09.2017) provides for a systematic transformation of the sector, the primary objective of which is the new high quality of education on all levels: ranging from the elementary school up to the higher education establishments. In particular, Article 6. Basic principles of state policy in the field of education and principles of educational activities and Article 19. The education of individuals with special needs are relevant to this research [11]. For this reason, a shortage of professionals, ensuring stable (sustainable) transition to the new system of teaching and providing educational services is a priority for securing standards for the implementation of education reform. That is why the main weaknesses of supporting education reform is a general scarcity of algorithms of retraining for key stakeholders in inclusive education.

This study refers to scientific research, which describes models supervision retraining of inclusive education professionals (C. Acedo, O. Akimova, S. Alila, I. Goodman, L. Ferguson, Z. Leniv, G. Lindsay, I. Hardy, T. Ryan, S. Symeonidou and others) considered consultative and collaborative models of team's support [1-10]. The meaningful idea of inclusion in New Ukrainian School, that teachers should be able to individualize and differentiate the educational process in terms of learning styles, abilities and additional needs of all children in education [12]. In addition to allowing more flexibility in school-wide curriculum development, inclusive education practices were made mandatory for children with special educational needs [2, 6, 5]. Therefore, the teacher needs the support of key stakeholders or collaborative team (administration, families, service staff, assistants and other specialists from the team) to guarantee this individualization of the educational process [7-8].

According to other researchers, a fundamental principle a successful program of retraining inclusive education professionals, one should take into account the world experience, and rely on the new paradigm of the European Education Area [5, 7, 8]. However, traditional forms leave the team approach during retraining and active training out of the account. That is to say, that outdated forms of retraining do not stimulate the school communities in Ukraine to fundamentally change their attitude towards their participation in the field of inclusive education according to European Standards. Consequently, the educational supervision for collaborative team is a dialogical process based on human values and understanding, and equal constructive relations between a supervisor and a recipient of supervision, which is targeted at accumulating the unique personal knowledge in a particular educational

* Corresponding author: katerynabondarr@gmail.com

situation and personal growth of both parties [1, 3, 4]. Thus, educational supervision creates a model retraining of the school's key stakeholders in inclusion.

The purpose of supervision is to increase the single-mindness of professional actions, upgrade qualification and improve the degree of satisfaction with the own work [3]. It means the subject of educational supervision is the analysis of the conscious and unconscious potential abilities of the key stakeholders. Educational supervision is possible in various forms [1]. The co-teaching group is a group form of the training for specialists-beginners (provides for the thematic focus) and intended for in-depth training and solving professional problems of key stakeholders [9]. Additionally, the co-teaching group carried out in a scheduled manner under the management of a supervisor. A thematic seminar is a group form of the work, in which the specialists of one profile of activities participate, for example, psychologists, and it is designed for discussing various professional problems, as well as carried out in a scheduled manner under the management of a supervisor [1, 9]. Interagency group of the supervisory work, the purpose of which is a group discussion of professional problems of the different team specialists, regardless of their experience, is conducted in a scheduled manner and under the management of a supervisor [1, 3, 9]. Additionally, educational supervision includes modeling and co-teaching the educational activities for children with special educational needs (SEN) for all participants of the team [4, 1]. Consequently, successful supervision should help the team members to understanding better the philosophy, main purpose, and practice of inclusive education, to find out about external and internal resources of activities of the collaborative team's participants in inclusive education.

Fundamentally, supervision, as an effective form of the collaborative and co-teaching model, is educating key stakeholders to solve difficulties and search for fresh ideas as well as effective educational solutions. According to the Standard Regulations on Supervision over Implementation of the Concept "New Ukrainian School", the educational supervision is a relatively new line of work with human resources for the Ukrainian educational system [11-12]. That is why there is a general scarcity of studies analyzing models of supervision in support of the Education Reform in Ukraine and the retraining of the school's key stakeholders in an inclusion.

The purpose of this study was an adaptation of the model of educational supervision to the Ukrainian educational system.

2 Method

The algorithm for conducting supervisory interagency groups includes four steps (Fig. 1) [3, 9, 10]. The first step is the report of the case from someone of key stakeholders and his/her formulation of a request. At this step, questions are asked by the group members to the presenter and aimed at clarifying the information about the child and educational situation are desirable. Then, the group discussion takes place, when the presenter is silent: the group members discuss professional actions without talking about the presenter's personal qualities.

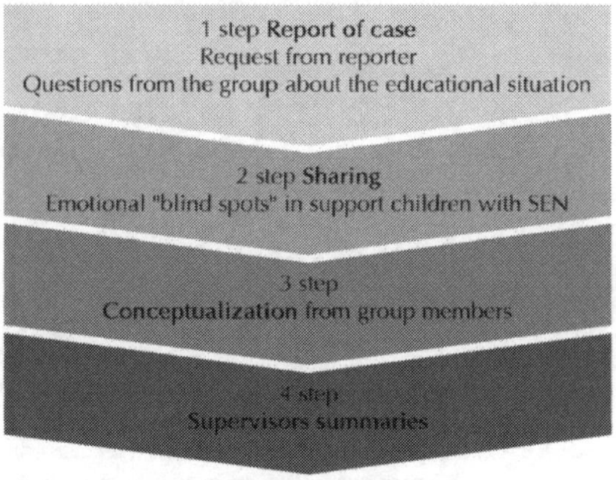

Fig. 1. Model of educational supervision.

In the second step, the group members share their feelings about the educational situation spontaneously. This step is important for establishing a safe atmosphere in the group and contributes to the awareness by the supervised "blind spots" in the work.

In the conceptualization step, the supervisor questions each group member about his/her concept of the origin of a problem, which the child with special educational needs has, and its remedy. No position is disputed or condemned by the supervisor and other group members.

In the fourth step, the supervisor summarizes the method of the presenter's work and proposes his/her concept of clarifying problems in the work with the child as well as optimal tactics of the work of options as effective educational solutions.

To prevent resistance of the group members and to distinguish themselves from the traditional methodical work, the following rules of work are introduced: 1) bringing his/her case for the consideration, the teacher has the right to determine what he/she wants from supervision; 2) the teacher forms himself/herself his/her request and expectation from the group and experts; 3) the "Stop" rule, which allows to interrupt the discussion at any time, or not to answer questions that cause your rejection for one or another reason, and protects against the disclosure of an unwanted information.

3 Implementation of the model of educational supervision

3.1 Background

The implementation of the model began with working meetings of the planning team (in January-April 2019). The work of planning team aimed at developing a concept for supporting inclusive education in Kryvyi Rih in cooperation between Deutsche Gesellschaft für Internationale Zusammenarbeit (GIZ) GmbH, Department of Education and Science of the Executive Committee of the Kryvyi Rih City Council and the

Kriviy Rih State Pedagogical University (KSPU), with the participation of the consultant Karsten Zehner. Structure of advanced training for educators in the premises of the KSPU was discussed: procedure for interaction with the education department, regulatory framework and stages of preparation of supervision.

Implementation of the model of educational supervision for developing inclusive education in the schools of Kryvyi Rih. Methods of the supervision groups are by the Standard Regulations on Supervision for Implementation of the Concept "New Ukrainian School" (developed by the Ministry of Education of Ukraine and the SSI "Institute of Education Content Modernization"). The basic idea of supervision – capacity development and practicing teaching skills for key stakeholders, who work with different categories of children with special educational needs in the schools of the city of Kryvyi Rih – "Introduction of inclusive education at the local level".

3.2 Summary

Empirical study has introduced a model of retraining through a socio-political and educational description of the current situation in Kryvyi Rih. That is why 124 key stakeholders from 20 schools with inclusive education were tested to plan the implementation model of supervision (teachers, tutors, staff, psychologists and parents from 7 districts of Kryvyi Rih were invited to participate). We consider the following to be the purpose of conducting supervision groups:

1) improve the quality of training for providing high-quality educational services in the context of inclusion for the teaching staff of general education institutions in the city of Kryvyi Rih according to the ideas of the New Ukrainian School and the new Law on Education;

2) using the team approach and principles of work of the supervision group, establish creative atmosphere for discussing the following problematic topics: specifics of identifying objectives and planning how to fulfill these in an individual development program; establishing learning behavior in children with SEN; autism spectrum disorders, etc.; difficulties in relationships with parents of children with SEN and involving them into the support team; organizational and managerial difficulties (structuring forms of education or additional educational services for children with SEN); personal problems of a specialist in connection with his/her professional situation;

3) present during the round-table discussion results of the work of supervision groups on inclusive education issues, held with the support of GIZ in Kryvyi Rih, get feedback and discuss issues and recommendations.

4) ensure during the regional round-table discussion gathering of ideas and discussion of opportunities for further extension, expansion, and institutionalization of the capacity development program in the field of inclusive education in Kryvyi Rih (further requirements, roles, and responsibilities, partner cooperation, ways of implementation, necessary resources).

3.3 Outcome of the planning team

The study was originally conducted on key stakeholders' experience. The first step of planning the implementation model of supervision was testing responses (key stakeholders from schools with inclusive education in Kriviy Rih). In the study, we relied on the questionnaire of responses' experience in inclusive education (124 persons, 120 – women, 4 – men) (Fig. 2).

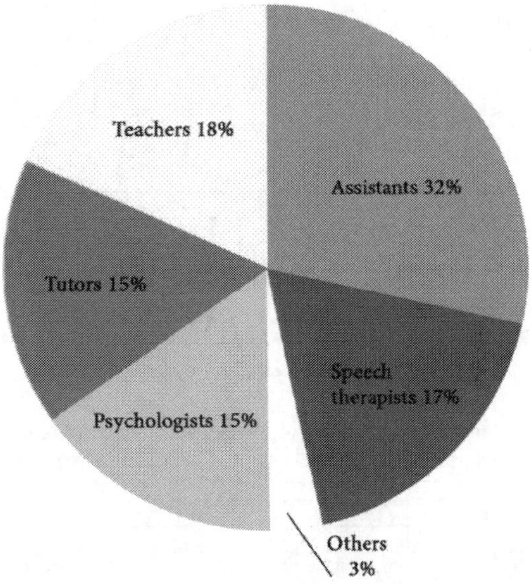

Fig.2. Distribution of professional groups among supervision participants.

The second step was the analysis of the answers to responses. For this evaluation, we focused only on the responses: an educational experience of working with children with SEN, background knowledge, emotional barriers in communicating, expectations from the retraining. According to the distribution of an educational experience of working with children with SEN, the selection is as follows 73% have experience of support children with SEN, 27% – specialists-beginners.

However, 80% of respondents answered that their knowledge of technologies, forms, and methods of education of children with special needs is not enough. Concerning the system retraining in special education, 64% have not improved their qualifications. It must be noted that only 14% of school team members have special education in the field of occupational therapy and special education (Table 1).

Table 1. Do you have special training in the field of inclusive education? (n=124).

Percentage %	Answer
14%	I have a special education
44%	I do not have a special education
5%	I received vocational retraining in the area
26%	I took further training courses in this area
12%	I did not take further training courses in this area

Most stakeholders consider their background knowledge insufficient and want to continue the process of further training (Tables 2-3).

Table 2. To what extent are you familiar with the developmental characteristics of the children with the different types of disabilities (Please, put "+" if your answer is "Yes" (n=124).

	I need additional education	I completed further training courses	I participated in the seminars	I know within a higher education program
Behavioral disorders	11%	23%	19%	17%
Speech impediments	15%	16%	15%	12%
Intellectual disabilities	14%	16%	20%	17%
Vision impairments	13%	6%	12%	17%
Hearing impairments	14%	10%	12%	17%
Musculoskeletal disorders	15%	16%	8%	15%
Autism spectrum disorder	20%	13%	15%	5%

Table 3. For which group of children with special needs do you want to learn a teaching practice in the first place? (n=124).

Percentage %	Answer
10%	For children with musculoskeletal disorders
13%	For children with sensory disabilities (hearing, vision)
9%	For children with intellectual disabilities
16%	For children with speech impediments
25%	For children with behavioral disorders
26%	For children with intellectual disabilities

Among the typical difficulties in interacting with professionals is the following – about 18% of respondents have emotional barriers in communicating with children with SEN, and for 16% of participants of the groups it is difficult to communicate with parents of these children, sometimes difficult for 4% of the respondents. School team members are generally willing to share their experiences, but some individuals do not care.

Consequently, a holding of workshops, the exchange of experience, participation in supervision groups prevails among the typical expectations for retraining of inclusive education professionals. The second place is taken by the improvement of methodological and procedural training in teaching children with SEN (Table 4).

It resulted in elaborating an understanding of supervision procedures, methods of material adaptation, and strategies for working with groups.

3.4 Adaptation of the model of education supervision

Adaptation of the model of educational supervision was held in Kryvyi Rih in April-November 2019. For this part of the research, we focused on the selection of educators from 20 schools (124 persons) on 15 groups. The following has been developed and given to team members to compile a case description:

- adapted version of the unstructured specialist's interview, in which she/he describes her/his case for supervision in a Balint group;
- protocol of types of interaction of the support team during the week;
- protocol of registering learning behavior;
- video presentation of components of learning behavior, working with social stories and visual simulation of learning rules of conduct in an educational institution;
- protocol of the weekly observation of the effectiveness of the material modification;
- standardized overseeing table for behavioral difficulties of children with SEN.

Table 4. What kind of help do you expect from the retraining of inclusive education professionals? (n=124).

Typical Answers	Percentage %	Rating
Methodological and procedural help in teaching children with SEN	36,45	2
Maintaining business documentation	3,85	5
Case studies, any practice-oriented training, experience exchange, supervisions	40,4	1
Detailed characteristics of specific categories of children with SEN, work of the support team, forms of cooperation with the teacher's assistant, diagnostics and correction technologies	13,5	3
Close collaboration with the inclusive and resource center or the employee of this center	5,8	4

It shall be noted that this is one of the first projects of this kind in Ukraine.

3.4.1 Interview

For the successful work on 1-2 steps of supervision, we offer the adapted version of the interview of the specialist, who describes the situation for the educational supervision (developed by authors). The interview contains 12 questions that evaluate the respondents' professional activity (professional and personal expectations). The study utilized the Interview to assess respondents' positions on professional expectations in inclusive supporting the child.

A pilot study with a small sample of teachers was conducted to establish whether questions were appropriate and understandable for the intended target group. The Interview provided indicators of the position of participants on two forms and six subdimensions of professional expectations: 1. professional expectations (6 items, Cronbach's α = 0.73); 2. personal expectations (6 items, 11, Cronbach's α = 0.70). For educators, psychologists, social employees (educators), teaching assistants (the positions are hereinafter referred to as "specialist"):

1. How satisfied is the specialist with his/her actions in the professional situation?
2. On what basis the specialist carried out the choice of his/her tactics of interaction with the child?

- How did he/she implement these tactics?
- What are the outcomes of such interaction?
- What could be done differently, if this opportunity were to happen?

3. What feelings and thoughts did the specialist have in the process of professional actions? For example, whether he/she felt irritation, anger, or frustration.

4. What body feelings did the specialist have during the difficult pedagogical situation? For example, muscle aches or abdominal discomfort.

5. What complicated (caused the emotional tension) the realization of professional activities, and at what point?

6. Who (in what "role") did the specialist feel he/she was in the process of professional actions? ("helping father (mother)"; equal, friend, "critical parent", another role).

7. Did the specialist have moments of the intense internal irritation to the child (one of the child's parents) in the process of interaction with his (her) and to what they related?

8. What did the specialist want to do in this case?

9. What factor, according to the specialist, is the problem related to?
- How exactly did I understand (imagine) the situation;
- What did I want in this situation;
- What did I feel in this situation;
- What did I do;
- What "should" I do?

10. What, according to the specialist, did the child (one of the child's parents) expect from the specialist's actions?

11. Are the emotional expectations of the child (one of the child's parents) equal to my abilities in this case?

12. How can one describe the problem that has arisen?
- Objective reasons (lack of experience, peculiarities of the child or its parents, etc.).
- Internal reasons (psychological impediments, severe emotional discomfort, frustration).

3.4.2 Cases

The result of supervision was mastering and intensification of application of theoretical knowledge and practical skills for analyzing typical complex cases of academic interaction in the context of inclusive education (school administration, child support team, specialists from the Inclusion and Resource Centre) (Table 5).

There were special cases of peculiarities of regulatory support of employment and public purchasing procedure, peculiarities of interactions with parents, announcement of the opening of an inclusive class, techniques for planning reasonable accommodation and universal design. According to the plan, during the introductory trainers analysed the procedure of supervision, the schemes of analysis of observations of the learning behavior, compiling social stories and practicing preparing an individual development program.

Consequently, the experience of conducting supervisory groups in Kryvyi Rih City revealed the typical requests in terms of implementing inclusive education (Table 4): 1) difficulties in understanding the problem of learning and development of the child with SEN; 2) difficulties in supporting and developing the child with SEN; 3) difficulties in dealing with the parents; 4) organizational difficulties (structuring of forms of education or additional educational services for children with SEN); 5) personal problems of the specialist in connection with the professional situation. That was also a supervision analysis of problems in children with autism spectrum disorders, including sexual education of such children, has been performed.

Table 5. Description of the inquiries of the Team specialists during supervisions (n=124).

Answers	Percentage %	Rating
Ambivalent perception of children with SEN by the parents of other children	8,0	4
What to do, if the child with autism spectrum disorders is distracted by background noise at the lessons	6,8	5
Algorithm of the contact with the "difficult" parents of children having SEN, who avoid responsibility	16,3	2
Children's protest against setting educational tasks and avoidance of performing tasks due to difficult speech problems	21,1	1
Refusal of educational tasks in the learning of the child with SEN	4	9
"Escape from learning activities" and immersion in the fantasy world of the child with SEN	3,3	11
Strategies for teaching natural sciences and mathematics for the child with SEN in a secondary school	3,7	12
Systematic school failures as a manifestation of spring exacerbation of psycho-neurological disease	3,8	10
Age-specific features of inclusive education for the child with hearing impaired	3,1	13
Work of the Team in the training material adaptation	6,1	6
Learning strategies in complex difficulties	10,1	3
Low attentiveness and fatigue in children with the asthenic syndrome	2,3	15
Professional burnout of specialists in the Team	4,4	7
Dilemma of transferring the child from the inclusive form of education to the individual one	3,0	14
The problem of the training material adaptation in a secondary school in severe dyslexia of the teenager.	4,0	8

3.4.3 Results

Analysis of the dynamics of the groups after the supervision based on the questionnaire of professional skills (developed by authors). The questionnaire contains 38 questions that evaluate the respondent's professional competences and skills in inclusive education on a scale of 1 (the skill is not mastered) to 5 (the skill is mastered). The questionnaire provided indicators of the position of participants on 5 scales of skills: 1. Supervision of complex cases (8 items, Cronbach's α = 0.70); 2. Knowledge of key features of authorities of the support team of a child with SEN (10 items, Cronbach's α = 0.69); 3. Work on the compilation and implementation of an individual development plan (IDP) (12 items, Cronbach's α = 0.71); 4. The resource/sensor rooms (8 items, Cronbach's α = 0.62).

Firstly, there was a positive growth of expertise concerning the following topics: "Supervision of complex cases" (M=4.31, SD=0.47). Secondly, the scale "Knowledge of key features of authorities of the support team of a child with SEN" (M=3.15, SD=0.57) indicates the heterogeneity of mastering skills about the team's work and its authorities. Therefore, practicing methods of coordination of the support team, purposes of compiling an IDP form a separate purpose of the supervision. Thirdly, the scale "Work on the compilation and implementation of an individual development plan (IDP)" (M=3.78, SD=0.57), indicates fluctuations in the mastering skills implementing an IDP. It is expedient to introduce a strategy for assessing the child's educational success, simplification/complication of the program in the supervision module. Lastly, the scale "The resource/sensor rooms" (M=2.47, SD=0.46) indicates fluctuations in confidence in the theory and practice of occupational therapy in the resource / sensory rooms.

The study utilized the content analysis of professional skills (developed by authors). Content analysis technique was used to categorize the answers of the listeners ("What did you like the most? Please describe" – Fig. 3):

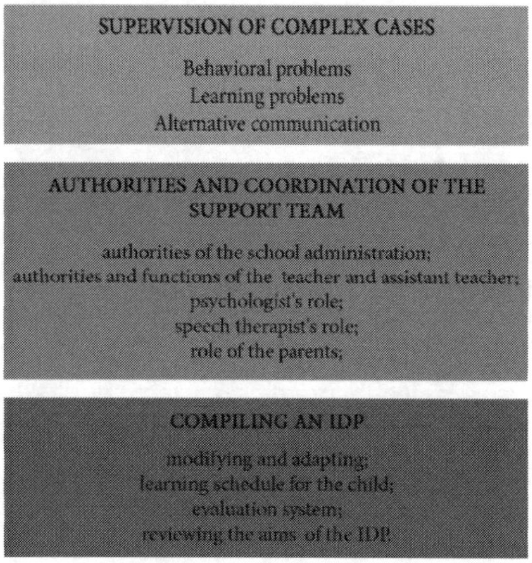

Fig. 3. Content analysis of answers from the respondents.

Cluster I "Supervision of complex cases": expanding the supervision module of problems in children with autism spectrum disorders, including sexual education; supervisions for team working with children with special language needs; supervisions for the frequency of meetings with children with special educational needs.

Cluster II "Coordination of the support team": video modeling of behavior; assembling groups according to the educational characteristics of children; psychodrama for key stakeholders for difficulties in establishing contact; peculiarities of interactions with parents.

Cluster III "Compiling an IDP": Practical exercises on the use of sensory equipment; a diagnostic protocol of the child's individuality; techniques for planning reasonable accommodation and universal design; case study.

Conclusion

Analysis of the system of organizational and methodological and informational retraining, implementation of the pilot project of advanced training courses in the form of supervision groups and training program for the team of psychological and educational support in inclusive education made it possible to offer a program of supervision (Fig. 4):

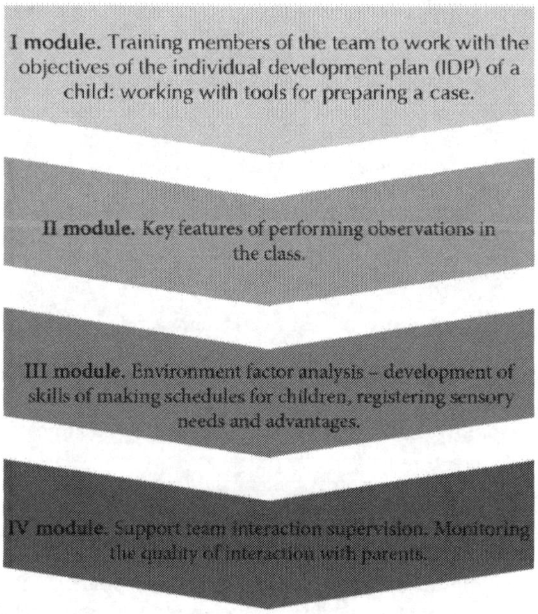

Fig. 4. Model supervision retraining of inclusive education professionals.

I module. Training members of the team to work with the objectives of the individual development plan (IDP) of a child: working with tools for preparing a case. Outline or plan of activities for a child within inclusive education in the school educational system.

Components of an individual development plan: available level of knowledge (academic knowledge) and functional skills – what are the success and achievements of a child when compiling an IDP; annual learning goals for children and how the school is going to monitor the progress; care and educational services, which will be rendered to the child. These can include special education, adjacent and auxiliary services, and a prolonged school year as well; time limits and conditions of the services: starting time, how often, for how long;

accommodations, that is changes and adjustments of the child's learning environment; modifications, that is changes of contents and essence of the child's learning plan; how will the child participate in the (final) standard tests; how will the child be included into the work of the general education class, how will she/he participate in different types of activities.

II module. Key features of performing observations in the class. Educational behavior. Stages of formation of educational behavior. Modifying materials for the classes about the level of maturity of the learning behavior: preparation of classes as a separate case. Modifying assessment.

Registration of child's behavior: detecting a motivator, prevention techniques, error-free learning strategies, incentive schedules, use of hints, task changes, change of hints (antecedent package).

III module. Environment factor analysis – development of skills of making schedules for children, registering sensory needs and advantages. Teaching how to introduce school rules for children. Social stories. Teaching school rules.

IV module. Support team interaction supervision. Monitoring the quality of interaction with parents.

In carrying out educational supervisions, the following methods have proven themselves: 1) case "A difficult child (its parents) for me", 2) development of skills of the non-violent communication, 3) group work to find solutions, 4) clarifying relations with the help of a supervisor, 5) training in the development of skills of observation and analysis of the child's behavior, 6) brainstorm.

References

1. C. Acedo, Preparing teachers for inclusive education. Prosp. **41**, 301 (2011). doi:10.1007/s11125-011-9198-2
2. O. Akimova, Osnovni aspekti inklyuzivnoyi osviti u pidgotovczi majbutnikh uchiteliv pochatkovikh klasiv (Main aspects of inclusive education in the education of future primary school teachers). Naukovi zapiski kafedry pedagogiki 35, 12–18 (2014)
3. S. Alila, K. Määttä, S. Uusiautti, How does supervision support inclusive teacherhood? International Electronic Journal of Elementary Education **8**(3), 351–362 (2017)
4. I. Goodman, M. Brady, M. Duffy, J. Scott, N. Pollard, The effects of "bug-in-ear" supervision on special education teachers' delivery of learn units. Focus on Autism and Other Developmental Disabilities **23**(4), 207–216 (2008). doi:10.1177/1088357608324713
5. L. Ferguson, International trends in inclusive education: The continuing challenge to teach each one and everyone, Eur. J. of SEN **23.2**, 115 (2008). doi:10.1080/08856250801946236
6. Z. Leniv, Kognitivno-kompetetnisniy component gotovnosti asistenta vchitelya do komandnoyi vzaemodiyi v umovah inkluziyi (The cognitive competency component of the readiness of the assistant to the team's cooperation in inclusion). Actual. pitannya korr. osvity **13**, 110–122 (2019)
7. G. Lindsay, Inclusive education: A critical perspective. Br. J. of Spec. Ed. **30**(1), 3–12 (2003). doi:10.1111/1467-8527.00275
8. I. Hardy, S. Woodcock, Inclusive education policies: Discourses of difference, diversity and deficit. Int. J. of Incl. Ed., **19**(2), 141–164 (2015). doi:10.1080/13603116.2014.908965
9. T. Ryan, J. Gottfried, Elementary supervision and the supervisor: Teacher attitudes and inclusive education. Int. El. J. of El. Ed. **4**(3), 563–571 (2017)
10. S. Symeonidou, H. Phtiaka, Using teachers' prior knowledge, attitudes and beliefs to develop in-service teacher education courses for inclusion. Teach. and Teach. Ed. **25**(4), 543–550 (2009). doi:10.1016/j.tate.2009.02.001
11. The Law of Ukraine "On Education" (2017), http://zakon3.rada.gov.ua/laws/show/2145-19. Accessed 25 Oct 2019
12. The Resolution of the Cabinet of Ministers of Ukraine "On Amendments to the Procedure for Organizing the Inclusive Learning in General Educational Institutions" (2017), https://zakon.rada.gov.ua/laws/show/588-2017-%D0%BF. Accessed 25 Oct 2019

How to achieve students' self-management in educational activity?

Liudmyla Bilousova[1], *Larysa* Kolgatina[1], and *Oleksandr* Kolgatin[2,*]

[1]H. S. Skovoroda Kharkiv National Pedagogical University, 29 Alchevskykh Str., Kharkiv, 61002, Ukraine
[2]Simon Kuznets Kharkiv National University of Economics, 9a Science Ave., Kharkiv, 61166, Ukraine

Abstract. The paper is devoted to the study of types of managing the student's educational activity. The educational discipline "Practicum of problem solving in informatics" for students of third year study, future teachers of informatics have been chosen for realising pedagogical conditions of computer-oriented management of students' educational activity. Progressive turn from direct management through co-management, subsidiary management to self-management was the main idea of designing the courseware. The information and communication educational environment has been based on the platform of learning management system Moodle. The Workshop elements of Moodle played the central role in management of students' educational activity. The results of our pedagogical observation and assessment showed the efficiency of suggested approach. Additionally, there were shown the lack of students' competency in time planning and introspection on the base of the experimental data.

1 Introduction

1.1 Statement of the problem

Informatisation of the educational process has led to the creation of information and communication educational environment in institutions of higher education and significantly influenced the goals, content, methods and means of students' educational activity, forms of its organization. The use of modern powerful computer tools for management tasks implementation in educational process means the transition to a new type of management – computer-oriented, which can provide the personalisation and not only serve the achievement of learning goals, but also help the student to become the active participant of such management, that is the subject of self-management. This task of introducing in education of innovative management practices, oriented on the use of information and communication technologies, remains relevant today.

1.2 Analysis of previous research

One of the most fundamental analysis of theoretical and methodological aspect according to the management of the independent learning activity of students of pedagogical higher educational institutions was suggested by O. V. Malykhin [1]. Recently, appropriate methodical systems have been introduced into the practice of the educational process to provide for computer-based management of students' educational activity. Information and communication educational technologies, especially, cloud technologies that transform education have been analysed according to results of the "Cloud Technology in Education" scientific conference by A. E. Kiv, V. N. Soloviev, S. O. Semerikov [2]. O. O. Lavrentieva, L. M. Rybalko, O. O. Tsys, and A. D. Uchitel [3] have analysed new methods of the organization of students' independent study activities together with the use of ICT and tools. Computer-based tools of supporting students' independent experimental activity in the process of learning quantum physics have been proposed in the work of S. P. Velychko and S. V. Shulga [4]. Management of students' educational activity was provided as support in instrument setup, measurements, results processing. K. V. Vlasenko, I. V. Sitak, O. O. Chumak [5] have designed and are developing an educational site "Differential Equations" to support students' educational activity. The site contains theoretical framework, practical classes, provides consultations on-line and via e-mail, testing, discussion cases, forum and provides support for the teaching of the course and solving practical problems of research character by students. S. O. Podlasov, O. V. Matviichuk, V. P. Bryhinets [6] have proposed elements of blended training in physics at a technical university on the basis of programmed learning (students study new material or fixate their knowledge) in the Moodle system with the help of the element *Lesson*. M. A. Kyslova and K. I. Slovak [7] developed methods of using the mobile learning environment in study of higher mathematics by future electromechanical engineers. These methods are based on the complex application of computer tools: using *Google Apps Education Edition* (texts, diagrams, links, videos); execution of practical tasks and research with developed models in cloud-oriented *GeoGebra* and *CoCalc* environments; application of *Drawings* for

* Corresponding author: kolgatin@ukr.net

generalization and systematization of concept connections, *Forms* for testing, *CoCalc* for task generation. Tools integration was provided with using *Classroom*, *Calendar* was used for scheduling training activities. O. O. Triakina, O. O. Pavlenko, N. P. Volkova, D. A. Kassim [8] analysed E-learning instruments for self-education and have suggested the ways of this tools implementation into professional training. Methods and technologies for the quality monitoring of electronic educational resources were analysed by H. M. Kravtsov [9]. A methodical system of computer-oriented management of independent work of future teachers in the process of learning computational methods (numerical methods) was developed in the works of L. I. Bilousova, O. G. Kolgatin, L. S. Kolgatina [10]. This system is based on the use of specially designed computational models in MathCAD environment and assumes learning management system Moodle to support for management of students' independent work. It should be noted that development of the educational process on the basis of its reorientation to students' self-management their own cognitive activity not only contributes to enhancing the autonomy of students, but also gives the education of personal significance. By determining the individual trajectory of educational and cognitive activity of each student on the basis of the maximum consideration of his individual and cognitive abilities, the necessary prerequisites for the formation of his skills of systematic and continuous professional self-improvement are created. Useful analysis in this direction were suggested by B. I. Kruk, O. B. Zhuravleva [11].

1.3 Objectives

The potential of computer-oriented management of students' educational activity is not fully realized, according to the above analysis. It actualizes the study of pedagogical conditions, the introduction of which improves the effectiveness of such management by contributing to the future teachers in obtaining better educational results and in acquisition of the active personal position in managing their own independent work.

The **aim** of this paper is in theoretical and practical study of pedagogical conditions of effective computer-oriented management of students' educational activity and progress from direct management through co-management and subsidiary management to self-management in information and communication educational environments.

2 Theoretical framework

There are various approaches to defining the concept of management in pedagogical systems in the psycho-pedagogical literature. Thus, M. Markov [12] views management as an organization of purposeful actions, I. B. Itelson [13] sees management as the actions that are directed to achieving a previously set goal. Yu. M. Korshunov [14] considers that management is the organization of a process that ensures the achievement of a predetermined goal. A. V. Filippov [15] considers management as the purposeful influence of the subject on the object and the change of this object as a result of influence. V. Nechaev [16] speaks about management as purposeful regulation of processes. In some studies, management is seen as an element of some system that connects all its elements and subordinates them to the goal. Thus, V. Yakunin [17] sees the essence of management in the interaction of the student and the teacher, which is carried out in accordance with the set goals and is aimed at activating the student's activity in the learning process and achieving the required results. We agree with all of the above statements, which highlight certain features of management and confirm the relationship between management and activities. It is also defined by the interpretative dictionary: "To manage – 1. To direct activity, work of someone, something; be led by someone, something; manage. 2. To direct the course of a process, to influence the development, the state of something" [18]. On the basis of this analysis of pedagogical research on the problem, the essence of managing student's educational activity is determined as realization of interaction of a student and a teacher, which is aimed at activating student's activity in the educational process and achieving the educational goals. As a result of this interaction, the social and cognitive experience of the student changes, which acquires the trait of independent, purposeful activity in order to become ready to solve future professional problems.

The development of information and communication technologies creates prerequisites for improving the efficiency of managing students' educational activity in modern higher education process. Given the new role of the teacher as a tutor, a moderator, who provides support to the student in choosing and building an individual educational trajectory, the new quality of management is seen in its variability, coordination of management actions with individual capabilities, needs and requests of the student. This management is directed to help the student to get knowledge and skills according to the curricula, but also to increasing involvement of the student in managing their own educational activity, in the progressive transition from direct management to co-management, subsidiary management and further to self-management. ICT-oriented management of student's educational activity is a multi-stage process (collection of information, statement of objectives, decision-making, implementation of the decision, monitoring and evaluation of results, adjustments) that is implemented with the use of appropriate ICT tools at each stage. Implementing student's self-management with the use of modern, powerful computer management tools means moving to a new type of management - computer-oriented, capable to provide higher quality of management. The new quality of computer-oriented management of student's educational activity is due to the following features:

- adaptability that is based on detailed data on the level of knowledge and skills of the future teacher required for independent work, as well as on the dynamics of their acquisition;
- flexibility that assumes gradual involving of a student in improving management of his / her own independent work through the transition from direct management to co-management, subsidiary management and self-

management on the base of analysis of the accumulated experience of using a certain type of self-management and data on its effectiveness;

- timeliness, which is provided by the opportunity to monitor the process of the task execution and the availability of communication resources, that allows timely and targeted assistance and advice to the student, based on the accumulation and analysis of data on the progress and effectiveness of his educational activity;
- transparency, which involves openness of requirements to the results of the educational activity, criteria for the evaluation, rating indicators of educational achievements of the future teachers;
- objectivity in making managerial decisions that is based on objective testing data and tracking the effectiveness of the student's educational activity.

On the basis of the analysis of new opportunities for managing the student's educational activity, pedagogical conditions for the effective implementation of the said management in the educational process have been substantiated:

- designing of information and communication educational environment, which contains variation educational-informative, instructive-methodical, software-instrumental, as well as communication resources for organization and support of the student's educational activity;
- using a system that automates the collection, accumulation and analytical processing of performance indicators of student's educational activity;
- ensuring the readiness of all participants in the educational process to implement computer-oriented management of the student's educational activity.

Only comprehensive application of all conditions ensures the effective management. The implementation stage of ICT-oriented student's educational activity management is the key stage, when the student actively takes part in this management as the person of educational process. Understanding the character of interconnections between the teacher and the student (Table 1) is very important to developing the flexible management based on different types of management (direct management, co-management, subsidiary management, and self-management).

Table 1. Activities of the subjects of the educational process at different management types on the stage of implementation of the decision.

Type of management			
Direct management	**Co-management**	**Subsidiary management**	**Self-management**
The teacher sets a task for the student	The teacher discusses a task with the student	The student chooses a task from a database	The students formulates a task and coordinates it with the teacher
The teacher sets the methods of the task execution	The teacher discusses the methods of the task execution with the student	The student chooses the methods of the task execution from suggested by the teacher	The student determines the methods of the task execution independently
The teacher suggests necessary resources to for the student	The teacher suggests necessary resources to for the student	The student chooses necessary resources from the given resource base	The student determines the necessary resources independently
The teacher gives the example of the correct operation sequence (detailed instruction)	The teacher gives the common schema of the operation sequence (framework instruction)	The student determines the operation sequence independently	The student determines the operation sequence independently
The teacher provides current correction of the task execution process	The teacher adjusts the process of completing the task, if necessary	The teacher adjusts the process of completing the task, if the student ask him for help	The student controls the task completing process independently
The teacher provides the student with current systematic help	The teacher helps the student, if necessary	The teacher helps, if the student asks	The teacher helps, if the student asks
The teacher gives the pattern of report to summarising obtained results. The student acts on the model	The teacher gives the plan of report to summarising obtained results. The student acts on the plan	The teacher gives the requirements to report and summarising obtained results. The student produces the analysis of obtained results independently	The student coordinates the form of report with the teacher and produces the analysis of obtained results independently

3 Methodology of empirical research

We have realised the above approach to management of students' educational activity in the practicum of problem solving in informatics for future teachers of computer science. 10 students took part in this work. The information and communication educational environment has been based on the platform of learning management system *Moodle*. It contains built-in communication resources as well as reference to educational-informative resources and instructive-methodical materials according to programming the basic algorithmic constructions for organization and support of the students' educational activity. The software-instrumental resources (*Eclipse* environment and the tools of common information technologies) were present at every students' computer. The leading information channel was the interactive lectures, where the elements of programs have been analysed in details. The educational activity of the students at this interactive lectures were managed directly, because the students did the notes in the form of parallel development the suggested and analysed algorithms as *Java* programs. Interactive parts of this lectures involved some students in co-management of educational activity, but some of them were passive and continue execute the tasks in direct management regime, using the ready fragments of code and orientating only at the teacher's commands in the time planning. The students' notes in the form of developed and tested programs became the instructional materials for management students' independent activity in problem solving.

There were created 5 *Workshop* elements of activity in the course in *Moodle* environment and suggested 5 series of individual tasks for each student according to the such topics: linear algorithms; branching; cycles; one-dimensional arrays; two-dimensional arrays. Every of this series contained three tasks of different levels. Student should execute one of this task for passing or all tasks for high grade. The first task assumed the direct (but distance) management of students' educational activity, because this task was very similar to the one was analysed at lecture. The second task assumed the direct or co-management. This task was based on some of analysed algorithms but was not fully similar. Students could solve this problem using only the lecture notes (direct management). But sometimes students needed in additional information for solving the problem. They could ask the teacher or colleagues personally or using the built-in tools of *Moodle* (co-management) as well as to use some additional information resources (subsidiary management). The third task was difficult and assumed using the algorithmic constructions that were not analysed at lectures. Students had to discuss this problem with the teacher (co-management) or independently use the additional information resources (subsidiary management). So, the students independently and intuitively made the decision on using some of above type of management of their own educational activity for each task according to their educational achievements and skills of independent activity.

Students uploaded the results of the tasks execution to the *Workshops*. It should be the correct program with the author's tests to prove it correctness. The second phase of students' activity in *Workshops* was to check and grade the works of the colleagues – assessment phase. Only the students, who have executed the first stage of the task and have submitted their works, could take part in the assessment. The assessment process is creative activity, but its management was direct, because the students assessed according to the simple instruction: +1 point, if the program is submitted and perform the required results; +1 point, if the program correctly work with the author's tests, +1 point, if the reviewer cannot suggest any tests to indicate the bugs. Grading of the assessment phase was produced automatically by comparison of student's given grades with other reviewers' grades for corresponding works. Teacher also took part in the assessment as a reviewer with weight coefficient of 10.

The teacher carried out pedagogical observation during the course. The students were suggested to design in classroom at fixed time three programs of different levels of difficulty as the final testing in the course. Students, who did not submit their works in time passed the task at additional time in the form of discussion of the results with the teacher personally. Such results were not analysed in this study.

4 Results and discussion

Correlation between students' final test results and students' average grades for submitted works has been evaluated to estimate the validity of our assessment tools (Figure 1). Coefficient of Pearson's correlation is 0.70 that is statistical significant at the 5% significance level.

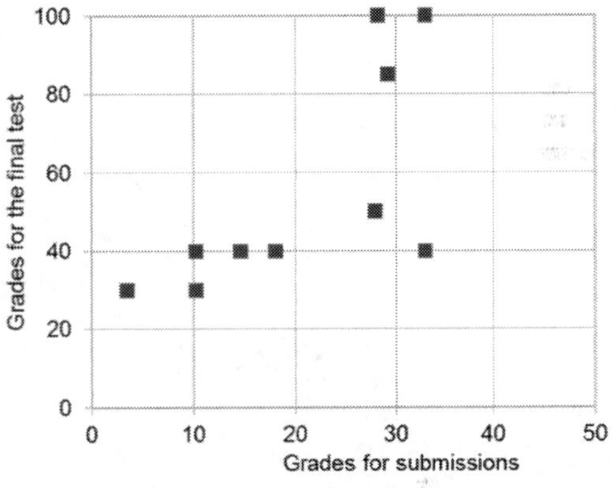

Fig. 1. Correlation between students' final test results and students' average grades for submitted works.

Correlation between the quality of tasks execution by students and their skills in assessment seems to be very good for 7 students (Figure 2), but 3 students with highest results of tasks execution did not take part in assessment, so we cannot prove this correlation statistically, the size of our sample is not enough. According to our pedagogical observation, some students did not take part in the assessment because of their mistakes in time planning.

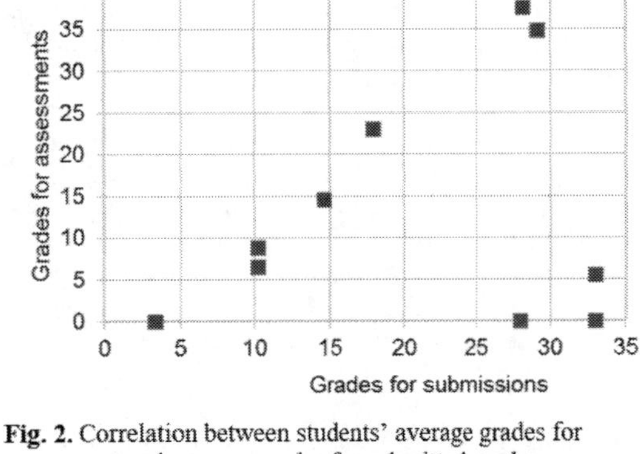

Fig. 2. Correlation between students' average grades for assessment and average grades for submitted works.

Not all students were able to maintain the proper pace of educational activity progressing from the first topic to the last (Figure 3). Some of them worked effective only in the begin, when direct instructions were full enough to execute some tasks. Because of low motivation and lack of elementary skills in self-management, they did not switch to co-management by their own initiative. These students did not submit some works in time and then passed the tasks at classes with personal participation of the teacher in the process of programming and time planning. So we should develop a mechanism for preventive diagnostic of students' skills in self-management and timely turn them to direct management of educational activity. Also, we see that direct management is an easier way of educational activity for some students. This way seems more comfortable for them. So, we should develop special methods to motivate this category of students for their progressing to self-management of the own educational activity. But other students actively used communication and additional sources to solve problems and did not decrease the level of submissions, when progressing to next, more difficult topic.

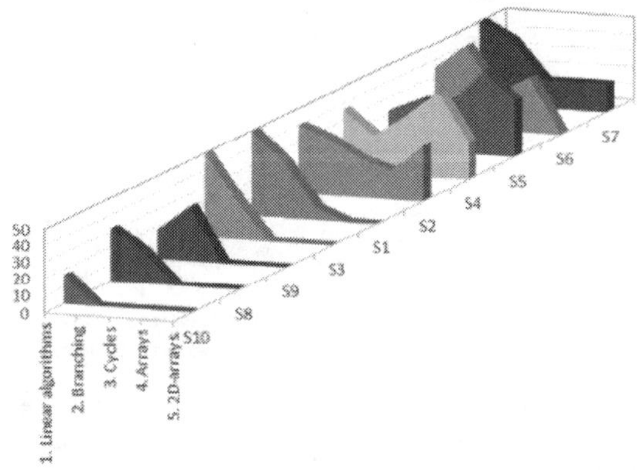

Fig. 3. Students' progressing from the first topic task (1. Linear algorithms) to the last (5. 2D-arrays). Students' names are shown as S1-S10. Vertical axis shows students' grades. The maximum possible grade was 50.

As expected, the most difficult for the students was the third task in each Workshop, the least difficult – the first task (Figure 4). Analysing the structure of student works according to given criteria (Figure 4), we can conclude that the most problem for students was not the development of the program but provident that it works correctly. The author's tests often were absent or incorrect. The program, if present, often was correct, but sometimes the reviewer's tests could find some bugs.

The analysis of the structure of students' works showed that the students' competency in introspection was not enough. In our opinion, the introspection is one of the leading elements of self-management competency. So the educational tasks should always content some sub-tasks on introspection.

Summarizing the result of our experience in combining of different types of management of students' educational activity, we can conclude that providing the above pedagogical conditions gave us possibility to improve the educational process in "Practicum of problem solving in informatics". Flexible management of students' educational activity with timely turn from the direct management to co-management and subsidiary management with return, when needed, supported the efficiency learning. In despite students' involving in practical labour out of the educational process, the goals of "Practicum of problem solving in informatics" were achieved.

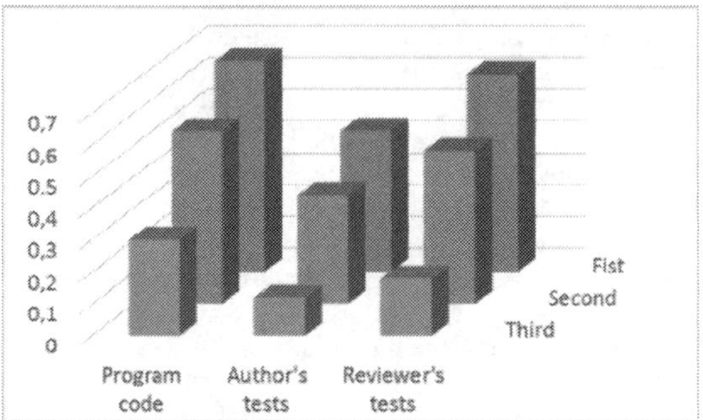

Fig. 4. Part of submissions that satisfy to the criterion (program code, author's test, reviewer's test) on each task (first, second, third). This part was calculated as average for all 5 Workshops.

5 Conclusions

Analysis of obtained experimental data in context of our theoretical framework has given the base for such conclusions:

- providing the pedagogical conditions of management of student's educational activity led to improving the educational process in "Practicum of problem solving in informatics";
- flexible management of student's educational activity with timely turn from the direct management to co-management and subsidiary management supported the efficiency learning;
- sometimes the type of management of student's educational activity should be timely turned back to direct management;
- there are two aspects in management of student's educational activity: content of activity choosing and time planning;
- students' competency in introspection is not enough, so the educational tasks should always content some sub-tasks on introspection.

Further work in the field of computer-oriented management of students' independent educational activity we see in developing new methods of students' progress from direct management through co-management and subsidiary management to self-management in information and communication educational environments, in introduction of these methods into various educational disciplines as well as studying pedagogical and psychological conditions to increase students' motivation for self-management of own educational activity.

References

1. O. V. Malykhin, *Orhanizatsiia samostiinoi navchalnoi diialnosti studentiv vyshchykh pedahohichnykh navchalnykh zakladiv: teoretyko-metodolohichnyi aspekt* (Management of the independent learning activity of students of pedagogical higher educational institutions: theoretical and methodological aspect). (Publishing house, Kriviy Rig, 2009)
2. A.E. Kiv, V.N. Soloviev, S.O. Semerikov, CTE 2018 – How cloud technologies continues to transform education. CEUR Workshop Proceedings **2433**, 1–19 (2019)
3. O. O. Lavrentieva, L. M. Rybalko, O. O. Tsys, A. D. Uchitel, Theoretical and methodical aspects of the organization of students' independent study activities together with the use of ICT and tools. CEUR Workshop Proceedings **2433**, 102-125 (2019)
4. S. P. Velychko, S. V. Shulga, ICT tools for support of students' individual work in the study of quantum physics. Information Technologies and Learning Tools **65**(3), 103–114 (2018). doi:10.33407/itlt.v65i3.2225
5. K. V. Vlasenko, I. V. Sitak, O. O. Chumak, Metodyka kompiuterno-oriientovanoho praktychnoho navchannia dyferentsialnykh rivnian bakalavriv z informatsiinykh tekhnolohii (Methods of computer-oriented practical study of differential equations by bachelors in information technologies). Cherkasy University Bulletin: Pedagogical Sciences **16**, 3–14 (2018)
6. S. O. Podlasov, O. V. Matviichuk, V. P. Bryhinets, Elementy zmishanoho navchannia fizyky v tekhnichnomu universyteti (Elements of blended learning in studying physics in the technical university). Information Technologies and Learning Tools **61(5)**, 151–161 (2017). doi:10.33407/itlt.v61i5.1741
7. M. A. Kyslova, K. I. Slovak, Method of using mobile learning environments in teaching mathematics of future electromechanical engineer (Metodyka vykorystannia mobilnoho navchalnoho seredovyshcha u navchanni vyshchoi matematyky maibutnikh inzheneriv-elektromekhanikiv). Information Technologies and Learning Tools **51**(1), 77–94 (2016). doi:10.33407/itlt.v51i1.1360
8. O. O. Triakina, O. O. Pavlenko, N. P. Volkova, D. A. Kassim, Usage of E-learning Tools in Self-education of Government Officers Involved in Global Trade Activities. CEUR Workshop Proceedings **2257**, 173–181 (2018)
9. H. M. Kravtsov, Methods and Technologies for the Quality Monitoring of Electronic Educational Resources. CEUR Workshop Proceedings **1356**, 311–325 (2015)
10. L. Bilousova, O. Kolgatin, L. Kolgatina, Computer Simulation as a Method of Learning Research in Computational Mathematics. CEUR Workshop Proceedings **2393**, 880–894 (2019)
11. B. I. Kruk, O. B. Zhuravleva, Analyz transformatsyi v ynnovatsyonnom pedahohycheskom obrazovanyy (Analysis of transformations in innovative pedagogical education). *Paper presented at III All-Russian scientific and practical Internet-conference "Innovative directions in pedagogical education"* (2010), http://econf.rae.ru/article/5138. Accessed 30 Dec 2019
12. M. Markov, *Teoriia sotsialnogo upravleniia* (The theory of social management). (Progress, Moskva, 1976)
13. I. B. Itelson, *Matematicheskie i kiberneticheskie metody v pedagogike* (Mathematical and cybernetic methods in pedagogy). (Prosveshchenie, Moskva, 1964)
14. Yu. M. Korshunov, *Matematicheskie osnovy kibernetiki* (Mathematical foundations of cybernetics). (Energoatomizdat, Moskva, 1987)
15. A. V. Filippov, Voprosy psikhologii upravleniia (Issues of Management Psychology). Psychological Journal **1**(2), 17–23 (1980)
16. V. Ya. Nechaev, *Sotsiologiia obrazovaniia* (Sociology of Education). (Izdatelstvo Moskovskogo universiteta, Moskva, 1992)
17. V. A. Yakunin, *Obuchenie kak protsess upravleniia: Psikhologicheskie aspekty* (Learning as a management process: Psychological aspects). (Izdatelstvo Leningradskogo universiteta, Leningrad, 1988)
18. *Novyi tlumachnyi slovnyk ukrainskoi movy* (New Interpretative Dictionary of the Ukrainian Language), vol. 4 (Akonit, Kyiv, 1999)

Soft skills, their development and mastering among post graduate students

Nataliia Volkova[1,*], *Nataliia* Zinukova[2], *Kateryna* Vlasenko[3], and *Tetiana* Korobeinikova[4]

[1]Alfred Nobel University, 18 Sicheslavska Naberezhna Str., Dnipro, 49070, Ukraine
[2]Alfred Nobel University, 18 Sicheslavska Naberezhna Str., Dnipro, 49070, Ukraine
[3]Donbass State Engineering Academy, 72 Akademichna Str., Kramatorsk, 84300, Ukraine
[4]Alfred Nobel University, 18 Sicheslavska Naberezhna Str., Dnipro, 49070, Ukraine

Abstract. The article deals with the issues of the content, assessment of the current state and the main tendencies in the development of the so-called "soft skills". At present soft skills are essential to have good job opportunities and university curriculum should be focused on developing and mastering them. The article analyses the results of the research of the soft skills development among post graduate students majoring in Professional Education of Alfred Nobel University, Dnipro. These data prove that postgraduate students realize the importance of mastering soft skills to make a successful career. Moreover, innovative teaching methods and techniques aimed at mastering soft skills are to be introduced into the educational process.

1 Introduction

Characteristic features of current situation are the presence of deep and rapid socio-economic, political, innovative and educational transformations in the life of Ukraine, the orientation of the country to integration the global world community, increasing the openness of the educational process, focus on the needs of the society and future employers. A challenging point nowadays is the importance of an individual which can actualize the need for highly qualified competitive scientific and pedagogical staff characterized by holistic high professionalism and personality culture, creativity, non-standard thinking, readiness to respond flexibly to changing conditions and solve professional pedagogical tasks creatively. These can be applied by new pedagogical thinking, intuition, ability to improvise, ability to establish and maintain contact with the interlocutor, to express their attitude to what is happening, to persuade and defend their view, to make decisions and manage their contacts, to reach goals and take on different roles, to maintain internal autonomy and encourage the partner to change behavior, to reach mutual understanding in different conditions of information exchange, etc. All mentioned above allows us to understand why the international education community emphasizes the orientation on vocational (in particular, pedagogical) education to the formation of soft skills and hard skills in their integrity.

Recent studies of the labor market indicate that the interest in soft skills is increasing. Most employers consider them as important as professional knowledge and skills. It is believed that the professional skills can be outdated but the soft skills are always relevant.

Literature review let us state that the scope of research on soft and hard skills is quite wide. For example, the focus of the classification of soft skills in the research of O. Abashkina [1], N. Zhadko and M. Churkina [24], K. Koval [13], Iu. Ivaniuk [16], A. Ivonina and others [11]. Several studies including papers by T. Ananieva [4], A. Zinchenko and M. Saprykina [23], K. Ananiadou and M. Claro [3] discuss the most required skills for the specialists in the 21-st century in many countries. The description of framework and design educational process are revieled in the studies by N. Zhadko and M. Churkina [24], A. Ivonina and O. Chulanova [11], N. Volkova [21], Ye. Yamburg [22], N. Vesnina [19]. Issues related to the new pedagogical format are analyzed in the works by S. Batsunov, I. Derecha, I. Kungurova and Ye. Slizkova [5].

So, today we observe the high popularity of using soft and hard skills in educational process, increase in their application not only for development among managers but also among the students of different specialties. At the same time we have to mark an obvious gap between theoretical claims and practical implementation of soft skils via hard skills into a formal educational context.

The objective of this article is to get acquainted with the soft skills system required by a university trainer in

* Corresponding author: npvolkova@yahoo.com

his or her professional activity and to highlight the opportunities for gaining and developing these skills during postgraduate study in the specialty 015 Professional Education.

2 Research methods

At the theoretical level, the article provides a retrospective analysis of the concepts of "soft skills" in domestic and foreign science. An empirical part of the research was carried out through the methodologies "Modified Questionnaire of Self-Actualisation Diagnostic" (A. Lazukin, in the adaptation of N. Kalin) [15]. To diagnose soft skills, a set of techniques was used, including the questionnaire "Measurement of Communicative and Social Competency" (V. Kunitsyna) [14], the methodologies for studying the motivation for university lecturers (Ye. Ilyin) [10], the diagnostic procedures of reflection level (A. Karpov) [12] and the methodology "Coping Behaviour in Stressful Situations" [6].

After theoretical analysis was completed, there were identified the most important issues:

1. Soft skills acquiring, their characteristics and distinctive features.
2. Advantages of acquiring soft skills for future university lecturers: social and communication skills, cognitive skills and personal qualities and components of emotional intelligence.
3. Possible ways of soft skills' integration into new educational context.
4. Soft skills acquiring among post graduate students in a new educational format.

Our analysis is also based on the survey carried out with a group of post graduate students of Professional Education specialty and from Alfred Nobel University, Dnipro. Forty-five post graduate students of the 1-st, 2-nd, 3-d and 4-th year of studying were interviewed. Each interview lasted from forty to sixty minutes. The detailed notes were taken and recorded. The group of experts was selected from academic staff from the Department of Pedagogy and Psychology of Alfred Nobel University, among whom there were 12 lecturers, senior lecturers and professors.

On the later stage, theoretical as described in scientific articles, projects, conference proceedings and web resources were compared and contrasted with the data from the experts and post graduate students' interviews, hence we could elaborate the all-round understanding of what it is necessary to start developing and mastering soft skills among post graduate students in the context of new pedagogical education format. As a result, final conclusions were made.

3 Results and discussion

First of all, we have to underline that there is an ongoing dialogue between employers and higher education institutions in different countries around the world to identify the so-called "skills of the future" that may become the key ones to the specialist. It is interesting to note that, in 2009, the OECD (Organization for Economic Co-operation and Development), which brings together countries around the world, presented a report entitled "21-st Century Skills and Competencies for New Millennium Students in OECD Countries" [3]. In 2015 the OECD published the results of a five-year study "Skills for Community Development. The Power of Emotional and Social Skills", which analyzed how education affects the personal success of learners and the development of society as a whole.

From the short review above, key results emerge that students' emotional and communicative qualities determine the life success no less than academic achievements. In foreign educational practice, emotional and communicative qualities are considered as "soft" skills that add to "hard skills" or technical, professional skills. Unlike the latter, which are usually well-measured, fairly robust, identified with specific designs, and included in job descriptions and learning outcomes, *soft skills* are versatile and essential for the successful professional. They are of vital self-determination of any person, regardless from the profession. These include the ability to communication, leadership, cooperation, diplomacy, establishment of the relationships; team-building and public skills; the ability to present ideas and to solve the open-ended tasks creatively. In today's world, these competencies are considered as an important educational outcome, along with professional competences.

Another promising finding was that eleven directives that foresee the introduction of new teaching and learning methods for the development of soft skills for EU university students by 2020 were adopted in 2018. The Europass of skills in 26 languages of the EU Member States has been introduced there. The analysts at the World Economic Forum have made a forecast that identified the ten key competencies that will be needed in 2020 [4], according to which the most important ones will be the ability to solve difficult tasks and critical thinking, the creativity and the ability to manage people, interaction skills and emotional intelligence, client orientation, the ability to negotiate and make decisions. From these findings it is clear that these competencies relate to the soft skills.

A similar pattern of results was obtained in 2017, when the officially adopted *soft skills classification* and its explanation called "European Skills, Competencies, Qualifications and Types of Occupation", including 1,384 skills that are in demand in the labor market was developed [7].

When investigating this issue of those international studies, it must be pointed out that this issue is also being researched in Ukraine. Thus, A. Zinchenko and M. Saprykina state that according to a survey conducted by the Development Center KSV during July-August 2016, such skills as teamwork, communication skills, and analytical thinking, the ability to learn quickly, flexibility, responsibility, initiative, competent written and oral language and emotional intelligence are the most in demand among Ukrainian employers [23].

In 2012 the National University of Kyiv-Mohyla Academy developed and presented the first comprehensive program "PROFI+" to prepare students for the labor market which provides gaining "soft skills" which is universal for all specialties.

We consider it essential to continue our analysis with defining soft skills as an important development element for university trainers. Planned comparison revealed that there are different approaches to the interpretation of soft skills and their classification, which is caused by the contextual approaches of researchers.

The analysis found evidence that in many contexts *soft skills* are treated or used as synonyms with such notions as employability skills, people skills, non-professional skills, key skills, skills for social progress (skills for social development), skills of the 21-st century, life skills.

Scholars take different approaches to defining the essence of *soft skills* or "flexible skills": the personal qualities of an individual, which allow to interact with other people more effectively and harmoniously [7]; communication and management talents [17]; skills that help you quickly find common ground with others, establish and maintain relationships, successfully present own ideas, be an effective communicator and leader [9] and others.

We can observe even more diversity in the approaches of such researchers as O. Abashkina, K. Koval, O. Sosnitskaya to the list of soft skills. Thus, K. Koval singled out the following: leadership skills and ability to work in a team, ability to teach and negotiate, ability to set and achieve goals, time management, purposefulness, presentation skills, effective communication skills, stress resistance, creativity, creative approach to solving problems and analytical skills, etc. [13]. As compared with the research by O. Sosnitskaya we can add to the list of soft skills such important elements as the ability to persuade, lead, manage, make presentations, find the right approach to people, ability to solve conflict situations [17]. O. Abashkina underlined the following skills as such competencies that provide personal dynamics (sense of responsibility, desire for achievement, self-confidence, high motivation), area of interpersonal relationships (contact, objective self-esteem, compassion and empathy for others), desire for success (self-giving, motivation to maintain status, tendency to systematize, initiative), endurance (resistance to criticism, resistance to failure, positive emotional attitude, firmness of life position, satisfaction with work) [1]. V. Shipilov gave the following elements to the list of soft skills: basic communication skills that help to develop relationships with people, support conversation, effectively behave in critical situations while communicating with others; self-management skills that help effectively control one's own state, time, and processes; effective thinking skills that help make life and work more systematic and the management skills that people need when they become leaders in any business process [18].

Iu. Ivaniuk in his tern, identifies the signs that one can say with certainty that a specialist has soft skills, if he is flexible, he is not afraid to take responsibility, establishes favorable relationships with people, knows how to work in a team, can teach others, is capable to set tasks for people quickly and accurately, motivating them for quality work, knows how to persuade people, manage their time with competence [16].

Based on the analysis of researchers' approaches to the classification of soft skills, we can distinguish the following *three groups*:

1. social and communication skills (communication skills, interpersonal skills, group work, leadership, social intelligence, responsibility, ethics of communication);
2. cognitive skills (critical thinking, problem-solving skills, innovative thinking, intellectual workload management, self-study skills, information skills, time management);
3. personal qualities and components of emotional intelligence (emotional intelligence, honesty, optimism, flexibility, creativity, motivation and empathy). It should be noted that, within each category and between categories, individual skills and qualities interact. It is clear that different types of soft skills are a priority for different activities.

Now let us consider to understand how important those soft skills are to the educators. The profession of the university lecturer is public, and therefore the educator, being a professional, must be a bright, intelligent and creative personality who is able to identify complex problems in vocational education, including research and innovation activity, and to offer some approaches to solve them, which involves deep analysis and creation of new holistic knowledge and / or pedagogical practice. However, nowadays current professional qualifications require a deeper, more practical and flexible knowledge within competencies. According to K. Koval, professional skills become somehow outdated, and "soft skills" remain always relevant [13]. Without claiming the completeness of the list, we consider that such skills are important for the university lecturers:

– *communication skills* (ability to listen to the interlocutor, to persuade and argue, to present themselves and their ideas; to work in a team in order to solve collaborative tasks, to establish relationships and maintain contacts, to communicate with the audience, to model relationships with stakeholders in the educational process – students, parents, educators, senior management, social partners; non-verbal communication skills and Internet communication culture, the ability to solve conflict situations, public speaking and leadership);
– *cognitive skills* (critical thinking, problem-solving skills, innovative thinking, intellectual workload management, self-study skills, information skills and time management);
– *management skills* (forecasting, the ability to form a group, the ability to form a team and a system of communication in a team, the ability to motivate team members and carry out formal and informal leadership, the ability to manage themselves and the audience);
– *strategic skills* (strategic planning, strategic decision making, the ability to work under risk conditions and delegate authority);
– *skills of self-organization* (the ability to set and achieve goals, skills of self-management, self-esteem, self-reflection and self-development);

– *emotional competences* (self-confidence, empathy, sense of responsibility, emotional intelligence, stress resistance, readiness to accept criticism and to respond to it adequately, honesty, optimism, flexibility, motivation and empathy).

It gives clearly better understanding that within each category and between categories, individual skills and qualities can interact. However, having one or the other skill is significant as an individual position, which is part of the requirements.

In addition, the university lecturer should be able to develop soft skills for future professionals, which will allow them to be successful and effective in their work due to a high level of self-organization, quality planning and effective communication.

We are aware that special disciplines are "responsible" for the development of "hard skills". They are "critical" in the short term and can be developed faster, with less effort and a guaranteed result (provided students have the motivation, the ability to study, etc.), as well as practically they do not incline to reverse development, otherwise, the soft skills more frequently can be developed spontaneously. They are "critical" in the long run, and their development is slower. They require more efforts, the achievement of the required level is not guaranteed ("limit" of competences, deep integration in the personality structure). We share the opinion of Ye. Yamburg that under specific conditions soft skills quite incline to reverse development [22].

Let us consider the signs of personality development presented in Figure 1.

You have an idea of development goals and a particular development plan

⇓

You know where and why you go and what you want

⇓

You analyze your actions and results, look for reasons for success and failure

⇓

You are ready to step out of your comfort zone and try not only what you are good at, but also some thing new, take the risk

⇓

You want to get feedback

⇓

You want to develop, to gain new experience, to grow professionally

Fig. 1. Signs of personality development (by Zhadko and Churkina, 2006) [24].

We are aware that the leading condition for ensuring the integrity in the development of soft skills is knowledge and understanding of graduate students' tasks and the context of their implementation. As practice shows, teaching different skills without relying on the values and priorities of professional activity does not lead to the expected changes. Realizing professional tasks allows you determining what skills are needed by the university lecturer, which skills will be used and which is not.

An experience analysis singled out two approaches to the soft skills formation today. The first one is to teach directly by introducing separate courses within the variant component of the curriculum. The second approach is based on the potential of all disciplines in conjunction with non-formal education and extracurricular activities. It is necessary to note that both approaches are common and widely spread at Ukrainian universities.

A solution to the problem under consideration is proposed in our approach implemented at the Department of Pedagogy and Psychology of Alfred Nobel University, Dnipro, Ukraine. The formation of soft skills of third-level (educational-scientific) level applicants in the specialty 015 Professional Education (by specialization) was facilitated by the training course "Professional-Pedagogical Communication", the content of which was organically combined with the traditional disciplines of the psychological-pedagogical cycle. This course contains *four training modules*: "Fundamentals of Professional Communication", "Verbal Professional Communication", "Non-verbal Professional Communication", "Computer Technology as a Professional Communication Tool". The basis of its development is the content of the textbook adapted for post graduate students [20].

The learning material of the *first* training module includes the consideration of the following issues:

– historical origins of professional-pedagogical communication;
– essence, types, functions and model of communication;
– communicative act as a unit of communication;
– the concept of "professional-pedagogical communication", its types, sources, features of interaction of entities of pedagogical communication;
– communicative knowledge, skills, experience, communicativeness of the university lecturer;
– information and its role in the university lecturer's professional communication, types of information, ways of its storage, transfer, free access and appropriate processing methods;
– the consideration of pedagogical communication as a leading component of communication of a university lecturer;
– the content of the stages of professional communication of the university lecturer;
– styles, lecturer's models of communication;
– barriers and constraints arising from the professional communication process;
– interpersonal relationships, social and psychological climate inside the university's academic staff;
– pedagogical conflicts: types and causes; resolving and preventing pedagogical conflicts. This training module has been entitled "*Fundamentals of Professional Communication*".

The structure of the *second* training module is based on the following issues:

– professional-pedagogical language and speech;

– the essence and functions of the lecturer's language and speech;
– speech activity and the lecturer's individual style of teaching.

This module contains the information about the rules, styles of modern language; the essence of verbal means of professional communication (external: oral and written; internal speech); expands on the concept: lecturer's culture and technique, lecturer-student interaction models of speech. It is entitled "*Verbal Means of Professional-Pedagogical Communication*".

The *third* basic module characterizes:
– non-verbal aspects of communication and their role in regulating relationships, establishing contacts;
– ratio of verbal and non-verbal means;
– revealing features of coding – interpretation of non-verbal behavior;
– non-verbal channels (facial expressions; touch; gesture; interpersonal communication space; visual interaction; intonation). It is entitled "*Non-Verbal Means of Professional-Pedagogical Communication*".

The *fourth* module contains teaching material covering the capabilities of the computer as a means of professional communication for the university lecturer; and dialogue in the system "man – computer", "man – computer – man". The issues of application in professional communication of information technologies, in particular computer technology, hypertext technology, multi- and hypermedia technologies, information-retrieval systems; about computer communications in off-line and on-line modes; difficulties, negative consequences of computer technologies introduction; distance learning, its essence, models and forms of interaction. It is entitled "*Computer Technology as a Professional Communication Tool*".

Focusing substantially on the activities of third-level (educational) 015 Professionals (specializations) on softskills development, we came to the conclusion that it was necessary to change the formats and technologies of interaction between university lecturers and post graduate students. The traditional classes are no longer of interest to a new generation of post graduate students. On the basis of the survey of post graduate students of Alfred Nobel University, Dnipro (45 persons) it was found out that they are fascinated by case studies (the situation analysis) (78.2%), micro group projects (67.2%), simulation game (simulation) (60.9%), gamification (55.8%), e-learning, m-learning (48.1%), facilitation technologies (55.8%).

The members of academic staff of the Department of Pedagogy and Psychology of Alfred Nobel University, Dnipro (12 persons) were asked to choose among the different teaching methods which they applied:
– "launching" of internal mechanisms of self-development, self-realization and self-improvement of personality in varieties of activity;
– changing stereotypes and realizing the need for personal change;
– reflection of one's own achievements in personal growth.

As a result, one can observe that the post graduate students have gained their own experience, which is the most powerful impetus for creative mastery of future professional activity. The interactive teaching methods used in the course are presented in the manual [21]. We would like to focus on some of them.

Through the implementation of *dialogical and discussion* methods of teaching (dialogue, debate, discussion, debriefing, brainstorming, lecture-dialogue, problematic lecture, lecture-analysis of a specific situation, lecture-consultation), each graduate student was given the opportunity to express his thoughts, to make up their own judgements in relation to the issues discussed; the educational process was transformed into mutual learning (collective and micro-group), cooperation of post graduate students and university lecturers as equal entities of learning on the basis of mutual understanding (forming a common field of partners, which unites their interests, allows to jointly consider problems in a specific situation), coordination (coherence, unity of action and effort), agreement (formation of common purpose, intentions, motivation of communication). This delivers significantly better results due to the use the techniques of *synectics* (synectics – "unification of heterogeneous elements"), "*Take a position*" (provides a demonstration of different thoughts on the topic under study), "*Fire on the lead*" (graduate students prepare problematic questions, the speaker gives them answers with evidence). Post graduate students were able to realize their individual, personal position, in principle, respect the opinion of another and, if necessary, to adjust their own position, which required the university lecturer to treat the post graduate student as a unique personality.

The implementation of *game teaching methods* involved modeling of professional situations, joint analysis and problem solving, which contributed to the development of their own values and cooperation. The attention was focused on: *business* (a form of modeling professional reality, imitation of specific professionally directed and conflict situations), *storytelling-role* (improvised play of a given situation), *situational* (presentation of a chosen to play a fragment of a professionally directed situation and presentation of post graduate students' professional or positional responsibilities to postgraduate students), *social and psychological* (games with imaginary situations, which involve the reproduction of difficult situations that were a real problem for a certain number of people; which stipulate obligatory comparison of the proposed solutions to the problem with actual decisions) games. As a result, each post graduate student could reflect on their own experience, individual characteristics, including those that serve as a source of barriers to communication and to discover alternatives to behavior in the proposed situations "try on" them and put them into practice. Among the techniques for implementing the aforementioned game methods were singled out the following: *reflection, duplication, advisory group, parallel, role rotation, support, interlocutor chair, mute assistant, replay and placement*.

The choice of *facilitation* methods (*World Cafe, Open Space, Kurt Levin Force Field Analysis, Anti-Brainstorming, Dynamic Facilitation*) is due to the fact that they have significant capabilities: increase the effectiveness of group decision making (situation analysis); identify options, alternative solutions to the problem; choose the best option), create and maintain a climate in the group (increase the level of involvement in the discussion process; stimulate the initiative; encourage personal responsibility for the process and result), provide the exchange of experience among participants and promote the personal development of the participants.

Superior results are seen for one more pedagogical technique. An important role was assigned to the work on the project, which makes it possible to use not only the intelligence, experience and consciousness of the post graduate students during academic process, but also their feelings, emotions, volitional qualities, promotes "*immersion*" in the teaching material, determination of the personality of their emotional and value attitude to it, improving the efficiency of assimilation, which gives the feeling of success. Project work involves mechanisms for storage and reproduction of information; transferring information to others; application of knowledge in variational situations; understanding cause-sequence relations, the interrelation between parts and the whole; presenting arguments and evidence, regrouping individual parts and creating a new whole, etc.

It is important to highlight the fact that post graduate students have the opportunity to acquire the skills of interaction, organization, punctuality, the ability to express their opinions clearly and concisely, to understand the essence of the problems of those with whom they communicate, creating the conditions for interaction and encouraging teamwork where it is appropriate. The effectiveness of the project method is that creative communication allows the post graduate student having realized themselves as a communicative personality and opening opportunities for accumulating communicative experience through self-realization in various communications, which are specific processes of information exchange.

In our research we would like to focus the technique of *training* (within the framework of the course "Self-Management and Self-Development of a University Trainer"), which can contribute to the development of soft skills to future university trainers.

Training as a form of study can reveal the full potential of the graduate students: the level and extent of their competence (professional, social, emotional, intellectual, leadership), independence, decision-making ability, interaction and more. Trainings has become an effective form of pedagogy of cooperation and developmental training:

1) new approaches (cooperation, openness, activity, responsibility, leadership);
2) new knowledge (intensive acquisition, refinement, knowledge sharing, knowledge building);
3) positive values, attitudes, ideals, motivation;
4) new skills (effective communication, self-control, self-management, leadership, teamwork, ability to provide and receive assistance, situation analysis and decision-making, etc.).

4 Conclusions and the prospects for further research

Having concluded our theoretical analysis, post graduates' and academic staff survey and having compared and contrasted data obtained, the results of the study demonstrate the following: the labor market today requires a person of comprehensive development, which would ensure the competitiveness of a specialist at the beginning of his professional activity.

The findings of the study are supported by the fact that in addition to professional knowledge and skills, a university lecturer must have a system of soft or social skills called "soft skills". Developing soft skills is not a bargain, but an objective requirement of the labor market. The high school should respond to these requests.

Results provide the basis to distinguish the following groups of soft skills important for future university lecturers:

1) communication skills;
2) cognitive skills;
3) management skills;
4) strategic skills;
5) self-organization skills and
6) emotional competences.

The acquisition of basic knowledge and mastering the basic skills of soft skills of third-level (educational-scientific) students in the specialty 015 Professional Education (by specialization) was facilitated by the training course "Professional-Pedagogical Communication", the content of which is organically combined with the traditional disciplines of psychological and pedagogical cycle. This course contains *four training modules* with the detailed content: "Fundamentals of Professional Communication", "Verbal Professional Communication", "Non-verbal Professional Communication", "Computer Technology as a Professional Communication Tool".

Through the implementation of dialogical and discussion methods of teaching using the techniques of synectics, game teaching methods involving modeling of professional situations, joint analysis and problem solving, our results can demonstrate the ability of post graduate students to reflect on their own experience and individual characteristics necessary to develop soft skills.

The findings provide a potential mechanism of the following techniques to implement the aforementioned game methods selected for training: reflection, duplication, advisory group, parallel, role rotation, support, interlocutor chair, mute assistant, replay and placement.

To develop soft skills in post graduate students it is necessary to choose some facilitation methods (World Cafe, Open Space, Kurt Levin Force Field Analysis, Anti-Brainstorming, Dynamic Facilitation).

Superior results could be seen for one more pedagogical technique, project work, which involves the "immersion" in the teaching material, determination of the personality of their emotional and value attitude to it and improving the efficiency of assimilation.

The implementation of interactive teaching methods was aimed at the development of soft skills in post graduate students. But in order to successfully develop the ability to apply these skills in professional pedagogical situations, post graduate students must practice them throughout their studies, as these skills tend to reverse. Today, soft skills not only add to hard skills and create new opportunities, they also contribute to the development and formation of professionalism. This is an issue for future research to explore in other fields of pedagogical studies.

References

1. O. Abashkina, *Soft skills: klyuch k karere* (Soft skills: The key to a career), http://elib.bsu.by/bitstream/123456789/165740/1/%D0%90%D1%85%D1%80%D0%B0%D0%BC%D0%BE%D0%B2%D0%B8%D1%87_%D0%9E%D0%9216_%D0%A71-018-022.pdf. Accessed 09 Nov 2019
2. M. Achramovich, *Soft skills kak tsennyiy komponent sovremennogo professionalnogo obrazovaniya* (Soft skills as valuable component of modern professional education*)*, http://elib.bsu.by/handle/123456789/165740. Accessed 11 Nov 2019
3. K. Ananiadou, M. Claro, 21st Century Skills and Competences for New Millennium Learners in OECD Countries, OECD Education Working Papers **41** (2009)
4. T. Ananieva, Desiat kompetentcii, kotorye budut vostrebovany v 2020 godu (Ten competencies in demand in 2020) (2016), http://tananyeva.com/desyat-kompetentsij-kotorye-budut-vostrebovany-v-2020-godu/. Accessed 12 Sept 2019
5. S. Batsunov, I. Derecha, I. Kungurova, Ye. Slizkova, Ped. Nauki **4** (2018)
6. J. Chen, A. Gilmore, N. Wilson at al., Cogn. and Behav. Pract., **24 (2)**, 226–244 (2017)
7. V. Davydova, Clushat, govorit i dogovarivatsia: chto takoe soft skills i kak ikh razvivat (Listen, speak and come to agreement: what soft skills mean and how to develop them) (2015), http://www.theoryandpractice.ru/posts/11719-soft-skills. Accessed 12 Sept 2019
8. ESCO Handbook: European Skills, Competences, Qualifications and Occupations (EC Directorate E, 2017). doi:10.2767/934956
9. Ye. Gaiduchenko, A. Marushev, Emotcionalnyi intellekt (Emotional intellect) (2019), http://l-a-b-a.com/lecture/show/99. Accessed 05 Nov 2019
10. Ye. Ilyin, *Motivation and Motives* (SPb, Piter, 2002), pp. 46–64
11. A. Ivonina, O. Chulanova, Yu. Davletshyna, Naukovedeniye **9** (1) (2017)
12. A. Karpov, *Psychological analysis of decision-making processes in activities* (Yaroslavl, Yaroslavl politetechnic insitute, 1985), p. 80
13. K. Koval, Visnyk of Vinnitsa Polytec. Instit. 2 (2015)
14. V. Kunitsyina, N. Kazarinova, V. Pogolsha, *Mezhlichnostnoe obshchenie* (Interpersonal communication). (Piter, Saint Petersburg, 2001)
15. N. Fetiskin, V. Kozlov, G. Manuylov, Modified Questionnaire of Self-Actualisation Diagnostic by A. Lazukin in the adaptation of N. Kalin *in Social and pedagogical diagnostic of personality and small groups development* (Moscow, 2002), pp. 426–433
16. Iu.Iu. Ivaniuk, What does soft skills mean and why are they so important in carrier? (2019), https://dnevnyk-uspeha.com/rabota-i-karera/chto-takoe-soft-skills-i-pochemu-oni-tak-vazhnyi-dlya-kareryi.html. Accessed 23 Вус 2019
17. O. Sosnitskaya, Soft Skills: "Miagkie navyki" tverdogo kharaktera (Soft Skills: "soft skills" of solid character) (2009), https://www.dw.com/ru/soft-skills-%D0%BC%D1%8F%D0%B3%D0%BA%D0%B8%D0%B5-%D0%BD%D0%B0%D0%B2%D1%8B%D0%BA%D0%B8-%D1%82%D0%B2%D0%B5%D1%80%D0%B4%D0%BE%D0%B3%D0%BE-%D1%85%D0%B0%D1%80%D0%B0%D0%BA%D1%82%D0%B5%D1%80%D0%B0/a-4837922. Accessed 02 Dec 2019
18. V. Shipilov, Perechen navykov soft-skills i sposoby ikh razvitiia (Soft skills list and the ways of their development) (2016), https://www.cfin.ru/management/people/dev_val/soft-skills.shtml. Accessed 12 Nov 2019
19. N. Vesnina, Bulletin of Alfred Nobel University. Philology **1**(17) (2019)
20. N. Volkova, *Professional and pedagogical communication* (Academiia, Kyiv, 2006), p. 256
21. N. Volkova, *Interactive technologies of teaching in higher school* (Alfred Nobel University, Dnipro, 2018), p.360
22. Ye. Yamburg, Russian Journal of Education, **15** (2013)
23. A. Zinchenko, M. Saprykina, *Skills for Ukraine 2030: business view* (USTON, Kyiv, 2016), p. 36
24. N. Zhadko, M. Churkina, Obuchenie hard skills i soft skills - v chem raznitca? (Teaching hard skills and soft skills - what is the difference?) (2006), https://hr-portal.ru/article/obuchenie-hard-skills-i-soft-skills-v-chyom-raznica. Accessed 12 Oct 2019

Developing critical thinking skills by students through active methods of the study of literature

Svitlana Kovpik[*]

Kryvyi Rih State Pedagogical University, The Department of the Ukrainian and World Literatures, Kryvyi Rih, 50086, Ukraine

Abstract. The paper deals with ways of developing critical thinking by means of literary works. It represents a number of analytical, searching, and cognitive tools aimed at activating thinking operations in the process of reading and interpreting fiction texts. It also focuses on the active methods of studying literary words (event analysis, fishbone structures, socionist analysis) which develop student critical thinking. The results of the research have proved that modern openness of literary education allows diversifying the ways and tools of studying linguistic and fiction "fabric" of the literary text.

1 Introduction. Analysis of recent research

Nowadays the methods of teaching the Ukrainian literature represent original and genuinely new ways of studying fiction works. Currently, philology teachers remark that students merely lose interest to reading literary works. Consequently such a tendency greatly affects the quality of their literary education. Therefore, the primary task of a teacher is to encourage modern students to read fiction texts. It is noted that at present students are inclined to interpret rather than analyze the literary work at the lesson. That is why there is a constant need to develop student analytical thinking skills while studying works of fiction.

Taking into account the fact that cognitive study of a literary work is carried out through its systematic, complex analysis (investigation) as well as motivated and purposeful interpretation (explanation), it should be mentioned that these processes are interrelated, but should not be observed as identical ones. They share a similar survey subject, it is a piece of fiction text, in particular. However, the subjects of study in these processes are significantly different. The subject of analysis focuses on the work of fiction which is considered as a whole for the sole purpose, while the subject of interpretation comprises the content of the literary work including all its meanings and senses.

It is well known that the purpose of analyzing the fiction text is the process of scientific cognition, splitting and evaluating the text "fabric", whereas the purpose of interpretation is interpreting and rendering the text content for better recipient (addressee) understanding. Hence, the task of an analyst is to study the literary work as an art, aesthetic, social system, through all consciousness levels in terms of all possible contexts and intertexts. The interpreter task is to help the recipient (addressee) understand the author's intention, content and meaning of the text. The analysis is carried out by means of literary criticism categories and regardless of the availability of judgments and conclusions. The interpreter takes into account the purpose of particular interpreting, the age and the intellectual level of the addressee, etc. These processes are different in the way of priority ranking and cognition scope: the analysis precedes the interpretation; the interpretation of the literary work is impossible without its previous analysis, but text cognition is not limited to its interpretation. Moreover, it should be noted that there is a difference at the level of their principles: the analysis of a fiction work is carried out on exact scientific understanding, while the interpretation is very often implemented on the basis of artistic taste and extra-literary factors. There are also some variations in the methods, ways, types and levels of analytical and interpretation procedures.

Reviews of recent research papers and surveys over the last decade clearly show that even guidance counselors intensify their efforts to solve an important problem i.e. "the adaptation of the most interesting and efficient forms for traditional and non-traditional ways of fiction work analysis at school" [1].

The World Economic Forum has stated that critical thinking is among the top three competencies for a young person in 2020. The above mentioned list also includes "problem solving skills" and "ability to communicate", which are formed by means of special methods at the literature lessons.

The human thinking process requires its constant training and improvement during his/her conscious life. Every person has a unique ability of thinking critically. Critical thinking is an exceptional process that combines a number of important thinking operations as following: analysis, synthesis, justification of evaluation, information summarizing. The development of critical thinking provides the improvement of person's skills to perceive the situation in the global and broad-scale way,

* Corresponding author: kovpiks@ukr.net

penetrating into the causes or alternatives of various phenomena, processes, etc. The uniqueness of critical thinking lies in the fact that a person develops specific skills to make balanced decisions. Scientists have already confirmed that this type of thinking is fundamentally different from creative, and intuitive types.

The development of critical thinking is not only an educational trend but also a social tendency. O. Pometun reasonably points out that critical thinking is "a human quality, the trait of his / her personality which is manifested not only due to the perception of information. This is a trait that manifests itself in the daily life, as we make many choices every day, and the more meaningful these choices are, the more developed critical thinking is" [1].

O. Slonovska in her research "Critical thinking – fashion chips or effective tools in training and knowledge obtaining by pupils at the humanity lessons?" ponders over the idea of using the term "critical thinking" in school practice "as the synonym of a well-known clear phrase "searching of literary work gist", based on a careful analysis of a particular literary work, character, clash, problem discussion through debating or chain way logical inferences. Obviously such systematic practice gives opportunities to achieve appropriate results and, above all, provide not intrusive and forced, but rather active, interesting and successful learning process" [2]. Views of O. Slonovska on functions of critical thinking are based on the need to develop the ability to analyze texts of fiction works. While analyzing Ukrainian literature texts at the lesson students learn to make logical conclusions, to give unbiased evaluation of character actions and behaviour. We share the opinion of O. Slonovska that critical thinking should be developed in every possible way among senior students, as it provides logical thinking development and psychology knowledge accumulation. Nowadays, critical thinking is an important skill of a modern person.

The purpose of the article is to reveal functional peculiarities of new active interdisciplinary methods of fiction studying, which are aimed at developing critical thinking among students.

2 Presentation of the main research material

It should be mentioned that the work of fiction is a result of unique creative thinking; therefore, it is very important for students at the literature lessons to analyze thoroughly and critically the actions, behaviour and the way of thinking of characters, giving them proper evaluation. The critical thinking complex includes a number of important mental processes: analyzing, comparing, synthesizing, hypothesizing, making assumptions, finding alternatives, making decisions. Critical thinking functions are to develop student's curiosity, establish cause-and-effect relationships between certain phenomena and processes, and form personal opinions. It also comprises the absence of mistake making fears, student readiness to dispute generally accepted truth, and logically comprehend important components of literary works.

Focusing on the development of student critical thinking, the teacher should consider the following steps in the structure of the literature lesson:

- the problem perception (this stage switches students on the problem, excites and provokes them to think, appealing to their previous knowledge. In terms of this phase, as well as during the whole lesson, it is important that the teacher speaks as little as possible and enables students to express their thoughts);
- the problem comprehension (this stage shows that students are entirely involved in the process of problem solving. It should be mentioned that the method of marking reading or INSERT means is of great use at this stage. This method allows students personally to determine already known information, to distinguish fundamentally new facts, to make notes indicating another way of thinking, to search for a better solution. It is important that the teacher puts series of free-answer questions rather than closed ones connected with the text. The teacher should not be afraid of free-answer questions, as they provide the development of student critical thinking);
- assessment or reflection (this stage is based on the consolidation of received information and new knowledge, restructuring ideas according to the new information, exchanging opinions with classmates about ways of problem solving).

As an instance, let us present a model lesson which applies the technology of critical thinking. It relies directly on the analysis of fiction works. At the beginning of the lesson, the teacher addresses to students, using the following motto of the critical thinking lesson: "You ask – you think!". This motto guides students in the way that putting the question stimulates their thinking process.

At the stage of comprehending the problem, firstly, the teacher invites students to identify the information that helps to reveal the meaning of the text. Among them there are the following: information about the history of work writing, the reasons for the involvement of cultural codes into the structure of the literary text.

Using the method of marking reading or INSERT suggests students to write down priori information, and separately make notes about the new facts. Afterwards they start to consider the ways of solving the conflict presented by the writer, and actively discuss it. Then students mark the text in such a way that the marks correspond to the following answers: "I knew it", "I knew but forgot", "New for me", "I thought different", "I want to know more".

Marking of the text can be submitted in the way of a table. The sample text marking is shown below (table 1).

Table 1. Marking of the text.

I knew it	I knew but forgot	New for me	I thought different	I want to know more	My own ideas
1	2	3	4	5	6

At the assessment and reflection stage, students express their thoughts based on summarizing information, which they have received, consolidating new knowledge, demonstrating how their knowledge has changed as new information has been acquired. Students share ideas about the ways of problem solving. Here it is necessary to adjust students' opinions of solving the problem to the way chosen by the writer. That is why nowadays it is very important to use critical thinking methods while studying the works of the Ukrainian literature, as they provide a brand new way of reading classics.

At this stage, fishbone technology can be applied. It allows students to answer a number of questions: "Why did the writer create the conflict in such a way?"; "Why did he focus the reader's attention on this or that event?"; "What is the essence of the writer's strategy?" Fishbone technology (fish skeleton) provides a fish-tail pattern. The determined problem is considered to be the head of the fish, the upper bones are the causes; the lower bones represent illustrative examples (facts and arguments). The tail gives the answer to the question. The author of this technology is Japanese professor Kauro Ishikawa, who has developed this technology for structural analysis of cause-and-effect relationships [3, 4].

Students write down this scheme using keywords or short phrases. While making the scheme students gradually find the answer to the question, which turns to be the tail of the fish. This scheme promotes better information memorizing, improves search and cognitive abilities, and develops logics and conclusion justifying skills.

Fishbone technology encourages students for groups work, develops critical thinking, visualizes the links between the cause and the effect, and ranks the facts in order of their importance.

Hence, we believe that the technology of critical thinking today should be actively implemented into the senior school practice of teaching the Ukrainian literature, because this technology has a number of important functions that contribute to the development of important thinking resources of modern youth:

- the independence of critical thinking makes to obtain ready knowledge not from textbooks, but to spend time on its acquiring from other sources;
- the process of critical thinking considers information as the starting point and not the destination one;
- critical thinking allows setting problems on the basis of student life experience;
- it carries out self-esteem of student activity;
- critical thinking develops skills of appropriate decision making;
- it helps to express thoughts in logical way and give convincing arguments.

It should be noted that the active methods of event analysis also deserve certain attention. These methods manifest a peculiar way of analytical and progressive studying the dynamics and the development of both separate events and sequence plot twists. This technology appeared in the 1960s, but today it has gained popularity mainly in political studies, less in conflictology and sociology. This method is primarily applied to investigate the development of purely military-political and strong socio-political confrontations, mass disruptions and it monitors the development of international events during certain periods. The following analysis is carried out according to a fairly established scheme: the subject-initiator of the event (who), the determinant of its development or the development of the event series as a complete "plot" of events.

Applying this technology in terms of literary studies is relevant, especially while studying events and event series in drama historical works.

At present in the Ukrainian literary criticism there is research dedicated to understanding the essence of specific events. The monograph "An Event in the Artistic System of Contemporary Ukrainian Prose: Morphology, Semiotics, Reception" (2010) by T. Grebeniuk confirms this statement. The researcher emphasizes in one of her study ("Event category in the receptive-communicative paradigm of the analysis of a literary work") that the event makes a basic, starting concept in distinguishing the literary genres, with the precise defining various literary terms, such as: "plot", "motive", "genre", "chronotope", "narrative", etc. She suggests that "because of its apparent empiricism and routine, this category is often perceived as extra-scientific, not worthy of professional analysis, and only in single theoretical works there are some attempts to represent it as a specific element of the literary work system" [5].

That means that T. Grebeniuk tries to comprehend the dynamics of events through the literary studies discourse, which is caused by paradigm shifts of scientific thinking, although the researcher does not specifically focus on the analysis of the event series in the literary work. Yu. Lotman gave his brief definition of a phenomenon called "event": "The text event represents the character who transfers the border of the semantic field" [6]. At the same time, the scientist emphasizes that the event is exceptional staying at the background of other components of macropoetics in the literary work, and it requires special study.

W. Schmidt interprets the event as the sequence of initial situations changing one another, whether external or internal and related to the character. In addition, the researcher defines the category "literary work eventuality" as a gradational peculiarity of events. The scientist identifies several conditions that should correspond to "narrative eventuality", which is marked by "factuality" (or reality), that is the obligatory connection between the events and their results [7].

"Narrative eventuality" has the following five major gradation relevant changes: eventuality increases if the change that has occurred is essential for this world; "unpredictability of an event", which indicates the increase of occasional properties; "consecutivity" shows that eventuality increases if changes have a significant impact on the process of thinking and consciousness of the character on the way of his subsequent actions and behaviour; "irreversibility" represents that the changes turn out to be definitely irreversible; "uniqueness" emphasizes that the changes can not develop otherwise.

Modern literary studies manifest several approaches to event and eventuality of literature activity comprehesion. Firstly, they are considered as "unfolding the plot in a fiction work that influences on the development of a story line, clashes, conflicts, the behavior of characters. It can be an action, an act, a chance. At the same time it is interpreted as an existent and one of the fundamental components in the story. Secondly, "plot event" is determined as the transfer of the character, "internal or external", that is crossing "the border, separating parts or spheres of the described space and the moment of literary time associated with the probable implementation of the planned events or vice versa its failure" [8].

In addition, N. Tamarchenko defines the concept "plot situation" as inseparable manifestation of time and space conditions. Moreover, it is stated that the transition from one situation to another is possible only in terms of replacing or transforming the character's conditions, changing his personal activity or shifting "activity" and "effectiveness" of external circumstances.

In drama works the event-creating situation helps to determine the relation between the counteracting forces, characterize the coincidence of conditions and circumstances, relationships and behavior of the characters. It also emphasizes the moment of conflict resolution.

It is supposed that the event situation of drama works can be represented by means of a number of different factors or a single replica of a character, who does not seem to generate an event, but noticeably prepare the specified situation, which can cause the event or even transform into it. If the plot of a play is the entire system of character clashes and intertwining of the story lines, then the main plot of the play is nothing but the system of events in its artistic reality. It forms the basis of its content, determines the essence and stages of a separate conflict or conflict systems. It may even make changes to its characters, etc.

It should be mentioned that playwrights P. Cornell and N. Boileau were the first who widely introduced the concepts of "event", "event development", "event exacerbation" in drama works, considering any fact, phenomenon or process of private or social life of a character as crucial and significant. Only then such situations can be determined and called events. Till the middle of the nineteenth century drama works showed concentric and holistic plots, in which all the actions of characters were subordinated to "the unity of actions", connected by one conflict, and the eventual chains were short and predictable.

The significant "liberation" of the processes and elements of micro- and macropoetics and composing poetics from these canons provided emergence of new poetic peculiarities and combinations of a play. It caused different types of transferring and reconstructing the time and space. It gave the way to brand new concentric systems of conflicts, and consequently to the accumulation or reducing of cause and effect relationships between the actions of characters and between the events of fiction works.

Certain events and event sequences of literary works start to be filled with new content, take new forms and formats, acquire new specific features and functions: from traditional and predictable features of plot dynamics to virtual-phantasmagoric and purely mystical shocks and shifts in the consciousness of individual characters, from fights and conflicts between characters to confrontations between ethnic groups and nations. Since the events and the event sequences, in particular, create peculiar psychofields, they not only allow the characters to reveal their nature to the recipient, but also determine the extent of their self-expression, requiring them to manifest their attitude to both certain events and event sequences. Moreover they can even become "an actant" (fr. actant – action move) events, that is why such play characters are typically "the engines" of event unfolding in literary works.

It should be also mentioned that the amount of events does not always depend on a number of participants, as it is often on contrary. The power and influence of the character can ultimately determine the following: the event in the life of one person may become the event of the whole nation and society, or vice versa, the event of the universal scale can change the life of one person. It is most likely that everything is settled due to the interdependence and interconnection of characters, people and events, the degree of intertwining of their goals and consequences.

The levels of conscious struggle and the nature of events and their participants, motives and conditions of events are not always the same. In this case much and / or even everything depends on the level of character consciousness, the level of the problem according to which the events or event sequences are completed, and the essence of social or natural events, etc.

Furthermore, literary critics divide the events of works into "setting" (events that give rise to fights, conflicts and confrontations), "major" (events during which these conflicts or confrontations are solved). However, not every single component of an event sequences is an equal and valuable part of its "through-going action", which can even begin with an insignificant event and be interrupted by several events. The ultimate event takes an extraordinary shape when it reveals resolving the crucial conflict in the end of the work.

In addition, acting as a plot, climax, and resolving in the play, events change not only the general nature of the "event development", but also the whole struggle of the characters, the general line of their behavior, being "catalyst" or "inhibitor" of the overall plot development. Probably this is the reason why some researchers consider event sequences to be the coincidence of circumstances that escalate into a certain situation, etc. Only when an event or event sequences become obsolete both in time and space, a significant retardation of events starts or the events of the work completely come to end.

As a result the event sequences of a literary work simultaneously make up its full structure, its main core (plot), and such a "turning point" of its development which may not be studied by the event analysis, but may be the main manifestation of genre-defining pathos, the ideological and aesthetic essence of the play.

It should be emphasized that certain events and event sequences in a drama play can also have double spatial and ambiguous temporal characteristics.

The main tasks of the event analysis method in literature lessons are the following:
- to teach students to explore the essence of each event and event sequences of the whole literary work, paying attention to their functions and the conditions of their development;
- to form the skills of analyzing all realistic and virtual events and characteristics that are in fact the main causes of character thinking and struggle, their emotions, moods, states and feelings, actions, and behavior, fight and lifestyle;
- to identify the main issues of the work that break the settled way of life of the character, events and event sequences, which attract to the work and rise the special interest of the recipients.

The series of analytical actions in accordance with the method of event analysis of a literary work include such steps:
1) determining the initiator of the event (character, forces of nature, phenomena);
2) identifying the actant character of the event (that is the one who moves, starts, aggravates the problem);
3) measuring the event magnitude in the work by a number of participants (micro-event (two participants) and macro-event (over two participants));
4) defying the influence of the event (an event from the life of one person, an event from the life of a nation, society);
5) defining the type of event: event-gesture, event-replica, event-facial expression, event-action, event-word, event-case, event-sit.

Lately, most emoji researchers are convinced that these characters have great hypertext potential. For instance, there have been attempts to translate the text of the novel "Moby Dick" by Herman Melville's using the appropriate emoji, called "Emoji Dick".

Today, emoji is not only a linguistic tool, but also a psycho-physiognomic means. By modeling this or that facial expression of an emoji, the author shows it the emotion that matches his / her level of sensuality. The emoji face performs an informative function; it tells the interlocutor the emoji author's response to the text which greatly differentiates written communication.

Therefore, by making the sense of poetry through a self-designed emoji body in Emoji-Maker, students have the opportunity to expand the range of emotions, moods and experiences that arise in the process of reading poetry and especially while its interpreting. Emoji face expression is a students' creative approach to understanding the essence of prose and poetry. After all, all human feelings are expressed not in words but facial expressions. So we offer emoji modeling technology that reproduces the artistic meaning of poetry.

Present day psychology, sociology and literary critics research shows the interest to the results of socionic achievements. Socionics is a new branch of knowledge that, since about 1970-ies predicts interpersonal relationships, helps to model relationships between representatives of different social groups, social institutions, and as a result, it allows the most effective implementation of personality socialization in the modern world.

Nowadays, the socionists have identified 16 types of personality, which they call sociotypes. In socionics, the sociotype is formed on the basis of knowledge about the psychotype and is connected with the social level of a person, the type and kind of his / her occupation, profession, and the level of intellectual development. Recently, pedagogy has become a new area of socionics in terms of team compatibility studying, training, family counseling etc.

At present, the main task of socialization at school is the process of thorough transmitting all the necessary knowledge and skills to a student for his / her successful entry into the society, as later students have to fulfill certain social roles. The socialization of modern schoolchildren is a key to their rapid adaptation in the society.

Sociologists distinguish the main mechanisms of socialization such as imitation, attachment to one of the family members, identification, competition awareness, etc. The difficulties of socialization are related to three main circumstances: the difference between the high levels of demands (the desire to become a hero, be famous, stand out) and a low social status, which is associated with certain age characteristics, as well as the divergence of an old style of paternity, which is oriented to the idea that for a mother her son or daughter always remains a child; new potential abilities of teenagers, made up by their psychophysiological adulthood; contradictions between the increased orientation towards their independence and dependence, which are enforced by the opinions and behavior of age-mates.

Socionists define several sociotypes. The first one is an intuitive-logical extrovert (inventors, innovators, producers of new ideas, thoughts etc.). Foreign fiction works depict this sociotype in the characters of Jacques Paganel (J. Verne "Children of Captain Grant"), Alice (L. Carroll "Alice in Wonderland"). Such people are marked by some awkwardness, absent-mindedness, but they are hardworking and good-natured, sincere and open, simple and creative, they are true friends who you can always rely on.

The next sensory-ethical introvert type includes mediators, peace-makers, and executors of someone's orders. Foreign fiction characters of this type are a soldier Schweik (J. Hasek), Sancho Panza and others. In the Ukrainian literature, it is illustrated by Nazar Stodolia's cobrother, Mykola (I. Kotliarevskyi "Natalka Poltavka"). Such characters stand out in society due to their desire to reconcile and cement relationships within the community; they act as intermediaries between the two opposing parties.

The following sociotype is an ethos-sensory lifelong enthusiast, optimist, promoter, salesman, salesman, manager implemented in such characters as: Moskal-magician, Mykola (I. Kotliarevskyi "Natalka Poltavka"), Santa Panko (Ostap Vyshnia "Som") etc.

The logic-intuitive introvert is represented by a fair analyst who is slow-moving, thoughtful, and difficult to enter the society. This type of personality can be noticed in the character of Mykhailo (M. Stelmach "Geese swans fly").

The intuitive extrovert is a teacher, mentor, educator, who constantly tries to be the first and it is manifested by Uncle Sebastian (M. Stelmakh "Geese swans fly"), Maurice Gerald (M. Reid "The Headless horse rider"), Margarita (M. Bulgakov "The Master and Margarita").

The logic-sensory introvert is an inspector, controller, and supervisor. This sociotype is represented in the foreign literature by the Constant Tin Soldier (H.-Cr. Andersen "The Constant Tin Soldier"), Malvina (A. Tolstoy "The Golden Key").

The intuitively-ethical introvert is a dreamer, a direct and desperate romantic, living on the principle of "the river flowed, flows and will flow". These are the people who as much as possible disconnect themselves from problems and well shown by the character of Ivan (P. Zagrebelnyi "B-en").

The sensory ethical extrovert is a patron. Artistic images in literature are the following: Lolita (V. Nabokov "Lolita"), Puss in Boots (Sh. Perot "Puss in Boots"). An important role in the study of sociotypes is given to the knowledge of social setting.

Thus, the socionic setting of a sociotype for a certain activity reveals an individual natural inclination to a particular profession, and the knowledge of his / her psychological type makes it possible to choose a proper profession, i.e. "related work". Each professional environment, in turn, requires a person to obtain certain psychological traits.

The information model of each psychotype allows determining the relations between the characters and identifying who will get along well and who will be constantly engaged in conflict relations. It is a crucial point for adaptation and it is the basis of the theory of interpersonal relations. The knowledge of the differences and similarities of psychic functions within the information model of the psyche allows accurately determining how the relationship between different psychotypes will develop. Different squares determine different relationships between individuals. Moreover, it can be useful for family planning and just in friendships matters. It is worth mentioning that a mental function is an ability of a person with particular skills to understand the information flow aspects. It is clear that mental functions are as numerous as the aspects; each aspect of the information flow corresponds to a certain mental function.

Practical psychology, mainly due to the attempts of psychologists K. Leonhard, A. Lichko, formed the idea of personality psychotypes, expressed through accentuation of characters, and developed corresponding typologies. It was noticed that these combinations of character traits are not countless, but they can be grouped into ten categories.

Psychologists emphasized that the character may be more or less expressed. The intensity of the character manifestation is revealed in terms of "norm", "accentuation", "pathology" or "psychopathia".

At the literature lessons the following model of determining the essence of each character psychotype to define their social compatibility is presented to students: the basic function identifies a person's occupation; the creative function focuses on the fact how a person implements his / her work; the role function shows person's role in a specific situation; the pain function covers complexes and vulnerabilities; the suggestive function allows a person to be influenced by someone; the activation function reveals the skills of self-realization; the supervisor function is responsible for receiving and processing information; the demonstrative function shows a person in action.

Therefore, knowledge of typical character traits and behaviors will help students in the future to understand a particular person, to know his / her unique personality. And this contributes to the rational use of human resources, reducing the level of conflict in small social groups and in society. It also provides the high level of self-realization of each person in those areas of life that are the most inherent.

Since socialization is an awareness of individual social experience, social connections, relationships; the lessons of the Ukrainian Literature, where students study sociotypes, are an important platform for students experience development. Using this experience, the students can meet new social requirements, and get ready for the transition to new social conditions. Psychologists and sociologists argue that individual socialization is a lifelong process, and the change of certain social principles or attitudes depends on changes in ideology, values and attitudes.

For better understanding of a character behavior model, students are offered the following algorithm of mental operations: to analyze intentions and plans of an action that precedes the action of the character; to identify the essence of the character's beliefs; to analyze the reactions, emotions of a character in terms of the words, and actions of other characters; to consider the reaction of the social environment to the behavior of a character; to determine how fast and why a character changes social settings; to understand how much the change in social attitudes influences his / her behavior in the future.

Consequently, lessons of the Ukrainian Literature, using the method of socionist analysis of literary works, help to assist student integration of into the society or various types of communities (groups, social institutions, and social organizations) due to studying the elements of culture, social norms and values that form socially significant personality traits. Lessons of the Ukrainian Literature make the process of socializing easy for students, and teach them to communicate with representatives of different social groups.

Taking into account that socialization is a long and multiaction process, it is worthwhile to adjust students to the fact that modern society is frequently evolving, and its structure is changing. At the same time, a person's life, age, attitudes, preferences, habits, rules of behavior, statuses and roles are constantly changing the life as well. By means of socialization, people fulfill their needs, opportunities and talents; establish relationships with other society members, groups, social institutions, organizations, which gives them more confidence.

3 Conclusions

Thus, the active methods of studying literary works proposed in the paper are clearly oriented to the development of personal critical thinking. Modern openness of literary education involves into the process of studying and analyzing the texts of fiction various psychology, conflictology, sociology tools, which contribute to the diversity of active working forms at the Literature lessons, and increase the interest to literature study.

We believe that active methods and technologies will be effective if there is their systematic usage. In our opinion, the following methods and techniques are active and productive: event analysis, INSERT reception, socionics character analysis of fiction works.

The diversification of using active methods at the literature lessons is a promising sphere for further scientific research, which will allow developing the system of modern methods of complex literary work studying in the future.

References

1. O. Pometun, *Osnovy krytychnoho myslennia* (Fundamentals of Critical Thinking). (Bohdan, Ternopil, 2010)
2. O. Slonovska, Horizons. **2**, 29-30 (2013)
3. K. Ishikawa, The Japanese Way **13**, 44–45 (1985)
4. K. Ishikawa, *Guide to Quality Control* (Asian Productivity Organization, Tokyo, 1976)
5. T. Hrebeniuk, *Podiia v khudozhnii systemi suchasnoi ukrainskoi prozy: morfolohiia, semiotyka, retseptsiia* (Event in the artistic system of contemporary Ukrainian prose: morphology, semiotics, reception). (Prosvita, Zaporizhia, 2010)
6. Yu. Lotman, *Structura khudozhestvennogo teksta* (Structure of the artistic text). (Iskusstvo, Moskva, 1970)
7. V. Schmid, *Narratologiya* (Narratology). (Slavyanskie kulturyi, Moskva, 2003)
8. Yu. Kovaliv, *Literaturoznavcha entsyklopediia* (Literary Encyclopedia). (Akademiia, Kyiv, 2007)

The research of the effect of different question forms and their combination variability on the results of testing

Antonina Hryvko*, and *Yurii* Zhuk

Monitoring and Assessment of the Education Quality Department, Institute of Pedagogy of the National Academy of Pedagogical Sciences of Ukraine, 52-D Sichovykh Striltsiv Str., Kyiv, 04053, Ukraine

Abstract. The purpose of the research was to identify and study the features of the effect of four questions forms and their combination in polymorphic test on the output indicators of secondary school students' testing. In the study, conducted in the subject field of studying native (Ukrainian) language, took part students of 8th and 9th-grades (N=164). The tools (criterion-oriented author's tests, questionnaire) and methods of results' analysis (factor analysis, dispersion analysis, etc.) used in the research process, made it possible to prove the existence of the effect of factor "question form" on the test scores (impact power is up to 35%), and also to determine combination of question forms that is characterized by the highest diagnostic accuracy. Results of a study indicate that it is advisable to construct a polymorphic linguistic test. The combination of the multiple-choice questions with one correct answer with using the text and constructed-response questions can ensure the clarity of output indicators of testing (with the least effect of question forms on them), in case evenly balance of these questions in the test. The disadvantages of the study are pointed out and prospects for their continuation and deepening, connected with further study of the optimal empirically valid test structure, are substantiated.

1 Introduction

1.1 Problem statement

The presence of discussed in the pedagogical community shortcomings of education quality evaluation through the use of test technologies on the one hand, and the use of test technologies to evaluate educational achievements of education seekers on the other, including for the further decision-making on the test subjects (high-stakes tests), updates researches related to the study of peculiarities of test technologies usage to objectively control students' educational achievements. Such studies should be aimed to determine the optimal questions forms and test structure in order to improve the method of the construction assessment tool in terms of functional and variation factors that affect the results of its application.

The purpose of the proposed article is to elucidate the study results of factors "question form" and "variability of combining different forms of questions in one test" in the context of effect study of a set of functionally-variational factors on the test results.

1.2 The rationale for the relevance of the study

Despite the bunch of studies, the problem of the effect of question forms on the test results remains relevant in the theory of testing. A number of scientific papers argue that the test with questions of close type and selected responses is easier than the open-ended test [1; 2; 3], which has led scientists to conclude that it is appropriate to combine questions of different forms in the test. In contrast to the situation about the different complexity of closed and open-ended questions and their corresponding impact on overall assessment results, some publications [4; 5; 6] highlight the results of studies that prove the equivalence of such questions in terms of statistical significance and reliability. In addition, some papers (for example, [7]) provide empirical information (correlation index between total scores from different subjects tests with construct-response (CR) questions and without them), which, according to the authors, is the evidence that conclusions for the test results will not change significantly, if you remove open-ended questions from the test or replace them with multiple-choice (MC) questions of the corresponding difficulty. M. Kastner and B. Stangl [8] explain the ambiguity of research results, comparing multiple-choice questions and open-ended questions, and the lack of consensus among scientists on the equivalence of diagnostic potential for such questions: the authors experimentally demonstrated the dependence of such studies results on the type of compared questions forms and scoring methods for evaluating tests and thereby questioned the reliability of some of them. The presence of different scientists' conflicting views on the problem of comparative assessment of the CR and MC questions importance makes it necessary to consider the few

* Corresponding author: av.hryvko@gmail.com

https://doi.org/10.1051/shsconf/20207504004

investigated aspects of this problem. In particular, the problem of measuring the effect of different questions forms of different types and different variants of combining them in the test on the output indicators of testing is not solved. It determines the relevance of studies, aimed at determining the optimal structure of the test, as well as finding out whether different question forms affect the test results of students performance who study in certain profile classes (physical, mathematical, humanities, etc.). The study presented in the proposed article is an attempt to solve the identified questions.

2 Theory

2.1 The theoretical basis of the research hypothesis

The formulation of the proposed study hypothesis (about the effect of functional and variational factors on the test results) is based on Bachman-Palmer theory of facets, developed in the subject field of linguistic testing [9], which proves the need to study different aspects of linguistic achievements evaluation of the tested students, including the coherence of results evaluation and factors, distinguished by researchers from the testing organization procedures and a set of characteristics toolkit. The authors of the theory divided these factors into 5 categories: (1) testing environment; (2) test structure (headings); (3) questions' characteristics; (4) characteristics of predicted responses; (5) relationship between questions and answers [9]. In accordance with the context of our research and on the basis of the assumptions based on the results of theoretical analysis of the scientific literature, we have distinguished functional and variation factors that affect on the results of testing: (1) uncontrollable – respondents' characteristics: personal (cognitive – knowledge and ability, affective – perceptive-emotional evaluation of the activity object – test questions) and group (in particular, class profile); (2) controlled – content and externally organized tests' constructs. The meaningful constructs of the test are, in particular, test questions content (displaying a fragment of the subject content in a test form) and the content of the test in general (optimally reduced reflection education content in the questions system [10]). Test content characteristics are always determined by the purpose of assessment (for example, thematic assessment, general determination of educational attainment levels of students at a particular stage of study; students selection by the level or orientation of their knowledge). As the subject content will always be wider than the test content [10], the issue of selecting the most relevant control material is one of the most problematic and requires the participation of experts. Selection of the test content for the final control is subordinated to the questions of determining levels of mastering basic constructs of the subject by the tested participants, which obviously determines the need for its compliance with the curriculum content and requirements of the standard. In this aspect, predictability of questions content impact on the results of their solution depends on externally organized constructs of the test.

Externally organized constructs of the test determine conditions, means and methods of ordering and expression of its content: setting characteristics (physical testing conditions, time of test questions and the whole test completion); general test characteristics (its specification) - purpose, structure, sequence and number of questions, instructions for the procedure of conducting and results evaluation; test questions format and form of predicted answers. The format means a method of initiating or carrying out certain actions that are accomplished by questions form, language design of the questions (formulation methods, volume and type of material), programming the method of processing material (determines the number of actions required to complete the questions). Forms of questions are ways of ordering its elements: closed (selective – MC question with one correct answer, with several correct answers, with the choice of the most correct answer, with the choice of true or false statement; the questions of establishing the sequence; the matching questions) and open-ended (constructed with a regulated response – add-on questions, freely-constructed – extended response questions). The response form is a method of answering defined by the form of a question. The choice of the appropriate form of the test questions is connected with the necessity to take into account the criterion of pedagogical measurement validity.

2.2 Analysis of scientific works on the topic of the research

An unreasonable choice of the test questions forms leads to a distortion of its content expression and complication of its understanding by tested students [11]. Therefore, scientists [12; 13] mostly associate the choice of test questions forms with the test complexity parameters, which are determined by psychometric test results. G. R. Hancock in his study compared MC and CR tests, designed according to Bloom's taxonomy levels. The author had experimentally demonstrated that the two formats measure similar cognitive constructs at the same taxonomic level but at different levels of complexity [14]. At the same time, researchers [15; 16; 17; 18; 19] believe that the structure of the test should contain different questions forms (polymorphic tests). Such a statement is in line with global trends in assessing academic achievement in an international context. Thus, in R. W. Lissitz et al. study [7], based on the analysis of the psycho-pedagogical literature, it is proved that, despite the possible equivalence of psychometric indicators of the test results, closed questions with the choice of the correct answer from a number of the offered ones cannot replace open questions with a constructed or freely constructed response, because of: first, in the process of solving MC and CR questions, different kinds of thinking activity are activated, and therefore its results are manifestations of different cognitive skills; secondly, the research presented experimental data, which show that some students perform better MC questions and worse – CR ones and others, on the contrary, more successfully write tests with CR questions than MC questions, so a test constructed from single format questions may be a reason of results deterioration of a particular tested category [7]. In this regard, as well as with the high level of questions diagnostic potential that involve a constructed or freely constructed form of response, in international benchmarking tests (PIRLS, PISA, TIMMS), most of the questions are presented in open forms. For the same reason, in Russia in the Unified State Examination in all subjects, since 2016, the share of expanded answer questions has gradually increased [20].

Consequently, the analysis of scientific works on the subject under consideration showed the existence of contradictions in the views of scientists regarding the possibility of using only one form of questions in the test and vice versa – the expediency of combining different forms of questions in accordance with the requirements of pedagogical measurements validity, as well as varying complexity / equivalence of MC and CR questions and the ability to adjust test results by constructing a test of a specific format. The relevance of provision concerning polymorphic tests for the final assessment of students' academic achievement confirms the need for empirical studies to substantiate the optimal structure of the test, which, accordingly, requires the study of the effect of the combination of different test questions forms on the numerical indicators of students' responses (test scores) to ensure the reliability of toolkit and pedagogical measurement.

3 Research method

The presented research is exploratory (in the study proposed a new approach and research methodology, developed by the authors of the article, also the results of their approbation are revealed). The article covers the results of the analysis of the linguistic tests, which at the same time determine the general tendencies and directions of further researches.

3.1 Study objectives

The proposed study is a component of scientific researches, *the purpose* of which is to determine the degree of functional and variational impact factors on the output test indicators (OI) – numerical values of students' answers to test questions (test scores) from the subject: Ukrainian language. *Factors under study:* since the proposed study considers OI as a manifestation of the impacts set of isolated factors ($OI = f\{F_1; F_2; F_3; F_n\}$), the following statistical hypotheses about the effect on the characteristics of the OI were tested during the analysis of the results: (1) "test questions form" factor (F_1); (2) combination of different questions forms in one test (F_2); (3) cognitive factor (F_3); (4) "profile specialization of class" factor (F_4); (5) affective factor (F_5).

3.2 Participants

The study was conducted in the subject field of Ukrainian language studying in the situation of the real educational process between students of 8th and 9th grades (N=164) of the following profiles: chemical-biological (CB), physical-mathematical (PM), economic (Econ). From the total amount of students: the number of 8th-grade students was 77, 9th-grade students – 87. The average age of the participants was M=14.04 (SD=0.65).

3.3 Instrument

In order to study the impact of the above factors, criterion-oriented tests from the subject "Ukrainian language" were developed in two variants (as expected, parallel in content and difficulty). In terms of content, the test questions offered to the lyceum students were consistent with the current Ukrainian language curriculum and provided an assessment of the most important knowledge and skills. Each variant of tests contained 4 blocks of different questions forms, aimed at testing students' mastering of the same studying constructs – knowledge and skills: MC questions, which provided the choice of one correct answer with using the text, hereinafter referred to as MC_{text} (block I), MC with the choice of one and several correct answers – $MC_{single+multiple}$ (block II), matching questions – MQ (block III), CR questions (open-ended questions that provided a student-designed response) (block IV). The tests included instruction on how to complete the questions each of the blocks. Such tests construction made it possible to compare the results of different questions forms with the same students (and not with different samples of students). To determine the consistency of the internal tests, their reliability was analyzed and based on Cronbach's alpha coefficient calculation [21] for each variant (0.641 for 1st variant and 0.673 for 2nd variant), as well as for each of the possible combinations of the described questions' blocks. This value ranges from 0.6 to 0.7, which makes it possible to conclude that the tests are suitable for usage in the research process.

To study the affective factor, a questionnaire was developed for students' attitudes toward different forms of the questions they solve, using the method of semantic differential (SD questionnaire) [22], by which, using a bipolar scaling procedure, students evaluated isolated blocks of test questions (by their forms).

As tests contained mostly the 8th-grade material, testing in the 8th-grades was carried out at the end of the school year (in one day), in the 9th-grades – at the beginning of the school year (in one day). Examinees were instructed to answer the questions of each form. The students took 45 minutes to complete the tests and the questionnaire in paper-and-pencil format.

According to M. Kastner and B. Stangl's study, MC and CR questions can be considered as interchangeable (which is an important prerequisite for comparing their results) if the Number Correct scoring rule is applied [8]. Thereafter, the proposed study obtained this rule, which allowed to take into account partial knowledge and not to reduce scores for incorrect answers.

3.4 Data Analysis

The test results were processed using statistical methods, such as variance analysis, factor analysis, etc. in MS Excel and STATISTICA software package.

According to the results of testing, compliance with the law of the normal distribution of both total and partial results is determined (for each questions block and in each sample). The characteristics of the comparative samples of the result indicators – numerical values of the test results of 8th and 9th-grades students – coincide with the level of 0.05 (according to the Wilcoxon-Mann-Whitney criterion (empirical value 0.8896, critical 1.96). Comparison of test results by 8th-grade students (N_1=77; M_1=7.32; SD_1=16.25) and 9th-grade students (N_2=87; M_2=7.80; SD_2=12.76) showed weak statistical power (0.056); a measure of effect d=0.12665 (that is weak) [23]. To check the statistical hypothesis of the equality of two samples, the Student's T-test criterion was applied, which showed that t (1,96) =.827. Accordingly, obtained statistics indicators confirmed the correctness of the hypothesis on the equality of the average general totality, which allowed to combine test scores, done by students of 8th and 9th-grades into a single sample (N=164; M=14.04; SD=0.65).

According to the results of the single sample, a multivariate dispersion analysis of the studied factors influence on them was carried out. The accuracy of calculations in each case exceeded 99% ($p<0.01$). The power of impact factors was determined by the method of J. Snedecor [24].

According to the results of processing verbal-numerical assessments on the SD questionnaire (analysis of averages, factor analysis), were identified dominant features (markers), expressing the generalized perceptual-emotional students' attitude to the proposed questions forms, and the profound factors which characterize the perception of different questions by the respondents. The results of the survey were compared with the OI of testing (test scores) in qualitative terms, given the different measurement systems and tools [25].

4 Main results of the research

The form of the test questions in the study was considered as an impact factor the output test indicators. The results of the variance analysis confirmed the hypothesis about the presence of this factor impact, whose power is up to 35%.

Comparative analysis of the test results of 8th and 9th-grades students on 1st and 2nd variants showed that the impact power of considered factor on OI varies depending on the texture of the test – for the 1st variant it is on average 45.5%, for 2nd – 27.8%. Besides, the tendency of a lesser expression of the impact power of the questions forms on OI in the second variant is observed in both 8th and 9th-grade students. Comparison of test results of students' groups by the profile of classes (CM, PM, Econ.) and educational level (8th and 9th-grades) revealed tendentious proximity of impact power indicators on OI factor "test questions form". This leads to the conclusion about the dependence of impact of profile studying on the test results from the subject field of assessment – in the subject field of Ukrainian language, indicators of the impact power of this factor on the results of different samples (classes of different profiles and educational levels) are tendency similar, which is explained by the presence of constant activity support in mastering and using the state language in the process of studying and everyday life.

At the same time, the analysis revealed that the impact power of the factor under consideration changes according to the variations of combining different questions forms in one test. It is determined that the highest diagnostic reliability, which (in addition to compliance with psychometric quality criteria – reliability, validity, etc.) defined by the smallest expression of the impact power on the test results (up to 10%), characterizes test, which combines the following forms of questions: (1) MC_{text} and $MC_{single+multiple}$; (2) MC_{text} and CR; (3) $MC_{single+multiple}$ and CR questions (Table 1).

Table 1. Calculation results of the effect of test structure on tests' scores

Combinations of questions forms (variants of the test structure)	**The impact power** (of the test structure on tests' score)	**Significance level of data***	**Test reliability** (α Cronbach)
I- II	8,45%	F(2,01)=16,14	.566
III-IV	17,30%	F(2,01)=35,32	.671
II-III	35,13%	F(2,01)=89,82	.6545
II-IV	9,02%	F(2,01)=17,25	.6431
I-IV	8,79%	F(2,01)=16,8	.6849
I-II-III	30,79%	F(1,80)=73,98	.6427
I-III-IV	20,35%	F(1,79)=42,89	.6526
II-III-IV	21,75%	F(1,79)=46,58	.6434
I-II-III-IV	21,27%	F(1,67)=45,31	.6373
I-III	32,89%	F(2,01)=81,37	.661
Note: N=164; *p<.0001			

As can be seen from the table, the most reliable and internally consistent combination of questions (α Cronbach's = .6849), which at the same time has a low level of impact on the test results (Impact power = 8.79%) is MC_{text} + CR questions.

An analysis of the cognitive complexity of the parallel questions offered to students in different forms [14] revealed the presence of an operational-activity component of complexity, which is determined by the number and nature of operations that need to be done in the process of completing a question. For example, compare a questions: to fill in gaps and an explanation questions, questions with the choice of one correct answer to the question that test *knowledge* of a certain theoretical construct, and questions that involve text reading or performing certain actions to select one correct answer and test *knowledge and application* of theoretical construct). Thus, in the proposed tests, the matching questions, which by cognitive complexity are identical to the parallel questions (with the choice of one and more than one correct answers), are characterized by cyclic repetition of operations and less variability of "extra" descriptors choice (than in the questions with the choice of right answer) that increases the probability of "guess" and decreases the diagnostic accuracy of the results. This thesis is supported by the analysis of the overall test results comparison (the level of questions' completion in each case is higher than the level of other blocks parallel questions implementation) with the results of the impact analysis of combining different questions forms in the test: in each of the test variants, which reduce the diagnostic accuracy of test results (Table 1), there are matching questions that are combined with other closed questions.

5 Conclusions, discussions and prospects of further researches

Generalization of the study results allows us to conclude about the presence of impact factor "questions form" on the test results (up to 35%). The variability of its power indicators shows the probability of a change in test result indicators according to the features of the external test organization. Variation of impact power indicators on OI factor "questions form" in the first and second tests' variants while maintaining the results of different samples of the general tendency of performing the comparable questions forms proves the existence of impact factor "linguistic design of questions": parallel in content and form, questions in two variants of the used in the study test were aimed at checking the assimilation of the same element of the educational material, but by means of different language-operating material (usage of different tokens, phrases and sentences to apply theoretical knowledge).

Presented results also lead to the assumption that CR questions should be included into the Ukrainian language test, which can increase the diagnostic accuracy of the test results, provided a certain ratio of CR and MC questions number. This assumption is confirmed by the consideration of an affective factor, which was analyzed by students' verbal-numerical evaluation of the proposed questions forms (the analysis is presented in a separate article [25]). According to the students' survey results, it is found out that the same questions are perceived by students in different ways depending on its form. Comparison of the affective factor analysis results with the testing results makes it possible to conclude that the form of the question is a sensory-afferent stimulus of the operative way of activity, which influences students' interest and motivation formation to perform the proposed questions. Considering the peculiarities of criterion-oriented assessment of students' level of mastery identified in the test components of educational material (irrespective of the group indicators level with subsequent results' ranking, as in the case of regulatory-oriented testing), this conclusion is significant in terms of test structuring and proves the feasibility of combining different forms and different cognitive complexity different questions in one test, and both for the formation of positive attitudes and motivational optima in the process of their implementation by students, and for testing and refining processes of test results.

Therefore, conclusions on the results of presented study in the subject field of linguistic testing confirm scientists' opinion who have studied external features of tests in other subject areas regarding the existence of effect of questions forms on the testing results (e.g., [26]), as well as different diagnosis power of MC and CR questions ([13], [3]) and expediency of constructing polymorphic tests (e.g., [18]) as a multidimensional toolkit for assessing knowledge of students with different abilities and cognitive models of cognition.

However, the dependence of OI on the test structure which was determined in the process of the proposed scientific study makes it necessary to take it into account in further scientific researches, aimed at deepening and explaining the presented results, which are preliminary and need clarification. Thus, the test used in the experiment, as noted earlier, consisted of 4 blocks that contained the same number of different questions forms, aimed at testing students' learning of the same educational constructs, but these questions were characterized by different degrees of cognitive and operating load, which made it impossible to evaluate dichotomous test performance and, consequently, made it difficult to compare different questions forms. Thus, we consider it desirable to carry out refinement studies on a larger sample: (1) using stem-equivalent individual tests, constructed from the questions of each of four investigated forms, which will give the opportunity to increase the number of comparable questions and increase results reliability (in the present study, questions of various forms were contained in one test that significantly limited the number of compared questions in each block of the test), as well as to compare the test results obtained by following different scoring rules ([27], [8]); (2) using tests, constructed from questions forms, the combination of which in one test, according to our study, has the least effect on test results (MC_{text} + CR). Conclusions about the impact of the test structure on the test results determine the following areas of further research: determination of the most appropriate ratio of different questions forms in the structure of the polymorphic test, the study of the problem of questions language-operational material as an impact factor of their understanding and evaluating in various subject areas.

The proposed study was performed for state funds in the framework of the investigation "Methodology of state final certification in the form of external independent evaluation of high school graduates" of the department of monitoring and quality assessment of general secondary education of the Institute of Pedagogy of the National Academy of Pedagogical Sciences of Ukraine.

References

1. I.P. Karpova, *Sravnenie otkrytyh i vyborochnyh testov* (Comparison of open and sample tests). Otkryitoe obrazovanie 3, 32–38 (2010)
2. A. Orlov, T. Ponomareva, O. Chukajev, S. Pazuhina, Tekhnologii otsenivaniya rezultatov obrazovatelnogo protsessa v vuze v kontekste kompetentnostnogo podkhoda (*Technologies for*

assessing the results of the educational process in a university in the context of a competency-based approach), 2nd edn. (Direkt-Media, Moscow, Berlin, 2017).
3. Ch.A. Melovitz Vasan, D.O. DeFouw, B.K. Holland, N.S. Vasan, Analysis of testing with multiple choice versus open - ended questions: Outcome - based observations in an anatomy course. Anat Sci Educ, **11**(3), 254–261 (2018)
4. D. Marengo, R. Miceli, M. Settanni, Test unidimensionality and item format: Do mixed item formats threaten test unidimensionality? Results from a standardized math achievement test. Testing, Psychometrics, Methodology in Applied Psychology **23**(1), 25–36 (2016)
5. F. Mozaffari, S. Mohammad Alavi, A. Rezaee, Investigating the Impact of Response Format on the Performance of Grammar Tests: Selected and Constructed. Journal of Teaching Language Skills **36**(2), 103–128 (2017)
6. S. Schladitz, J. Groß Ophoff, M. Wirtz, Effects of different response formats in measuring educational research literacy. Journal for educational research online **9**(2), 137–155 (2017)
7. R. W. Lissitz, Xiaodong Hou, Sh. Slater, The Contribution of Constructed Response Items to Large Scale Assessment: Measuring and Understanding their Impact. Journal of Applied Testing Technology **13** (3) (2012), http://www.jattjournal.com/index.php/atp/article/view/48366. Accessed 21 Jul 2019.
8. M. Kastner, B. Stangl, Multiple-Choice and Constructed Response Tests: Do Test Format and Scoring Matter? Procedia – Social and Behavioral Sciences **12**, 263–273 (2011)
9. L.F. Bachman, A.S. Palmer, *Language Testing in Practice: Designing and Developing Useful Language Tests*, 1st edn (Oxford University Press, New York, 1996)
10. V. Avanesov, *Soderzhaniye testa i testovykh zadaniy* (The content of the test and test items). (2007), http://testolog.narod.ru/Theory61.html. Accessed 21 June 2019
11. V. Avanesov, *Problemy kachestva pedagogicheskikh izmereniy* (The problem of the quality of pedagogical measurements) (2008), http://testolog.narod.ru/EdMeasmt2.html#_ftn23. Accessed 6 July 2019
12. K. Djatlova, T. Mikhaleva, Issledovaniye vliyaniya raznoobraziya form testovykh zadaniy na statisticheskiye kharakteristiki testov (Investigation of the influence of the forms variety of test tasks on the statistical characteristics of the test). Voprosyi testirovaniya v obrazovanii 4, 65– 75 (2006)
13. M.E. Martinez, Cognition and the question of test item format. Educational Psychologist **34** (4), 207–218 (1999)
14. G.R. Hancock, Cognitive Complexity and the Comparability of Multiple-Choice and Constructed-Response Test Formats. The Journal of Experimental Education **62/2,** 143–157 (1994)
15. J.C. Alderson, *Assessing Reading* (Cambridge University Press, UK, Cambridge, 2000)
16. B. Clay, *Is This a Trick Question? A Short Guide to Writing Effective Test Question* (Kansas Curriculum Center, 2001).
17. T. Hudson, *Teaching Second Language Reading* (Oxford University Press, Oxford, 2007)
18. A. Kan, O. Bulut, D.C. Cormier, The Impact of Item Stem Format on the Dimensional Structure of Mathematics Assessments. Educational Assessment **24**(1), 13–32 (2019)
19. H. Wainer, D. Thissen, Combining Multiple-Choice and Constructed-Response Test Scores: Toward a Marxist Theory of Test Construction. Applied Measurement in Education **6**(2), 103–118 (1993)
20. O.A. Reshetnikova, Chto okazyvayet vliyaniye na izmeneniye kontrolnykh izmeritelnykh materialov dlya gosudarstvennoy itogovoy attestatsii? (What influences the change of control measuring materials for state final certification?). Pedagogicheskie izmerenija 2, 5–9 (2016)
21. C. Hoyt, Test reliability estimated by analysis of variance. Psychometrika **6**, 153–160 (1941)
22. Ch.E. Osgood, G. J. Suci, P. Tannenbaum, *The Measurement of Meaning* (University of Illinois Press, Champaign, 1957)
23. J. Cohen, *Statistical power analysis for the behavioral sciences*, 2nd edn. (Lawrence Erlbaum Associates, Hillsdale, 1988)
24. G.W. Snedecor, *Statistical Methods Applied to Experiments in Agriculture and Biology* (Collegiate Press, Ames, 1937)
25. A.V. Hryvko, Yu.O. Zhuk, Vykorystannia zasobiv IKT u protsesi eksperymentalnoho doslidzhennia emotyvno-otsinnoho stavlennia uchniv do riznykh form testovykh zavdan z ukrainskoi movy (Using the means of computing technologies in the process of experimental research of the students' emotive-estimate relation to different forms of test tasks in Ukrainian language). Information Technologies and Learning Tools **70**(2), 285–297 (2019). doi:10.33407/itlt.v70i2.2621
26. C. Jonick, J. Schneider, D. Boylan, The effect of accounting question response formats on student performance. Accounting Education **26**(4), 291–315 (2017)
27. E. Lesage, M. Valcke, E. Sabbe, Scoring methods for multiple choice assessment in higher education – Is it still a matter of number right scoring or negative marking? Studies in Educational Evaluation **39**, 188–193 (2013)

Secondary data analysis in educational research: opportunities for PhD students

Liubov Panchenko[1,*], and *Nataliia* Samovilova[2]

[1]National Technical University of Ukraine "Igor Sikorsky Kyiv Polytechnic Institute", 37 Peremohy Ave., Kyiv, 03056, Ukraine
[2]Luhansk Taras Shevchenko National University, 1 Gogol Sq., Starobilsk, 92703, Ukraine

Abstract. The article discusses the problem of using secondary data analysis (SDA) in educational research. The definitions of the SDA are analyzed; the statistics of journals articles with secondary data analysis in the field of sociology, social work and education is discussed; the dynamics of articles with data in the Journal of Peace Research 1988 to 2018 is conducted; the papers of Ukrainian conference "Implementation of European Standards in Ukrainian Educational Research" (2019) are analyzed. The problems of PhD student training to use secondary data analysis in their dissertation are discussed: the sources of secondary data analysis in the education field for Ukrainian PhD students are proposed, and the model of training of Ukrainian PhD students in the field of secondary data analysis is offered. This model consists of three components: theory component includes the theoretic basic of secondary data analysis; practice component contains the examples and tasks of using SDA in educational research with statistics software and Internet tools; the third component is PhD student support in the process of their thesis writing.

1 Introduction

In the modern digital globalized world, we see a large data flow from different sources and large datasets. That's why it's important to prepare future researchers for a secondary data analysis with new computer tools and technologies.

Secondary data is collected by someone other than the researcher and with another purpose. During the secondary research authors may draw data from government documents, scientific papers, statistical databases and other sources.

The relevance of this direction is indicated by a number of initiatives. For example, The Secondary Data Analysis Initiative [1], developed in 2019, aims to deliver high-quality high-impact research through utilising existing data resources created by the ESRC and other agencies in order to address some of the most pressing challenges facing society.

Secondary data analysis is a promising area in the field of educational sciences, but it is scarcely presented in PhD research in the pedagogy field in Ukraine.

1.1 Problem definition

The purpose of the article is to establish the features of the secondary data analyses in educational research and how it is presented in scientific articles of authoritative journals, conference proceeding and program courses for PhD students.

1.2 Analysis of recent research and publications

The methodology of using secondary data analysis in scientific research has received wide recognition in the global scientific community [2-9].

J. Sobal discussed the problem of teaching secondary data in the field of sociology [2]. E. Smith analyzed the pros and cons of using secondary data analysis in educational research [3-4]. T. P. Vartanian presented advantages, disadvantages, feasibility, and appropriateness of using secondary data analysis with focus on social work [5].

"Practical Methods for Secondary Data Analysis" course program for students of School of Public Health (University of Minnesota) is presented in [6]. The course emphasizes practical approaches to pre-statistical data processing and analysis with Stata statistical software on a PC with a MS Windows operating system.

T. Logan recent work about practical iterative framework for secondary data analysis in educational research deserves attention [7].

V. Sherif discussed the problem of evaluation preexisting qualitative research data for secondary analysis [8]. M. P. Johnston describes secondary data analysis for qualitative and quantitative data in the field of libraries research [9].

The paper of J. Carter and others [10] focuses on the World Bank data and presents the usage of socioeconomic secondary data to develop quantitative skills of social science students in UK university.

Analysis of scientific sources shows that in Ukraine SDA is not sufficiently used in education in general, and in the training of Pedagogy majors PhD students in particular.

* Corresponding author: lubov.felixovna@gmail.com

2 Results of the study

2.1. SDA methodology analysis

What is the definition and essence of secondary data analysis?

J. Sobal notes that any data which have been collected for "another purpose and later reanalysed may be seen as secondary data" [2, p.480]. P. Vartanian says, that "secondary data can include any data that are examined to answer a research question other than the question for which the data were initially collected" [5].

We agree with E. Smith and others, that secondary data analysis is a research methodology that has the potential to greatly impact greatly educational research [3]. We share also the opinion of J. Sobal that secondary data analysis, "the reanalysis of machine-readable data, is one of the great supplements to traditional teaching methods, especially for teaching research methodology and statistics" [2]. The training in using SDA is especially important for PhD students because they are preparing to become both researchers and university teachers.

There are different methods of using SDA. We can use SDA in isolation with the purpose of re-assessing data set with a new research question. The other path is the combination of two or more data sets for investigation of the relation between the variables in those data. We can also combine secondary data analysis with primary data analysis.

Secondary data can be numeric or non-numeric or qualitative data. Qualitative secondary data include data retrieved second hand from interviews, ethnographic accounts, photographs, documents, conversations and other.

The list of sources of numeric or quantitative data that are suited to secondary analysis would include: population census, government surveys, cohort and other longitudinal studies, administrative records and other regular or continuous surveys, university and college records, author websites and other.

Secondary data can be restricted or public; it can arise from direct (biomarker data) and indirect observation (self-report).

Analysis of scientific sources shows [11] that SDA is a wide field, related to literature search and Internet search, literature review, cross-national research, demographics data, qualitative and quantitative data analysis, comparative research etc. (Fig. 1).

The scientists presented a wide list of examples of large secondary datasets for educational and social sciences research [12]:

- Common Core of Data (CCD)
- Current Population Survey (CPS)
- Early Childhood Longitudinal Study (ECLS): Birth (ECLS-B) and Kindergarten (ECLS-K) Cohort
- General Social Survey (GSS)
- Head Start Family and Child Experiences Survey (FACES)
- Monitoring the Future (MTF)
- National Assessment of Educational Progress (NAEP)
- National Education Longitudinal Study (NELS)
- National Household Education Surveys (NHES)
- National Longitudinal Study of Adolescent Health (Add Health)
- National Longitudinal Survey of Youth (NLSY)
- National Survey of American Families (NSAF)
- National Survey of Child and Adolescent Well-Being (NSCAW)
- National Survey of Families and Households (NSFH)
- NICHD Study of Early Child Care and Youth Development (SECCYD)
- Programme for International Student Assessment (PISA)
- Progress in International Reading Literacy Study (PIRLS)
- Trends in International Mathematics and Science Study (TIMSS)
- U.S. Panel Study of Income Dynamics (PSID): Child Development Supplement (CDS).

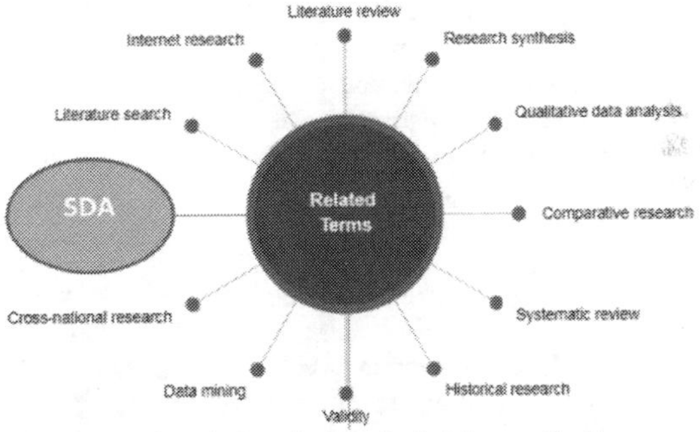

Fig. 1. Secondary data analysis and related terms (by Sage Method Space) [11].

According to T. Vartanian, an excellent archive for educational datasets, is the International Archive of Educational Data [13]. Here, we will find datasets and online tools to examine a wide range of educational surveys.

We can add some Ukrainian resources for this list. The first one is the Ukrainian Center for Education Quality Assessment. It offers a service through which you can analyze the results of external independent evaluation, taking into account different indicators. There are data sets from 2015-2019 [14]. Our sociology students used this data to compare the ZNO results of their region with another region, Kyiv, all of Ukraine in social statistics classes and in course papers.

The second source we presented in our work [15]. We offer our PhD students the survey data from Ukrainian teachers [16-17] for analysis. In 2017, the Ukrainian Association of Educational Researchers conducted the All-Ukrainian monitoring "Teaching and Learning Survey on Principals and Teachers of Secondary Education Institutions" (based on the TALIS methodology [18]). 3,600 teachers and 201 school principals from 201 schools, representing all regions of Ukraine, took part in the study. According to the OECD policy the results of the study, are open and accessible. This year we can use the data of a new wave of TALIS-2018 and conduct the comparative research with different countries.

The third source is a population census in Ukraine. We use data bases that contain Ukrainian census data since 1959 [19]. For example, one of the tasks is related to building and comparing the gender-age pyramid of the population of Ukraine at different years and includes searching for the relevant, data, building the pyramid using standard diagram building Excel tools, using SPSS tools (Chart Builder, Histogram, Population Pyramid), and using pyramid package of R environment. The second task is related to the calculation of child care and grandparent care load coefficients, visualizing of their dynamics, and includes an introduction to the demographic passport of Ukraine [19].

In Demographic and Social Statistics / Education page on the State Statistics Service of Ukraine (http://www.ukrstat.gov.ua/) we can find some educational statistics about:

- Preschool educational institutions (1990-2018)
- Secondary education schools (1990-2018)
- Vocational schools (1990-2018)
- Institutions of higher education (1990-2019).

Also the Women and Men / Demographic and Social Statistics / Education page presents gender data about:

- Pre-school education in 2017
- Secondary education schools and vocational schools in 2017
- Institutions of higher education in 2017
- Indices of gender parity among students of educational institutions of Ukraine

What are the advantages of using secondary data? We can save time and money; those datasets are ideal for use in classroom examples, course projects, master's theses, dissertations and supplemental studies; data may be of higher quality and more representative.

The disadvantages of using secondary data are: data may not facilitate particular research question; information regarding study design and data collection procedures may be scarce; data may potentially lack depth; may require knowledge of survey statistics and methods which is not generally provided by basic graduate statistics courses.

Scientists list [20] the following important steps in the teaching SDA.

1. Develop student's research question
2. Identify a secondary data set
3. Evaluate a secondary data set
   - What was the aim of the original study?
   - Who has collected the data?
   - Which measures were employed?
   - When was the data collected?
   - What methodology was used to collect the data?
   - Making a final evaluation
4. Prepare and analyse secondary data.

It is useful to correlate these steps with use SDA in isolation, with the combination two or more data sets and to combine secondary data analysis with primary data analysis.

What software is used for SDA? We can use the software specifically developed for analysing complex survey data [12]. It is generally free, but may lack flexibility and be only useful for initial data analysis. The examples of such tools are: PowerStats (http://nces.ed.gov/datalab/), Data Analysis System (DAS)(http://nces.ed.gov/das/), AM Statistical Software (http://am.air.org/). Also we can use general purpose software that can account for complex sampling. These tools are usually commercial and cost a lot. (except R). They are generally syntax-based, more flexible. Examples of such tools are: SAS (certain analyses require SUDAAN add-on), Stata, SPSS, Mplus and other.

In R environment there is a special package called "survey" [21]. The package is oriented on analysis of complex survey samples and provides the following features: summary statistics, two-sample tests, rank tests, generalized linear models, cumulative link models, Cox models, log linear models, and general maximum pseudo likelihood estimation for multistage stratified, cluster-sampled, unequally weighted survey samples. Also, we can use variances by Taylor series linearization or replicate weights, post-stratification, calibration, and raking. There are two-phase subsampling designs, graphics, PPS sampling without replacement; principal components, factor analysis. So, the students need substantial training in order to be able to use this package.

The next section discusses how the secondary data analysis application is displayed in the articles of scientific journals, as well as the maintenance of the article by data sets.

2.2. Presenting secondary data analysis and quantitative methods in the journal article

The British Scientist E. Smith [4] explores the use of quantitative methods in educational research and the use of numeric secondary data analysis.

She reviewed the published output of eight well-regarded journals in the fields of Education, Sociology and Social Work over a seven-year period (Table 1). Those journals were:

In the Education field

- British Educational Research Journal
- Oxford Review of Education
- Research Papers in Education

In the Sociology field

- British Journal of Sociology
- Sociology
- Sociological Review

In the Social Work field

- British Journal of Social Work
- International Social Work

Table 1. The number of papers using secondary data analysis and quantitative methods (E. Smith [4, p. 327])

Journal	Secondary data analysis	Quantitative methods	Total papers
Education journals	80	192	627
Sociology journals	89	119	706
Social work journals	33	181	683
All journals	202	492	2016

About one quarter of all the papers (24 %) that were reviewed by E. Smith used some form of quantitative method, of these around 42% presented secondary data analysis. The use of quantitative methods changed from 31% of papers in the 'Education' journals, 27% in the 'Social work' journal, and 17% in 'Sociology' (Fig. 2).

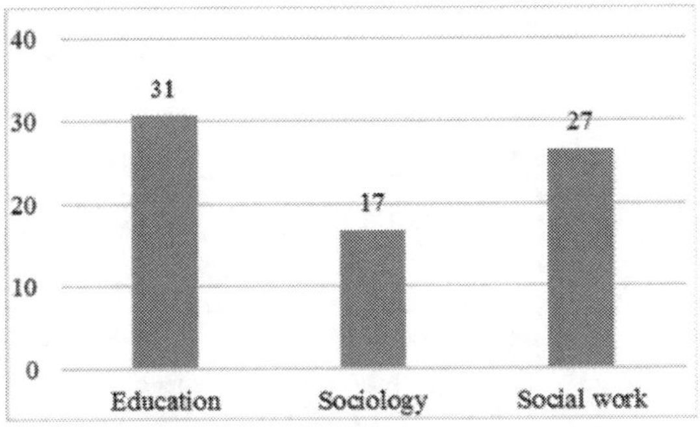

Fig. 2. Percent of papers with quantitative methods from total papers. Built by author with data from [4, p.327]

Less than 10% of all papers reviewed involved some analysis of secondary data. In the 'Sociology' journals the majority (75%) of numeric papers did make use of secondary data, including the data from surveys such as the National Child Development Study, the British Family Resources Survey, the Labour Force Study and the European Values Survey. In 'Education' journals, 42% of the papers which used numeric methods involved the analysis of secondary data (Fig. 3).

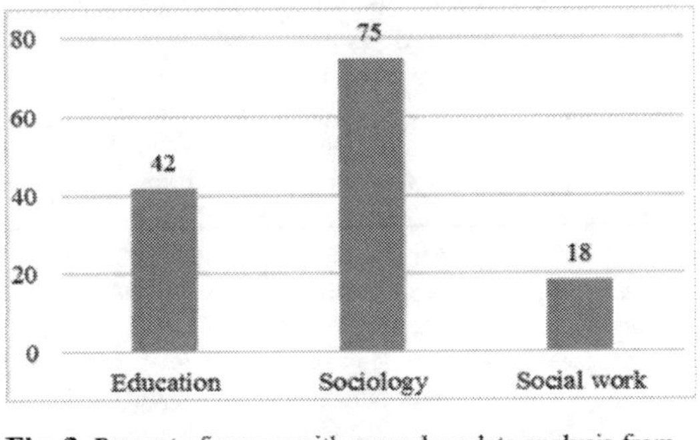

Fig. 3. Percent of papers with secondary data analysis from paper with quantitative methods. Built by author with data from [4, p.327]

The vast majority of articles made use of school performance data; some others authors used studies such as the Youth Cohort Study, the 1958 British Birth Cohort Study and administrative data produced by the Higher Education Statistics Agency [4].

We are going to perform a secondary statistical analysis for this data. The research question is: "Are publications of the three education journals significantly different in using SDA?" To compare the journals we used the statistical Fisher criterion φ^*, which estimates the significance of differences between the percentages of two samples that have an effect of interest to the researcher. The data for calculations for two journals are given in the Table 2.

Table 2. Comparison of publications of two educational journals using SDA (calculated with data from [4]).

Journals	Secondary data analysis, yes		Secondary data analysis, no		Total
	n	%	*n*	%	
British Educational Research Journal	34	12,4	240	87,6	274
Oxford Review of Education	30	13,6	190	86,4	220
Total	64		430		494

The empirical value of Fisher's criterion $|\varphi^*|$ is 0,403, which does not exceed the critical one 1,64, so these journals do not differ significantly in terms of the proportion of articles that use the SDA. Similar results were obtained when comparing the other two pairs of the educational journals.

We also analyzed the conference proceedings of UERA (Ukrainian Educational Research Association). The aim of the UERA is to promote the development of scientific competence of the researchers in Education field, to raise the quality of educational research in order to influence the educational system and the society (uera.org.ua). The discussion of Third UERA Conference "Implementation of European Standards in Ukrainian Educational Research" (June 21, 2019) was held in the following networks: Educational Research Potential for Developing Education in Ukraine; Practical Application of Educational Research for Pre-Service Teacher Training Reform in Ukraine; Academic Integrity and European Ethical Standards in Educational Research [22]. 62 articles were submitted to the conference. Among them, 3 articles contained a secondary data analysis, and 14 – a primary quantitative analysis. Articles with secondary analysis accounted for about 5% of the total number of articles, and articles with quantitative methods – for about 23%.

2.3 Journal articles with data: Journal of Peace Research

One of the trends in the social and behavioral sciences is to support the idea of reproducible research, as a result of which the author publishes, together with the publication, research data, scripts for their processing, support tools and files. This data can be the useful source of secondary analysis.

Consider the example of the Journal of Peace Research [23], how to publish reproducible research on peace and conflict. The journal is guided by the principles of access to data and transparency of research [24], which means that research authors, editors, publishers, and professional associations seek to increase the reliability and openness of various studies by publishing the authors data.

We obtained the following statistics about the number of articles with data in 1984-2018 (Table 3).

An analysis of the dynamics of the number of articles with data published in the journal since 1984 (Table 3, Fig. 4) shows that, unlike one article in 1984, readers received an average of 52 articles in the last three years (2015-2017).

Table 3. Statistics about articles with data in Journal of Peace Research

Year	Number of articles with data	The average number of articles with data in a single issue
1984	1	0,2
1998	10	1,7
1999	22	3,7
2008	28	4,7
2014	41	6,8
2015	52	8,7
2016	55	9,2
2017	49	8,2
2018	45	7,5

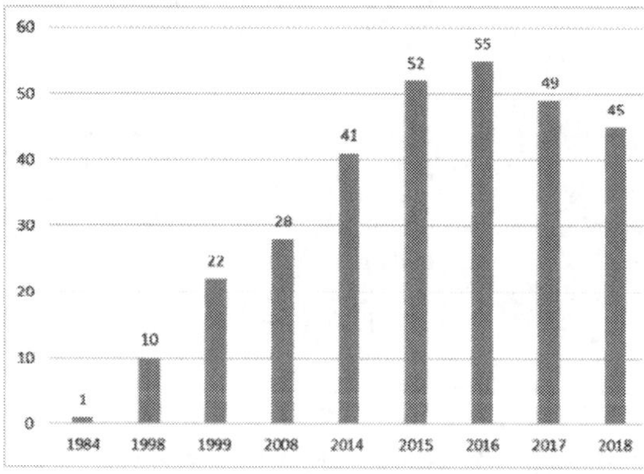

Fig. 4. The dynamics of the number of articles with data in the Journal of Peace Research

Fig. 5 shows dynamics of average number of articles with data in a single issue. There are 8-9 articles with data on average per issue. Note that unlike articles in the journal, access to the data provided by the authors is free.

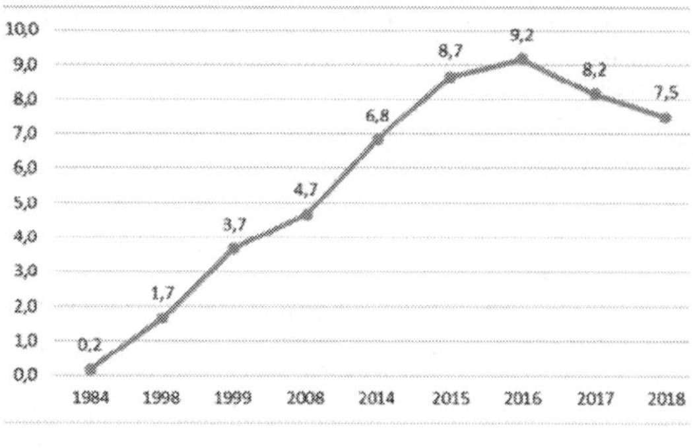

Fig. 5. The average number of articles with data in a single issue

Consider, for example, one of the 2018 articles which provides the contents of the files (Fig. 6). We can see that the authors of the article provided numerous files as support for their article [25]. They also offered a detailed explanation of the software needed to reproduce the calculations; which packages should be installed etc. R 3.4.1 and Stata 13.1 software versions are required to reproduce this study; the following R packages need to be installed: "MatchIt", "dplyr", "ggplot2", "haven", "readr", "xtable", "tidyverse", "RStata". Among the files that accompany the article are .txt, .csv text files; scripts R, Stata; html files; Stata (.dta) and R (.rda) data files.

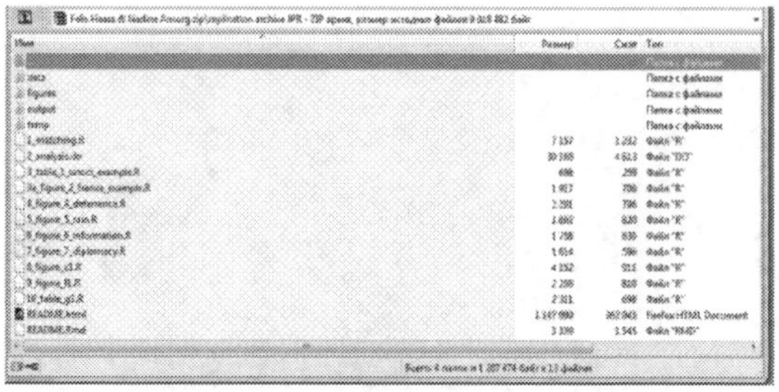

Fig. 6. Files to article "Better peacekeepers, better protection? Troop quality of United Nations peace operations and violence against civilians" [25]

Thus, the magazine promotes the publication of reproducible research, which has the following advantages: increased reliability of research and quality of data analysis; expanding further research on this topic; examples and standards for teaching future researchers, in particular in the field of peace and conflict studies [26].

2.4 Secondary data analysis courses for master and PhD students

We analyzed courses on SDA for masters and doctoral students. We will discuss the following two examples.

The course 'Practical Methods for Secondary Data Analysis' from the School of Public Health (University of Minnesota) emphasizes practical approaches to pre-statistical data processing and analysis with Stata statistical software, advantages and limitations of several national data resources are discussed [6]. The course is designed for three credits. The course goals are:

- Better appreciation the steps required to take a raw data set and produce an analytic data set.
- Appreciation the vast number and variety of existing data resources potentially useful in the field of public health investigations, as well as methods for finding and exploiting these data.
- Understanding basic and moderately advanced data structures, including the binary and hexadecimal number systems, flat-files, relational and hierarchical data resources.
- Ability to read into Stata at least moderately complex ASCII data.
- Ability to exploit existing data conversion software.
- Awareness of central issues in complex sampling designs.
- Familiarity with U.S. Census data.
- Awareness of the National Health Interview Survey.

Course 'Use of electronic archives of social data' from Taras Shevchenko National University of Kyiv [27]. The purpose of the course is to familiarize the students with the possibility of using in the research and analytical work of electronic archives of social data, including archives of the results of sociological research (quantitative and qualitative); data from statistical agencies; global (international) indexes and their ratings of countries, cities, regions; data from national and international non-governmental research organizations, etc. The purpose is also to teach students to search data in electronic archives; to acquaint students with the peculiarities of preparation for analysis of data obtained from archives, with the specifics and methods of secondary data analysis; provide basic knowledge of data management planning in empirical sociological projects and preparation of own research data for placement in electronic archives of social data.

An analysis of the content of these courses showed that not all topics related to the SDA (Fig. 1) were reflected in their programmes.

In addition, we have not found such courses for masters and doctoral programs in the field of pedagogical sciences in Ukraine.

3 Conclusion

Advanced informational technologies have made data resources more accessible and easier to research. Modern initiatives about open access data provide wide opportunities for researchers. The most important initiatives are: UK Data Service (UKDS), Office for National Statistics, Organization for Economic Cooperation and Development (OECD), World Bank.

So, it is important to prepare future researchers for a secondary data analysis using new computer tools and technologies. This is especially true for PhD students in the field of education. They should search, analyze and interpret educational statistics in the framework of their dissertations.

This model of this training may consist of three components. Theory component includes the theoretic basic of secondary data analysis, strength and weakness of this methodology. Practice component contains the examples and tasks of using SDA in educational research with computer tools (specialised and general). These two components are implemented in lectures, seminars and independent work in courses on research methods and courses on quantitative methods. The third component is implemented as PhD student support in the process of writing a dissertation work and includes consultations, seminars and peer reviews.

In our opinion, the course of research methods need to contain a mandatory unit about SDA. The further development of the study is integration of secondary data analyses in the courses of research methods for PhD students in the field of Education in Ukraine and building the model of their support on the stage of thesis writing. This model can be structural and content [28] or structural and functional.

References

1. Secondary Data Analysis Initiative (SDAI) - open call, https://esrc.ukri.org/funding/funding-opportunities/secondary-data-analysis-initiative-sdai-open-call (2020). Accessed 20 Feb 2020
2. J. Sobal, Teaching with Secondary Data. Teaching Sociology **8**(2), 149–170 (1981). doi:10.2307/1316942
3. E. Smith, Pitfalls and promises: the use of secondary data analysis in educational research. British Journal of Educational Studies **56**(3), 323–339 (2008). doi:10.1111/j.1467-8527.2008.00405.x
4. E. Smith, *Using Secondary Data in Educational and Social Research* (Open University Press, Maidenhead, Berkshire, 2008)
5. T.P. Vartanian, *Secondary Data Analysis* (Oxford University Press, New York, 2011)
6. Practical Methods for Secondary Data Analysis (2017), http://www.sph.umn.edu/site/docs/syllabi/Syllabi/2017/Fall/PubH-6617.pdf. Accessed 30 Dec 2019
7. T. Logan, A practical, iterative framework for secondary data analysis in educational research. The Australian Educational Researcher **47**, 129–148 (2020). doi:10.1007/s13384-019-00329-z
8. V. Sherif, Evaluating Preexisting Qualitative Research Data for Secondary Analysis. Forum Qualitative Sozialforschung / Forum: Qualitative Social Research, **19**(2), Art. 7 May (2018). doi:10.17169/fqs-19.2.282
9. M.P. Johnston, Secondary Data Analysis: A Method of which the Time Has Come. Qualitative and Quantitative Methods in Libraries **3**(3), 619–626 (2014), http://www.qqml-journal.net/index.php/qqml/article/view/169. Accessed 21 Mar 2020
10. J. Carter, S. Noble, A. Russell, E. Swanson, Developing statistical literacy using real-world data: investigating socioeconomic secondary data resources used in research and teaching. International Journal of Research & Method in Education **34**(3), 223–240 (2011). doi:10.1080/1743727x.2011.609553
11. SAGE Publications, MethodSpace - Connecting the Research Community (2020), https://www.methodspace.com. Accessed 21 Mar 2020
12. N. Koziol, A. Arthur, An Introduction to Secondary Data Analysis. CYFS (2011), http://r2ed.unl.edu/presentations/2011/RMS/120911_Koziol/120911_Koziol.pdf. Accessed 30 Dec 2019
13. Archive of Educational Data, http://www.icpsr.umich.edu/IAED/index.html (2019). Accessed 30 Dec 2019
14. Open data ZNO 2019 (2019), https://zno.testportal.com.ua/yearstat/uploads/OpenDataZNO2019.7z. Accessed 30 Dec 2019
15. L. Panchenko, Methodology of using structural equation modeling in educational research. CEUR Workshop Proceedings **2393**, 895–904 (2019), http://ceur-ws.org/Vol-2393/paper_411.pdf. Accessed 30 Dec 2019

16. Access the full survey by TALIS methodology raw data (in SPSS): TEACHERS_DATA, https://drive.google.com/open?id=1bzh6U7MnOaFSt_1CV1BsQndCuLX_WpWt. Accessed 20 Feb 2019
17. Questionnaires (in ukr): TEACHERS_Questionnaire, https://drive.google.com/open?id=1L6SHvqpMAGPzeLkp8Ksb9E-KXdnkd0sd. Accessed 20 Feb 2019
18. TALIS - The OECD Teaching and Learning International Survey (2019), http://www.oecd.org/education/talis/. Accessed 20 Feb 2019
19. L.F. Panchenko, Training Sociology Students in Computer Analysis of Demographic Processes and Structure. Information technologies and learning tools **65**(3), 166–183 (2018). doi:10.33407/itlt.v65i3.2034
20. Bank danykh (Census) (2019), http://database.ukrcensus.gov.ua/MULT/Database/Census/databasetree_uk.asp. Accessed 30 Dec 2019
21. Package 'survey' (2020), https://cran.r-project.org/web/packages/survey/survey.pdf. Accessed 21 Mar 2020
22. *Implementatsiia yevropeiskykh standartiv v ukrainski osvitni doslidzhennia* (Implementation of European Standards in Ukrainian Educational Research), ed. by S. Shchudlo, O. Zabolotna, L. Zahoruiko. III International Scientific Conference of Ukrainian Association of Educational Research (21 Jun 2019). (TzOV "Trek-LTD", Kyiv–Drohobych, 2019)
23. Journal of Peace Research, http://journals.sagepub.com/home/jpr. Accessed 30 Dec 2018
24. Data Access & Research Transparency, https://www.dartstatement.org. Accessed 30 Dec 2018
25. F. Haass, Better peacekeepers, better protection? Troop quality of United Nations peace operations and violence against civilians. Journal of Peace Research **6** (2018)
26. L.F. Panchenko, *Prozorist doslidzhen z pytan myru ta konfliktiv: Journal of Peace Research* (Transparency of Peace and Conflict Studies: Journal of Peace Research), in *The Great Wars - the Great Transformations: Conflicts and Peace in the 20th and 21st Centuries.* Materials of IX International Scientific and Practice Conference (Kyiv, 26-27 Nov 2018). (Kyiv, 2018), pp. 87–89
27. Program course 'Use of electronic archives of social data' (Taras Shevchenko National University of Kyiv), http://www.soc.univ.kiev.ua/uk/courses. Accessed 30 Dec 2019
28. I. Lovyanova, K. Vlasenko, A. Krasnoschok, D. Dmytriiev, R. Shponka, Modeling of ICT Competence Formation of Would-be Mathematics Teacher. Information Technologies and Learning Tools **74**(6), 186–200 (2019). doi:10.33407/itlt.v74i6.2421

Professional training of future music art teachers for inclusive education: theoretical and technological aspects

Svetlana Sysoieva[1], *Natalia* Ovcharenko[2,*], and *Olga* Chebotarenko[3]

[1]National Academy of Educational Sciences, Department of General Pedagogy and Philosophy of Education, Kyiv, 04053, Ukraine
[2]Kryvyi Rih State Pedagogical University, Music Education, Vocal and Choir Conducting, Kryvyi Rih, 50086, Ukraine
[3]Kryvyi Rih State Pedagogical University, Music Studies, Instrumental and Choreographic Training, Kryvyi Rih, 50086, Ukraine

Abstract. The article is devoted to the actual problem of contemporary art education – to improve the professional training of future music art teachers for children's inclusive education. Its purpose is to present the results of theoretical understanding of the developmental and healing potential of music, to develop technological support in future music art teachers' professional training for children's inclusive education. According to the results of the study, the basic concept of the study was defined as: "professional training of future music art teachers for inclusive education", which is seen as an educational process aimed at ensuring students' readiness to provide a system of educational services to persons with special educational needs in their future professional activity, and it involves mastering their inclusively oriented musical and pedagogical knowledge, skills and personal qualities. To improve the professional training of future music art teachers for children's inclusive learning, the findings of the latest research in the field of art and music-pedagogical theory concerning the harmonizing and music-therapeutic influence of music on the personality of a child in need of educational inclusion were used; criteria of musical works selection for children with psycho-physiological disabilities have been identified; there was developed and substantiated developmental and corrective technique of training singing that can be used in inclusive classes of different age, gender and cognitive abilities of children. Developmental and corrective technique of training singing includes innovative content of children's teaching, comprising a specially selected vocal repertoire; lesson and extracurricular forms of training; specific teaching methods for students with special educational needs.

1 Introduction

The humanistic tendency of the modern world society is to increase attention to persons with special educational problems. The Ukrainian state, which has chosen a course on European integration, intensively develops and implements reforms aimed at the development of national inclusive education, as evidenced by such documents as: Law of Ukraine "On General Secondary Education" (2010), Law of Ukraine "On Education" (2017), the concept of the National Strategy for the Development of Inclusive Education for 2020-2030 (2019), which emphasize the importance of training highly qualified teachers able to provide quality services in the light of human educational needs. In this context, there is a growing need to address the problem of improving the professional training of future music art teachers for the children's inclusive learning to be competent in the use of specific characteristics of musical art to shape their outlook, motivation, cognitive, emotional and volitional spheres; educational as well as correctional and developmental music techniques for children's teaching and socialization. At the same time, the theoretical understanding of the developmental and healing potential of music art in relation to the technological provision of future music art teachers' professional training in inclusive music education in general secondary education is poorly understood and needs special study.

2 Literature review

The influence of musical art on the spiritual and psycho-physiological spheres of the man has always been the focus of thinkers, scholars, musicians, performers as well as musicians-educators. In the time of increasing the negative effects of socio-political, environmental, man-made influence on the personality of a person, it is especially important to take into account the modern achievements of philosophy, art, music pedagogy, not only in revealing the essence and functions of musical art in the harmonization of the inner world of personality (H. Hesse [1], A. Samoilenko [2], S. Shushardjan [3]), and also in the problems of music influence on successful socialization of the person in the environment, which requires special educational needs, that is – educational inclusion (E. Ryabinina [4], S. Fadeeva [5]).

Analytical review of modern scientific sources on the problems of inclusive music education revealed that Ukrainian scientists thoroughly investigated in their scientific works the following: conceptual foundations of inclusive education of Ukraine (A. Kolupayeva [6],

* Corresponding author: shvager77@gmail.com

M. Poroshenko [7], O. Taranchenko [6]); main provisions of music-educational inclusion (O. Oleksiuk [8], O. Lyashenko [9]); theoretical and methodological principles of correction direction of musical activity (Yu. Bondarenko [10]); specifics of professional activity of music teacher in the conditions of inclusive education (A. Polychronidi [11]); the current state of professional training of future teachers of music art for educational inclusion (N. Ovcharenko [12]); music and therapy techniques and methods for working with children who have special educational needs (N. Bazima [13], N. Kvitka [14], O. Moroz [13]).

Inclusive music education for children with different educational needs, including autism, gender and physical problems is widely represented in a number of foreign researchers' studies (M. Hairston [15]; V. Kinsella, M. Fautley, S. Gray [16]; N. Sarrazin [17]; M. Winebrenner [18]).

However, the study of domestic and foreign experience made it possible to state that the theoretical and technological aspect of preparing future music art teachers training for children's inclusive teaching in the main instrument and vocal class has not yet been the subject of a separate research.

Therefore, the objective of our study is to shed light on the results of theoretical understanding of the developmental and healing potential of music art, the development of technological support for the professional training of future music art teachers to children's inclusive education.

3 Research methodology

To achieve this goal, the following methods were used: theoretical analysis for the study of scientific works on this problem; specification for consideration of basic concepts of the research; analysis, generalization for the study of the harmonizing potential of musical art; classification for the distribution of music material into three groups according to age categories (for junior, middle and senior school age students); the rationale for determining the criteria for selecting composer works for artistic inclusion; theoretical substantiation for the development of developmental and corrective technique of teaching singing and its basic principles; synthesis and structuring to determine the content, forms, methods of professional training of future music art teachers for children's inclusive teaching.

4. Results of discussion

4.1 Inclusive education and inclusive learning

The solution of the problem of future teachers' training for children's inclusive learning involves defining basic concepts of the research as: "inclusive education", "inclusive learning", "persons with special educational needs", "professional training of future teachers of music art". In defining the concept of "inclusive education", legislative documents emphasize that it is a complex process of ensuring equal access to quality education [19]. At the same time, in the researches of modern scientists, and we hold on to this position, inclusive education is considered as "a continuous process aimed at creating an inclusive educational environment based on the principles of tolerant attitude, respect for the individual characteristics of the child and non-discrimination" [7, p. 17].

The concept of "inclusive learning" is interpreted as a system of state-guaranteed educational services based on the principles of non-discrimination, respect for human diversity, effective involvement and inclusion of all its participants into the educational process [16]. Such training is needed by persons who: cannot attend school due to health condition; living in villages and towns (when the number of students in a class is less than 5 people); reside in an area of armed conflict, in a temporarily occupied territory of Ukraine, or in settlements in the territory of which state authorities temporarily do not exercise or fully exercise their authority; are in emergency situations of a natural or man-made nature; have high educational potential and can finish school fast; are foreigners or stateless persons [17]. In defining the concept of "person with special educational needs" we subscribe to the opinion of M. Poroshenko that such is "a person who needs additional permanent or temporary support in the educational process in order to secure his right to education" [7, p. 237].

The concept of "professional training of a specialist" is understood as a continuous controlled process of acquiring a personally subjective experience of professional activity, which allows to systematically and holistically perceive reality and act on the basis of humanistic value orientations, embedded in modern educational concepts of continuity, professional training in higher educational institutions for the formation of professional competence of the individual and successful careers taking into account the modern requirements of the labor market [22, p. 251]. We consider the professional preparation of future teachers of music arts for inclusive learning as an educational process aimed at ensuring students' readiness to master the basics of inclusive learning, as well as competences aimed to use musical and pedagogical knowledge and skills in the process of inclusive learning.

The difficulty of organizing this training is to teach and satisfy the educational needs of all children in music lessons. According to N. Kvitka's diagnostic study, there were three categories of children with special needs identified [14, p. 2]: the first category – students with typical cognitive development, who have sensory, speech, musculoskeletal disorders, and who can study according to the program of institutions of general education, for them, only in a special case, educational materials need adaptation; the second category is students with a slight decrease in cognitive development who have sensory impairments, mental retardation or severe speech impairment, and who may also study under the general education program, but the program material may not be acquired; and have considerable difficulty in learning the curriculum material. Awareness of such categories is necessary in the professional training of future music art teachers for children's inclusive learning, as it enables them to approach each child in a more differentiated manner and apply specific methodological techniques.

4.2 Instrumental music art as a means of harmonizing personality

The complexity and inconsistency of the modern world introduces a certain disharmony into the soul of a child, which forces one to close oneself from it, escaping into its own created world. In order for this illusory world not to completely swallow the child, having lost touch with reality, there are efforts required to overcome these boundaries, to find ways for internal harmonization and to build a "dialogue of consent" instead of a "dialogue of the deaf" [2].

Such an important connecting and harmonizing beginning is classical music, which, according to H. Hesse, is "an extract and embodiment of our culture, because it is the most clear, most characteristic, most expressive gesture. In this music, we own the heritage of Antiquity and Christianity, the spirit of cheerful and brave piety, unsurpassed chivalrous morality" [1, p. 34]. The selection of musical material becomes very important, which, on the one hand, will be the "standard" of harmony, purity, that is, those ethical and aesthetic qualities that make up the essence of the "ideal" world; on the other hand, it will be in tune with the soul of the child, his ideas about this world. Only in this case, through familiarization with musical creativity (perhaps due to performance), will there be a "dialogue of consent" [2], as a result of which a state of harmonious fusion with the world will be achieved.

Music therapy today is one of the most rapidly developing areas; as S. Shushardjan notes, this direction uses "more than 50 different musical and acoustic methods and techniques for correction of mental and physical health, disease prevention, social rehabilitation and creative development of a person" [3, p. 7]. As S. Fadeeva rightly claims, music is a very effective tool in psychiatry, as it well regulates emotional states, causing either peace, calm, or excitement. "The greatest effect of exposure to music is achieved when the patient performs it himself (plays the instrument, sings, and dances to the music). Active experience of music through performance is important for the restoration of motor abnormalities, singing is necessary in the treatment of respiratory diseases" [5, p. 18].

Many authors believe that the harmonization of music is achieved due to its nature, which includes vibrational, rhythmic and melodic spheres. S. Fadeeva notes that music can convey: moods (the whole spectrum, from positive ones – joy, fun, tenderness, to negative ones – despondency, anxiety); intellectual and volitional processes (determination, thoughtfulness, inertness, lack of will, seriousness, etc.); generalized properties of the phenomena of reality (strength, lightness, orientation, breadth, spatiality, etc.); characteristics of movements (fast, moderate, slow, lethargic, elastic, gusty, etc.) [5, p. 21].

The musical world can become the connecting link that establishes a connection between the real and inner world of the child. According to E. Ryabinina, "the actualization of musical qualities as aesthetic is therapeutically significant because it opens up channels for regulating the patient's conditions, ordering and studying the effects on them through aesthetic experience" [4, p. 206]. Thus, through music (performance), you can harmonize your inner world, and, on the other hand, find a way to interact with the external environment (communication methods).

According to many researchers, the selection of music depends on a number of reasons. It is necessary to take into account the emotional state, and the level of intellectual development, and the degree of disposition and perception of music, and age, and much more. In the work of each composer, you can find works that open up a special harmonious, ideal (illusory) world, devoid of conflicting acuity, contradictions and suffering. Researchers talk about a special harmonizing function of the works of W. Mozart and J. Brahms, but this list can be increased.

As S. Fadeeva notes, "most scholars prefer academic music of the European tradition of the era of classicism, the style of classic baroque (I. Bach), or romanticism (J. Brahms), music related to the folk tradition (I. Stravinsky), music depicting the sounds of nature" [5, p. 23]. On the one hand, music should be understandable and in tune with the soul of the child; on the other, it should carry new information and feelings for a stronger emotional effect.

In this case, it is important to select musical works in accordance with the inner sensations of the child, so that through a complete merger with the composer's language, performing contact with the public is possible. According to the definition of E. Ryabinina, and we completely agree with her position, music therapy is based not so much on the acoustic impact, but on the potential of the psycho-aesthetic effect of music. "Of particular importance is musical communication, the capabilities of which are fully realized by such most common forms as live instrumental and vocal performance" [4, p. 210].

Music can also have a beneficial effect on the elimination of the child's depressive states, which can be caused by the imbalance between the processes of nervous excitation and inhibition. S. Fadeeva notes that in such cases, "the goal of music therapy is to regulate the activity of the central nervous system: along with an increase in its activity to a certain level, the activity of the organism as a whole improves" [5, p. 16]. However, the process of influencing music on the nervous system is very complex. As noted by E. Ryabinina, music therapy cannot be reduced only to the organization of playing music and fixing the beneficial effects of this process on the well-being of a person. "The fundamental importance of the aesthetic qualities of music is connected, first of all, with its inherent ability to create strong, and therefore therapeutically effective, experiences. It is characteristic that in the treatment of post-traumatic stress disorders, the world leader among the options for the direction of musical therapy, not pleasant sounding fragments are used (like musical "relaxation", where ordered sounds are not always music), but, more often, work with patients is carried out through their beloved music" [4, p. 207].

Musical material can have explicit programmability (in the form of headings, epigraphs, poems preceding the play (for example, in the cycle Tchaikovsky's "Seasons"), paintings that inspired the image, etc.) and hidden (which is transmitted through the genre, style, stylistics, composer program, tonality, etc.). In the case when the program is not prescribed, you can use a synthesis of arts (literature, poetry, painting, sculpture, cinema, etc.), which greatly enriches the imagination and enhances the perception of music.

Of particularly great help are the composer's comments, in which he reveals the features of musical works, the history of their creation, thoughts and feelings that accompanied the composer's process of creating these plays. It is such a material that significantly deepens the semantic and emotional perception of musical creativity [23]. The child's response to a piece of music, on the one hand, depends on the level of his or her emotional culture; on the other, from the value system embedded in it (ethical, spiritual, aesthetic, cultural, etc.). According to S. Fadeeva, music can become "a decisive factor in the development of the axiosphere of a child's personality, helping to cultivate value-semantic relations to oneself and the world" [5, p. 33].

The issues we have raised about the harmonizing and healing effects of music on a child's personality are particularly important in the professional training of future music art teachers for inclusive student's learning. Indeed, the generalizations we make allow us to select the necessary music material and distribute it into three groups according to the age categories of children (for the younger, middle and older age students) who require special educational services. For elementary school students, the material should be selected regarding two major semantic spheres – lyric and play, where light and harmonious, conflict-free atmosphere prevail. An example of the lyrical sphere are the following works: P. Tchaikovsky – "Mom", "Sweet Dream" ("Children's Album"); S. Prokofiev – Morning, ("Music for Children"); G. Sviridov – "Lullaby", "Gentle request", ("Children's Plays Album "); B. Filtz – "Doll Cradle" ("Yavoriv Toys"). Another, playing sphere, can be represented by such works as: P. Tchaikovsky – "Playing horses", "March of wooden soldiers" ("Children's album"); S. Prokofiev – "The Procession of grasshoppers" ("Music for Children"); G. Sviridov – "Hopper", ("Album of plays for children"); B. Filtz – ("Hand-drawn nozzle", "Cuckoo" ("Yavoriv Toys").

For middle school-aged children, it is possible to offer works with a more diverse figurative sphere, but which also does not contain contrast-conflict comparisons. To reflect the lyrical figurative sphere you can use such works as: P. Tchaikovsky – "Morning Prayer", ("Children's Album"); R. Schumann – "Dreams"; S. Prokofiev – "Rain and Rainbow", "The Moon Walking over the Meadows" ("Music for Children"); G. Sviridov – "Music box", "Before bed", ("Album of plays for children"); B. Filtz – "Magic Bow", "Yavor's Box of Memories" ("Yavoriv Toys"). Another, playing sphere, can be represented by such works as: P. Tchaikovsky – "Waltz", "New doll", "Nanny's fairy tale" ("Children's album"); S. Prokofiev – "Walk" ("Music for Children"); G. Sviridov – "Rain", ("Album of plays for children"); B. Filtz – "Colorful fiddle", "Merry Horse" ("Yavoriv Toys").

For high school students, it is possible to use works both curricular and genre structured (preludes, saraband, pavana, lullaby, etc.), taking into account the fact that children already have certain life and music experience. The following works can be suggested to depict the lyrical sphere: S. Rachmaninov – Preludes D-dur op.23, G-dur op.32; Musical moment Des-dur No.5; R. Schumann – Romance Fis-dur; K. Debussy – "Moonlight" ("Bergamas Suite"), prelude "Girl with flax-colored hair"; M. Ravel – "Pavana". Another, playing sphere, can be represented by the following works: M. Ravel – "Water game"; P. Tchaikovsky – "Horse Play" ("Children's Album"); S. Rachmaninov – Musical moment C-dur No.6; B. Filtz – "Horse-drawn cart" ("Yavoriv Toys"); G. Sasko – Blues, Ragtime ("Playing Jazz"), Jazz-Waltz.

In the process of inclusive education, it is necessary to take into account the characteristics of the child's psyche (either prone to affectation, or, on the contrary, prone to autism and isolation). In the first case, the musical repertoire should include both fairly vivid play pieces and lyrical ones. It is necessary to approach children with special care who are prone to autism, as they are very vulnerable and fragile, and musical works should not cause rejection. Among the criteria for the selection of composer works, we can name the following ones: conflict-free and harmonious images; the dominance of the major sphere; melodic tune and euphony (cantellation); lack of sharp-sounding modern harmonies and rhythms. The following works can be mentioned as examples of such pieces: J. Bach – Prelude F-dur (Volume 2 WTC), Aria from the "Goldberg of Variations", Aria from the orchestral suite No. 3, Bach-Gounod "Ave Maria"; V. Mozart – part 2 of the sonata G-dur No. 5; L. Beethoven – part 2 of Sonata No. 8 op. thirteen; F. Liszt – Consolation of Des-dur; F. Chopin – Nocturne Des-dur; C. Saint-Saens – The Swan; R. Schumann – Romance Fis-dur; S. Rachmaninov – Prelude G-dur; S. Prokofiev – Transience b-moll; V. Silvestrov – "Children's Music" 2.

Some of the sonatas by M. Clementi and V. Mozart, as well as dance pieces (for example, K. Debussi – "Golliwogg's cake walk" from the cycle "Children's Corner") can be offered as moving active games.

We offer the music material for future music art teachers to master at lessons in basic musical instrument in order to use them in inclusive classes in the process of musical perception. Students need to understand the inclusive potential of both instrumental and vocal music during their professional training. Therefore, in the vocal class, students, being trained for children's inclusive learning, should master teaching singing techniques aimed at establishing intellectual, emotional and physiological relationships, that is, to harmonize them. This is the developmental and corrective technique of teaching singing that can be used in inclusive classes of different age and gender children.

4.3 Technological bases of inclusive vocal training

The technique developed by us involves the education of students in inclusive classes, which were integrated into the first and second categories classes, according to N. Kvitka [14], since children of the third category need the professional help of psychologists. The purpose of this technique is socialization, formation of musical culture, development and correction of psycho-physiology of children by means of vocal art. Developmental and corrective technique of singing training includes initial, basic and effective stages of application. At the *initial* stage, the content, the choice of forms and methods of teaching, corresponding to the educational needs of young, middle, and senior school students, are formed. At the *secondary* main one – the content of educational material, forms, methods, consequently, according to age are applied. At the *final* effective stage – students' educational achievements are revealed, the results of technique implementation are analyzed, positive and negative moments in its application are determined.

This technique is based on the principles of: harmonization, which ensures the interconnection of spiritual and physical, intellectual and emotional; nature correspondence, which provides education of the child according to his / her gender, age, individual intellectual and psycho-physiological characteristics; cultural responsibility, which ensures that in the process of vocal training the world and national cultures are taken into account; accessibility, which in the process of vocal training implies resistance to the existing level of development of students; systematic nature, which ensures systematic learning of the students' material; positivity, which involves creating a situation of success and a willingness to support children in the vocal learning process.

The content of this technique is the educational vocal repertoire, the selection of which depends on the degree of intellectual impairment, complex disorders of psycho-physiological development of children who are taught in an inclusive class, as well as their age characteristics. Thus, we have selected vocal works for the preparation of musical perception of children, especially for elementary school students: Ukrainian carols and shchedrivkas, M. Leontovich "Dudaryk", Y. Stepovy "Lullaby", M. Lysenko Fragments from the opera "The goat-dereza", K. Stetsenko from the opera "The Chanterelle, the Cat and the Cockerel", B Filtz fragments from the "Forest opera", the Neapolitan folk song "Santa Lucia" by V. Mozart "Longing for spring", J. Brahms "Lullaby", etc.; to perform folk songs and works of composers, which are filled with a bright, joyous outlook, have a playful and choreographic potential and do not contain dramatic conflicts, contrasting musical images. Such works, in particular, are: folk songs "Come out, get out, ladybug", "Shchedryk", "Hryts, Hryts, come to work", "Podolianochka", "Halya went around the garden", "Beautiful dance hopachok", "Oh, winter is over", "They lived at their grandmother's", "Gray cat", "Oh, there's rye on the hill, a bunny sits ", etc.; children's songs by composers: M. Lysenko "Song of the Fox", V. Kosenko "Pastoral", V. Podvala "Musical riddles", A. Filipenko "Cheerful musician", "Chanterelle Took a Violin " and "Play with a teddy bear", B. Filtz "Bells Ringing", "We were in the woods", V. Vermenich "Grandfather Frost's Invitations" and others.

Mastering the content of students' musical training using the designated song repertoire should take place both in the form of lessons – in the lessons of music art, and at the extracurricular time – on special vocational developmental and correction training sessions, where it is advisable to use group and collective forms for teaching children by means of art. We propose that in inclusive classes the offered works of music are to be introduced in music lessons, since the content of the programs available in the schools is instructive and educational, but not corrective. During extra-curricular activities, the teacher has more opportunities in the selection of vocal works, taking into account the individual characteristics of children.

Of particular importance in the successful implementation of our proposed technique are singing teaching methods developed by vocal educators, adapted for us to work with children with special educational needs, and the author's vocal development and vocal-correction methods. Having diagnosed and analyzed the cognitive and vocal abilities of children, it is important to choose appropriate teaching methods of singing and, if necessary, methods of social, intellectual, psycho-physiological correction.

Thus, during passing the main secondary stage of technique at the beginning of vocal training of inclusive class students in elementary school, it is important to systematically and gradually apply the methods that are preliminary to singing in the system of musical arts lessons and at special vocal trainings:

1) method of game "musician-actor", which is aimed at invoking the mimic-emotional reaction of students to the word, and then – to vocal music, in the combination of words and music. The essence of the method lies in the fact that children are called a word (adverb) and they must express it through facial mimics. It is important to use emotionally positive mood words as an example: gentle, joyful, lyrical, inspired, uplifting, dreamy, insightful, delicate, sensitive, and more. In the future, children are invited to respond to facial expressions by listening to a vocal work, identifying internally the emotions it evokes;

2) method of playing a game "musician-conductor", such a method is aimed at developing children's reaction to the rhythm of a musical work. Children should stand up, a vocal recording is played to them, and they must move their hands down, a conductor gesture, showing the strong beats of vocal music;

3) method of a game "conductor-actor", which is aimed at making students respond with gesture and facial expressions to the heard vocal music. Children listen to the vocal work on the record, get acquainted with it, and then stand up or sit down; they are played a recording of the work and they try through gestures and facial expressions to show what the music tells. Students who are able, if desired, can dance or move freely in singing. It is necessary to direct children's attention to the correspondence of movements and gestures to the essence of music;

4) method of "improviser-singer" and "composer-singer", which is aimed at developing children's creativity. Children are given a creative task to make some changes to the tune the teacher suggested and then to sing it. Further the tasks become more complicated: it is necessary to create the beginning, middle or final part of a melody, come up with a verbal text to it and then

sing it. A more complex task is to create a basic soundtrack to accompany a song using music computer programs;

5) articulation gymnastics (by V. Yemelyanov), which provides the opportunity to coordinate the work of the articulation apparatus with the help of such exercises [24, p. 161]: make a tongue circular motion between lips and teeth with the mouth closed in one direction and then in the opposite direction; rest your tongue on upper lip, lower lip, right cheek, left cheek; to tongue, changing the shape of the mouth, listening to different clicks and invite children to click equally, creating unison; raise the upper lip, exposing the gums and giving a smile; massage your face with your finger tips starting from the roots of the hair on the forehead down to the neck with a circular kneading movements; touch the face by tapping with the tips of bent fingers; put your index fingers on the nose and control the wrinkled nose; at the same time lift the muscles under the eyes and wrinkle the nose with wide open eyes and raised eyebrows; massage the jaw and jaw joints with your fingers. Such gymnastics should be held only under the teacher's supervision;

6) breathing exercises, which promote the development of the breathing apparatus mastery of the ability to calm the nervous state through the control of the singer's breathing. The practice of singing has gained considerable experience in the use of exercises for the development of singer's breathing. However, we do use some of them that, in our opinion, give an initial idea of the work of breathing in the singing process. The gymnastics includes a set of training methods:

– initial breathing exercises are aimed at awareness of the types of breathing: clavicular – undesirable, chest – possible, lower-rib-diaphragmatic – most comfortable in singing and three stages of singing breathing: inhalation, delay and exhalation (K. Linklater [25], S. Riggs [26]). Children are encouraged to stand up, with their hands put to their sides, to breathe in peacefully and to bring air into the upper part of the lungs to raise their shoulders, then imagine that they exhale a candle, hold their breath, exhale slowly (clavicular type). The next exercise – children inhale, filling the breast with air, hold, then exhale in the same way (breast type). Furthermore, the following exercise is trained – students inhale as if they are sighing, while inhaling they fill the lower part of the lungs to feel that a ball has formed from the air in the stomach part, hold it, and then exhale slowly. Students should be told that such a ball should be kept while singing. The exercises should repeat, setting a task for the students to compare which type of breathing allows more air in the lungs, and the exhalation is longer. If children find it difficult to master the lower-diaphragmatic type of breathing, they are invited to lie down and put their hands on the stomach to breathe so that their hands are lifted. And then, transfer the feeling to a standing position;

– the exercises referred to by the castrate singer Farinelli and then by Manuel Garcia [27], who suggested breathing in as slowly as possible for several seconds to develop breathing; exhale very slowly; keep lungs as full as possible; keep as empty as possible;

– the exercise that has a filtering and coordinating effect on the respiratory apparatus, using a simple wind instrument – nozzle, harmonica, etc. Inhaling through the instrument, make a long sound out of the instrument, pulling up the stomach;

– the exercise that strengthens the muscles of the respiratory apparatus: hands should be placed onto the sides, breathe in deeply the air to fill the lower and middle part of the lungs, and then utter the voiceless consonants, such as "p", "s", "t", " tʃ", four times each, pushing the diaphragm closer to yourself;

– health-improving breathing exercises (by G. Strelnikova [28]), which should be used for diseases of the upper and lower respiratory tract, sound disorders, in particular: turn your head to the right and to the left, while doing short, noisy, quick breaths each time. Calm your breath down through your mouth. Tilt your head right and left, back and forth, inhaling sharply through your nose, exhale effortlessly. Then, turn your hands in front of your breasts in turns inhaling sharply and exhaling slowly. The exercise rate should be – 1 breath per 1 second;

– breathing exercises with sound, which are used at the completion of preparatory exercises for the development, coordination, healing of the respiratory system: using the lower-rib-diaphragmatic type of breathing slowly inhale, hold the breath, and at a comfortable pitch of the sound, sing in unison vowels "uh", "oh" as long as possible; then gradually introduce simple exercises with syllables, words, poems instead of the sounds;

6) phonopedic exercises for making children aware of the process of sound formation (by V. Yemelyanov [24]), aimed at feeling the attack of sound: it is necessary to imagine as if you warm your hands in the cold, holding your palms at the mouth and warming them with breathing and making a sound "hhhhhh", and then add the sound "a" to it, first at a comfortable pitch for children and then at a certain pitch defined by the teacher to get "hhhhaaaa"; another exercise "Wave", which enables students to imagine the motion of the wave, to understand the process of sound formation, the dynamics of sound, and involves the alternation of exhalation or strobass and vowel "a" with the strengthening of the dynamics, which looks like: hhh HHHA HHHA; an exercise for the awareness of the boundaries of the voice range and the ability to switch from register to register: sing the lowest note of the voice on the vowel "ah", using portamento (from Italian – sliding), move the voice to the highest note on the vowel "uh";

7) phonetic exercises are used in working with children who have speech disorders, they are based on the diagnosis of sound detection with each child or syllables that sound most casual. In the sound production of such sounds the voice apparatus is free. As evidenced by scientific and methodological researches (D. Ogorodnov [29], A. Yakovlev [27]), most often, the most convenient sound for children is the loud "uh", which makes it possible to naturally coordinate the work of the singing apparatus. At the beginning of the work, the exercises should be performed with the vowel "uh" on one sound, using the strokes legato, non legato, staccato, then – reflecting the sound of the cuckoo, in the distance, the designated vowel should be accompanied with other sounds to the singing trying to preserve its previous sensations.

These exercises create a solid basis for mastering vocal repertoire by children. The skill of a music teacher to work with students in vocal training using group and collective forms that facilitate their artistic communication and personal communication is crucial. Therefore, mastering the content, forms and methods of children's inclusive learning by the students requires constant self-improvement and the search for new musical works, forms, methods of vocal practice with students.

For future music teachers, it is also important to master the ability to analyze the learning outcomes, identify the positive and negative points in applying it; try to carry out diagnostics and control for identification of pupils' educational achievements, results of developmental and corrective work, which is the final effective stage of the technique.

According to the results of the study, we conducted a pedagogical experiment in 2018–2019, engaging 36 students from the Arts Department of Kryvyi Rih State Pedagogical University along with 24 students from inclusive classes in general secondary educational institutions in Kryvyi Rih. All of the participants were divided into 2 groups: an experimental group, made up of 17 people and a control group – 19 people.

We conducted a students' survey to determine their own level of professional competence in inclusive education, which showed that: 84% of future teachers of the experimental group and 82% of the control one are not ready for inclusive learning activities; 12% of students in the experimental group and 14% of the control group are not fully confident in their knowledge and skills to teach children in inclusive classes; only 4% of students in the experimental group and 4% of the control group feel ready for this activity.

Subsequently, educational material for educational artistic inclusion was introduced into the educational process of the experimental group and the vocal class: specially selected pieces of music with accompanying didactic material in the main musical instrument class; developmental-correcting technique of singing teaching in vocal class. The control group students did not study the innovative material. Future music art teachers applied their knowledge and skills in the teaching practice, for six academic weeks, with 24 students enrolled in inclusive general secondary education classes.

At the end of pedagogical practice, another survey was conducted, which revealed that: 22% of future teachers of the experimental group and 80% of the control are not ready for inclusive educational activity; 14% of students in the experimental group and 16% of the control group are not fully confident in their knowledge and skills to teach children in inclusive classes; 64% of the students in the experimental group and 4% of the control group feel ready for this activity. The results of the experimental study are shown in Table 1.

The experiment showed a significant increase in the level of future teachers' readiness with high and medium levels of readiness for inclusive learning and a significant decrease in students with low levels of readiness for inclusive learning. Whereas the level of students' readiness of the control group has hardly changed.

Table 1. Dynamics of readiness levels of future music art teachers for children's inclusive learning

Levels	groups		groups		change criteria	
	initial stage of experiment		final stage of experiment		expe-rime-ntal	con-trol
	experi-mental	con-trol	experi-mental	con-trol		
Low (non-developed)	0,84	0,82	0,22	0,80	–62	–2
Medium and high (developed)	0,16	0,18	0, 78	0,20	+62	+2

The students of the experimental group revealed an rise in children's interest in music and after-school extracurricular lessons, an increase in students' academic progress, an improvement in children's communication skills, and a good mood due to the perceived music and singing lessons.

5. Conclusions

1. Based on the results of the theoretical analysis, it is determined that professional training of future music art teachers for inclusive education is an educational process aimed at ensuring students' readiness to master the basics of inclusive learning, as well as competences designated to apply music-pedagogical knowledge and skills into the process of inclusive learning.
2. The criteria for the selection of musical works for children in need of special educational services are revealed: non-conflict, harmony of images, dominance of the major sphere, melodic tune and gentle sounding, cantillation, the absence of sharp modern harmonies and rhythms.
3. The developmental and corrective technique of singing training for children with different educational needs is substantiated, which includes the innovative content of children's education – specially selected vocal repertoire; lesson and extracurricular forms of training; specific methods of teaching children.
4. It has been experimentally proved that the level of future music art teachers' readiness for inclusive learning in the experimental group is much higher than that of the control group students'.

References

1. H. Hesse, *Igra v biser* (The Game of Beads). (Folio, Kiev, 2017)
2. A. I. Samoylenko, *Muzyikovedenie i metodologiya gumanitarnogo znaniya. Problema dialoga* (Musicology and methodology of humanitarian knowledge. The problem of dialogue). (Astroprint, Odessa, 2002)
3. *Muzyikalnaya terapiya segodnya: Nauka, Praktika, Obrazovanie* (Music Therapy Today: Science, Practice, Education), Materialy Mizhnarodnoi naukovo-praktychnoi konferentsii, Moskovskiy pedagogicheskiy gosudarstvennyiy universitet, 22-23 marta, Moscow, 2019

4. E. F. Ryabinina, *Problemy ekstremalnoi ta kryzovoi psykholohii* **14**(III), 203–210 (2013)
5. S. A. Fadeeva, *Muzyikalnyiy dizayn v obrazovatelnom uchrezhdenii* (Musical design in an educational institution (N. Novgorod, 2011), p. 95
6. A.A. Kolupaieva, O.M.Taranchenko, *Inkliuzyvna osvita: vid osnov do praktyky* (Inclusive education: from the basics to practice). (ATOPOL LLC, Kyiv, 2016)
7. M. A. Poroshenko, *Inkliuzyvna osvita* (Inclusive education: a textbook). (Agency Ukraine LLC, Kyiv, 2019)
8. O.M. Oleksiuk. *Visnyk Pedahohiky* **5**, 5–12 (2001)
9. O. D. Liashenko, Inkliuzyvna mystetska osvita: innovatsiina problema XXI st. (Inclusive art education: an innovative problem of the 21st century), in *Materialy Mizhnarodnoi naukovo-praktychnoi konferentsii*, Universytet im. B.Hrinchenka, Kyiv, 16–17 zhovtnia, 2014
10. Yu. A. Bondarenko, Dissertation, National Pedagogical Dragomanov University, 2017
11. A. Polikhronidi, Molod i rynok **8** (139), 111–114 (2016)
12. N. Ovcharenko, Education: Modern Discourses **2**, 55–62 (2019)
13. N. V. Bazima, O. V. Moroz, Lohopediia **3** (2013)
14. N. O. Kvitka, *Dydaktyko-metodychne zabezpechennia diialnosti asystenta vchytelia v zahalnoosvitnikh navchalnykh zakladakh na predmetakh khudozhno-estetychnoho tsyklu*, (Didactic and methodological support of teacher's assistant's activity on subjects of artistic and aesthetic cycle in general comprehensive educational institutions) (2014), http://lib.iitta.gov.ua. Accessed 28 Jun 2019
15. M. Hairston. Strategies for Working with Special Needs Students in the General Music Classroom (2013), http://www.nafme.org/strategies-for-working-music-classroom. Accessed 24 Sep 2019
16. V. Kinsella, M, Fautley, S. Gray. Musical Inclusion. (2018), http://www.open-access.bcu.ac.uk/6478/. Accessed 5 September, 2020
17. N. Sarrazin. Music and the Child (Open SUNY Textbooks, New York, 2016)
18. M. Winebrenner, Teaching Kids with Learning Difficulties in Today's Classroom: How Every Teacher Can Help Struggling Students Succeed. (Free Spirit Publishing, Minneapolis, 2014)
19. Nakaz MON, 912, 01.10.2010 (MES Order, 912, 01.10.2010), https://mon.gov.ua/ua/npa/prozatverdzhennya-kontseptsii-rozvitku-inklyuzivnogo-navchannya. Accessed 20 December, 2020
20. Zakon Ukrainy "Pro osvitu" (Law of Ukraine "On Education") (2017), https://zakon.rada.gov.ua/laws/show/2145-19. Accessed 28 Nov 2019
21. Nakaz MON Ukrainy 8, 12.01.2016 (Order of the Ministry of Education and Science of Ukraine 8, 12.01.2016), https://legislation.com/ucraini-mon/nakaz-vid-12012016. Accessed 24 December, 2020
22. S. O. Sysoieva, in *Osvitolohiia: vytoky naukovoho napriamu* (Edelveis, Kyiv, 2012), pp. 213–237
23. O. V. Chebotarenko, Naukovi zapysky 169, 164–167 (2018)
24. V. V. Emelyanov, *Razvitie golosa. Koordinatsiya i trening* (Voice Development. Coordination and training). (Planeta muzyiki, Sankt-Peterburg, 2010)
25. K. Linklater, *Freeing the Natural Voice. Imagery and Art in the Practice of Voice and Language* (Drama Publishers, Hollywood, 2006)
26. S. Riggs, Singing for the Stars: a Complete Program for Training Your Voice (1992), https://kupdf.net/download/seth-riggs-singing-for-the-stars_59a85746dc0d602b1e568ee0_pdf. Accessed 28 December, 2020
27. M. Garsia, *Sovetyi po peniyu* (Tips on singing). (Planeta muzyiki, Sankt-Peterburg, 2014)
28. M. N. Schetinin, *Dyihatelnaya gimnastika Strelnikovoy* (Breathing Gymnastics by Strelnikova). (Eksmo, Moskva, 2016)
29. D. E. Ogorodnov, Muzyikalno-pevcheskoe vospitanie detey v obscheobrazovatelnoy shkole (Musical and singing education of children in the secondary school). (Muzyichna Ukraina, Kiev, 1981)
30. A. V. Yakovlev, *Fiziologicheskie zakonomernosti pevcheskoy ataki* (Physiological patterns of the singer's attack). (Muzyika, Leningrad, 1971)

Suggestive techniques in music-performing training of future music art teachers: methodological aspect

Viktoriia Mishchanchuk

Kryvyi Rih State Pedagogical University, Department of Musicology, Instrumental and Choreographic Training, Kryvyi Rih, Ukraine

Abstract. The article deals with the use of suggestive techniques, which by influencing the conscious and subconscious sphere of future music arts teachers through the use of verbal and non-verbal, external and internal means of influence, will contribute to the disclosure of their inner potential, the development of creative thinking as well as the realization of their abilities into the practical activities. The author defines various spheres of the suggestive techniques use, in particular, in the process of future music teachers training and provides own interpretation of the concept of "suggestive technique in the music-performing training of future music teachers". In addition, the author offers methodological developments aimed at conducting creative music-performing activities – suggestopedic self-training of self-regulation skills along with suggestopedic method of forming the skills in artistic interpretation of a musical work, which is presented as a chain of suggestopedic classes, where each stage corresponds to a certain interpretation phase. Furthermore, it is noted that the developed methodological toolkit with the complex involvement of suggestive techniques with a number of methods, tools and approaches to art education along with the varied application of these methods in the process of students' professional training, methods of performing reliability formation and innovative methods, provides to carry out pedagogical influence on independence as well as performing activities effectively. Admittedly, the introduction of typical suggestopedic forms of classes (concert sessions, classes involving related arts, class performances, storytelling, play sessions) into the methodological toolbox brings additional aesthetic coloring into the interpretative processes, which, in combination with the globalization of the learning material, have led to the reconsidering of musical works artistic interpretation techniques on the whole.

1 Introduction

The processes of globalization and technologies integration in the educational field, and in the professional training of future music teachers in particular, are primarily aimed at finding and integrating traditional and innovative principles, methods, tools and forms of learning, developing and substantiating new approaches to the training of highly qualified, competitive, capable of self-discovery, self-development and self-realization of their abilities and opportunities in a creative activity specialists. Therefore, the use of suggestive techniques in the process of students' professional preparation, allowing exerting influence on the conscious and subconscious sphere of the personality through the use of verbal and non-verbal, external and internal means of influence, will contribute to the holistic perception of the world picture through a combination of both logical and imaginary mechanisms; the development of creative thinking, the discovery of the internal potential of a person as well as the achievement of the professional skill heights.

2 Literature analysis

It should be stressed that the analysis of scientific sources showed that in the field of philosophy and psychology, research on conscious and unconscious mechanisms of suggestion were conducted by A. Adler, M. Burno, J. Murphy, B. Porsnev, S. Rubinstein, P. Simonov, D. Uznadze, Z. Freud, K. Jung and others. Thus, S. Freud believed that the man's mind comprises two areas – the conscious and the unconscious. According to the researcher, the unconscious is deprived of any logic, is emotionally unstable, does not face external reality and lives to a greater extent in the past than in the present and acts on its own imaginary principles [1]. The suggestion issues were reflected in the fundamental writings of such prominent psychologists as T. Akhmedov, G. Goncharov, K. Kulikov, I. Monakhova, I. Schulz, etc. From the perspective of V. Kulikov, suggestion is a psychological influence, the exercise of which occurs through speech and non-verbal means and is characterized by reduced personality orientation [2]. The researcher G. Goncharov carried out a historical analysis of the origin of suggestion, examined the periods of suggestion, its varieties, teachings and their representatives [3]. In terms of our study, it was important to refer to the developments provided by the scientists R. A. Berk, S. Goncharov, G. Lozanov, S. Palchevsky, L. Tarasova, T. Chubukova, and others in

* Corresponding author: victoriya.n.m@gmail.com

the field of suggestive techniques appliance in the educational process. The scientist G. Lozanov made a research into the problems of suggestology and developed a method that allows using suggestion to reveal the inner potential of the individual in the process of foreign languages learning [4]. Researcher S. Palchevsky modifies G. Lozanov's suggestopedic system and creates his own model of AECreS (Art, Emotionality, Creativity, Spirituality), the basis of which lay the psychological-physiological as well as didactic-methodological foundations. S. Palchevsky considers that the main goal of the pedagogical process is the development of the child's creative abilities, which should take place in specially created conditions of psychological comfort [5, 163-164]. The researcher R. Berk points out that the use of suggestive methods, means, forms of conducting classes (active and passive), developed by the eminent Bulgarian scientist G. Lozanov, plays a significant role in the students' learning process. Thus, the scientist is considering the use of musical art in the educational process with the purpose of aesthetization, creation of a corresponding emotional atmosphere in the classroom, improving efficiency along with rapid learning of the educational material. R. Berk draws attention to the issue of proper musical material selection in accordance with the tasks set by the teacher, students' characteristics and interests [6].

The problem of the use of suggestive means and suggestive techniques in the process of future music art teachers' training was reflected in the studies of E. Ekonomova, G. Erzhemsky, A. Kozyr, L. Masol, N. Morozova, V. Petrushina, O. Polyakova, T. Reisenkind, O. Rudnitskaya, V. Fedoryshyn and others. O. Polyakova's scientific work deals with the use of both verbal and non-verbal suggestion in the preparation of future music teachers, which are defined as four main streams – verbal, visual, tactile interrelation between the teacher and the student in music and performing classes, and - language of music. Owing to the lined out streams, the teacher affects the subconscious of students with the purpose of conveying and acquiring educational information [7].

Regarding the definition of "suggestive technique", L. Tarasova believes that this means allows carrying out the educational process in the state of wakefulness based on emotional suggestion [8, p. 216]. According to N. Morozova, the use of suggestion in the pedagogical process, namely in music lessons, contributes to a more effective implementation of any actions (learning the material, speed of memorization, etc.). The scientist believes that during the suggestion implementation, the suggerend experiences an altered state of consciousness (trance). This condition is characterized by a decrease in the function of the left hemisphere and the opening of access to the sphere of the unconscious, which leads to the disclosure of human creativity, awakening intuition as well as the development of its hidden resources of sensory systems. N. Morozova proposes that the future music art teachers would use the following suggestive methods in the lessons, namely: joining, choosing without choice, incorporation, psychological anchor or breaking the stereotype, which will increase the concentration and stability of attention and better assimilation of musical material, expanding a person's range of interests [9].

Such authors as A. Kozyr and V. Fedoryshyn emphasized the importance of using suggestion in future music professionals' training. The scientists claim that having a teacher an arsenal of suggestive methods and means, their use in the educational process will open up the students' internal potentials and will promote the suggestive abilities of future music teachers. Thus, these abilities will help the future specialist to create a favorable positive atmosphere of learning, solve complex creative tasks and exert emotional and volitional influence on the team in order to achieve the desired results [10]. Researcher N. Mukhorina notes that a modern teacher should have the latest techniques allowing him to influence children, but it is only possible if they have acquired certain suggestive skills. In her research, the scientist develops a set of techniques for suggestive skills formation, defines organizational and psychological and pedagogical conditions contributing to the formation of this phenomenon. N. Mukhorina notes that the use of suggestion by the teacher allows building subject-subject relations between the participants of the educational process, promotes activation of self-regulation and self-education processes with the younger generation, regulation and correction of their behavior as well as formation of their views and attitudes to the surrounding world [11]. T. Reisenkind's research indicates that the use of suggestive technique in the process of professional training of future music teachers allows them to influence their emotional-figurative sphere and thereby leads to an understanding of their own "I", stimulating the intuitive sphere, creating vivid interpretations of musical works [12]. While studying the use of suggestive techniques in music-performing training of future music teachers, we have come to the conclusion that the use of these techniques "... involves the organization of a psycho-dynamic complex process based on the mutual influences of partners at the conscious and subconscious levels in terms of psychological comfort; develops and forms certain musical and artistic representations of the personality, not inherent before, to create and realize original concepts of musical works in the performing process" [13, p. 66].

We believe that in order to achieve impregnable results in the process of students' music-performing training using suggestive techniques, it is important to develop the creativity of future music teachers, and thus provide for the formation of *self-regulation skills as an effective tool for performing creative music and performance activities, interpretation activities, in particular.*

The result of such a creative music-performing activity, in our opinion, is an ability to independently create an innovative, original product of a certain activity – a unique artistic image of a musical work that arises in the process of its interpretive processing. It is the complex involvement of suggestive techniques with a number of methods, tools and approaches to artistic education (artistic management, verbal, visual-and-demonstrational aids, etc.) and the varied application of these methods in the process of students' professional preparation that have effective pedagogical impact on the students' independence, originality of their music-performing activities overall as well as their creative and interpretive activities as future music teachers in particular.

Correspondingly, considering the question of artistic teaching methods use in the process of future music teachers' music-performing training, in our study, we refer to the research by G. Padalka, who carried out their classification as follows: "by sources of transmission and nature of perception of artistic information; according to the nature of the artistic activity; in accordance with the nature of artistic tasks by stages of study; depending on the tasks of developing the students' personal artistic properties" [14, p. 178]. The first group of methods includes verbal, demonstrational-visual aids (visual) and artistic-creative (practical). The scientist emphasizes that the essential role in art education is given precisely to verbal methods (conversation, storytelling, explanation, discussion, current commentary, verbalization of works of art content), resulting in expanding the students' theoretical and methodical knowledge in the field of musical performance, directed at students' inner understanding of the essence of artistic images, obtaining valuable counselling on working on the image of a musical work, as well as on their activation to independence, expressing their own opinions and their own position on the understanding of the latter [14].

Obviously, in the course of music-performing training of future music teachers, it is advisable to use demonstrational-visual methods of artistic teaching as well. According to G. Padalka, this group is represented by the following methods: demonstrations of works of art and verbal explanations through artistic illustrations. The significance of the former method "... is to provide an example of the artistic result the student strives to, in providing a kind of standard for the study of educational material" [14, p. 183]. The latter method, the one of artistic illustration of verbal explanations, involves the demonstration of musical works with the indispensable presentation of certain theoretical positions regarding the given examples or reasoned own opinion and its presentation to the audience. An important feature of this method is the professional performance of a musical work. It should be noted that in the process of students' music-performing training, it is important to use the method of artistic management, which involves "multiple repetition of artistic actions in order to achieve perfection of their performance, fixation at the level of automatism" [14, p. 188], which will facilitate the formation of music-performing skills with the future specialists. The effectiveness of the method depends on the teacher's use of methodological and psychological recommendations.

The methodological support for future teachers of music art training with the use of suggestive techniques also involves the use of various creative tasks aimed at the development of musical interpretation skills, self-regulation of their psycho-emotional processes and states while working on a musical piece and the implementation of its interpretive concept in the music-performing process. With regard to suggestive teaching methods, we propose the use of an autogenous training method developed by I. Schulz. This method involves the use of self-suggestion formulas in a state of rest for psychological self-regulation. The autogenous training helps future music teachers to overcome the negative phenomena (difficulties in learning a piece of music, presenting a composition to a listening audience, etc.) and reduce neuropsychiatric overload [15]. The implementation of this method and the achievement of the desired results depends on the formulated self-guidance formulas. According to E. Coue, they should be short, simple and not forced. Since the formula is addressed to our subconscious "I" and is perceived as an order for execution, it does not necessarily have to be true. You should also consider the positive component of the formula and try to avoid the "not" particle in its formulation. Also, the process of suggestion must take place without any willful efforts [16, p. 323].

Our study also dwells upon the integration of such suggestive methods that involve the use of verbal paths of information transmission based on a combination of linguistic, paralinguistic and extralinguistic means. Scientist L. Masol attributes these methods to suggestive methods of the linguistic group, which also includes paralinguistic and extralinguistic ones, and it is aimed at creating emotional situations of success, surprise or delight. The peculiarity of these methods is the use of intonation, figurative and timbre palette of voice, appropriate pauses and whispers to influence the students [17, p. 136]. In our opinion, it is also important to follow a certain rhythm and pace of speech, to use a dynamic scale to perceive, memorize, and reproduce educational information. Among the suggestion methods, we also propose to use the immersion method, which will allow future teachers of music art to dive deep into the world of musical images of a given musical piece, its history of origin, to learn about stylistic features of the composer, etc., as well as to create vivid concepts for music pieces and present them to a listening audience. In our study, we also support the use of aesthetic therapy techniques (involvement of various arts) to aesthetize the learning process, overcome nervous tension, and master the submitted material. In our opinion, the introduction of suggestopedic forms of classes (concert sessions, classes with the involvement of related arts, class performances, storytelling, play sessions), providing the involvement of our methods, contribute to the formation of original interpretations of music pieces.

3 Methodological research results

In the study, the strategy of formation of self-regulation skills in the process of music-performing training was designed to direct students through suggestive techniques to find an algorithm for self-organization and self-regulation of their own actions in the process of practical music-performing activity: during both ongoing individual and group activities as well as during the announcement of the results of music and performance

 https://doi.org/10.1051/shsconf/20207504007

training progress – at the credit-testing, exams, concert events in terms of higher education institutions.

The possession of self-regulation tools is a very important factor in the formation of music-performing readiness of future music teachers. After all, the developed skills of self-regulation determine the performer's reliability in revealing the results of the cooperative creative activity carried out by the teacher and the student.

Unfortunately, the pedagogical observations made during the exam sessions, modular control, concert events, etc., prove the absence of self-regulation skills in most students in practice not only in stressful situations when announcing results of their artistic and performing training, but even during ongoing individual and group activities. This is due, first of all, to the students' inability to psychologically adjust to the reproduction and presentation of the artistic image of the musical work to the audience and to overcome the excitement during the concert performance.

Scientists A. Kozyr, G. Padalka, D. Yunyk argue that there is a direct correlation between the formation of self-regulation skills in the process of creative music-performing activity and reliability in communicating its results to the listener. According to D. Yunyk, "... the performing reliability of the artists of musical art is not an innate but acquired integral personal characteristics, which is manifested in the unmistakable, accurate performance of musical works in the regular and emotional conditions of the stage activity" [18, p. 125].

Based on the above statement of the researchers, it should be noted that for the effective performance of students' music-performing activities it is not enough to provide even careful, painstaking practice of learning end educational repertoire of the courses "Basic instrument", "Additional instrument", "Voice production", "Choral conducting", etc. After all, in the conditions of emotional stressful situations of the concert performance, the practice done often does not matter. As a rule, the implementation of the interpretative concept of creating an artistic image of a musical work, its operational and technical support, makes up no more than 60%, and sometimes scores 20-30% of the performance of a musical work in the regular conditions of a classroom individual lesson.

Thus, these observations allow us to state the need for purposeful formation of self-regulation skills in the process of music-performing activities as a tool for immunity forming against the stressors. Therefore, the formation and development of these skills were carried out through specially developed methodological tools, which contain methods of performing reliability formation, innovative methods, as well as modern techniques of suggestive influence.

The Toolkit for Performance Reliability includes:
- creative tasks for self-regulation of musical memory in its sensory, short- and long-term aspects, based on the methods of conscious selection of "valuable" musical information by activating arbitrary attention; methods of logical structuring of musical material; methods of selective performance of fragments of musical works; methods for ranking pieces of music according to the difficulty level of the memorizing them, etc.;
- creative tasks for self-correction of emotional and logical evaluation of intermediate and final results of music-performing activity;
- creative tasks for self-tuning for the implementation of the interpretative model of musical works;
- creative tasks of self-control during the implementation of artistic interpretation [13].

In order to form self-regulation skills in the process of music-performing training, we also created suggestopedic auto training of self-regulation skills, the basis of which was the suggestopedic method of autogenic modification. The application of this methodological support provided the creation of appropriate psychological and pedagogical conditions, namely:
- "students' psychological confidence as for the possibility of 100% reporting to the listener the results of artistic and performing training based on the development of self-regulation skills of their own actions;
- permanent provision of positive emotional support of the specified skills formation process;
- suggesting students' sensation/sense of their own rich intellectual and volitional potential" [19, p. 100].

During the training, the regulation of students' psychological state occurs first under the guidance of the teacher, and in the process of mastering the tasks of auto-training gradually transforms into self-regulation on the basis of enhancement of suggestion activity directed at itself.

The advantage of suggestopedic auto-training is that in the process of conducting it, the psychological-pedagogical goals of the teacher-suggester and the student-suggerend are completely the same. The success of the self-suggestion process determines the level of self-regulation, and, which is very important for the effectiveness of the training itself, the student, in real time, is enabled to self-manage his or her own attention, memory and emotional manifestations.

In fact, an instrument of suggestopedic self-training for the formation of self-regulation skills in the process of music-performing training is lexical self-suggestion formulas [19, p. 100]. The use of these lexical formulas promotes concentration of attention, control of emotional state, actualization of intellectual activity, etc. So, we suggest to get acquainted with the suggestopedic self-training of self-regulation skills in the process of music-performing training on the basis of autogenic modification:

1. Identification of problematic situations in terms of self-regulation and specification of a positive set of achievements.

Task 1.1. Make a list of psychological and pedagogical problematic situations on the separate yellow cards that arise in the course of ongoing music-performing activities during the individual lessons of the "Basic Instrument", "Additional Instrument", "Orchestra Class", etc., and during the exam-concert performances.

Task 1.2. Rank the cards with the problematic situations recorded on them according to your personal importance value of solving them.

Task 1.3. In accordance with each of the ranked problematic situation, write a description of your own psychological and physiological state on the red cards, preventing the full realization of the artistic image of the musical work creation, its creative interpretation, the operational and technical side of the performance. Describe your own condition according to the emotional, intellectual and physiological-sensory components. (For example: "Before performing a musical element that contains a double note invoice, there is a timidity and tension, accompanied by the thought "Still does not work" and a feeling of "stiffness in hands").

Task 1.4. On green cards, use a lexical self-suggestion formula to create a comprehensive positive description of psychological and physiological states that is completely opposite to that described on the red cards. (For example: "Before performing a musical element that contains a double-note invoice, there is a boldness and looseness, accompanied by the thought" I am sure I will play brilliantly "and a sense of "flexibility and ease of movement of hands"

The lexical self-suggestion formulas contained in comprehensive positive green card descriptions are to meet the following requirements:

1) punctuality, conciseness, capacity that promotes better perception and engaging in consciousness (short lexical formula: "I'm calm" works better than "I'm always calm when performing double notes";

2) affirmative positivity (the lexical self-suggestion formula should not contain any objection that may exacerbate the problem; for example, the lexical formula: "I'm not afraid to forget the text during a speech" can reinforce the fear; the lexical formula "I remember everything" works better);

3) individual orientation (the lexical self-suggestion formula works only at the personal level; if a certain key phrase helped solve the problem earlier, it can be used for autogenic modification; for example, if the word "Basta!" helped once overcome the fear of stage, then it can be engaged during the process) [13].

2. Autogenic modification: self-suggestion of a positive set of self-regulation skills.

Task 2.1. Compiling all cards, make a table containing 3 columns: yellow, describing the problematic situations, red, describing the negative set of feelings, and green, describing the positive complex containing lexical self-suggestion formulas.

Task 2.2. In calm, focused state, work with the table as follows, starting with the first, easiest task:

1) look at the yellow card with the record of the problem situation;

2) symbolically remove a red card from the table describing the negative set of feelings;

3) being concentrated and focused, repeat the description of the positive set of sensations containing lexical self-suggestion formulas;

4) after completing the first task, move on to the next one.

3. Autogenic modification in the mode of relaxopedic training: self-suggestion of a positive set of self-regulation skills in the state of relaxation.

Task 3.1. In case of major difficulties, such as the total inability to recollect sheet music during public performance, work with an autogenic modification using a card table in Relaxopedic Training under the teacher-suggester's guidance. The systematic combination of autogenic modification with the state of relaxation provides self-regulation in order to resolve the problematic situation in the process of music-performing training.

To measure the formation of self-regulation skills of future music teachers in the process of music-performing activity, we have used the modified methodology of N. Kuzmina and V. Ginetsinsky "Diagnostics of skill mastery of self-regulation methods in the process of music-performing activity as well as during concert performance" [20]. The carried out diagnosing of future music teachers allowed us to state that the students acquired the ability to self-regulate in the process of professional training, which facilitated the concentration of their attention on the creation and presentation of the integral emotional-figurative content of the artistic work, performing technical and artistic tasks, regulating inner psycho-emotional processes in order to realize the interpretative concept of musical composition in the performing process.

We believe that in the process of music-performing training of future music teachers, it is important to develop the ability to create an independent and original interpretation of musical works, which fully attests to the degree of development of students' creative and thinking qualities.

Thereby, O. Scholokova insists on the necessity of overcoming reproductive and imitative tendencies in the interpretation of musical works. The researcher notes that "... every interpretation of a work of art begins with one's own, individual interpretation, i.e., proceeds from perception" [21, p. 75].

At the same time, one cannot deny the leading role of teachers' guidance in such subjects as "Basic Musical Instrument", "Orchestral Class", "Choral Class", "Voice Performance", musical-theoretical disciplines on the formation of musical interpretation skills in the process of music-performing training. We have specially created a suggestopedic technique of forming the skills of artistic interpretation of musical works. The *goal* of this methodology is to bridge the gap between the reproductive and independent ways of performing artistic interpretation of musical works in the context of the synthesis of rational and intuitive in the process of students' music-performing training.

The goal has been achieved by introducing characteristic *suggestopedic* forms into the methodological toolkit, namely: *concert sessions, classes involving related arts, class performances, storytelling, play sessions* and more. These suggestopedic forms have introduced additional aesthetic coloring into the interpretative processes, which in combination with the globalization of teaching material have led to the reconsidering of musical works artistic interpretation techniques on the whole.

The methodology is based on the basic principles of suggestopedia developed by G. Lozanov, namely:

– "joyful and spontaneous concentrative calmness;

– dynamic, structured and hierarchical globality – the part in the global and the global in the part; the global as a part and the part as a global; and

– desuggestive set-up with the aim of the spontaneous freeing of the reserve capacities of the mind and brain" [4, p. 66].

The *essence* of the methodology lies in *the globalization of the educational material regarding the basic laws of the interpretation process, which allows to begin to elaborate a piece of music as a whole, to summarize its figurative and artistic content, generalizing its nature, without focusing on the first stage of the interpretation process in detail.* However, in the further interpretative elaboration, the individual details should be considered as certain elements of one musical integrity.

The *specific characteristics* of the methodology include the following:

1) "embeddedness" in the above suggestopedic forms of teaching methodical tools of artistic teaching: verbal methods (conversation, current commentary, verbalization of the content of musical works), demonstration methods (demonstration of musical works, artistic illustration of verbal explanations, etc.); methods of artistic management, varied development of artistic material, creation of artistic images; music-pedagogical discussions, art debate, etc.;

2) the availability of aesthetic means of the interpretation process or artistic means of art, which are the natural component of creating artistic interpretation in the context of a system-holistic approach: the means of music, choreography, painting, fine and applied arts, photography, cinema, etc.;

3) the orientation of methodological support for the creative interpretation skills formation on the creativity of the interpretation process in the context of originality and independence:

- the role of the teacher is presented solely in providing the most favorable conditions for the students' interpretive activity;

- the interpretative processing itself is carried out by students independently.

Thus, the suggestopedic method of forming the skills in artistic interpretation of a musical work, is presented as a chain of suggestopedic classes, where each stage corresponds to a certain interpretation phase of the process. This method is assigned for such disciplines as "Orchestral Class", "Choral Class", during which there are constantly occurring problematic situations in the implementation of students' interpretive skills. At the same time, the methodology can be adapted and applied during individual lessons in "Basic Instrument", "Additional Instrument", "Concertmaster Class", etc.

Stage I. Music study context. Content: comprehensive and substantiate introduction of students to the global historical and theoretical information about the historical era in which the author of a musical work under interpretation lived and worked, about the creativity of the designated author, corresponding artistic direction, his or her artistic and stylistic features. The suggestopedic form – a concert session. The lesson lasts for an academic hour, which is very convenient for both individual and group classes in the current conditions of music-performing training of future music teachers.

The concert session requires the careful preparation of the teaching assistant. During the preliminary preparation for the concert session, the teacher selects an auxiliary piece of music or a medley of several works that belong to the same era, the same artistic direction as the main piece of music planned for interpretation. It is also possible to use a piece of music as an ancillary by the same author as the main work. Suggestopedic requirements for the selection of ancillary work:

- the work of music should be calm, contemplative;

- the piece of music is to contain artistic and stylistic features characteristic of the artistic direction in which the main composition is created;

- the duration of the ancillary musical work should exactly coincide with the duration of the sound of the educational material planned for teaching;

- the performance of the ancillary piece must be perfect.

During the preparation for the concert session, the teacher-suggester will listen to the ancillary piece of music several times in order to determine figurative changes of mood, nuance, tempo, dynamics, etc. After that, he distributes the educational material on historical and theoretical information about the era of creation and author of the main musical work, the corresponding artistic direction, characteristic style features, etc. in accordance with the characteristics of the sound of the ancillary musical work. The teacher-suggester should train the nuances of the sound of the voice in the process of teaching the teaching material in accordance with the nuances of the work. These are the factors to determine the quality of learning information by the students.

Methodological requirements for conducting a concert session:

1) the behavior of the teacher-suggester at the beginning of the concert session corresponds to the beginning of the concert: solemn and restrained;

2) teaching of learning material begins after the sound of a musical introduction or after the presentation of the first musical theme of the work with full breath: the intonation of the text is sublime, but depending on the nature of the sound of the ancillary music work;

3) the phrases pronounced by the teacher-suggester must coincide with the musical phrases; any change in timbre, strength, dynamics of voice is caused by changes in music: joyful, confident intonations correspond to the major system; melancholic, lyrical – minor;

4) the beginning of the speech falls on the first strong part of the musical phrase, and the end - on the last weak; if there is a corresponding malfunction in the speech, it is necessary to wait for the end of the musical phrase and quietly continue the production with a strong fate of the next musical phrase;

5) the rate of speech depends on the agogics of the ancillary musical work [13].

Thus, a properly conducted concert session allows students not only joyfully and without strain to absorb the educational information of the musical-theoretical content necessary for a competent, well-grounded interpretative process at the global level, but also empirically, by means of the perception of the ancillary work, to confirm educational information on the embodiment of the composer's creative intent in the characteristic stylistic features of the musical language of a particular era.

After the concert session, the teacher invites students to work independently for a week (before the second session) to find and study music-theoretical, musical-historical literature in the context of continuing the global acquaintance with the era and artistic direction in which the main musical composition was created.

Stage II. Music-operating context. Content: on the basis of demonstration and comparison of different performances of the main work be able to formulate a detailed definition of its character, as well as an analysis of the means of musical expression (musical form, melody, harmony, metro-rhythm, texture, dynamics, strokes, etc.) in creating a defined character. Suggestopedic form – lessons with the involvement of related arts. Duration – an academic hour.

Engaging with related arts, as well as a concert session, requires careful teacher's preparation. Classes engaging with related arts are conducted using multimedia tools - interactive whiteboards, monitors, tablets – depending on the equipment at the audience's disposal.

The teaching material is assembled by the teacher as follows: students are offered several different options of performing the main work under interpretation. Each version of the performance is accompanied by a display of works of painting, sculpture, architecture, etc. of the relevant cultural era. The aforementioned show is purely illustrative in nature and does not need to be focused on. The suggestopedic meaning of the illustrative display is the aesthetic coloring of the process of musical perception of the main work, the understanding of its figurative and artistic content in the context of the globalized perception of the corresponding historical and musical style. The duration of this part of the class is 30 minutes, during which students are to listen to at least 3 variants of the performance of the main work, in order to prevent the creation of stereotypes in further interpretive processing.

During the next 15 minutes, students should spend on a thorough, generalized definition of the character, figurative content, genre, and form of a piece of music that must be done on a hearing basis in writing.

The homework that students receive after engaging in related arts is to independently write a musical-theoretical analysis of the main work in accordance with the requirements of the course "Analysis of Music Works". The use of sheet music is required for homework. Means of musical expression are analyzed solely in the context of a generalized, detailed definition of the nature and artistic content of a piece of music on the basis of a globalized understanding of the genre-style features of the musical direction in which the work was written.

Stage III. Analytical and pedagogical context. Content: assessing the level of musical and performance complexity of the work, as well as revealing its musical and pedagogical potential. Suggestopedic form –an interactive game lesson.

The lesson lasts 45 minutes. Applying interactive games such as "Student as a leader of an orchestral collective", "Artistic debate in the style of television talk shows", "Scientific and practical conference on artistic interpretation of musical works", "I am the moderator of the round table", etc., students are to analyze a number of music-performing tasks concerning the interpretive creation of an artistic image of a musical work, to specify the musical-performing means of maintaining the integrity of the form, to detect the point of the "Golden Cross", to reproduce the musical style, inherent to a particular musical direction, to plan nuances and dynamics, etc. in the context of the author's design. They are also to determine the pedagogical potential of a musical work in order to form students' interpretive and performing skills on the basis of the collective formulation of conclusions.

After the interactive play, students are given a written homework assignment, a creative assignment "Annotation-interpretation of a musical work", in which the author's two recommendations for the interpretation of the musical work are added to the two previously completed assignments. The completion of the specified creative task can serve as a basis for the preparation of module control testing, credit or examination in the subjects of "Orchestral class" or "Choral class". It should be noted that the use of this method provides practical preparation of students for the exam in the field of creating an annotation of a polyphonic piece of music, which is a mandatory requirement for its compilation.

As the study shows, we tested the method of formation of artistic interpretation skills of a musical work by the method of I. Grinchuk and O. Burska "Diagnostics of Musical and Performing Thinking: Instrumentation Kit" [22, p. 202-221], which showed the positive dynamics of raising the level of formation of certain music art skills in the process of music-performing training of future specialists.

4 Conclusions

Thus, the analysis of scientific sources on the problem of the use of suggestive techniques in music-performing training of future music teachers allowed us to provide our own definition of the concept and to create the author's methods of music-performing training of future music professionals using suggestive techniques in the educational process, namely: suggestopedic auto-training for formation of self-regulation skills in the process of music-performing training, suggestopedic method of formation of skills of musical work artistic interpretation. The developed methodological toolkit of techniques is based on the combination of forms, methods and means of suggestive influence with the forms, methods and means of artistic teaching. Performing creative tasks to ensure performing reliability, creating appropriate psychological and pedagogical conditions and lexical self-suggestion formulas during self-suggestive auto-training for self-regulation skills formation has contributed to the development of creative and intellectual capabilities of individuals, improving the performance of musical and performing activities by updating and self-correcting the inner state of music specialists. Moreover, the introduction of regular suggestopedic forms of classes (concert sessions, classes involving related arts, class performances, storytelling, play sessions) into the methodological toolbox of the methods of musical work artistic interpretation skill formation, the availability of bringing aesthetic coloring into the process of interpretation and teaching classes conducting contribute to students' reserve opportunities revelation aimed at increasing the volume of learning material, deep immersion into the imaginary world of performing works, creating and implementing original and unique interpretations of musical works in the performing process.

Thus, the proposed author's methodological developments in future music teachers' music-performing training using suggestive techniques in the educational process opens new perspectives of students' personal and professional progress and contributes to their reaching of professional skill pinnacles.

References

1. S. Freud, The ego and the id, ed. by J. Strachey (W.W. Norton, New York, 1962)
2. V.N. Kulikov, *Psihologiya vnusheniya* (Psychology of Suggestion). (IvGU, Ivanovo, 1978)
3. G.A. Goncharov, *Suggestiya: teoriya i praktika* (Suggestion: theory and practice). (Prosveshenie, Moskva, 2008)
4. G. Lozanov, *Suggestopaedia-desuggestive teaching. Communicative method on the level of the hidden reserves of the human mind, Summary of Lectures 1995–2005 Vienna* (2005), http://www2.vobs.at/ludescher/pdf%20files/suggestopaedia%20-%20book.pdf. Accessed 5 Mar 2020
5. S.S. Palchevskyi, *Suhestopedahohika: novitni osvitni tekhnolohii* (Suggestion education: new technology). (Kondor, Kyiv, 2005)
6. R.A. Berk, Use of Technology and Music to Improve Learning (2008), https://images.pearsonassessments.com/images/NES_Publications/2008_12Berk_13012_1.pdf. Accessed 5 Mar 2020
7. E.S. Polyakova, *Psihologicheskie osnovy muzykalno-pedagogicheskoj deyatelnosti* (Psychological foundations of musical pedagogical activity), 2nd edn. (BGPU, Minsk, 2005)
8. O.M. Piekhota (ed.), *Osvitni tekhnolohii* (Educational technologies). (A.S.K., Kyiv, 2001)
9. N.V. Morozova, Vestnik kafedry YUNESKO "Muzykalnoe iskusstvo i obrazovanie na protyazhenii zhizni" **4**, 57–67 (2016)
10. A.V. Kozyr, V.I. Fedoryshyn, *Vstup do akmeolohii mystetskoi osvity* (Introduction to artistic education acmeology). (Vyd-vo NPU imeni M. P. Draghomanova, Kyiv, 2012)
11. N.B. Muhorina, Dissertation, Kolomna State Pedagogical Institute, 2003
12. T.Y. Reizenkind, *Dydaktychni osnovy profesiinoi pidhotovky vchytelia muzyky v peduniversyteti* (Didactic foundations of vocational training of music teacher at a university). (Vydavnychyi dim, Kryvyi Rih, 2006)
13. V.M. Mishchanchuk, Dissertation, National Pedagogical Dragomanov University, 2015
14. H.M. Padalka, *Pedahohika mystetstva (Teoriia i metodyka vykladannia mystetskykh dystsyplin)* (Art Pedagogy (Theory and methodology of teaching artistic disciplines)). (Osvita Ukrainy, Kyiv, 2008)
15. I. Schultz, Übungsheft für das autogene Training: Konzentrative Selbstentspannung (Georg Thieme Verlag, Stuttgart-New York, 1983)
16. T. Ahmedov, *Prakticheskaya psihoterapiya: vnushenie, gipnoz, meditaciya* (Practical psychotherapy: suggestion, hypnosis, meditation) (AST, Torsing, Moskva, Harkov, 2005)
17. L.M. Masol, *Khudozhno-pedahohichni tekhnolohii v osnovnii shkoli: yednist navchannia i vykhovannia* (Artistic and pedagogical technologies in primary school: unity of education and upbringing) (Drukarnia Madryd, Kharkiv, 2015)
18. D. Yunyk, *Teoriia ta metodyka mystetskoi osvity. Naukova shkola H. M. Padalky* (Theory and methodology of art education G. M. Padalki Scientific School), 2nd edn. (NPU imeni M. P. Draghomanova, Kyiv, 2011), pp. 121–150
19. V.M. Mishchanchuk, Academic notes. Series: Pedagogical Sciences **170**, 97–101 (2018)
20. A.A. Karelyn (ed.), *Psihologicheskie testyi* (Psychological tests), vol. 2 (VLADOS, Moskva, 2003)
21. O.P. Shcholokova, *Osnovy profesiinoi khudozhno-estetychnoi pidhotovky maibutnoho vchytelia* (Fundamentals of professional future aesthetic preparation of a future teacher). (Vipol, Kyiv, 1996)
22. I. Hrinchuk, O. Burska, *Problemy muzychnoho myslennia: teoriia i metodyka rozvytku. Dialektyka muzychnoho lohosu ta eidosu* (Problems of Musical Thinking: Theory and Methods of Development. Dialectics of music logos and eidos). (Pidruchnyky i posibnyky, Ternopil, 2008)

The concise associative and dissociative dictionary as a resource for accelerated teaching of a foreign language: experimental testing of the theory

Nataliia Slukhai[1], *Liudmyla* Slavova[1,*], and *Anastasiia* Flanagan[2]

[1]Taras Shevchenko National University of Kyiv, Institute of Philology, 14 Shevchenko Blvd, 01601 Kyiv, Ukraine
[2]Upgrades Language School, Skeppargatan 27, 11452 Stockholm, Sweden

Abstract. The paper outlines the results of experimental testing of the effectiveness of the principles of analogy, cognitive effects and the frame-cluster approach as a basis for a concise associative and dissociative dictionary, which is seen as a means of accelerated teaching of a foreign language based on the knowledge of the mediator language (here: English). It is a proven fact that analogy (including association) is a universal principle of thinking which enables one to correctly identify lexical units of the language of one group through the mediator. The correctness of involving the dissociative principle has been confirmed experimentally: it is based on memorizing the uncommon and different. Accelerated, interest-driven teaching of a foreign language based on associative and dissociative principles takes into account the resources of cognitive effects, which are cognitive illusions not controlled by consciousness. We have substantiated the expediency of distributing vocabulary in the proposed dictionary according to the frame-cluster principle, in which the learner constructs a situationally relevant phrase from cluster-organized associated forms. Repeated use of lexemes in communication provides a correction of errors over time. The experiment confirmed that such a dictionary can satisfy the needs of the modern information society, where multilingual communication is in demand.

1 Introduction

Today's information space in the context of globalisation and widespread penetration of virtual technologies is characterized by its high intensity, multisemioticity, intertextuality and collage character. The world has changed, and these changes have called forth new challenges for the information society, especially in the field of mastering foreign languages: politics, business, and the social sphere, showing, in particular, a sharp increase in the number of immigrants. Language for them is the main component in adaptation in society. Finally, cultural needs of contemporaries call for new high-speed technologies of accelerated teaching of a foreign language, corresponding to the high pace of modern life.

The purpose of the concise associative and dissociative dictionary based on the principle of analogy is to satisfy this need by providing the opportunity to master the resources minimally sufficient for communication in the private sphere in one or two weeks, and communication in fields of human activities in 3-9 weeks. The principle of analogy rests on the resources of the mediator language; thus, English can be a stepping stone to a fast and high-quality teaching of other Germanic languages.

Mastering a great number of lexical units within a short period of time quickly depletes short-term memory resources; however, the support of cognitive effects (cognitive illusions) allows the learners to use the potential of suggestive learning and maintain their interest in mastering the language; ensure that a learner is "in the flow", which is defined as a state of complete involvement in the process of learning with the dominants of joyful anticipation, pleasure, full engagement (as understood by Csikszentmihalyi [1]).

In a dynamic world the passive learning of lexical units becomes an outdated practice. The concise associative and dissociative dictionary is a teaching tool based on the frame-cluster principle and pragmatically focused on the learning of vocabulary integrated into a communicative situation. Reproduction of words in the structure of repeated communicative situations will ensure the preservation of the vocabulary learned in long-term memory.

New ideas in the field of foreign language learning are promising only under the condition of experimental confirmation of theoretical constructs: the advantages and disadvantages of their implementation are noticeable only in practical application.

* Corresponding author: slavovall16@gmail.com

Thus, the relevance of this research is preconditioned by the need for experimental confirmation of the possibility of high-speed situation-oriented, interest-driven foreign language learning in the context of requirements of a modern, dynamic and multicultural information society.

The purpose of the paper is to test the possibility of accelerated teaching of a foreign language involving a mediator language on the basis of using the principle of associative and dissociative teaching; activating reserves of cognitive effects; a frame-cluster approach to mastering the language.

The latest large-scale experience in successful high-speed interest-driven learning is suggestopedia as "the science of targeted use of means of suggestion in various forms and stages with pedagogically reasonable purpose, in particular for expanding the learners' mnemonic capabilities" [2, p. 5]. Suggestopedia emerged as a response to dissatisfaction with extensive methods based on training at the expense of time, involving an increasing amount of technical means and expanding training material. These methods all included the grammar and translation method of the 18th century, the directive method of the 19th century, Harold Palmer's method (the 20s and 40s of the 20th century), the audio-lingual method (Ch. Fries, R. Lado, the 1940s and 60s), the audiovisual method (P. Gouberina, P. Rivenc, the 1940s and 1960s). Suggestopedia generated intensive methods which are based on acceleration of learning per time unit due to the appeal to the reserve mental resources of a person. Broadly defined, suggestopedia is based upon the use of natural suggestion resources as a manifestation of spontaneous speech communication ("Silent Way" of Caleb Gattegno; "Natural approach" presented in works of Stephen D. Krashen and Tracy Teller (its variety became the method of Total Physical Response, known as J. Usher's method), drama pedagogy, "Situational Language Learning"; Hi TECH; the project or case method described by J. Dewey and W.H. Kilpatrick; the tandem method). In a narrower sense, suggestopedia is based on the use of suggestion resources outside the rational control of a person (this is the suggestive method of Georgi Lozanov and the emotional semantic method of Galyna Kitaygorodskaya, the method of verbal image emotional-volitional control of a person's state (Georgiy Sytin); the suggestocybernetic training method after Viacheslav Petrusinskii; but also hypnopedia, rhythmopedia, relaxopedia, Freire pedagogy and others. All these methods are based on the cognitive approach to language and human thinking (the history of suggestopedia is fully considered in [3, pp. 170-183]). All these methods rely on teaching by using a typical or atypical dictionary – either encyclopedic or linguistic – dictionary as an extract of speech or "dissolved" in a discourse, visualized or sounding, curtailed to "dots" of a consciously unconscious associative-verbal network or expanded and emotionally expressively enhanced dictionary, suggested or imposed onto the mind, with the added possibility of it assuming a virtual form.

The history of dictionary making relevant to foreign language learning abounds in promising solutions, as does the history of the above-mentioned training methods. There are different types of dictionaries compiled in order to satisfy various needs. Dictionaries assist users to achieve a set of goals; correspondingly, there are distinguished: encyclopedic dictionaries; etymological dictionaries; monolingual dictionaries; bilingual dictionaries; thematic dictionaries; spelling dictionaries; terminological dictionaries, etc. Among monolingual, bilingual, and multilingual dictionaries the most widely-spread are bilingual ones.

The most typical reason for consulting a dictionary is finding out what a word means. In this case, bilingual or explanatory dictionaries are used. In bilingual dictionaries the source language is a foreign one for the user and the target language can be either the user's native one (or the one better known in comparison with the source language) [4]. In such type of dictionaries we are supposed to find the meaning of the source language entry. L.V. Ščerba differentiated between a translation dictionary and an explanatory dictionary, arguing in favour of the latter [5].

The main idea of a bilingual dictionary is to provide a perfect target language equivalent for each source language headword. The classification of equivalents (cognitive, translational, explanatory, and functional) was described in works of Zgusta [6], Adamska-Sałaciak [7, 8], etc.

At the end of the twentieth and the beginning of the twenty-first century there was observed a tendency in the sphere of dictionary compiling from introspection-based to corpus-based. Mainly electronic corpora (e.g. Sketch Engine, Mova.info, etc.) have been involved. Online bilingual dictionaries are gaining popularity, and new lexicographic projects appear, e.g. Norwegian and Swedish Lexin project, French Dictionnaire ARTES project. The tendency of applying information technologies has also been traced in Slavic lexicography [9, 10, 11, 12, 13, etc.].

Bilingual dictionaries differ in number and type of items included, amount of information in a single headword, and the target audience. The last factor predetermines the scope and directionality of a dictionary [14].

We'd like to focus our attention on bilingual dictionaries serving educational purposes, i.e. bilingual learners' dictionaries [15, 16, 17, 18, etc.]. They are suitable for learners of all levels: from beginners to advanced students. The role of dictionaries in the process of learning a foreign language cannot be overestimated. The idea is to suggest a relevant dictionary that will facilitate that process.

Zöfgen writes, "It is therefore all the more surprising that in a market saturated with language learning material a bilingual dictionary developed along the lines of pedagogical lexicography and especially designed for the foreign learner is [...] still not available" [18, p. 2888].

Intensive learning of a foreign language is possible if the cognitive approach is applied, the so-called cognitive effects are involved [19]. There has already been a successful attempt to compile the Associative Thesaurus of English [20]; The Edinburg Associative Thesaurus and the Associative Thesauri of Russian and English languages based on a psychological perception of consciousness units linked in the human psyche [21].

As for the present situation with the bilingual learners' dictionaries, it is quite vivid that there is practically a small variety of dictionaries that can boost the learning process. There is an urgent need in bilingual dictionaries specially designed for learners, especially those who learn other kindred languages like English, Swedish, and German.

If comparison of languages at the stage of their development is typology, then at the stage of their perfection it is primarily comparison of their pictures of the world created with the help of languages [22, p. 66].

It is necessary to bring truly innovative approaches into the vast field of dictionary-making.

The intensification and complication of the information flow of modern civilization requires finding new principles and teaching methods, offering new types of dictionaries in order to simplify and speed up teaching.

The concise associative and dissociative dictionary we offer as a resource for accelerated teaching of a foreign language is suitable for learners with good knowledge of the mediator language and is governed by three principles: using the capabilities of associative and dissociative thinking; reserves of cognitive effects; and the distribution of lexis among frame clusters simulating simple communicative situations ("I buy bread" is made up of the subject, action, object clusters). Swedish has been chosen as the language to be learned. With over 9 million speakers it is the most popular Germanic language spoken in Scandinavia, it is quite relevant to use the associative and dissociative method to master this language. English is chosen as the mediator language with a view to its global status with the highest percentage of speakers in the world.

The associative and dissociative principle is based on unequal components: the associative principle dominates as a reflection of the universal law of analogy. Analogue thinking is both the product and the tool of human social development. It is as old as society that has realized its human nature. Fundamental analogy is a consistent property of consciousness [23], i.e., after M. Heidegger, "analogy, understood as the main feature of being, outlines completely certain possibilities and ways of producing the truth of this being inside existence "correspondence, thought, as the fundamental characteristics of the Being of whatever is, furnishes the pattern for very specific possibilities and modes of setting the truth of this Being, in whatever has being, into the work" [24, p. 143]. Thus, analogy is characteristic of speech activity and can be observed on all levels of the language structure. The logic and linguistic content of the analogy axis is the relationship of identity (including pseudo-identity), similarity, analogy proper and close or remote association. The midpoint of the axis is constituted by the logical linguistic relations of similarity, but that is only by convention, since any point on the axis may occupy a central position depending on the focus of attention. Therefore, the analogy axis may also be called the axis of similarity, identity or association (explained by N. Slukhai [25, pp. 260-264]).

Dissociative thinking, which mirrors the associative, is helpful for memorizing. Lexemes chosen on the basis of dissociation should take up to ten times less volume in the learner's dictionary and to the highest degree meet the principle of dissimilarity and even exotic character for the learner's linguistic consciousness; they must evoke great interest and be placed in the von Restorff cognitive effect zone, as one exploits the cognitive effect reserve of the unusual, different from the norm, interesting, and exotic.

Cognitive effects are a by-product of human creative thinking. They allow using resources of the subconscious, not controlled by the mind; opening up the resources of cognitive zones that are not covered by rational thinking. The role of cognitive effects, such as distortions and illusions, in associative and dissociative teaching can hardly be overestimated; the algorithm of these effects is described in [19].

The frame principle of organizing the material allows the learner to quickly get the pragmatic effect of language proficiency. It is well known that grammar in the human mind is lexicized, it is represented by words that appear in "pre-speech readiness". This makes it possible to easily place analogue forms in the utterance structure, reflecting the typical frame structure. Modern thinking is highly anthropocentric, placing the "I" in the subject actant position of the frame, we assume after mentioning the action ("want") and a range of objects ("bread, butter, milk"), which are easy to memorize when placed in a typical frame position. The best way to learn vocabulary is when a lexical unit is related to a given frame. Such an approach enables a learner to retain words in long-term memory. The idea that the meaning of words is best characterized in terms of experience-based schematization of events and objects in the speaker's world [26, p. 1320] has been proved in the FrameNet project operated by the International Computer Science Institute (Berkeley, the USA). It brings pleasure to learners when they can use language in basic situations, even if it is just the result of the cognitive effect of surface knowledge (the Dunning-Kruger effect). Maintaining interest in a language and being able to see one's own progress facilitate further learning and help quickly achieve basic proficiency.

The theory of accelerated teaching of a foreign language with the help of a concise associative and dissociative dictionary has been substantiated [19], and applied in a teaching practice. However, it still needs experimental verification. Any theory supported by experiments turns out to be an important step either in confirmation or rejection of its outcomes. But as practice shows, experiments usually prove its efficiency. The theory of frame semantics has been applied to the construction of the online Multimedia Encyclopedia of Japanese Mimetics [27]. The results of the experiment proved the efficiency of the frame-semantic approach to mimetic semantics.

An experimental verification of these provisions is presented in the results of this paper.

2 Methodology

Material of the research consists of the following experimental data: 56 extended questionnaires based on the language material of the sample in the amount of more than 3,000 lexemes from the online resource hosgeldi.com ("The Swedish language: lessons"), reduced to about 1000 lexemes of the concise associative and dissociative dictionary. The main method is experiment, supplemented by cognitive analysis, the ranking of lexemes according to associative and dissociative principles, degree of usage in speech, frame and qualitative characteristics; observation and description.

A free associative experiment was conducted in student groups majoring in translation (translation from English and other Western European languages) in their 3rd or 4th year of the Bachelor's program or 1st and 2nd year of the Master's program. The experiment was carried out at the Institute of Philology, Taras Shevchenko National University of Kyiv, from November 10 to December 10, 2019. The number of students is 56. The number of responses received to each question varied between 47 and 56.

The aim of the experiment was to establish whether the principle of associative and dissociative teaching has any effect on accelerated learning of the language, and if so, then to what extent and in which groups of lexemes; whether cognitive effects are effective (based on one: the von Restorff effect); and whether the frame-cluster distribution of material influences the process of memorizing the words.

The associative experiment was carried out in the form of a questionnaire; the duration of the experiment was limited to two academic hours.

The questionnaire included two batteries of questions. The first dealt with some identifiers of the linguistic personality and potential linguistic competence of students (1. Your age and gender. 2. Level of English proficiency; 3. How long have you been learning English? 4. What other European languages are you proficient in?).

The second battery of questions concerned the identification of resources and the influence of associative and dissociative thinking of students; cognitive effects and the frame-cluster distribution of words to help students in their accelerated learning of Swedish with English as the mediator language.

Two batteries of questions were given to students, whose proficiency in English was at the level of upper-intermediate. Such level of proficiency allows identifying based on the method of analogy on average 37% of the lexemes randomly offered by the electronic resource of the website www.hosgeldi.com. However, the analogy fixed in consciousness between the lexemes of the Swedish and English languages has different degrees of manifestation, which was shown by the test data in the questionnaires.

Carrying out the experiment among the students, the authors of this technique deliberately left out of sight a great number of lexemes which pose zero translation difficulty with a high level of coincidence of the form and meaning of correlative lexemes (*att kopiera, att decorera, att dansa, att notera, att spendera, att reparera*). It is obvious that there is no difficulty in translating them; the number of these lexemes reaches 10% of the virtual source dictionary.

The associative thinking of students was measured by tasks for translation from the unfamiliar language (Swedish) into English (language proficiency is not lower than Intermediate). The translation was expected to be made on the basis of association by form and place in the analogy paradigm, the inner form (image).

Tasks of low, medium and high difficulty were proposed. For example, the lexemes *att hata, att sitta, att gå upp, att landa, att sälja* were placed in the low translation difficulty level. The lexemes of low difficulty were distributed according to the frame positions of subject, object, action/state, instrument, time, space, number, color (1st column) or not distributed (the 2nd column). The aim was to establish whether the identified position of the lexeme in the frame structure makes the analogue translation easier, e.g., subjects: *jag, en man, en son, en dotter, en bror*; objects: *bröd, mjölk, ett misstag, en vinge, ett bröst*; actions: *att gå, att komma, att höra, att svälja, att ha*; states: *livlig, arg, törstig, drucken, trött*; instruments: *en panna, en hammare, en kniv, en yxa, en maskin*; time: *morgon, nu, en dag, efter, en natt*; space: *här, nära, Syd, Väst, en ocean*; number: *få, en, två, tre, fyra*; color: *röd, grön, blå, vit, grå*.

In the next task the students were encouraged to translate vocabulary with the low degree of difficulty, not arranged by clusters: *att förbjuda, att dricka, att möta, att bita, att blåsa, att glida* and the like.

In the task of the medium level of difficulty there were offered lexemes that were not distributed according to frame position (correlating with similar ones in the previous task with lexemes of the low level of difficulty of identification, lexemes of medium level of difficulty of analogue translation demonstrate the depth of human analogue thinking), e.g. *att kyssa, att slicka, att följa, att ljuga, att viska*.

The task of a high level of difficulty contains lexemes that are difficult to identify in the analogue paradigm, i.e. those that are not suitable for the associative teaching methods. Lexemes that are difficult to identify include those ones the interpretation of which requires special knowledge, though their analogue potential is quite high: *att möta* (to meet), *att låta* (to let), *att tro* (to trust).

 https://doi.org/10.1051/shsconf/20207504008

3 Research results

The analysis of the answers given by the students has allowed us to establish the following correlations: the depth of associative and dissociative thinking does not depend on age or gender but does depend on the level of English proficiency (it was easier to detect analogue connections for those who had Upper Intermediate or higher). In the case of Swedish, it also depends on knowledge of other Germanic languages (in particular, it is productive to know both English and German; while knowledge of Romance languages does not influence the successful search for analogue connections).

Lexemes which had been sorted by clusters according to typical frame positioning were translated more successfully than those that had not been sorted (see Fig. 1).

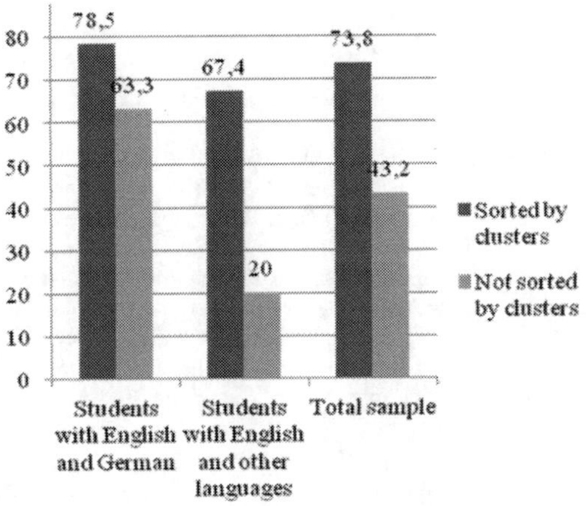

Fig. 1. The difference in translation done by: all students; those who know English and German; those who know English and other languages (except German), of analogue forms of low degree of translation difficulty sorted by clusters and not.

As it was expected, students translated lexical units displaying a high and medium degree of analogy, i.e. a low degree of translation difficulty (combined when calculating) more successfully than lexical units with a low degree of analogy. The correlation between proficiency in various languages and the rate of correct answers was as described (see Fig. 2).

Students' dissociative thinking was measured by two tasks: to translate lexemes with a high level of identification difficulty (i.e. the low level of associativity mentioned above) and to translate lexemes not distributed in the frame structure, indicating ethnically relevant phenomena that usually cause increased interest in studying the language: *Midsommar* – Midsummer, *Jul* – Christmas, *Nyår* – New Year, *Kanelbullens dag* – Cinnamon Bun's Day, *fikastund* – coffee break, *Tjugondag Knut* – St Knut's day, *Alla Helgons Dag* – All Saints' Day, *sambo* – cohabiting partner, *särbo* – partner but living apart (the words were given without translation).

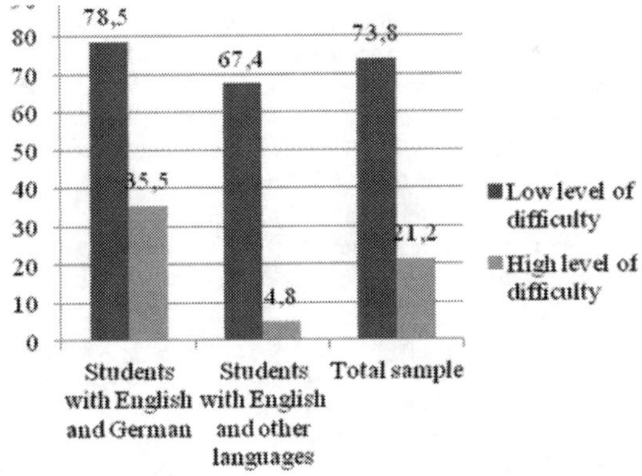

Fig. 2. Comprehension of low difficulty word sets versus high difficulty ones with words grouped by clusters (per cent)

They include lexical units of ethnoculturally relevant phenomena (modern and historical, profane and sacral), ethnoexotisms, including an unusual visual form (*mormor* – maternal grandmother) or their sounding (*Hur mår du?* – How are you? including homophones: *sex* – six, *bra* – good) and ethno-culturally resonant deviations concerning average regional norms, and not necessarily negative (*nusvenskar* – new Swedes). The level of difficulty of the task (in relation to the answers received) was rated by students as 4.1 (on a scale from 1 to 5). The level of difficulty in perceiving the words of a dissociative block (lexemes of a high level of difficulty in translation and ethnoculturally relevant lexemes) in connection with the range of learned languages is displayed in Figure 3.

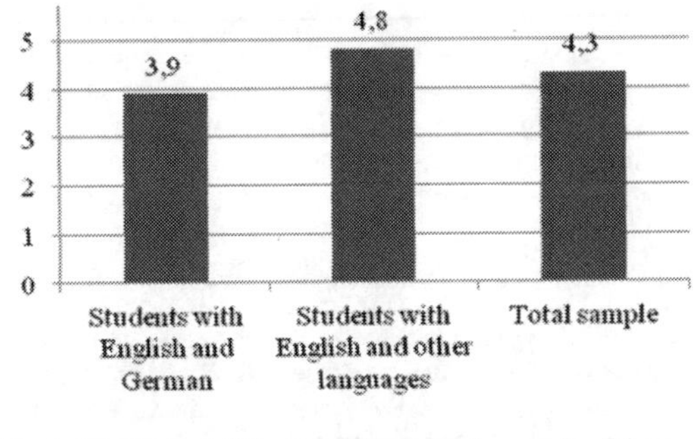

Fig. 3. The level of difficulty while perceiving words of the dissociative group by all students; students with competence in English and German; in English and languages other than German (in points on a scale of 1 to 5).

The dissociative group as a rule includes ethnoculturally relevant lexemes which usually elicit increased interest among recipients of the new culture and are usually perceived differently from non-transparent dissociative lexical units.

They are expectedly difficult to understand (see Figure 4). However, they attract special attention of students interested in understanding a new linguoculture. Accordingly, the aim of the next task was to establish if it is possible to use the increased interest of students in mastering ethnocultrally relevant information to accelerate learning. When presenting a translation followed by explanations of ethnoculturally relevant information, students displayed a higher degree of involvement and easily reproduced the meanings of lexemes when asked during the follow-up assessment a week later. To assess the success of dissociative training students were also asked to answer the question whether it was fun and interesting to translate ethnically relevant lexemes, and if so, to give the reason (the aim of the task was to find out whether the von Restorff cognitive effect works in the field of teaching a foreign language; or the isolation effect, which makes the different element in a range of more similar ones easier to memorize [3]). The answers were the following: yes (without explanation – 7); yes, because it is fun and informative; because it is unusual due to belonging to a different culture while denoting something familiar; because it is a different designation within the Germanic group; it is extremely interesting because it activates associative thinking and causes one to look for hidden regularities in the lexical system; it is very interesting to face the "dissociative challenges"; it brings satisfaction when you give a correct answer; it is exciting to translate ethnically relevant lexemes; it is always exciting to implement knowledge in new fields; it is truly enticing because it challenges your knowledge and logic; it is fun to look for the inner form of the words; difficult but interesting because it activates translation skills, because it widens your horizon and enriches your background knowledge in culture; it is interesting to compare the degree of similarity between English and Swedish words; it is interesting to find associations based on knowledge of other known languages. However, there were answers 'no' that amounted to 8% (it does not provide ground for the development of linguistic intuition; it is difficult; complicated because it is not possible to see correspondence; lack of knowledge is not compensated by the associations). Presumably the efficiency of cognitive effects while studying the dissociative layer of lexical units will increase if the visual representation of the image is provided: this task is easily solved in a digital version of the associative and dissociative dictionary.

In the questionnaire the students were also asked to find an explanation for their ability to translate the words of the unknown language. As a hint students were offered three answers: the form, the inner form of the word, and attraction of analogue paradigm (rated on an ascending scale from 1 to 5). At the same time, the form (basically connected with the meaning) and the attraction of the analogue paradigm, presented to students as two answer options, in fact, demonstrate one phenomenon.

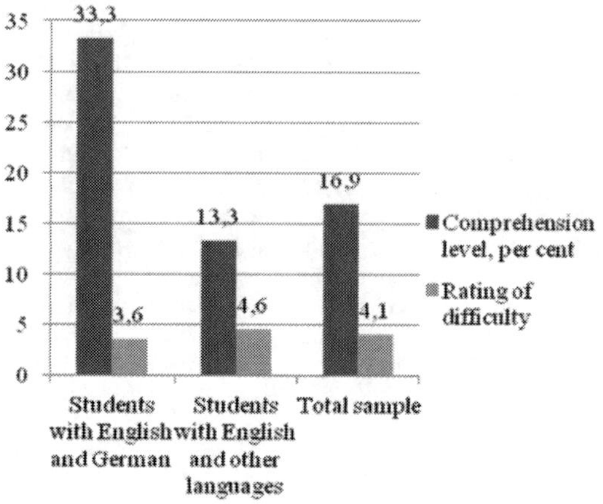

Fig. 4. Comprehension of the ethnic relevant lexics.

The total number of points for "the form" is 147, for "the inner form" – 134 and "attraction of the analogue paradigm" – 207. These results demonstrate the students' awareness of the impact of analogue paradigm. Qualitative results are even more revealing: the maximum influence of the analogue paradigm (rated with 5 points) was indicated by 41 out of 56 students; the maximum importance of the form was perceived by 5 students; the maximum importance of the inner form was indicated by only 4 students; others found it difficult to answer. The qualitative results of these responses are correlated with the qualitative results of training students majoring in this specialty, which is likely to indicate that the higher the students' linguistic competence, the higher their orientation potential in the analogue paradigm. However, taking into account that only 4 students rated the influence of the analogue paradigm with two points, and only three students gave one, it can be argued that almost all students are aware of the high influence of the analogue paradigm.

The next task was to transcribe 5 lexemes that were easily visually identifiable, but difficult audibly, all supplied with translations: *att gäspa* – to gape, *att sjunga* – to sing, *tjock* – thick, fat; *att knäfalla* – to kneel, *ett hus* – a house. The purpose here was to determine whether the visual analogue paradigm indeed plays a leading role in accelerated language learning, and losses in the communication channel with distorted speech production should be recognized as acceptable. As expected, students transcribed the words with distortions of the sound image, to a greater or lesser extent they resembled sounding of English words (like *ett hus – ət ys*; *att gäspa – æt gəspə* (sic), etc.). However, the sound image of a word reproduced by an inexperienced speaker with distortions will still be adequately perceived by an experienced speaker. Several attempts to pronounce the word or form with distortions and subsequent external corrections will help come to the correct pronunciation according to the rules; however, these losses in communication might be viewed as acceptable in view of the significant gains in terms of accelerated language learning.

4 Conclusions

Thus, the experiment confirmed the potential of the concise associative and dissociative dictionary as a method for accelerated foreign language teaching with reference to the mediator language, namely: the correctness of reliance on analogue (associative) forms of low and medium levels of perception difficulty (a zero level of perception difficulty will bring definitely positive results); the relevance of involving dissociated forms when it comes to ethnically relevant phenomena that cause obvious interest; the effectiveness of cognitive effects; efficiency of lexical units distribution in an associative and dissociative dictionary according to the frame-cluster principle. In this case, both digital and printed versions of the associative and dissociative dictionary are better to compile according to clusters, enabling the learner to build situationally relevant utterances out of the associative forms: *Jag vill äta*; *Jag vill ha en röd hatt*; *Jag måste köpa bröd och mjölk*; *Jag vill ha en kopp kaffe*; *Jag måste gå*; *Ge mig bröd och fisk*. Any English-speaking person can build these phrases having the associative and dissociative dictionary, which determines the prospects of this project. It is advisable to supplement the digital version with visual images of ethnically relevant phenomena.

Based on the results of the experiment, however, the following drawbacks of the proposed dictionary can be predicted:

1 Accelerated language learning with the help of a concise associative and dissociative dictionary requires knowledge of the mediator language at the level of upper-intermediate (preferably two languages).

2. The part of the dictionary that will be learnt will correspond to a strong anthropocentric regulator, placing the main emphasis on daily language and, partly on media and arts spheres, but will be much less relevant for professional communication.

3. Only part of the dictionary subordinated to the associative and dissociative principle can be mastered and practically applied due to the suggested technique, i.e. less than 50%.

4. It can be assumed that there will occur distortions in the pronunciation of separate sounds or parts of words until they are situationally corrected.

5. There can also be some grammatical distortions, which will be corrected in spontaneous communication.

Despite these restrictions, it has been experimentally confirmed that a concise associative and dissociative dictionary as a resource for accelerated teaching of a foreign language is able to satisfy the needs of today's information society, where there is an acute demand for personal multilingual communication.

References

1. M. Csikszentmihalyi, *Flow: The Psychology of Optimal* Experience (Harper and Row, New York, 1990)
2. S.S. Palchevskii, Dissertation, Institute of Pedagogy of National Academy of Pedagogical Sciences of Ukraine, 2003
3. N.V. Slukhai, *Suggestiya i kommunikatsiya: lingvisticheskoye programmirovaniye povedeniya cheloveka* (Suggestion and communication: linguistic programming of human behavior). (Kyiv National University Press, Kyiv, 2012).
4. A. Adamska-Sałaciak. In *International Handbook of Modern Lexis and Lexicography,* ed. by P. Hanks, J.-M. de Schryver (Springer, Berlin, Heidelberg, 2015). doi:10.1007/978-3-642-45369-4_6-2
5. L. Ščerba, Towards a general theory of lexicography (trans: Farina, D.). Inter. J. Lexic. **8**(4), 305–349 (1995). doi:10.1093/ijl/8.4.314
6. L. Zgusta, *Manual of lexicography* (Academia, Prague; Mouton, The Hague, 1971)
7. A. Adamska-Sałaciak, *Meaning and the bilingual dictionary: The case of English and Polish* (Peter Lang, Frankfurt-am-Main, 2006)
8. A. Adamska-Sałaciak, Examining equivalence. Inter. J. Lexic. **23**(4), 387–409 (2010). doi:10.1093/ijl/ecq024
9. M. Banko, *Z pogranicza leksykografii i językoznawstwa* (On the borderline of lexicography and language study). (Wydzial polonistyky UW, Warszawa, 2001)
10. P. Zmigrodzki, *Wprowadzenie do leksykografii polskiej* (Introduction to Polish lexicography). (Wydawnictwo Uniwersitety Śląskiego, Katowice, 2009)
11. V.A. Shyrokov, *Linhvistychni ta tekhnolohichni osnovy tlumachnoyi leksykohrafiyi* (Linguistic and technological basis of interpretative lexicography). (Dovira, Kyiv, 2010)
12. V.A. Shyrokov, *Kompiuterna leksykohrafiya* (Computer lexicography). (Naukova dumka, Kyiv, 2011)
13. Z. Saloni, Perspektywy polskiej leksykografii jednojęzycznej (Outlook of Polish monolingual lexicography). Poradnik językowy **7**, 1–18 (1996)
14. A. Adamska-Sałaciak, In *The Bloomsbury companion to lexicography*, ed. by H. Jackson (Bloomsbury Academic, London, 2013), pp. 213–231
15. A. Adamska-Sałaciak, in *English Learners' Dictionaries at the DSNA 2009*, ed. by I.J. Kernerman, P. Bogaards (K Dictionaries, Tel Aviv, 2010), pp. 121–137

16. R.H. Gouws, Monolingual and bilingual learners' dictionaries. Lexikos 14, 264–274 (2004)
17. J. Tomaszczyk, In *Lexicography: principles and practice*, ed. by R.R.K. Hartmann (Academic Press, London, 1983), pp. 41–51
18. E. Zöfgen, in *Dictionaries: an international encyclopedia of lexicography*, ed. by F.J. Hausmann et al., vol. 3 (Walter de Gruyter, Berlin, New York, 1991), pp. 2888-2908
19. N. Slukhai, S. Slukhai, A. Flanagan, Kognitivnyye effekty kak resurs analogovogo obucheniya: printsipy sozdaniya assotsiativno-dissotsiativnogo slovarya (Cognitive effects as a resource for analog learning: principles for creating an associative-dissociative dictionary). Science and education: a new dimension. Philology VII(62), I. 211, 49–53 (2019)
20. G. Kiss, *An associative thesaurus of English* (University of Edinburg, Edinburg, 1972)
21. N. V. Ufimtseva, The associative dictionary as a model of the linguistic picture of the world. Procedia – Soc. Behav. Sc. **154**, 36–43 (2014). doi:10.1016/j.sbspro.2014.10.108
22. V. M. Alpatov, *Istoriya lingvisticheskikh ucheniy* (History of Linguistic Doctrines). (Jazyki slavjanskoj kultury, Moscow, 2005)
23. O.O. Taranenko, *Polisemichnyy paralelizm i yavyshche semantychnoyi analohiyi* (Polysemic parallelity and semantic analogy). (Naukova dumka, Kyiv, 1980)
24. M. Heidegger, *The questions concerning technology and other essays* (Harper Perennial, New York, 1977)
25. N. Slukhai, Dissertation, Taras Shevchenko National University of Kyiv, 1996
26. *Recent developments with focus on electronic and computational lexicography*, ed. by R.H. Gouws, U. Heid, W. Schweickard, H. E. Wiegand. Dictionaries. An international encyclopedia of lexicography. Suppl. vol. (Mouton De Gruyter, Berlin, 2013).
27. K. Akita, in *Cognitive-functional approaches to the study of Japanese as a second language*, ed. by K. Kabata, K. Toratani (Mouton De Gruyter, Berlin, 2016), pp. 139–168

Interactive methods of training for enhancing communication of IT-professionals

Svitlana Symonenko[1,*], *Nataliia* Zaitseva[1], *Viacheslav* Osadchyi[2], and *Margaryta* Vynogradova[1]

[1]Dmytro Motornyi Tavria State Agrotechnological University, Department of Foreign Languages, Melitopol, 72312, Ukraine
[2]Bogdan Khmelnitsky Melitopol State Pedagogical University, Department of Informatics and Cybernetics, Melitopol, 72300, Ukraine

Abstract. The paper deals with the issue of enhancing communication of IT-professionals using interactive training methods. It is emphasized that this problem is urgent for IT-education, which has the quickest rate of the content change, formation of new knowledge fields, shift in types and volumes of communication and interaction. Regulations, guides and qualifications frameworks of European countries and Ukraine in terms of requirements to communication of specialists in different fields have been analysed. The concepts of interactive teaching, interactive teaching methods and activities, and their advantages have been studied. It has been noted that interactive methods can be successfully used in teaching general and special subjects for future IT-professionals. Main peculiarities of certain interactive teaching methods have been described. Some practical ideas of the project method, brainstorming, the gamification method, the debate use in the educational process with the focus on the English language at higher educational institutions for enhancing better communication of IT-professionals have been presented. It is emphasized that the meticulous combination of interactive methods and preparation fosters comprehension at the sufficient level both in use of language aspects and in live communication.

1 Introduction

Modern researchers and practitioners are in the search of the most efficient forms, methods and approaches of the vocational training of specialists in different fields. The search is urgent for IT-education, which has the quickest rate of the content change, formation of new knowledge fields, shift in types and volumes of communication and interaction. The issue of enhancing communication of IT-professionals is even more relative.

Importance of successful communication has been proved by a series of rules, guidelines, procedures, frameworks. The analysis of regulatory acts of European countries has shown that the communicative competence is one of the vital and professional necessities. The European Qualifications Framework for Lifelong Learning includes eight key competences: 1) communication in the native language; 2) communication in a foreign language; 3) mathematical competence and basic competences in science and technology; 4) information competence; 5) ability to learn; 6) social and civic competences; 8) a sense of initiative and entrepreneurship; 9) cultural awareness and expression.

The document emphasizes that all competencies are equal since each competence contributes to the successful life in the knowledge society; competences can be partially matched and combined; many topics can be used in the European Qualifications Framework for Lifelong Learning, as they play a significant role in key competencies: critical thinking, creativity, initiative, problem solving, risk assessment, decision making and constructive management of emotions [1].

The necessity of constant competence improvement and development is confirmed by ISO (the International Organization for Standardization) which has issued ISO 10015:2019 "Quality management – Guidelines for competence management and people development" document where certain crucial modifications in comparison with the ISO 10015:1999 version have been introduced. The authors of the document offer specific guidelines for establishing, implementing, maintaining and improving systems for competence management and people development in order to get positive outcomes of their work. The authors insist that companies, employers or organisations are to encourage people to get or improve competences "by creating learning and training opportunities with circumstances to deploy the outcomes that have been acquired" [2].

National qualifications frameworks of certain European countries (Bulgaria, Germany, Scotland and others) include definite descriptors of communicative competence. Descriptors of 5, 6, 7 and levels of the National Qualifications Framework of Ukraine [3] in the "Communication" category are defined as: interaction, cooperation with a wide range of persons (colleagues, managers, clients) for professional or educational activities; bringing information, ideas, problems, solutions and own experience in the field of professional activity to specialists and non-specialists; ability to formulate a communication strategy effectively;

* Corresponding author: asimonenko@ukr.net

comprehensive and univocal reporting own conclusions, knowledge and definitions which substantiate them to specialists and non-specialists, particularly to learning people, application of foreign languages in their professional activity; communication in the dialogue mode with broad scientific community and public in a definite field of scientific and/or professional activity. It is obvious, that the contents of communicative competence descriptors have been complicated from one level to another in the document, as specialists are to develop their competences while studying or working.

The PMBOK (A Guide to the Project Management Body of Knowledge, PMBOK Guide) which is considered to be the set of terminology and guidelines for project management [4] emphasizes the importance of communication while working on a project which is in turn an inevitable demand for any high qualified specialist. The guide details three project management processes that are key to successful project communication: planning (developing a communications management plan), execution (management of communications) and control and monitoring (control of communication).

The purpose of the paper is to determine the interactive methodological approaches and forms of training for enhancing communication of IT-professionals, choose academic disciplines which foster communication of students, and present real examples of interactive activities which have been used.

2 Interactive methods of training at higher educational institutions

The most promising and most efficient training forms and methods are interactive forms of educational process organization, technologies and teaching methods at practical classes which can be successfully used in order to form and develop the communicative competence. The practitioner R. Knapen [5] defines interactive teaching as "instructing the students in a way they are actively involved with their learning process". The author insists on the existence of numerous ways of such involvement: teacher-student interaction, student-student interaction, audio, visuals, video application, hands-on demonstrations and exercises. The researcher gives the following classification of interactive teaching activities: 1) interactive activities for speech encouraging; 2) individual student activities; 3) student pair activities; 4) student group activities; 5) interactive game activities.

The Ukrainian researcher O. Sichkaruk [6] gives the detailed description of interactive teaching methods in his work: lectures with conversations or discussions, problematic lectures, seminars-discussions, "question-answer" seminars, discussions with provocative questions, consultations, conversations, round tables, brainstorming, group-specific situations, business games, role-playing and didactic games, business stimulations, projects, and panel exercises.

The use of interactive teaching methods, according to L. Naseikina, initiates such an organization of the educational process, which cannot be carried out in the cognitive process, because interactive teaching starts stimulating the cognitive activity of students and involves each participant in the mental and behavioural activity [7].

It is important to note that interactive learning methods are based on communication, which is not characterized by the exchange of thoughts or knowledge, but on the basis of such exchange the intellectual activity is stimulated. In the same way new knowledge is born, the following abilities are enhanced: to use theoretical knowledge in a particular situation, to combine knowledge from different fields, to take into account the opinion of another person, to work collectively to solve a problem, to take someone else's experience, to take responsibility for the decision made, for the results of their own actions, to manage and obey own ambitions for the sake of the common goal [6].

As professor S. Sysoieva [8] states, the following interactive technologies will be appropriate during practical training in the professional disciplines: the project method, the training, the cooperation in cooperation technology, the method of discussions, and the game method.

Interactive methods are considered by V. Petruk [9] to be helpful in forming the ability to communicate both as a person and a specialist. Students engage in a relationship with each other, resulting in the reproduction of a conflict situation accompanied by natural emotional tension. This leads to increased interest in the learning process. The ability to directly test different situations, promotes further active discussion of a particular problem by students.

As the English language is the humanitarian discipline and suggests communication of education agents even with application of interactive methods and forms, multiple factors should be taken into account when developing a syllabus. Each topic of the course should be looked into in terms of the interactive method which is the most appropriate for the participants, chosen timing, supposed equipment, and the venue. The most efficient language-related interactive methods are presented in Table 1.

In our research we proceed from student massive involvement in communication and creative activities, from timing which requires minimum previous self-work and ninety minutes of English classes at the university.

3 Application of interactive methods of training for enhancing communication of future IT-professionals

Having analyzed scientific works, frameworks and guidelines, we have identified the main academic disciplines which foster the development of communication skills for IT-professionals: "Web technologies", "Software development", "Group Dynamics and Communication", "Ukrainian for professional purposes", "English for professional purposes", and "Business English".

In the course of studying the "Group Dynamics and Communication" discipline future IT-professionals aim to form skills of interpersonal communication, developing students' personal and business skills to

 https://doi.org/10.1051/shsconf/20207504009

work effectively in partnership with colleagues, in project teams, and in the further professional activity in interaction with customers and users of software products [10]. The discipline is aimed at the development of such competences, defined in 5-6 levels of the National Qualifications Framework of Ukraine in the "Communication" category.

Table 1. Language-related interactive methods.

Examples / Aspects	Creative tasks (essays, reviews)	Games (business games, role plays)	Use of human resources (expert lectures, tours)	Use of new material (instructive debates, multimedia lectures)	Solving tasks (brainstorming, case analysis)
Main agents	students	students	invited experts, guides, teachers	teachers, lecturers, invited experts	students
Timing	90 minutes, independent learning	90 minutes, independent learning	sufficient timing	90 minutes, previous independent learning	90 minutes
Venues	university classroom	university classroom	university classroom / appropriate alternative	university classroom / lecture hall, computer lab	university classroom
Support materials in English	individual / group tasks, legends, terms	legends, terms, rules, timing	leaflets, brochures, introductory handouts	lecture notes, presentation files, handouts, audio / video files	cases, individual / group tasks, SWOT schemes
Equipment / devices	appropriate for the task	appropriate for the game	wide screen, microphone, computer	wide screen, microphone, computer	white board, stationary

The purpose of learning the "Ukrainian for professional purposes" discipline [11] is to develop the communicative competence of future IT-specialists, to gain communicative experience, to develop the skills of optimal language behaviour in the professional field: influence on the interlocutor through the skillful use of a variety of linguistic means, mastering the culture of monologue, dialogue and polylogue; perception and reproduction of professional texts, mastering vocabulary and terminology of their speciality, choice of the communicatively justified language.

The "Ukrainian for professional purposes" discipline is aimed at developing such competences, which are defined as common ones in the Tuning project [12]: the ability to put knowledge into practice; ability to communicate, including verbal and written communication in Ukrainian and at least one of the common foreign languages; ability to develop project and program documentation that meets the regulatory documents.

When teaching special disciplines, through which students develop software projects and work as a team [13; 14], interactive teaching methods are also applicable. A set of methods can be applied, namely: the project method, the cooperation method (small groups), brainstorming, the case method, and the debate.

At the very beginning of the course, teachers of the department of foreign languages at Dmytro Motornyi Tavria State Agrotechnological University give students of the "Computer Science" speciality commonly known reasons for learning English. At the beginning of the second term of studying English students participate in a debate on the same matter. The goals of the debate are:

1) comparing students' perspective on English learning necessity at the beginning of the course and after six months of being students in their major;
2) generating effective critical thinking;
3) comprehending significant and complex ideas about participants' future profession;
4) stimulating team work and developing negotiation skills;
5) training participants how to articulate own position regarding learning English.

At the first stage, two teams are imposed the attitude by their teacher (either willingness or neglecting learning a foreign language). After the panel discussion, when both teams are given scores for each reason they represent and prove right as a group each student may shift their ground and join the counterpart team. The third stage includes individual acts of speech when each speaker represents their real vision of English necessity in the modern world, for their personal and professional prospect.

The scores are awarded either, especially when speaker's personal experience in communicating with English speakers, using English for task solving or resource exploration is provided. From previous experience it is obvious that "pro"-group is always a victorious team. In spite of fair pointing to difficulties in language acquisition (like a lack of free time because of English home assignments, advisable money investment in educational resources, different personal educability levels, different backgrounds of school English courses), the winner usually scores over their rivals due to numerous arguments about learning English as an integral part of the prospective career path as an IT-specialist.

The project method is based on the idea that reflects the essence of the concept of "project", its pragmatic focus on the result that can be obtained when solving a particular problem. The result can be seen, comprehended, applied in the field. The solution of the problem becomes the nature of the project activity [8].

The main feature and advantage in the process of forming the communicative competence of future IT-professionals is that the final result should be a real product, but not an imaginary project. Students' efforts should be directed towards collection, analysis, processing of real information, taking into account the real situation in the country or in a certain industry, social sphere, based on the existing rules, laws and regulations [6].

The practical use of the project method in the process of forming the communicative competence of future IT-professionals in the study of professional disciplines can be explained by the example of the "Web-technology" discipline concepts:

1. Idea: development, self-development of the communicative competence of future IT-specialists.

2. Goals:

- to learn how to use knowledge in a practical activity that simulates future professional activity;
- to acquire new professional competences or to develop already existing professional competences of a student, in particular, the communicative competence;
- to achieve high results in educational research work by performing project activities;
- to develop teamwork skills.

3. Content: covers the theory and practice of creating online resources.

4. Main groups of tasks:

- educational tasks – mastering the technology of project activity, theoretical knowledge of the discipline and the ability to apply it in practical activities, learning to find the best ways to create a project, to master communicative skills;
- scientific tasks – developing the ability to analyze scientific and methodological sources, Internet sources, the ability to summarize scientific facts and advanced pedagogical experience;
- methodological tasks – developing the ability to design project development activities, plan and design their project activities, apply communication skills in teamwork.

5. General requirements for the project:

1) independent and authentic work;

2) relevance of the project, i.e. its relevance to the current state of information technology;

3) external attractiveness and performance accuracy;

4) conformity of the development tools with the discipline studied;

5) content-richness (appropriate volume and quality of materials);

6) following the specific requirements of the project (the requirements for the webpage construction, functionality, content volume and quality must be met).

6. Possible project topics: "Development of a personal website", "Development of a thematic site", "Development of the website of the department", "Development of the site of the university", "Creating a blog", "Development of an online store" and others.

7. Competences of students to be improved: personal-intellectual, communicative, creative, social, information-communication, and research competences.

8. Forms of organization of students: simple project – one student; complex project – 2-3 students.

9. The role of a teacher in project-based learning: compiles project topics, advises students, defines reporting forms, assessment criteria, delivery time, technical design, and project defense forms.

10. Project defence forms: presentations, reports, demonstrations.

It is necessary to note, that in order to be successful, all communication components in projects must function effectively and communication barriers must be overcome to the maximum. Developing a communication plan can be one way of overcoming communication barriers and building successful communication.

The American researcher S. Rajkumar [15] proposes four steps for effective communication between a project manager, a project team, and stakeholders while working on the project. According to the author, the first step is to determine the communication requirements, i.e. to determine the information required for each project participant, to calculate the number of communication channels and to determine the communication time. The second step is to determine 5W (Why, What, When, Where, Who, and 1H (How)), in other words, to identify the recipients of the messages (Who), information types (What), communication timeline (When), environment (Where), causes of communication (Why), and communication channels (How). The third step is to identify and take into account the environmental factors of the organization that influence the communication process, such as organizational culture and structure, project management information system, etc. The last but not least important step is to determine peculiarities of organizational processes that affect communication, such as standards and policies unique to the organization, organizational instructions, work instructions and criteria for evaluating the effectiveness of others.

The scientist A. Zub [16] insists that a communication plan is an integral part of the project plan. It may include a plan for collecting information that identifies the sources of information and how to obtain it; an information sharing plan that identifies consumers of information and how information is delivered; a detailed description of each document to be received or transmitted, including the format, content, level of detail and definitions used; plan for putting into operation various types of communications; methods of updating and improving the communication plan. The communication plan can be formalized and detailed depending on the needs of the project.

Within the project there is a need for different types of communication: internal (within the project team) and external (with the management of the customer, external organizations, etc.) communication; formal (reports, requests, meetings) and informal messages; written and oral communication; vertical and horizontal communication.

All the mentioned above states that student projects can and should be interactive and professionally managed in order to be successful. Let us consider the project which is based on both communicating with different agents and activities with the PMBOK Guide.

The PMBOK Guide may become a very effective means to improve students' communicative skills and to enhance the research activity within an interactive project – planning by students their research activity for an academic year. The participants of a study group (the undergraduates and postgraduates who major in Computer Science at Dmytro Motornyi Tavria State Agrotechnological University) were familiarized with the PMBOK Guide at the end of the spring term. They were given group and individual assignments for the

https://doi.org/10.1051/shsconf/20207504009

summer holiday. As a group, 8 participants should have interacted and cooperated to perform the following tasks:

1) to negotiate in order to come to an agreement with their group supervisor and with each other about the schedule of group sessions acceptable for everyone;

2) to find online 10 conferences and paper calls, to contact the organization committees (in writing or online) and to make sure that every participant's research field corresponds with the search results and can be represented as a conference report or a thesis;

3) to collaborate in order to plan in advance the summarizing discussion (its form, venue, participants, technical support and handouts provided) at the end of the academic year.

The individual assignment supposed a researcher to establish correspondence with a potential scientific supervisor and to represent their written approval of student's future line of investigation and timing.

The outcomes of the mentioned above interactive project at the end of the winter term have been:

- having successfully planned definite group and individual activities due to the latest practices in project management;
- training of peer communication within the study group;
- teaching students to address themselves in adequate manner to experts in their field of research and official committee representatives and to maintain academic correspondence both in writing and orally (online);
- academic and official English language acquisition through acquiring language patterns;
- providing students' awareness of communication strategy transferability between Academic English and Business English.

In current project activities presented above, the communication competence is being effectively enhanced. Therefore, we consider the project method to be the most appropriate and adequate method to correspond the specifics of ICT. Moreover, in the project activity, future IT-professionals are able to fully demonstrate communication skills and develop communication skills, especially in the case of combination of the project method with other interactive methods.

The issues of building successful communication and overcoming communication barriers are key issues in the modern world of employment, where workers and employers can play different roles, change these roles, learn new methods and technologies, and communicate effectively regardless of their new roles and functions. Moreover, they should be able to build relationships of any type and within any hierarchical organization model. Modern higher education curricula should include special courses, modules, seminars, lectures and psychological training to prepare future professionals to work in different circumstances for effective communication, to form readiness to overcome various communication barriers, to develop communication strategies to be ready to function effectively in the workplace.

Both in the course of project work and in the process of solving software development problems the collaborative learning method can be used. It is the method aimed at developing certain skills and competences, mastering concepts, academic and professional knowledge provided by the curriculum, or organizing project activities with further discussion. In order to build the communicative competence of future IT-professionals, it is important that this method is focused on group goals and the success of the entire group, which can only be achieved through the independent work of each member in simple interaction with other members of the same group when working on a problem, a question to be studied [8].

Brainstorming is a common practice in the software industry. In the concept of software project management, brainstorming is the first stage in the development of a software product. Its purpose during training sessions is to activate students' communicative and intellectual activity, which aims at expressing ideas for solving a specific problem, proposing ways to solve a certain problem, as well as freeing from the inertia of thinking, overcoming stereotypes in solving a creative problem, accumulating ideas for solving the proposed problem [6].

Brainstorming is widely used within planning research and vocational training events jointly with students. Along with bright ideas for projects engaging equally instructors and students, student view and ideas matter for improving learning and educational resources and tools.

The examples of realization of such spontaneously contributed ideas are improvement of the department websites. At the beginning of 2018, students of Dmytro Motornyi Tavria State Agrotechnological University majoring in Computer Science were asked to skim through newly adopted corporate design of the department of foreign languages website within the university domain and to generate ideas and ways to increase traffic to the webpage of the Department of Foreign Languages [17]. In the evaluation session several attracting visitor website headers have been developed and some menu items have been deleted. The content selection strategy has also emerged from that brainstorming session and nowadays it tends to be less entertaining and more progressive.

The website of the Informatics and Cybernetics Department within the domain of Bogdan Khmelnitsky Melitopol State Pedagogical University [18] has also been created considering the results of brainstorming session. The ideas of using the template with automated scroll to given anchor when clicking on a header or a link and also displaying smart popups connected to the department instructors photos presenting brief information about their positions and scientific interests have arisen from collaboration with students majoring in Computer Technology and Digital Design.

Consequently, combining and improving mutual brainstorming ideas have ensured easy-to-use website navigation, attracting content and present-day layout, preferred by students themselves.

The case method is widely used in vocational education because situational exercises (a real description of the situation that occurred in professional practice) offered to students can be created to master the profession. The situation (case) can be offered for student analysis as a task for independent individual work, and can be used in the organization of group classroom work [6]. For example, when studying the "Web technology" discipline students are given a situational exercise about breaking an institution's website, detailing the structure of the site, technology of its creation, symptoms of problems (in which browser, at what stage of work, under what conditions, and what just happened). Having processed the data reported by the teacher, students are divided into groups or work individually to hypothesize about the cause of the failure and the means to remedy it. The results of the work of groups and strategies from individual students are evaluated by the instructor. It is concluded who was closest to the correct definition of the problem and ways of solving it, what was wrong and what was right. In this way, situational exercise can be triggered by brainstorming or discussion. However, students can practically demonstrate computer-assisted troubleshooting.

The rival method regarding student interest is the gamification method because it is one of the most powerful motivation factors for learning languages. Business English provides lots of opportunities to combine professionally based and business-like situations within one role game. Postgraduate students of the 'Computer Science' speciality of Dmytro Motornyi Tavria State Agrotechnological University were given legends to dramatize a job interview.

The main task was to discuss professional skills required by the job. The main challenge was that according to the legends both the employer (a big financial consortium) representative and candidates (IT-professionals with different personal and professional background) had some issues to conceal.

The English teacher was a participant of the game either, but their legend was "a trade union representative", who was monitoring that neither party's rights were violated and that inadequate, inappropriate questions (about candidates' private life, health issues, financial background) were not asked. In the case of facing a difficulty with the usage of English or discussing a candidate's hard and soft skills, the trade union representative could be appealed to for a brief prompt. The teacher did not support any of the parties and did not articulate their evaluation of the process. Student improvisation and spontaneity in the legend development and extension were welcome unless they violate the game rules.

The participants had to communicate in adequate manner (politely, business-like and completely honest). Nevertheless, each party had to reveal the other's concealed facts (gaps in candidates' CVs; disadvantages of the workplace) through questioning and to make a reasonable decision. The employer's representative had to choose the most suitable supervisor for their IT-team, and each of the candidates had to give arguments for their willingness or refusal to take the job.

The outcomes of the game include students' progress in:

- consolidation of oral and written speech token and IT terminology;
- including learned language patterns for Business English in spontaneous speech;
- improving skills in analyzing oral speech and applicant papers, deducing language pattern meaning from context, inferring and predicting;
- raising awareness of work ethic and business communication norms;
- increasing student motivation to be engaged in successive role games.

Having implemented several interactive methods from the methods described above the Business English Essentials for Software Engineers guide for senior students of IT-specialities has been composed. The guide is intended for extending reading, writing, and speaking skills with the accent on job-related situations: searching for a job, applying for a job, job interviews etc. Every unit has an active vocabulary list, an authentic text, numerous exercises, samples of documents, infographics and useful phrases for discussions. The information for the course book is retrieved and adapted from modern Internet resources.

In addition to the guide a distant learning course on Business English for IT-specialists [19] and a course on English for Computer Science students [20] have been developed. Every unit comprises a vocabulary list, an authentic text on the corresponding topic, useful grammar references, videos and audios with tasks, vocabulary and grammar tests. The most used questions are Drag and Drop into Text, Embedded Answers (Cloze), Matching, Multiple Choice, Selecting Missing Words, Short Answer, Description, and Essay. The course includes the compulsory self-testing unit to check student outcomes at the end of the term. In order to facilitate communication in English, a chat has been created in the course in which it is possible to ask questions, leave comments, answer questions, and share useful information All the students who participated in interactive class activities have proved the comprehension level of course material (76% average).

4 Conclusions

The interactive methods are inevitable in modern education: when meticulously prearranged, their implementation provides background for student's confident participating, supports developing and enhancing available skills and abilities (both academic and professional), delivers opportunities for students to follow up independently in their personal self-development and vocational mastering. The most preferred by students interactive methods (the project method, gamification, brainstorming) are the most effective since they are positively perceived, allow students to contribute most of the input into the method implementation terms, encourage students' improvisation and spontaneity. As our research shows, the thorough selection of interactive methods and preparation for English classes fosters comprehension at the sufficient level both in use of language aspects and in live communication. The prospects for future research can be development of the methodology of application of interactive teaching methods for enhancing communication of IT-professionals at higher educational institutions and experimental examination of the methodology.

References

1. Recommendation on Key Competences for Lifelong Learning adopted by the European Parliament and the Council in December. (Official Journal of the European Union, 2006), https://eur-lex.europa.eu/LexUriServ/LexUriServ.do?uri=OJ:L:2006:394:0010:0018:en:PDF. Accessed 08 January 2020
2. ISO 10015:2019 Quality management – Guidelines for competence management and people development. (International Organization for Standardization, 2019), https://www.iso.org/obp/ui/#iso:std:iso:10015:ed-2:v1:en. Accessed 20 December 2019
3. Pro zatverdzhennia Natsyonalnoii ramky kvalifikatsii (On National Qualifications Framework). (Legislation of Ukraine, 2020), https://zakon0.rada.gov.ua/laws/show/1341-2011-%D0%BF?lang=uk. Accessed 29 October 2019
4. *PMI. A guide to the project management body of knowledge.* (Project Management Institute, Newtown Square, Pennsylvania, 2017), pp. 82–86
5. R. Knapen, 20 interactive teaching activities for in the interactive classroom (2018), https://www.bookwidgets.com/blog/2018/06/20-interactive-teaching-activities-for-in-the-interactive-classroom. Accessed 20 December 2019
6. O. Sichkaruk, *Interaktyvni metody navchannya u vyshhij shkoli* (Interactive teaching methods at higher school). (Takson, Kyiv, 2006), pp. 19–20
7. L. Naseikina, Realizatsiya kompetentnostnogo podhoda pri formirovanii professionalno-znachimyih lichnostnyih kachestv studentov-programmistov (The implementation of the competence-based approach to the formation of professionally significant personal qualities of software engineering students). Bulletin of OSU 9(158), 34–41 (2013)
8. S. Sysoieva, *Interaktivni tekhnologii navchannia doroslykh* (Interactive technology for adult learning). (EKMO, Kyiv, 2011), pp. 120–121
9. V. Petruk, Dissertation, Vinnitsa National Technical University, 2008
10. P. Lozynskyi, Vikoristannia kompiuternoi programy Skype na zaniattiah inozemnoi movy (Application of Skype at foreign language classes). Bulletin of Ivan Franko Zhytomyr State University 1, 230–235 (2015)
11. T. Andreieva, *Ukrainska mova (za profesiinym spriamuvanniam)* (Ukrainian for professional purposes). (Kyiv, 2012)
12. J. González, R. Wagenaar, Universities' contribution to the Bologna process. (Publicaciones de la Universidad de Deusto, Bilbao, 2008), pp. 27–32
13. V. Kruglyk, V. Osadchyi, Formirovanie kompetentnosti v oblasti programmirovaniya u budushchikh inzhenerov programmistov (Developing Competency in Programming among Future Software Engineers). Integration of Education 23, 587–606 (2019). doi:10.15507/1991-9468.097.023.201904.587-606
14. V. Osadchyi, V. Kruhlyk, H. Chemerys, K. Osadcha, in ICTERI'2019: Integration, Harmonization and Knowledge Transfer, ed. by V. Ermolayev, F. Mallet et al. 15th Int. Conf. ICTERI 2019, Kherson, Ukraine, June 12-15, 2019. Vol. II, p. 17–28
15. S. Rajkumar, in PMI® Research Conference: Defining the Future of Project Management, Washington, DC, 2010
16. A. Zub, Upravlenie proektami (Project management). (Yurayt Publishing House, Moscow, 2018), pp. 280–285
17. Kafedra inozemnykh mov (Department of foreign languages) (2020), http://www.tsatu.edu.ua/im/. Accessed 05 January 2020
18. Visokoyakisna vyshcha osvita IT fakhivtsia u Melitopoli (High-quality higher education for IT-specialists in Melitopol) (2020), http://inf.mdpu.org.ua/. Accessed 05 January 2020
19. Business English for IT-specialists course (2020), http://nip.tsatu.edu.ua/course/view.php?id=1507. Accessed 05 January 2020
20. English for Computer Science students (2020), http://nip.tsatu.edu.ua/course/view.php?id=1507. Accessed 05 January 2020

Mathematical preparation of students for their professional self-realization in modern innovative society

Mariia Astafieva, *Dmytro* Bodnenko*, *Oksana* Lytvyn, *Volodymyr* Proshkin, and *Oleksii* Zhyltsov

Borys Grinchenko Kyiv University, 18/2 Bulvarno-Kudriavska St., Kyiv, Ukraine

Abstract. The article analyzes the problems of mathematical university preparation of specialists in the context of their professional self-realization in the modern labour market. The structure of mathematical competence of students is considered, its most important components are substantively described (the ability to ask and answer questions in and with mathematics, the ability to deal with mathematical language and tools). As a result of the first (ascertaining) stage of pedagogical experiment allowed to establish the level of students' mathematical competence formation according to the specified structure was established. It is proved that students of both mathematical and non-mathematical majors have poor mathematical knowledge and skills for solving problems that require their complex application, in particular, modeling problems in the context of future professional activity. The basic approaches to formation of mathematical competence of students are substantiated: research-oriented teaching; interdisciplinary approach and practical orientation; use of digital learning environments and related tools. The methods of implementation of these approaches are disclosed in the authors' elective discipline for the students of mathematical specialty "Methods of formation of students' mathematical competence for the future successful career". The continuation of the research will be the introduction of the discipline above to the educational process which contains the next monitoring of future teachers' readiness to use indicated approaches in their professional activity.

1 Introduction

Today's innovative society needs professionals of a new type, able to act responsibly, actively and productively in the conditions of social change, to carry out work activities with a high degree of interdisciplinarity and adaptability, while being ready to learn throughout life and to do it quickly and effectively. It is proved that one of the key qualities of such a specialist is a mathematical competence, as indicated by a number of international and Ukrainian regulatory documents (Recommendation 2006/962/EC of the European Parliament and of the Council on basic competences for lifelong learning of 18 December 2006 [1], The EU Framework Program for Renewed Key Competences [2], the Law of Ukraine "On Education" [3], the reform of the Ministry of Education of Ukraine "New Ukrainian School" [4], measures for the implementation of STEM education in Ukraine [5], etc.).

Today's tasks (and the ones of the near future) require mathematical competence from specialists at both general (extra-curricular) and subject levels, which is caused by the expansion of the scope of mathematics, methods of mathematical modeling. A specialist in a particular subject area is often unable to formalize a task, select, and apply appropriate mathematical methods to solve it. Mathematician - to understand the essence of the real problem, to determine the characteristics of the process (phenomenon, object) being modeled, to assess the adequacy and effectiveness of the model developed. Therefore, their cooperation is needed, and, thus, qualitatively different mathematical preparation of the first and the second. These challenges of time have led to a change in the educational paradigm - the transition from a knowledge-based to a competent approach in teaching, the intensification of research and practice in this direction.

The issue of mathematical competence formation nowadays is the subject of research of many Ukrainian and foreign scientists. Here are just a few studies. S. Rakov [6] highlights the issues of training mathematics teachers on the basis of a competent approach using digital technologies. N. Tarasenkova [7] outlines the theoretical foundations of the competence approach in teaching mathematics. B. Jaworski and J. Matthews [8] explore the problems of teaching mathematics at engineering majors, offer ways to improve students' conceptual understanding of them, particularly through research and partnerships, so that they can flexibly use mathematics in engineering contexts [9, 10]. M. Niss, member of the PISA Expert Group on Mathematics, outlines the characteristic features of mathematical competence and argues that the key problems of mathematical preparation at all educational levels, from school to university, are the same and are addressed through mechanisms of competence-based learning [11]. The authors of this article substantiate some approaches to the formation of critical thinking as an important component of students' mathematical competence [12,

* Corresponding author: d.bodnenko@kubg.edu.ua

13], consider the problems of teaching mathematical modeling to students of different specialties, theoretical and methodological foundations of mathematical modeling [14-16].

However, despite numerous studies, there is a "paradox of relevance": modern society is interested in specialists in high-tech industries, and young people are reluctant to choose educational programs with a strong mathematical component or do not receive sufficient mathematical training at the university.

The results of the international project ERDF (European Foundation for Regional Development) "Cross-border network of adaptation of mathematical competences in socio-economic development (MatNet)" point to certain problems of university's mathematical training of specialists implemented in the period 2011-2013 by the Agrarian University of Latvia and the University of Šiauliai (Lithuania), during which the role of mathematics in the modern labour market in the context of modern education was studied [17]. In particular, surveys of more than 300 employers and employees in Latvia and of more than 200 – in Lithuania indicate that 22% of those surveyed consider mathematics to be a set of formulas to remember, and 20% say it is a meaningless game of numbers by the rules devised by mathematicians. Regarding the role of mathematics in professional activity, 52% of respondents said that mathematics is needed and widely used in their professional activity, 48% – that their profession does not require a deep knowledge of mathematics, it is enough to be able to perform arithmetic calculations and find percentages. 59% of respondents believe that mathematical thinking helps to solve real professional problems, and this indicator, despite the fact that it is not too high, indicates the great value for the professional activity of mathematical thinking, since one who does not possess such thinking could not. evaluate its value. Significantly, more than 61% of respondents are sure that employees with good mathematical preparation achieve greater success in their profession and are highly valued by the employers.

As for employers 'and employees' assessment of mathematics teaching at university, only 50% of respondents consider that teaching mathematics at university was interesting and useful, and 62% confessed that they did not understand mathematics and tried to learn by heart the theorems and formulas. More than 65% said that mathematical knowledge helped them to understand and learn the specific disciplines of their subject area in one way or another, and 76% are convinced that mathematics develops logical thinking.

There is no reason to believe that if such a study was conducted in Ukraine, the results would be fundamentally different. Therefore, there is a need to overcome the existing contradiction between modern market inquiries, the needs of a person in successful professional self-realization and the real state of mathematical preparation of future specialists.

This encourages further research by scholars and investigations by practitioners. One of the directions and evidences of such activation is purposeful scientific researches, international projects on mathematical preparation of students with the purpose of enabling their professional self-realization in the modern innovative society, studying and creative introduction of positive foreign pedagogical experience, which is realized at the Department of Computer Science and Mathematics of Borys Grinchenko Kyiv University, both within the complex scientific topic of the department and through international projects on the problems of teaching and learning mathematics at high school.

The aim of the article is to draw up and develop approaches to the formation of mathematical competence of youth for a successful career in an innovative society, using the European experience.

2 Research Methods

The achievement of the aim of the research was facilitated by the use of a set of appropriate methods: analysis of scientific literature in order to establish the state of elaboration of the problem under study, determination of the categorical and conceptual apparatus of the research; synthesis, generalization, systematization, modeling for theoretical substantiation of approaches of formation of mathematical competence of youth for successful career in an innovative society; empirical: diagnostic (talk, testing), statistical (Pearson criteria) to assess the level of mathematical competence.

The research was performed within the framework of the complex scientific topic of the Department of Computer Science and Mathematics of Borys Grinchenko Kyiv University "Theoretical and practical aspects of the use of mathematical methods and information technologies in education and science", DR No. 0116U004625. The experimental base of the research is Borys Grinchenko Kyiv University.

3 Results and Discussion

Based on the ideas of a competent approach from the project "KOM: Competencies and the Learning of Mathematics", implemented in Denmark [18], and taking into account the results of our own experience of forming students' mathematical competence, we are trying to answer the key question: "How to teach mathematics to students to receive the best possible mathematical preparation at the university in order to help them in their future professional activity?"

The Danish project KOM, led by Mogens Niss, professor of mathematics at Roskilde University, was implemented to create a platform for reforming mathematical education in Denmark, from school to university. The ideas and results of this project have become widespread in European countries.

The summary documents of the KOM project define a person's mathematical competence as the ability to formulate, apply, and interpret mathematics in a variety of contexts (both mathematical and non-mathematical). In other words, it is knowledge, understanding, in particular, also understanding of situations in which mathematics, its methods play or may play a certain role, and the compulsory ability to act in the context of knowledge, competencies, skills, formed ways of thinking and values. The Project's recommendations highlight 8 components of mathematical competence, which are divided into two groups (Table 1) [18].

Table 1. Structure of mathematical competence.

The ability to ask and answer questions in and with mathematics	**The ability to deal with mathematical language and tools**
1. Mathematical thinking 2. Mathematical reasoning 3. Posing and solving mathematical problems 4. Mathematical modelling	5. Representing of mathematical entities 6. Handling mathematical symbols and formalism 7. Communication in, with and about mathematics 8. Making use of aids and tools

Let's describe each of the components briefly.

1. Mathematical thinking, which involves:
- awareness of the types of questions specific to mathematics, the ability to ask such questions and understand what type of answer can be expected;
- understanding the scope and limitations of mathematical notions, concepts, methods;
- ability to distinguish different types of mathematical expressions;
- ability to abstract and generalize.

2. The ability to give a reason mathematically means:
- ability to trace and evaluate the chain of arguments;
- understanding what mathematical proof is and the ability to distinguish strict proof from heuristic reasoning;
- ability to reveal the main idea in argumentation (proof);
- possession of deductive method of proof.

3. Posing and solving mathematical problems involves not only identifying problems and formulating different types of "pure" and applied problems and solving either their own or formulated by others problems, but also the ability to search for and find different approaches and ways of solving them, evaluate the pros and cons of each.

4. Mathematical modelling involves:
- ability to analyze the properties of existing models, to assess their compliance with the criteria of correctness and adequacy;
- translation and interpretation of elements of mathematical model in terms of real model;
- the ability to create mathematical models in a given context.

5. Representation of mathematical entities implies:
- interpretation of mathematical essence in different ways (analytical, graphic, visual, dynamic, schematic, tabular, symbolic, verbal-descriptive, etc.);
- understanding the strengths and weaknesses of each interpretation;
- ability to choose the most appropriate interpretation, according to the aim.

6. Handling mathematical symbols and formalism is:
- understanding, deciphering and interpreting symbolic mathematical expressions;
- understanding of the nature and rules of formal mathematical systems;
- translation of mathematical texts written in natural language into formal language.

7. Communication in, with and about mathematics is:
- understanding of the texts of mathematical content, represented in written, visual, oral or other form;
- ability to present certain mathematical results in different forms and at different levels of theoretical and technical accuracy.

8. The use of aids and tools, in particular, digital ones means:
- awareness of the existence and properties of different techniques and tools for mathematical activity, understanding of their capabilities and limitations;
- the ability to reflectively use such techniques and tools.

In order to find out the real state of the students' mathematical competence formation, a testing (ascertaining stage of pedagogical experiment) was conducted at the Faculty of Information Technology and Management of Borys Grinchenko Kyiv University in September – October 2019. A total of 39 students of different years of study of specialties "Finance", "Economics", "Management", "Computer Science", which provides the study of mathematical disciplines ("Higher Mathematics", "Probability Theory and Mathematical Statistics", "Econometrics", etc.), as well as 20 students of different years of study of the specialty "Mathematics" took part in the study. Such groups division was caused the need of determination the level of readiness mathematical competency by students of different specialties.

According to Pearson's criterion χ^2, this division of students into two groups is comparable to equable as $\chi^2_{emp} < \chi^2_{cr}$, where $\chi^2_{emp} = 5.492$ is the empirical value of the criterion, $\chi^2_{cr} = 6.635$ is the critical value of the criterion. In addition, 8 teachers of mathematical disciplines the Department of Computer Science and Mathematics of Borys Grinchenko Kyiv University were involved in the study. For the determination of students' mathematical competency paying attention to its components (see the table 1) the tasks were prepared [19]. The results of testing are shown in Table 2.

Table 2. Formation of mathematical competence of students.

No	Components of mathematical competence	Gr.	Level, %		
			Low	Medium	High
1.	Mathematical thinking	A	40.0	45.0	15.0
		B	41.0	48.7	10.3
2.	Mathematical reasoning	A	35.0	50.0	15.0
		B	30.8	56.4	12.8
3.	Posing and solving mathematical problems	A	40.0	50.0	10.0
		B	35.9	56.4	7.7
4.	Mathematical modelling	A	45.0	45.0	10.0
		B	33.4	58.9	7.7
5.	Representing of mathematical entities	A	25.0	60.0	15.0
		B	38.4	51.3	10.3
6.	Handling mathematical symbols and formalism	A	40.0	50.0	10.0
		B	25.6	66.7	7.7
7.	Communication in, with and about mathematics	A	35.0	50.0	15.0
		B	38.4	51.3	10.3
8.	Making use of aids and tools	A	35.0	55.0	10.0
		B	35.9	56.4	7.7

The table includes the notation: A – a group of students of specialty "Mathematics", B – a group of students of non-mathematical specialties. As the results of the experiment show, the percentage of mathematics students with a high level of development of mathematical competence is slightly higher than the percentage of the same students in other specialties. This excess is not statistically significant, after all

$$\chi^2_{emp} = 0.266, \chi^2_{cr} = \begin{cases} 14.067, p \leq 0.05 \\ 18.475, p \leq 0.01 \end{cases} \quad (1)$$

Exactly students with high level of mathematical competence, as practice shows, are able to correctly carry out logical reasoning, competently build the proof of mathematical facts, demonstrate the ability to use fundamental mathematical regularities in solving theoretical and applied mathematical tasks and problems, recognize mathematical structures in other (non-mathematical) theories, etc.

Mathematics students are better at solving mathematical problems, but just like other students are not good enough at using mathematics to solve applied "life" problems (students are practically unable to comprehensively solve the problem, to reveal its scientific essence in the professional field, not capable in analyzing, verifying, evaluating the completeness and reliability of information in the course of professional activity, produce new ideas, creative approaches to their implementation, etc.). Therefore, math does not serve as a basis for students to develop a successful career.

As a result of conducting a testing and on the basis of discussions with expert teachers, we outlined the problematic field of research, the ground of which lies in insufficient level of mathematical competence of students, and also confirmed the factors that hinder the successful (effective) teaching of mathematics at the university:

- insufficient basic mathematical knowledge, which makes it impossible to study the proper mathematics courses at the university (the critical state of school mathematics education in Ukraine is attested by the results of the External Independent Evaluations (EIEs) of recent years and the recently released PISA-2018 report [20]);
- superficial and formal mastering of mathematics, which in turn causes a weak ability to apply it in solving vital and professional problems;
- low interest of students to study mathematics at university;
- poor general education and research skills (or total lack of them) at a large proportion of students.

Considering that, as our research has shown, the levels of formation of most components of mathematical competence are approximately the same for students of different specialties, we have come to the conclusion that the way to improve the situation is not due to changes in the content of mathematical training, programs of mathematical disciplines (each specialty has its own program), but due to changing teaching approaches, improving teaching methods. The traditional technological approach to education (defining a mandatory standard minimum of knowledge, skills, acquaintance students with them, standardized testing of the ability to reproduce this knowledge), which prevails today in higher education, not only Ukrainian, has proved to be not very effective and inappropriate to the time requirements. Since mathematical competence means the ability to act, it is obvious that its formation is possible only in the course of active performance.

Therefore, to improve the motivation for learning, to deepen the conceptual understanding of mathematics and the ability to apply it, to develop general educational and research skills, which will inevitably lead to an increase in the actual level of mathematical knowledge and mathematical competence in general, active approaches to learning are required. The most effective of them are:

- research-oriented training;
- interdisciplinary approach and practical orientation;
- use of virtual learning environments and appropriate tools.

We would prove our choice.

By research-oriented training, we mean a set of pedagogical techniques and tools for the development of an individual's research competence, which implies the ability to identify a problem, formulate a research problem and find ways to solve it. Research-oriented learning, as opposed to reproductive learning, is aimed at developing students' experience of self-seeking and constructing new knowledge, ways of action, their application in different situations and conditions, developing the ability to creative activity, certain values.

Obviously, the real practice of teaching has never been pure reproduction. It is equally impossible to imagine a study based on research only. We are talking about creating such conditions for a student under which he / she would take not just an active position in the educational process, but an initiative position as well, not only master the proposed material (this also requires activity), but also master the methods of the world's comprehension, making an active dialogue with him / her, seek for the answers and ask himself/herself new questions. A number of significant differences between reproductive (informative) and research-oriented teaching of mathematics are summarized by us in Table 3.

Table 3. Comparison of reproductive and research-oriented approaches to mathematics teaching.

No	Reproductive (informative) training	Research-oriented learning
1.	The teacher informs the students the theoretical material provided by the program (definition of concepts, formulation and proving of theorems, the essence of certain methods, problems, ideas and ways of solving them)	The teacher "brings" students to concepts, facts, ideas, creating situations in which they, on the basis of observations, independently establish the characteristic features of mathematical structures, express hypotheses, participate in the proving
2.	Mathematics emerges as a unitary and complete system of consistent, non-questionable, authoritative information	The presentation of a certain theory is followed by consideration of alternative points of view, different approaches, analysis of their advantages and disadvantages
3.	The main purpose of the practical classes is the formation of standard	The share of reproductive exercises and tasks is negligible; higher-level

No	Reproductive (informative) training	Research-oriented learning
	manipulative abilities and skills, as well as the ability to act on a model, algorithm, instruction; exercises and tasks that illustrate lecture theory prevail	tasks that create the conditions for students to come up with ideas, offer alternative approaches and ways to solve, discuss, persuade and oppose, evaluate, interpret, and so on prevail
4.	Emphasis on individual learning activities (each solves the task independently)	Promotion and organization of collective research work

Here are some examples that illustrate these alternative approaches.

Example 1. Formation in students of the concept of continuous at the point of function of one variable ("Mathematical analysis", first year of study).

Reproductive (informative) approach.
The teacher reports the definition of a continuous function at a point: "A function $f(x)$ is called continuous at a point x_0 if its limits at a point x_0 is equal to a value at this point, i.e. $\lim_{x\to x_0} f(x) = f(x_0)$".

A research-oriented approach. **Case task.** The teacher shows the image (Fig. 1), indicates that only in Fig. 1a) a graph of a continuous at x_0 point of a function is shown, and proposes to find out the characteristic signs of a continuous function at a point and suggest its definition.

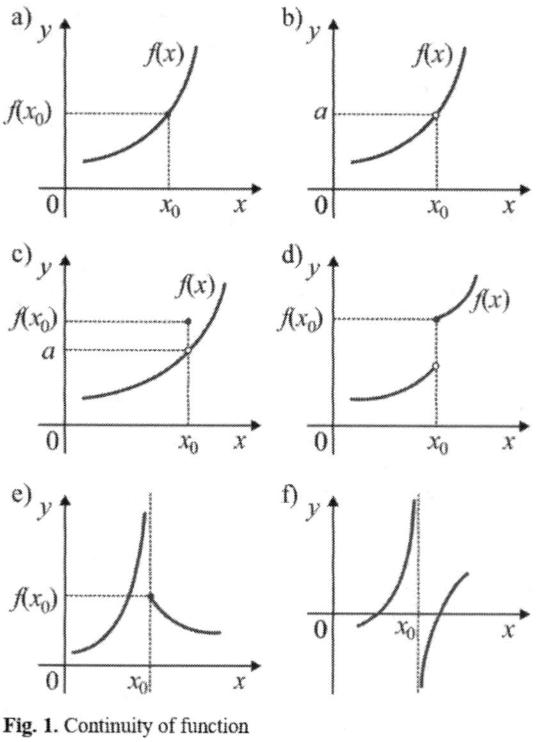

Fig. 1. Continuity of function

The benefits of the second scenario to form not only a deeply conscious notion of continuous function at the point, but also the ability to observe, analyze, based on previously acquired knowledge (in this case, the boundaries of the function), to identify essential features, to generalize, etc. – are obvious.

Example 2. Formation of students' notion of non-native elements in extended Euclidean space as a model of projective space ("Projective Geometry and Image Methods", second year of study).

Reproductive (informative) approach.
The teacher introduces the concept of projective space and formulates the basic axioms of projective geometry. Give examples of models of projective space.

A research-oriented approach.
Teachers and students analyze Euclid's fifth postulate in Hilbert's formulation: "Let l be an arbitrary straight line and A be a point that does not belong to it; then in the plane defined by point A and line l, it is possible to draw not more than one line that passes through A and does not intersect l". Replacing it with negation of that statement, students come to the following conclusions: 1) it's possible to draw at least two straight lines that pass through A and do not intersect l (Lobachevsky's geometry); 2) it's not possible to draw any that passes through A and does not intersect l, and further – arbitrary lines on a plane always have a common point.

The example illustrates understandable for students and logical introduction to projective geometry. After explaining the concept of a non-peculiar element, the teacher emphasizes that the extended Euclidean space is only one of the models of projective space. Finding other models of projective spaces is also an interesting research challenge that we recommend to tackle with students. This scenario allows students to integrate different approaches to reflecting the world in geometry. Possibilities of projective geometry for the development of abstract and figurative thinking were also successfully used.

Research-oriented learning (in it's any organizational model) aims at gaining new knowledge by students (about objects, phenomena, processes, relationships, methods, actions) through the direct experience of purposefully designed experience.

Research-oriented learning can be implemented as a creative combination of different methods. We consider the most basic of them are the following three: problem-based learning, inquiry-based learning and project-based learning. Each of them has its own peculiarities. For instance, the aim of problem-based learning is *application of theoretical knowledge*, the aim of inquiry-based learning – *gaining of a new (for a student) knowledge*, the aim of project-based learning – *project implementation*.

Despite the peculiarities and differences of each of the methods (which we will not dwell on in detail), let us note their common features:

1) transfer of the initiative in the organization of his educational knowledge to a student himself; the teacher in this process is not the main source of knowledge and directives, but a manager, coordinator and partner;
2) reflective activity on both intellectual and emotional levels;
3) collective search and research work; synergy;
4) enhancement of intrinsic motivation for learning;

5) skills development, development of research procedures are a separate, independent didactic goal.

Therefore, only a research-oriented approach to learning develops cognitive abilities and provides a deep understanding of the essence of processes, cause and effect, and other relationships. With the constant improvement and widespread use in all subject areas of computer mathematical programs, these are the qualities of a specialist, rather than a simple, even very broad, set of certain knowledge and technical (operational) skills that cannot be formed once in a lifetime, will be more and more demanded in the labour market.

An important and necessary condition for successful and effective mastering of mathematical disciplines by students is a clear answer to the question "why study it?". For non-mathematics students, this should be in the context of seeing mathematical knowledge as a compulsory basis for the study of professional training disciplines in future and professional activity in various subject areas. For mathematicians, on the contrary, there must be an understanding of how a mathematician can become an indispensable specialist in the fields where it is required to have a sophisticated mathematical apparatus to analyze processes and implement the latest technologies in production, science, economics, etc.

We see the implementation of cross-curricular links traditionally through a system of interdisciplinary and applied problems and the use of mathematical modeling (for example, T. M. Zadorozhnaya, Y. M. Krasyuk, Rita Ferri, Nicholas Mousoulides [21, 22]). And also through some specific forms of classes in the so-called Centers of competences and within the framework of student's teaching (vocational) practices.

Let us dwell on the latter. The main purpose of creation of the Centers of competences at the University is to introduce new forms and methods of teaching that will meet the request of the parties concerned to change higher education towards greater practical orientation, the development of independence of educational recipients, the formation, in addition to professional, social competencies. The centers provide solutions not only to practical tasks but also to problems that are not always well known and require knowledge and skills acquired during the learning of different disciplines.

For example, within the discipline "Computational geometry and computer graphics", mathematics students study the geometric foundations of computer graphics using the Autodesk 3ds MAX suite, using knowledge of analytical, differential and projective geometries, and image theory. Studying of logical programming by computer science students and building of expert systems requires knowledge of mathematical logic and mechanisms of logical inference (methods of logical inference).

An interesting experience is to conduct a specialty training for students in the first year of study of the specialty Mathematics and Computer Science in the hackathon format.

Hackathon as a technology competition is an example of project-based learning where teams work on an idea, design, product prototype, and present solutions within a proposed topic. The event ends by the presentation of its project by each team to the jury. During the hackathon, student teams create different devices of the Internet of things. The work is very intensive: workshops for participants, distance courses, selfstudy of the issue and, in fact, the hackathon. The company representatives evaluate the development. This form of work allows students to master a professional competencies (finding and developing solutions with rapid prototyping technologies; creating circuits with electronics, encoders and actuators; design and construction of mathematical and computer models of the device, their calculation and implementation in the form of a bulk model using different materials; building and programming devices with embedded systems and the Internet of things) and general (soft skills), as well, (systematic approach to problem solving; creativity and creation; team collaboration and management; self-presentation, effective presentation and protection of the created product; entrepreneurship, self-management, stress resistance and others).

One of the components of mathematical competence is the use of digital technologies: virtual learning environments, web-oriented educational portals, computer mathematics systems (CMS), cloud services for communication and collaboration, tools for presentation and multimedia materials, services for monitoring activity etc. In addition, their introduction into the educational process enhances the effectiveness of research-oriented and interdisciplinary and practically oriented approaches to learning.

For example, to find and visualize solutions of mathematical problems, properties of geometric objects, illustrations of regularities, relations, correlations, transformations, we most often use CMS GeoGebra.

Here is an example of a mathematical experiment in using the GeoGebra system to study Pappus's projective theorem: let A, B, C be three points on one line, D, E, F - be three points on another line. Let the three straight lines AE, BF, CD intersect the three straight lines DB, EC, FA, respectively, at points P, Q, R. Then the points P, Q, R lie on one straight line (Fig. 2).

Students, by constructing the described configuration in GeoGebra and changing the position of its elements, formulate themselves the hypothesis that the points P, Q, R belong to one straight line (Figs. 2a, b, red straight line), and remain on it regardless of the position of given points at straight lines or given straight lines.

One can continue the study by considering the two straight lines as a degenerate second-order curve. Then, considering the non-degenerate oval curve with students, we come to the formulation of Pascal's theorem.

The approaches of formation of mathematical competence of students outlined by us require, in turn, modernization of the content, forms and methods of preparation of future teachers of mathematical disciplines at higher education establishments.

As an example, we take an optional discipline for Master's students of specialty "Mathematics" developed by us – "Methods of formation of students' mathematical competence for the future successful career".

The main idea that we plan to implement in the course of teaching this discipline is to familiarize students with the latest methods of teaching mathematical disciplines at the university, which based also on using a positive European experience.

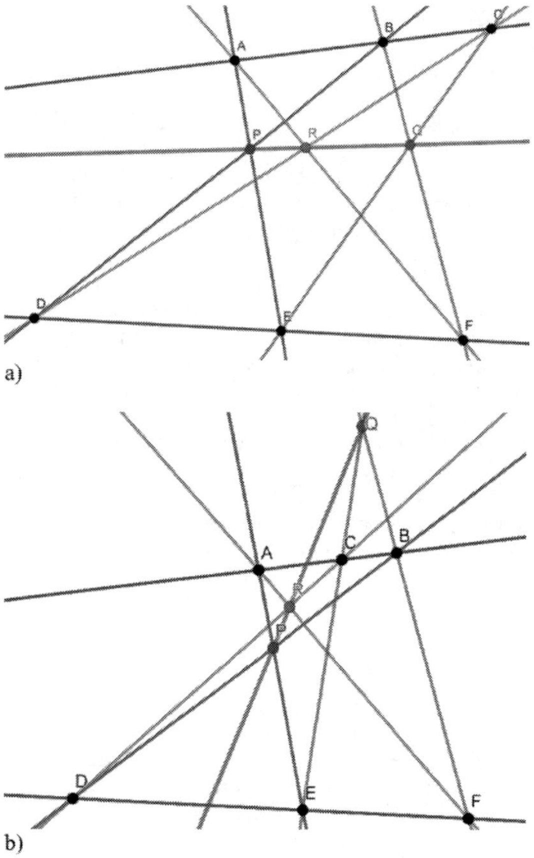

Fig. 2. Pappus configuration at different elements positions

The main blocks of theoretical knowledge:

1. Transformation of mathematical education to ensure a successful career for the youth in an innovative society.
2. Research-oriented learning is the basis of a competence approach and a means of achieving quality mathematical education.
3. Implementation of cross-curricular relations in the study of mathematical subjects for the formation of conceptual and systematic mathematical knowledge.
4. The role of practical problems in different subject areas and mathematical modelling in the formation of professional qualities of future specialists.
5. The use of network services and cloud-based learning technologies in the process of learning mathematics.
6. STEM-education and its implementation in Ukraine. Use of STEM-technologies for formation of key and subject competencies.

A lecture was and remains the main form of theoretical studies. The didactic purpose of the lecture is to introduce students to the scientific problem (task), to reveal the main issues of the topic, to solve the given problem, to focus on the most difficult (problematic, contradictory) moments, to prepare students for further selfstudy work. In order to achieve this goal, it is necessary to achieve educational interaction "lecturer – student", aimed at mastering of educational material. Properly organized lecture makes passive learning impossible. In such a way, the basis of the students' actions for the further formation of knowledge is laid, and this happens when students directly communicate with the teacher. Therefore, by virtue of influencing the student's personality, the lecture cannot be replaced by any of the newest learning tools or sources of information with which he or she is able to work independently.

Active and fruitful interaction in the system "teacher – training material – student" can be significantly enhanced by the use of modern digital technologies. In our view, an effective use of this is a multimedia lecture. It does not simply allow the computer to transfer part of the features of the lecturer. It provides students with a better understanding of the material through a combination of verbal and visual perception. In addition, a multimedia lecture enhances learning motivation, making the lessons emotionally engaging and interesting. An example of one frame of a multimedia lecture on "Formation of mathematical competence in students of non-mathematical specialties" of our optional course is shown in Fig. 3.

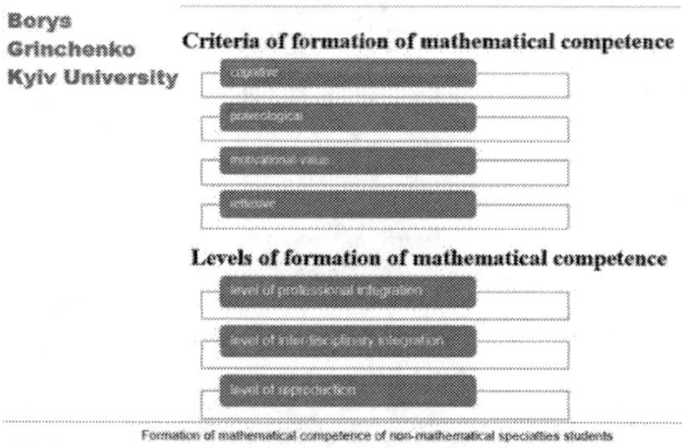

Fig. 3. Fragment of the lecture "Formation of mathematical competence of non-mathematical specialties students"

The seminars are provided in an interactive form. These include, in particular, discussion seminars, research seminars, business games, analytical seminars, interdisciplinary seminars. For example, the course on the topic "Transforming Mathematics Education to Ensure a Successful Youth Career in an Innovative Society" is planned to take the form of a discussion seminar, where future teachers will learn how to make discussion tolerately: express their thoughts accurately and convincingly, listen and hear the position, different from their own, actively defend their point of view, oppose reasoningly.

Seminar-research on the topic: "Formation of mathematical competence in students of mathematical specialty" provides within the limits of research training work in groups, collective search for the solution of professional tasks and problems. The main purpose of the lesson is to form students' concept of the content of the

subject mathematical competence, to understand its importance for the professional activity of mathematician; development of practical skills of work of the teacher for formation of mathematical competence at students of mathematical specialties. Students are offered the following tasks:
1. Investigate the specifics of mathematical competence in mathematics, what structural components distinguish it from the mathematical competence of specialists in other spheres.
2. On the basis of comparative analysis of Ukrainian and European practices, to find out the content, forms and methods, technologies of mathematical education of students of mathematical specialties, optimal for the formation of mathematical competence in them.
3. Develop a fragment of a practical lesson on a given topic (prepare a case; plan a business game or other type of training or task) in mathematical discipline for bachelors in Mathematics.

Of particular interest to students are the lessons in the form of business games. So it is planned to hold the project "Project Method as STEM Technology".

At the analytical seminars, students have the opportunity to try themselves in four professional roles:
- participant – performer of specific actions in a certain mathematical field;
- expert – a person with a high level of knowledge in a narrow field of mathematics, conducts examination and is responsible for the accuracy and completeness of analysis, validity of conclusions in accordance with the tasks of examination;
- moderator – the person who coordinates the group work participants;
- seminar facilitator for the overall management of the seminar.

Taking into account the principle of interdisciplinarity in teaching as one of the leading in the conditions of modern higher education, we plan interdisciplinary seminars, which are jointly held by different teachers. For example, such seminar is "Networking Services, Computer Mathematics Systems, and Cloud-Based Mathematics Learning Technologies." To prepare for this seminar, students are offered a homework assignment – to prepare a project "Using Computer Mathematics (Cloud Service) in Professional Activity".

4 Conclusions and prospects for further research

1. An analysis of the scientific literature and the results of European research projects has shown that students' mathematical competence is the key to their successful professional self-realization and the demand in the modern labour market in an innovative society. The most important components of mathematical competence, which are conventionally divided into two groups, are considered and substantively characterized:

I. The ability to ask and answer questions in and with mathematics,

II. The ability to deal with mathematical language and tools.

2. As a result of conducting the first (ascertaining) stage of pedagogical experiment at the Borys Grinchenko Kyiv University (September, October 2019), a problematic field of study is identified, which consists in the insufficient level of students' mathematical competence formation, and also the factors that inhibit the successful study of mathematics are confirmed: insufficient school mathematical preparation for studying university courses in mathematics; superficial and formal learning of mathematics, which does not allow to apply it effectively in solving vital and professional problems; low interest of students in teaching mathematics; poor general education and research skills in a large proportion of students.
3. Effective approaches to the teaching of mathematics at the university are substantiated: research-oriented teaching; interdisciplinary approach and practical orientation; use of digital learning environments and related tools. In particular, examples of the difference between reproductive (informative) and research-oriented teaching of mathematics are described and illustrated. A form of realization of cross-curricular communication as a students' learning practice, namely hackathon, is demonstrated. Examples of the use of digital technologies for mathematical experiment are given.
4. Outlined approaches to teaching mathematics are the main content of the developed elective discipline for the Master's program students of the specialty "Mathematics" – "Methods of formation of students' mathematical competence for the future successful career". The implementation of this discipline during the second (formation) stage of pedagogical experiment, followed by an assessment of the willingness and ability of future teachers to apply these approaches in their professional activities will be the subject of further scientific research.

References

1. UE, Union Europea, Recommendation 2006/962/EC of the European Parliament and of the Council, of 18 December, on key competences for lifelong learning (2007)
2. European Commission, Proposal for a council recommendation on key competences for lifelong learning (2018)
3. The 10 skills you need to thrive in the Fourth Industrial Revolution (2017).
4. The New Ukrainian School. Conceptual Principles of Secondary School Reform (2016). https://mon.gov.ua/storage/app/media/zagalna%20serednya/nova-ukrainska-shkola-compressed.pdf. Accessed 27 Oct 2016.
5. Action Plan for the Implementation of STEM Education in Ukraine for 2016-2018 (2019), https://drive.google.com/file/d/0B3m2TqBM0APKQmc4LUd2MmVFckk/view. Accessed 05 May 2016.
6. S. Rakov, *Matematychna osvita: kompetentnisnyi pidkhid z vykorystanniam IKT* (Mathematical Education: A Competent Approach Using ICT). (Fact, Kharkiv, 2005).
7. N. Tarasenkova. Mathematics at home school **11**, (2016)

8. B. Jaworski, J. Matthews. Developing teaching of mathematics to first year engineering students. Teaching Mathematics and Its Applications **30** (2011)
9. B. Jaworski< Research practice into/influencing mathematics teaching and learning development: Towards a theoretical framework based on co-learning partnerships. Educational studies in mathematics **54** (2003)
10. B. Jaworski, Theory and practice in mathematics teaching development: Critical inquiry as a mode of learning in teaching. Journal of Mathematics Teacher Education **9** (2006)
11. M. Niss, Competencies and educational description. Education **9** (1999)
12. M. Astafieva, D. Bodnenko, V. Proshkin, Information Technologies and Learning Tools **71** (2019)
13. M. Astafieva, D. Bodnenko, V. Proshkin, in *Poster of the 15th International Conference on ICT in Education, Research and Industrial Applications. Integration, Harmonization and Knowledge Transfer*, Kherson State University, Ukraine, 12-15 June 2019
14. Ya. Goncharenko, Economical and mathematical methods and models in the system of training students of economic majors. Didactics of mathematics: Problems and Investigations **36** (2011)
15. C. Michelsen, Mathematical modeling is also physics interdisciplinary teaching between mathematics and physics in Danish upper secondary education. Physics Education **50** (2015)
16. R. Zulkarnaen, Why is mathematical modeling so difficult for students? AIP Conference Proceedings 2021, 060026 (2018). doi:10.1063/1.5062790
17. A. Zeidmane, Development of Mathematics Competences in Higher Education Institutions. International Journal Of Engineering Pedagogy **3** (2013)
18. M. Niss, The Danish "KOM" project and possible consequences for teacher education. In *Educating for the future: Proceedings of an international symposium on mathematics teacher education*, Malmo University, Sweden, 5-7 May 2003.
19. The tasks for the determination of students' mathematical competency. https://docs.google.com/forms/d/e/1FAIpQLScZUqLp9EXn35NFtXgHgtf_heQvrPYHNB8LNmcjuKZbj4KA9A/viewform. Accessed 08 Jan 2016
20. PISA 2018 Results PISA, OECD Publishing, Paris I: (2018), http://www.oecd.org/pisa/. Accessed 03 Dec 2018.
21. T. Zadorozhnia, Y. Krasiuk, Physical and Mathematical Education **3** (2017)
22. R. Ferri, N. Mousoulides, in *Abstracts of the Tenth Congress of the European Society for Research in Mathematics Education, Dublin City University*, Ireland, 1–5 Feb 2017

Enhancing interest in research activities in mathematics students in teacher training universities

Kateryna Vlasenko[1,*], *Olha* Rovenska[1], *Iryna* Lovianova[2], *Oksana* Kondratyeva[3], and *Vitaliy* Achkan[4]

[1]Donbass State Engineering Academy, Department of Mathematics and Modelling, Kramatorsk, Ukraine
[2]Kryvyi Rih State Pedagogical University, Department of Mathematics and Methods of its Training, Kryvyi Rih, Ukraine
[3]Bohdan Khmelnytsky National University of Cherkasy, Department of Mathematics and Methods of Teaching Mathematics, Cherkasy, Ukraine
[4]Berdyansk State Pedagogical University, Department of Mathematics and Methodology of Teaching Mathematics, Berdyansk, Ukraine

Abstract. The article looks into the issue of developing an interest in mathematics students' research activities in the teacher training universities. The study is dedicated to the feasibility of involving the 5E Instructional Model to the organization of students' scientific research during the workshop on the Approximation Theory and Fourier Series. The research considers the results of the survey among students who helped to evaluate their emotional state during the workshop. To collect the data we used the tool of express evaluation of positive and negative emotionality the Differential Emotion Scale. The article discusses the positive influence of the environment developed through the 5E Instructional Model on students' emotional state and forming their interest in scientific research while organizing workshop classes. We have grounds to conclude that there is the efficiency of implementing workshops based on the 5E Instructional Model. The index reduction of students' negative emotions encouraged their activity during the workshop and the improvement of interest in research activities.

1 Introduction

The determining factor of training a competitive teacher is the organization of his/her scientific activities that allow the teacher to create independently new means of pedagogic activities, create new ideas and approaches that correspond to the changeable requirements of the present time. I. F. Yarullin, N. Bushmeleva, I. Tsyrkun [1] called the organization of the student's scientific activities one of the mechanisms to form their research competence. The scientists: A. ScHoenfeld [2], R. Terner [3], A. Vintere and A. Zeidmane [4], J. Proulx [5], B. Koichu and A. Pinto [6] determine the research competence as one of the components of mathematics teacher's professional competence that guarantees further research activities of the graduate in mathematics. Regarding this, the issue of organizing research activities of pre-service mathematics teacher stays constantly open in pedagogical studies.

Many authors have emphasized the necessity to support active students' research activities. J. Lithner [7] pointed out that international tendency in mathematics education is acquiring mathematical knowledge not only in terms of context but in terms of getting skills connected with carrying out mathematical research. C. Bonwell and J. Eison [8], quoted in E. Fallon, S. Walsh, T. Prendergast [9], stated that students have to do more than just listen. They have to read, discuss and do research on the problems. Scientists emphasized the importance of organizing intensive students' activities, forming their positive attitude to research projects. This idea is agreed with the conclusions made by P. Hernandez-Martinez and P. Vos [10], that described the critical state of the question to form students' interest in research activities. In their work K. Jones, L. Black and A. Coles [11] confirm that at every level of university students' training to mathematics teacher's activities, it is necessary to form their creative thinking and their investigative skills. Scientists emphasize that the organization of students' research activities during their training encourages the development of research competence, necessary both for solving practical problems and for being able to adapt fast to changeable conditions of the modern time and master their skills constantly.

We agree with L. Yore [12], who states that forming an interest in research activities is the first stage in the development of research competence during mathematics learning. We also took into account the ideas of T. Dreyfus [13], who considers research activities during Mathematics learning as a natural part of the educational process, which is directed at forming research competence among students. In order to organize such activities, scientists offer to use special courses dedicated to special scientific researches in the priority areas of modern mathematics. This fact is evidenced by the opinion of the mathematical community (I. Yarullin, N. Bushmeleva, I. Tsyrkun [1], I. Biza, V. Giraldo, R. Hochmuth, A. Khakbaz, C. Rasmussen [14], N. Telegina [15]) about the significant potential in the researches on forming a

* Corresponding author: vlasenkokv@ukr.net

positive attitude to students' research activities using the materials of different mathematical branches. In scientists' opinion, the usage of interesting mathematical theories encourages students to get more meaningful education of theoretical materials, facts, and methods of solving mathematical problems and it allows getting particular experience. We can also meet the confirmation of this opinion in the works by A. Matejko and D. Ansari [16], S. Sevinc and R. Lesh [17], who investigated the organization of research activities related to particular branches of mathematics. The idea caught on, that is why guided by the conclusions made by the above-mentioned scientific researches we decided to research the creation of interest among students of mathematics departments in research activities through organizing a workshop on Approximation Theory. The choice of this branch results from its extensive use in practice. This is explained by the fact that the modern stage of science and technology development is characterized by the use of a considerable amount of information. As experience shows this tendency will only enhance in the future – the development of computer science, telecommunication, and registration equipment lead to steady growth of the data amount. Therefore, tools and methods of their processing and analysis are growing. The creation of a single methodical approach based on general mathematical principles is actual for several tasks such as to get, model, register and process data. The series finds a mass use as a tool to represent a considerable class of functions, carrying out analytical transformations, approximate calculations in many applied tasks. Algorithmic and computer software that is created on their basis is characterized by high universality and is included in computer and hardware – computer complexes of different purposes, which is confirmed by the numerous researches by H. S. Malvar [18], F. Dedus, S. Makhortykh, M. Ustinin [19], A. Pankratov [20], O. Novikov, O. Rovenska [21].

The article is aimed at describing an introduction of a seminar in Approximation Theory and Fourier Series searching for a method to implement a workshop on that was developed to enhance interest in research activities in mathematics students in teacher training universities.

2 Method

At the first stage of the research, we used the survey method to evaluate students' interest in research activities. We used the Differential Emotions Scale by C. Izard [22] to survey students.

The relevance of involving this methodology to evaluate students' interest in research activities is proven by the researches where the direct dependency between the subject's interest in cognitive activities and their emotional state during its implementation is emphasized. Since the feeling is a dynamic component of the emotion (J. Panksepp [23]) and two psychobiological processes connected with it – fascination and individuation (S. Langer [24]), motivating, managing and informative functions of feelings allow them to capture or simplify and organize the thing that can become (especially in difficult situations) a great number of impulses in concentrated cognitive processes. During 2015 – 2019 we surveyed master's degree students of physics – mathematics departments of Kryvyi Rih State Pedagogical University and Berdyansk State Pedagogical University. 49 master's students took part in the survey (17 male students and 32 female students aged from 20 to 28). The use of the online survey, first through Google form, posted on the Internet, and then, moved to the forum of the platform «Higher School Mathematics Teacher» [25] had an advantage in comparison to the survey on paper as it encouraged the respondents' frankness and prevented missed questions.

According to the chosen methodology, we selected Likert scale to evaluate each of the basic emotions 1 – 'feeling is completely absent'; 2 – 'feeling is slightly expressed'; 3 – 'feeling is moderately expressed'; 4 - 'feeling is strongly expressed'; 5 – 'feeling is fully expressed'. At the beginning of the research, the most significant (> 9 points) positive emotion related to the experience of research activities was 'interest', negative – 'shame' and 'fear'. Students usually face the last two emotions while learning mathematics.

Students believe that the key problem of learning mathematical theory is the absence of the connection between the theory and practice and abstract character of the subject.

During the second stage of the research, we developed the subjects of the workshop on Approximation Theory.

The workshop program consists of six classes.

1. The history of the development of approximation theory and Fourier series.
2. The ways of periodic function classification.
3. Approximation methods that are based on matrix series summing.
4. Main tasks of approximation theory: approximation of individual function, class approximation, precise and asymptomatically precise ratio.
5. Examples of researches by subject.
6. Examples of using approximate aggregates in computer complexes of broad purpose.

The workshop was aimed at the formation of students' interest in research activities through their implementation into the real process of using series in applied tasks.

The workshop was held for a group of 7 – 8 students twice a month for three months. Every class included two hours of classwork and three hours of extracurricular work. The classes were held by the prominent teachers of mathematics departments who took part in the development of the workshop and looked for the method, the implementation of which would encourage the formation of students' interest in research activities during the workshop.

With the help of the Deductive Approach to Content Analysis, we determined the possibility to involve the 5E Instructional Model. During the selection of resources for the analysis of the model usability, we were oriented at those representing the usability of the 5E Instructional Model while learning. Among them we can name TeachThought [26], Lesley [27], The National Academies Board on Science Education [28], Alberta Education [29] (Table 1).

Table 1. Analysis of the resources that represent the 5E Instructional Model usage.

Resources	Used while learning a subject	Model features	What are the efficiency grounds
Teach Though	Biochemistry and Molecular Biology Education, Mathematics	Joint activities	The solid knowledge foundation through an active part
Lesley University	Mathematics, Life sciences	Constructing knowledge based on experience	Possibility for the full cycle of education
The National Academies Board on Science Education	Biological sciences	Structure and sequence of education are directed at creating a challenging situation	Integration of learning activity with laboratory experience
Alberta Education	Librarianship, Work with information	Student's involvement in metacognition; encouragement of critical and creative thinking	Focus on achieving defined learning outcomes in different subjects

We also found out what the purpose of using the 5E Instructional Model by other scientists was. Ping-Han Cheng, Ya-Ting Carolyn Yang , Shih-Hui Gilbert Chang, Fan-Ray Revon Kuo [30] noted the efficiency of using the model stages to increase the motivation of students' learning. L. Duran, E. Duran [31] describe the use of this model in programs of professional development in education. S. Supasorn, V. Promarak [32] see the use of the 5E Instructional Model as an efficient method of improving students' understanding of natural processes. In conclusions of scientific researches done by R. W. Bybee, A. Gardner, J. A. Tailor, P. Van Scotter [33], A. Abdi [34], E. T. Ong [35] we also find the confirmation of the 5E Instructional Model efficiency in improving students' achievements in science. This model consists of the stages: Engagement, Exploration, Explanation, Elaboration, and Evaluation. Every stage provides a specific pedagogical function and encourages teaching processes from the teacher's side and forming a better understanding of scientific knowledge, abilities, and skills from the student's side.

We developed the recommendations that were implemented during workshop classes in order to use the stages of the 5E Instructional Model during the organization of workshop classes on Approximation Theory.

Engagement. At this stage, the teacher has to determine what students already know regarding the concept that is considered and what kind of knowledge they still need. In order to master new educational material, it is necessary to help students to revise mathematics sections such as Algebra, Mathematical Analysis, Functional Analysis, Function Theory. Moreover, at this stage, the teacher is only a consultant who helps students to prepare short reports encouraging students' interest and motivation. For this purpose, the teacher presents the actuality of the researches dedicated to learning approximate features of approximation methods that are generated by certain transformations of partial sums of Fourier series and allow building the sequence of trigonometric polynomials that would equally coincide for any function (Table 2).

Table 2. Recommendation for the teacher on organizing the state of Engagement.

Appropriate	Inappropriate
– encourage students to raise their questions; – offer to compare their ideas with others.	– read the lecture; – give definitions to terms; – explain or give tasks.

Exploration. The stage is directed at strengthening students' activities regarding knowledge and skills. At this stage, students can revise the tasks that use the methods of approximation theory on special subjects that they learn. As a rule, students cite examples of tasks on periodic signal approximation in the theory of control engineering, pattern recognition, nondestructive testing, etc. Students can discuss and write down approximation methods in every particular case. The teacher is only a consultant who offers students such research methods as observation, hypothesis generation, forecasting. Students' communication and work in groups without the direct teacher's involvement are encouraged. equally coincide for any function (Table 3).

Table 3. Recommendation for the teacher on the organization of Exploration.

Appropriate	Inappropriate
– encouragement of search for several ways to solve the problems; – comparison of ideas; – self and mutual survey	– use of traditional explanation; – implementation and involvement of a great amount of terminology.

Explanation. At this stage, students can describe their point of view regarding the search for solving extreme problems of approximation theory. After this, the teacher has to introduce common terminology and acquaint the students with the general scheme of researching integral images of trigonometric polynomials variations that are generated by linear methods of summing Fourier series, from periodic functions. Generating students' new ideas on methods of approximation improvement, their comparison with the ideas of the previous stage is possible. At this stage, the teacher also has to prevent possible mistakes while explaining misconceptions that could arise at the stage of engagement and exploration. During the classes of this stage, the teacher involves interactive methods and presentations for mathematical modeling of periodic processes (Table 4).

Elaboration. After getting an explanation about the research main scheme regarding integrated images of trigonometric polynomials variations on the classes of periodic functions it is important to involve students in further research activities. Further work includes significant analytical calculations connected with exact and approximate methods. Starting from the integral image students can learn asymptotic behavior of exact upper bounds of trigonometric polynomials variations from periodic functions to infinity. The stage is aimed at helping students to develop a deeper understanding of general methods of mathematical analysis and the use of approximation processes in practical tasks. Students can carry out additional researches, develop new approximation methods, exchange ideas and use acquired research experience to integrate approximation theory in practice. It is possible to create algorithmic and computer algorithmic products based on built methods (Table 5).

Table 4. Recommendation for the teacher on the organization of Explanation.

Appropriate	Inappropriate
– teacher's explanation; – expression of the ideas using generally accepted terms; – idea review and forming new ones.	– forming a great amount of terminology; – focus on independent work.

Table 5. Recommendation for the teacher on the organization of Elaboration.

Appropriate	Inappropriate
– understanding enhancement through strengthening the ideas acquired by experience; – use of algorithms that are close to new situations; – grounds for conclusions; – support of forming student's proper ideas	– development of the ideas that do not have a connection with previous experience; – generating a great number of ideas without deepening in the essence of the theory.

Evaluation. Evaluation is considered to be a permanent process during which the teacher only observes the students and supports during report presentations, idea introduction, and question tasks. The use of peer assessment is relevant. Such a form of evaluation can be complemented by students' self-assessment of their level. During the classes of this stage, the teacher involves interactive methods and presentations for mathematical modeling of periodic processes (Table 6).

Table 6. Recommendation for the organization of the concluding stage.

Appropriate	Inappropriate
– evaluate the progress in general in comparison to the initial level; – evaluate the ability to use approximate methods to solve complex problems; – give students feedback regarding the feasibility of their ideas; – encourage questions that enhance a deeper understanding of the influence of individual function features on the approximation order.	– evaluate single facts and separate elements of approximation theory; – offer a survey in a test form.

Using the 5E Instructional Model does not oblige the teacher to strictly follow the implementation of five stages. If necessary, it is possible to repeat the stages of Explanation and Elaboration several times.

3 Results

During the preparation stage, we selected the target type as a selection strategy, because the selection had to include the students who have a high achievement level in mathematical branches. By high level, we understand the absence of the final mark «satisfactory» and lower following the national 4-level scale «unsatisfactory», «satisfactory», «good», «excellent» for each of the subjects «Algebra», «Mathematical analysis», «Functional analysis» and «Function theory». The target selected analysis provided us with sample size n=49 of students that represents 23% of the general number of master's degree students of the first year during 2015 – 2019. At the stage of organizing data collection, we used the tool of express-evaluation of positive and negative emotional state the Differential Emotion Scale (Izard [22]), which ensures diagnostics of a wide range of emotional states. Each of the ten basic emotions (x_i, i=1,2...,10) is represented by three independent changeable 5-character scales for factors that describe emotional states. The points on every scale correspond to the level of emotional feedback and can be in total from 3 to 15 points. The stage of data analysis of every profile implies the selection of significant (>9 points) emotions, creation of 'emotion profile', determination of the dominant emotional state.

At the beginning of the research, the most significant positive emotions regarding the experience of research activities is 'interest', negative – 'shame' and 'fear' (Table 7).

Table 7. Distribution of significant emotions at the beginning of the survey Font styles for a reference to a journal article.

Emotion	Number of students who have this emotion as dominant (>9 points)	Comparison with the general number of students
Interest	32	65,3%
Fear	45	91,8%
Shame	27	55,1%

While processing every profile we defined the indexes of emotional states that characterize the level of subjective students' emotional attitude to the present experience of research activities. The Index of positive emotions and Index of critically negative emotions could range from 9 to 45 points, the Index of anxio-depressive emotions ranged from 12 to 60 points. We defined that the positive emotional state turned out to be dominant among 69,4% of students, a strong level (> 36 points) of expressing positive emotional state was marked only among 6,1% of respondents. Also, a distinct (from 29 to 36 points) level of positive emotional state was fixed among 10,2% of students. Other students (53,1%) showed moderate (from 20 to 28 points) and weak (< 20 points) level. So, most students' attitude to the research process can be mainly characterized as positive. However, this positive attitude is weakly expressed, unstable and cannot ensure the proper motivation in overcoming difficulties that inevitably arise during research activities. This fact plays an important (if not the most important) role in the failure of attempts to involve an unprepared student to research activities in any area, including mathematics.

The dominant critically negative emotional state regarding the present experience of research activities was fixed among 12,2% of respondents, half of whom had a strong (>32 points) or distinct (from 25 to 32 points) level. It is important that among all the students who had the critically negative state as dominant, the factor "Dull" took no less than 4 points, and, accordingly, made the greatest contribution to the calculation. It testifies a stereotype regarding the complexity and absence of interest in research activities among young people. We considered this aspect while searching for methods of workshop implementation.

As mentioned above, the emotions 'fear' and 'shame' were detected as significant among 91,8% and 55,1% of respondents. These emotions are included in the third group of emotions that determine the anxio-negative emotional state of the subject regarding the experience of research activities. Despite this fact, the given state is dominant only among 18,4% of students. It demonstrates that these two emotions influence the formation. 4,1% of respondents have strong (> 30 points) level of emotional state, distinct (from 21 to 30 points) - 10,2%, moderate (from 12 to 20 points) and 4,1% of respondents - weak (< 12 points). Such a noticeable selection of two emotions in the general image of the emotional state confirms the idea that fear and shame prevent students from implementing their interest in the research process and take an active position while conducting research.

The repetitive survey was carried out after finishing the workshop. The distribution of significant emotions after taking a workshop is represented (Table 8).

Table 8. Distribution of significant emotions after taking a workshop.

Emotion	Number of students who have this emotion as dominant (>9 points)	Comparison with the general number of students
Interest	44	89,7%
Surprise	18	36,7%
Fear	39	79,5%
Shame	5	10,2%

Interest turned out to be a significant positive emotion among 44 students. We can note that the number decrease in students who had shame as a significant negative emotion is well seen – 17 respondents. At the same time, the number decrease of students who had fear as a significant emotion is minor – 6 students (Figure 1).

Despite this fact it is impossible to claim that this emotion in the context of the given research is badly adapted. The profile analysis of respondents' emotions shows the decrease of fear expression to varying degrees among 77,5% of students. The presence of surprise among the significant emotions, as well as interest, which is included in the positive group, is predictable.

More detailed analysis of the feasibility of implementing a workshop that was carried out using the index calculations of students' emotional states. We detected the increase of students with the dominant positive emotional state up to 81,7%, where 63.2% of respondents had a strong and distinct level. At the beginning of the workshop, the same indicator was 16,3%. Thus, we managed to form a stable positive attitude to research activities among more than half of the workshop participants.

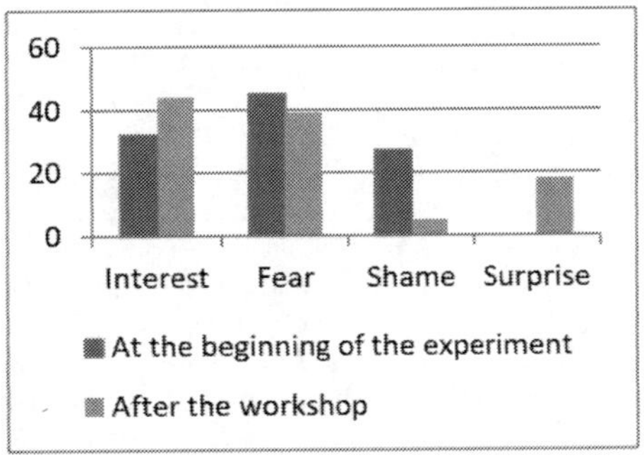

Fig. 1. Distribution of significant emotion.

The number of students who have a critically negative emotional state as dominant remained at the level of 12.2%, though the qualitative structure of this subgroup changed. In our opinion, it is connected with a greater amount of working practice in small groups during workshop classes in comparison to individual work. As teachers pointed out certain students perceived such format negatively.

The dominant anxio-negative subject's attitude to experience of research activities after taking a workshop was fixed among 6,1% of students. Among them 4% of respondents have moderate and 2,1% - weakly expressed level of emotional state. The comparative analysis of the students' number regarding dominant emotional states is displayed (Figure 2).

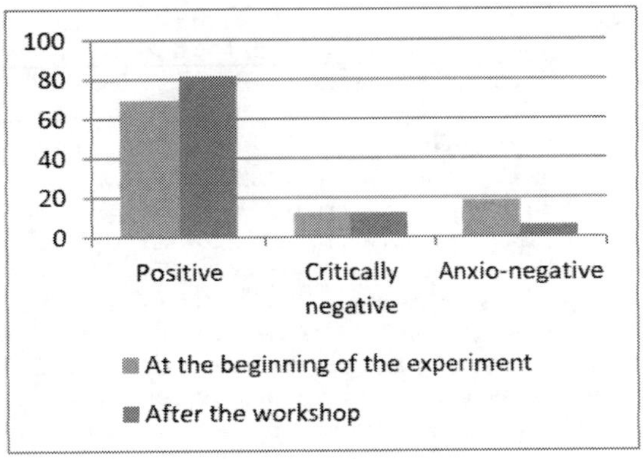

Fig. 2. Distribution of dominant states.

The analysis of the results proved that creating the environment based on the 5E Instructional Model during the scientific workshop where students did not feel negative emotions to research activities encouraged the increase of their interest in research activities.

4 Discussion

Searching for the ways of forming interest of master's students of pedagogical universities in research activities we faced the researches done by W. Sandoval and V. Reiser [36], M. Rocard [37]. The scientists point out that in order to form students' impression of the real world it is necessary to show them how to organize their activities as real scientists do, experiencing the process of learning and knowledge grounding. Fallon et al offered to seek the possibilities to organize students' research activities through method selection and forms of a learning organization that influences active students' involvement.

Traditional educational methods, which are oriented at the teacher, don't provide an active involvement of students to research activities (L. D. Yore [12], J.-L. Lin [38], K. Vlasenko K. et al [39]). The scientists emphasize the importance of searching for educational models that encourage the strengthening of students' learning activities. The Deductive Content Analysis Method helped us to choose the 5E Instructional Model as the foundation of developing a scientific environment of students' education.

The efficiency of the 5E Instructional Model to encourage students' research activities is proved in the works by L. Duran and E. Duran [31], R. Bybee, N. Landes [40], S. Supasorn, V. Promarak [32], P.-H. Cheng, Y.-T. Carolyn Yang, S.-H. Gilbert Chang, F.-R. Revon Kuo [30]. Also, we support the opinion by K. Vlasenko, O. Chumak, I. Sitak, O. Chashechnikova, I. Lovianova [41], who believe that learning has to be built so that students can research, explain, extend and estimate their progress, and the introduction of ideas assumes students' awareness of the reason or necessity of their use. Such an approach is completely agreed with the structure, functioning, and consistency of the 5E Instructional Model.

M. Alshehri [42] believes that while organizing research activities it is necessary to direct students to the main models of subject matters. For the pre-service mathematics teacher, one of the key subject matters is Approximation Theory, its broad influence on the modern state of innovation and technology development is widely known. The research is aimed at searching for ways of implementing a workshop on Approximation Theory to form students' interest in research activities. The main research result testifies that the use of the 5E Instructional Model influenced efficiently the formation of students' positive attitude towards research activities. Within this model, the involvement of the workshop on Approximation Theory encouraged the increase of the level of expressing students' positive emotional state (particularly interest, surprise increase) and decrease of anxiety level. These results are agreed with the conclusions by D. Coulson [43], E.-T. Chin, F.-N. Lin [44], A. Abdi [34], N. Jung [45], E. Ong [35], who studied the connection between interest growth and persons' emotional state. This justifies the use of methodology Differential Emotions Scale by Izard [22] during the experiment.

5 Conclusion

The actuality of involving students to research activities in education arises from the fact that research competence is considered as one of the components of the professional competence of the mathematics teacher. Enhancing students' interest in research content and research activities during the studies also requires the use of a model that implies complete students' awareness of the importance of the research problem. The Deductive Approach to Content Analysis helped us determine the possibility to involve the 5E Instructional Model to the organization of workshop classes on Approximation Theory, point out the characteristics of the model and parameters of its efficiency. The model stages provide the formation of a better understanding of scientific knowledge, abilities, and skills from the students' side. We developed the content of workshop classes on approximation theory according to the model stages. Relying on the analysis of the present recommendations on the model use in learning different subjects, we offered the recommendations on the organization of workshop classes, It should be noted that the course should be provided following the stages Engagement, Exploration, Explanation, Elaboration, and Evaluation following certain recommendations that encourage students' interest in research activities.

Forming a positive attitude to research activities is the first step to the development of the research competence of pre-service mathematics teachers. The analysis of works on the connection between the person's interest and emotional state allowed formulating the most important positive and negative emotions that are connected with the experience of the research activities. The results of calculating the indexes of students' emotional states proved that the creation of the environment based on the 5E Instructional Model where students do not feel negative emotions to research activities encourages emotional state and interest in research activities.

The perspectives of future research involve the creation of the courses that use Inquiry-based approaches with the purpose of further research on forming research competence among students of pedagogical universities.

References

1. I. Yarullin, N. Bushmeleva, I. Tsyrkun, Mathematics Education **10**(3), 137–146 (2015)
2. A. ScHoenfeld, *in Proceedings of the Fourth International Congress on Mathematical Education*, California, 1983
3. R. Turner, Research developments. A **5**, 24 (2010)
4. A. Vintere, A. Zeidmane, Research in mathematical competence in engineers' professional activities engineering for rural development. Paper presented at the 13th International Scientific Conference Engineering for Rural Development, University of Agriculture Faculty of Engineering, Jelgava, 29–30 May 2014
5. J. Proulx, For the learning of mathematics **35**(3), 25–27 (2015)

6. B. Koichu, A. Pinto, Canadian J. of Sc., Math. and Techn. Ed. **18**(1), 68–85 (2018)
7. J. Lithner, Mathematical reasoning and familiar procedures. Int. J. of Math. Ed. in Sc. and Techn. **31**(1), 83–95 (2000). doi: 10.1080/002073900287417
8. C. Bonwell, J. Eison, *Active Learning: Creating Excitement in the classroom*, ASHE-ERIC Higher Education Report No. 1. Washington, DC: George Washington University (1991)
9. E. Fallon, S. Walsh, T. Prendergast, An Activity-based Approach to the Learning and Teaching of Research Methods: Measuring Student Engagement and Learning. Irish Journal of Academic Practice **2**(1), 2 (2013). doi:10.21427/D7Q72W
10. P. Hernandez-Martinez, P. Vos, Why do I have to learn this? A case study on students' experiences of the relevance of mathematical modelling activities. J. ZDM Mathematics Education **50**, 245–257 (2018). doi:10.1007/s11858-017-0904-2
11. K. Jones, L. Black, A. Coles, Marking 21 years of Research in Mathematics Education. Research in Mathematics Education **21**(1), 1–5 (2019). doi:10.1080/14794802.2019.1592336
12. L.D. Yore, Electronic Journal of Science Education **5**(4), 1–7 (2001)
13. T. Dreyfus, M. Artigue, D. Potari, S. Prediger, K. Ruthver (eds.), *Developing Research in Mathematics Education Twenty Years of Communication, Cooperation and Collaboration in Europe* (Routledge, New York, 2019)
14. I. Biza, V. Giraldo, R. Hochmuth, A. S. Khakbaz, C. Rasmussen, in *Research on Teaching and Learning Mathematics at the Tertiary Level. ICME-13 Topical Surveys* (Springer Open, Cham, 2016), pp. 1–32
15. N. Telegina, The Use of Project Activity in Teaching Mathematics. Eurasia J. Math., Sci Tech. Ed. **15**(8) (2019). doi:10.29333/ejmste/108439
16. A. Matejko, D. Ansari, Contributions of functional Magnetic Resonance Imaging (fMRI) to the Study of Numerical Cognition. Journal of Numerical Cognition **4**(3), 505–525 (2018). doi:10.5964/jnc.v4i3.136
17. S. Sevinc, R. Lesh, Training mathematics teachers for realistic math problems: a case of modeling-based teacher education courses. ZDM Mathematics Education **50**, 301–314 (2018). doi:10.1007/s11858-017-0898-9
18. H.S. Malvar, *Signal processing with lapped transform* (Artech House, Norwood, 1992)
19. F. Dedus, S. Makhortykh, M. Ustinin, Application of the Generalized Spectral-Analytic Method in Information Problems. Pattern Recogn. Image Anal. **12**, 429–437 (2002). doi:10.1134/S1054661806030175
20. A. Pankratov, Spectral analysis for identication and visualization of repeats in genetic sequences. Pattern Recogn. Image Anal. **19**(4), 687 (2009). doi:10.1134/S105466180904018X
21. O. Novikov, O. Rovenska, Approximation of classes of poisson integrals by repeated Fejer sums. Lobachevskii J. Math. **38**(3), 502–509 (2017). doi:10.1134/S1995080217030209
22. C. Izard, *Differential Emotions Theory* (Springer, Boston, 1977)
23. J. Panksepp, Damasio's error? Conscious. Emot. **4**, 111–134 (2003). doi:10.1075/ce.4.1.10pan
24. S. Langer, *Mind: An Essay on Human Feeling* (Jonhs Hopkins University Press, Baltimore, 1967), p. 512
25. K. Vlasenko, Higher School Mathematics Teacher (2019), http://formathematics.com. Accessed 22 Feb 2019
26. Lesley University Online, What Is The 5E Model? A Definition For Teacher (2019), https://www.teachthought.com/learning/what-is-the-5e-model-a-definition-for-teachers. Accessed 22 Feb 2019
27. Lesley University, Empowering Students: the 5E Model Explained (2019), https://lesley.edu/article/empowering-students-the-5e-model-explained. Accessed 17 Oct 2019
28. R. Bybee, The BSCS 5[e] Instructional Model and 21st century skills (The National Academic Board on Science Education, 2009), https://sites.nationalacademies.org/cs/groups/dbassesite/documents/webpage/dbasse_073327.pdf. Accessed 5 Oct 2019
29. J. Branch, D. Oberg, *Focus on Inquiry Glossary*. (Alberta Learning, Alberta, 2004)
30. P.-H. Cheng, Y.-T. Carolyn Yang, S.-H. Gilbert Chang, F.-R. Revon Kuo. IEEE Transactions on Education **59**(2), 147–153 (2016)
31. L. Duran, E. Duran, The Science Education Review **3**(2), 49–58 (2004)
32. S. Supasorn, V. Promarak, Implementation of 5E inquiry incorporated with analogy learning approach to enhance conceptual understanding of chemical reaction rate for grade 11 students. Chem. Educ. Res. Pract. **16**, 121–132 (2015). doi:10.1039/C4RP00190G
33. R.W. Bybee, J.A. Taylor, A. Gardner, P. Van Scotter, J.C. Powell, A. Westbrook, N. Landes, The BSCS 5E Instructional Model: Origins, Effectives, and Applications (2006), https://media.bscs.org/bscsmw/5es/bscs_5e_full_report.pdf. Accessed 28 Nov 2019
34. A. Abdi, The effect of inquiry-based learning method on students' academic achievement in science course. Universal Journal of Educational Research **2**(1), 37–41 (2014). doi: 10.13189/ujer.2014.020104
35. E.T. Ong, 5E Inquiry Learning Model: Its Effect on Science Achievement among Malaysian Year 5 Indian Students. International Journal of Academic Research in Business and Social Sciences **8**(12), 348–360 (2018). doi:10.6007/IJARBSS/v8-i12/5017

36. W. Sandoval, B. Reiser, Explanation-driven inquiry: Integrating conceptual and epistemic scaffolds for scientific inquiry. Science Education **88**, 342–375 (2004). doi:10.1002/sce.10130
37. M. Rocard, Science Education NOW: A renewed Pedagogy for the Future of Europe (2015), http://ec.europa.eu/research/science-society/document_library/pdf_06/report-rocard-onscience-education_en.pdf. Accessed 25 Oct 2019
38. J.-L. Lin, Eurasia J. Math., Sc. Techn. Ed. **10**(5), 415–426 (2014)
39. K. Vlasenko et al., Training of Mathematical Disciplines Teachers for Higher Educational Institutions as a Contemporary Problem. Universal Journal of Educational Research **7**(9), 1892–1900 (2019). doi:10.13189/ujer.2019.070907
40. R. Bybee, N. Landes, The American Biology Teacher **52**(2), 92–98 (1990)
41. K. Vlasenko, O. Chumak, I. Sitak, O. Chashechnikova, I. Lovianova, Espacios **40**(31), 1–15 (2019)
42. M. Alshehri, IOSR Journal of Research & Method in Education (IOSR-JRME), **6**(2), 43–48 (2016)
43. D. Coulson, *BSCS Science: An inquiry approach – 2002 evaluation findings* (PS International, Arnold, 2002)
44. E.-T. Chin, F.-L. Lin, A survey of the practice of a large-scale implementation of inquiry-based mathematics teaching: from Taiwan's perspective. ZDM Mathematics Education **45**, 919–923 (2013). doi:10.1007/s11858-013-0546-y
45. N. Jung, How emotions affect logical reasoning: evidence from experiments with mood-manipulated participants, spider phobics, and people with exam anxiety. Front. Psychol. **5**, 570–562 (2014). doi:10.3389/fpsyg.2014.00570

Selection criteria for cloud-oriented learning technologies for the formation of professional competencies of bachelors majoring in statistics

Olga Gavryliuk[1], *Tetiana* Vakaliuk[2,*], and *Valerii* Kontsedailo[3]

[1]Institute of Information Technologies and Learning Tools of the NAES of Ukraine, Department of Cloud-Oriented Systems of Education Informatization, Kyiv, 04060, Ukraine
[2]Zhytomyr Polytechnic State University, Department of Software Engineering, Zhytomyr, 10005, Ukraine
[3]Easygenerator, 3012 KN Rotterdam, Netherlands

Abstract. This article scientifically substantiates the criteria for the selection of cloud-oriented learning technologies for the formation of professional competencies of bachelors majoring in statistics, as well as presents the results of expert evaluation of existing cloud-oriented learning technologies by defined criteria. The criteria for the selection of cloud-oriented learning technologies for the formation of professional competencies of bachelors majoring in statistics, were determined: information-didactic, functional and technological. To implement the selection of cloud-oriented learning technologies for the formation of professional competencies of bachelors majoring in statistics, and effective application in the process of formation of relevant competencies, the method of expert evaluation was applied. The expert evaluation was carried out in two stages: the first one selected cloud-oriented learning technologies to determine the most appropriate by author's criteria and indicators, and the second identified those cloud-oriented learning technologies that should be used in the educational process as a means to develop professional skills Bachelor of Statistics. According to the research, the most appropriate, convenient and effective cloud-oriented learning technologies for the formation of professional competencies of future bachelors of statistics by the manifestation of all criteria are cloud-oriented learning technologies CoCalc and Wolfram|Alpha.

1 Introduction

The European integration processes, change, and development of the educational system of Ukraine creates new requirements for the training of specialists in almost all spheres of human life. The formation of general competencies is the basis of general education, and the formation of professional competencies of future specialists is carried out in the process of education in higher education institutions (HEI). Traditional learning is out of date and needs updating, replenished with new technologies, forms, means, and is confirmed in the text of the National Doctrine of Educational Development that "continuous improvement of the quality of education, updating its content and forms of organization of educational process; development of the system of continuous education and training throughout life; introduction of educational innovations, information technologies" [1].

An important achievement in the field of education has been the creation of open education platforms based on the implementation of the principle of the functioning of cloud technologies; comprehensive updating of training technologies, methodological support, and content of distance and e-learning based on the introduction of information and communication technologies (ICT); introduction of new forms and methods of teaching based on cloud-oriented technologies, Web 2.0 technologies, services of electronic social networks [2].

Formation of professional competencies of specialists, including the future bachelor of statistics, is carried out during the training at HEI, and the use of the latest information and communication technologies is an important key element in this process. That is why one of the leading areas of qualitative training of specialists in the requirements of today is the application of cloud technologies, and in the educational process – cloud-oriented learning technologies (COLT).

2 Analysis of recent research

Research on evaluating the effectiveness of ICT learning has largely highlighted the problem of evaluating learning outcomes.

The analysis of existing ICTs, criteria and indicators of their selection were analyzed and highlighted in the works of such scientists as V. Yu. Bykov, O. A. Galchevska, V. M. Demianenko, O. S. Golovnia, K. R. Kolos, G. P. Lavrentyeva, L. A. Luparenko, O. M. Spirin, et al.

* Corresponding author: tetianavakaliuk@gmail.com

In particular, a team of authors (V. Yu. Bykov, O. M. Spirin, L. A. Luparenko) considered open web-oriented systems for monitoring the implementation of scientific and pedagogical research results [3]. O. S. Golovnia in her works investigated the virtualization software in the training of UNIX-like operating systems and identified the criteria and indicators of their selection [4]. V. M. Demianenko, G. P. Lavrentieva, M. P. Shyshkina give methodological recommendations on the selection and use of electronic tools and resources for educational purposes [5]. K. R. Kolos has developed criteria for selecting components of a computer-oriented educational environment for a postgraduate teacher education institution [6]. O. M. Spirin offers criteria for external evaluation of the quality of information and communication training technologies [7].

The use of cloud technologies in education has dedicated their works by such scholars as E. I. Ablialimova, O. G. Glazunova, O. V. Korotun, S. H. Lytvynova, L. M. Medzhitova, M. V. Marienko, Z. S. Seydametova, S. O. Semerikov, A. M. Striuk, S. N. Seitvelieva, Yu. V. Tryus, V. M. Franchuk, M. P. Shyshkina, and others.

In particular, the problem of developing a methodological system for the use of a cloud-oriented environment in the training of databases of future computer science teachers was investigated by O. V. Korotun [8]. The question of designing a cloud-oriented educational environment of a comprehensive educational institution by S. H. Lytvynova was also investigated [9]. Several teams of authors have considered cloud technologies in learning at different intervals [10; 11; 12; 13]. At the same time, the question of research into the use of cloud technologies in training future bachelor of statistics and the development of appropriate criteria and indicators of selection have not been sufficiently studied.

The **purpose** of the article is to define criteria and establish appropriate indicators for the selection of cloud-oriented learning technologies to shape the professional competencies of bachelors majoring in statistics.

3 Results

Research on the implementation of cloud-oriented learning technologies to shape the professional competencies of future professionals is being actively pursued by various researchers. As this research is aimed at COLT to shape the professional competencies of future Bachelor of Statistics, it is important to identify, by a certain set of criteria, the most effective, convenient and relevant cloud-oriented learning technologies to be used in the educational process of HEI.

To begin with, we will define the term "criteria", since this definition is presented differently by different researchers.

In encyclopedic reference publications, the concept of "criterion" is defined as "a trait, a basis for evaluation, taken as a basis for classification" [14].

In "Vocational Education: A Dictionary" ed. N. G. Nychkalo, the criterion is called "the criterion for evaluating something, a means of verifying the truth or falsehood of a statement" [15].

V. N. Bahrii in his research, argues that the criterion is "a standard against which to evaluate, compare a real pedagogical phenomenon, process, or quality by reference" [16].

R. V. Torchevsky notes that "in the most general form, the criterion is an important and defining feature that characterizes the various qualitative aspects of a particular phenomenon under study, helps to clarify its essence, helps to specify the main manifestations. In this regard, the indicator is a quantitative characteristic of this phenomenon under study, which makes it possible to conclude on the state of statics and dynamics" [17].

In a short terminological dictionary I. M. Dychkivska term "criterion" is given as "an indicator that characterizes the property (quality) of an object, the evaluation of which is possible using one of the measurement methods or the expert method" [18].

Under the selection criteria of COLT for the formation of professional competencies of future bachelors of statistics, we will understand such features, qualities, and properties of cloud-oriented technologies that are required for their effective use in the educational process to form the professional competencies of future bachelors of statistics.

An expert evaluation method was used to implement the selection of the COLT for the formation of the professional competencies of future bachelors of statistics and for effective application in the process of forming the corresponding competencies. According to the purpose and objectives of the method, the corresponding COLT is numbered in ascending or descending order based on a separate trait, by which further ranking is made. It should be noted that the peer review was carried out in two stages.

In the first stage, experts were asked to evaluate 8 COLT that could be used in the process of forming the professional competencies of future bachelors of statistics.

20 experts of different profiles were invited to the expert evaluation procedure, among them officials of the State Statistical Service of Ukraine and the State Treasury in Zhytomyr, employees of banking institutions, employees of commercial financial institutions.

A point scoring system was used in the study [19]. According to the aforementioned evaluation system, for the number of N COLT, the maximum possible estimate of N is given to the most significant in the use of COLT and 1 to the least significant. The results of the assessment are presented in the form of a table, where the columns indicate the hotline number and the fields the expert number. The COLT name card is presented in alphabetical order (A to Z), to prevent psychological clues that could affect the outcome of the assessment.

To determine whether there is an objective agreement between experts, calculated concordance coefficient Kendall's W [20] by the formulas:

$$W = \frac{S(d^2)}{S_{max}(d^2)} = \frac{12 \cdot S(d^2)}{m^2(n^3-n)} \quad (1)$$

$$S(d^2) = \sum_{j=1}^{n} d_j^2 \quad (2)$$

$$d_j = S_j - 0{,}5 \cdot m \cdot (n+1) \quad (3)$$

$$S_j = \sum_{i=1}^{n} R_{i,j} \quad (4)$$

Where:

S_j – is the total rank of the j-th indicator (it should be noted that this is the main parameter for evaluating the significance of the indicator);

$j = 1, 2, 3, \ldots, n$; n is the number of indicators;

m – number of experts;

$R_{i,j}$ – is the rank of the j-th indicator, determined by the i-th expert.

Applying formulas (1) to (4) for the sake of calculations, obtain a certain value W based on the experimental data. If the results of the calculations differ significantly from zero, this means that there is an objective agreement between the experts (if W=0, it is considered that there is no correlation between the ranking of experts, at W=1, the rankings are completely identical) and the total ranks are quite objective.

The results of the peer review are presented in Table 1.

Table 1. Ranking Cloud-Oriented Learning Technologies for the formation of the professional competencies of future bachelor of statistics.

COLT / Expert number	CoCalc	Excel Online	GeoGebra	Google Sheets	MapleCloud	Scilab	WebMathematica	Wolfram|Alpha
1	6	4	2	1	3	5	7	8
2	6	5	1	2	3	4	8	7
3	8	1	2	3	4	5	7	6
4	5	3	2	1	4	8	7	6
5	5	2	1	4	3	6	7	8
6	6	1	5	2	3	4	8	7
7	8	2	3	1	5	4	7	6
8	5	3	1	2	4	6	7	8
9	6	1	4	3	2	5	8	7
10	7	1	2	3	4	8	5	6
11	7	3	2	4	1	6	5	8
12	5	2	3	6	1	4	8	7
13	8	1	2	3	4	5	6	7
14	6	4	1	3	2	5	8	7
15	7	4	1	3	2	5	6	8
16	5	3	2	4	1	6	8	7
17	8	2	1	3	5	4	7	6
18	7	1	2	3	4	8	5	6
19	4	3	2	1	8	7	5	6
20	7	4	1	2	3	6	5	8
S	**126**	50	40	54	66	**111**	**134**	**139**
d	36	-40	-50	-36	-24	21	44	49

The result was selected COLT 4: CoCalc, Scilab, WebMathematica, Wolfram|Alpha.

After calculating (by formulas 1–4) based on the experimental data presented (see Table 1), obtained a coefficient of concordance W = 0.71. Since the value obtained is non-zero, there is an objective agreement between experts.

In the second phase of the study, another group of specialists was recruited to evaluate the most significant COLT according to certain criteria. It is worth noting that the second stage involved 15 specialists of different profiles, namely: teachers, heads of departments and deans of faculties of higher education institutions of Ukraine, having experience and related to the professional training of future bachelors of statistics, employers (Department of Statistics in Zhytomyr region, Department of the State Treasury Service of Ukraine in Zhytomyr, Main Department of State Tax Service in Zhytomyr region, heads of state and commercial banks, managers financial companies), which worked directly with the selected COLT and could objectively evaluate them according to the degree of manifestation of each criterion.

The manifestation of each of the presented criteria was evaluated for each of this COLT. To this end, experts have been asked to evaluate its performance using the scale shown in Table 2.

Table 2. Scale bar for evaluation of the relevant criteria

Scores	Evaluation of the indicator
0	the indicator is missing
1	the indicator is partially available (not available more than available)
2	the indicator is more available than not available
3	the indicator is completely available

The indicator will be considered positive if the arithmetic mean of these points is at least 1.5. If more than half (50%) of the indicators of the relevant criterion are negative, then the criterion is defined as insufficiently developed. In the case of:

- when 50-55% of the indicators of the criterion are positive, the criterion is characterized as critically manifested;
- if 56-75% of the indicators of the criterion are positive, then the criterion is characterized as sufficiently manifested;
- if 76–100% of the criterion indicators are positive, then the criterion is characterized as highly manifested [19].

An analysis of existing cloud-oriented learning technologies to shape the professional competencies of future bachelors of statistics [21] has made it possible to identify the criteria and relevant indicators of these cloud-oriented learning technologies:

- information-didactic: information support; coverage of various sections of mathematics and statistics; graphical presentation of results; teamwork on the project; ability to apply programming knowledge;
- functional: user-friendly interface; free of charge; accessibility; multilingualism;
- technological: cross-platform; integration with other cloud services; adaptability.

The results of the peer review of each of the selected criteria and relevant indicators will be discussed in more detail.

The **information-didactic criterion** characterizes the information and didactic component of cloud-oriented learning technology and is based on the laws of assimilation of knowledge, skills, and competences, namely:

1.1. the indicator "information support" characterizes the presence of a description of the use of the tool, examples, or the presence of a section of assistance;

1.2. the indicator "coverage of various sections of mathematics and statistics" characterizes the possibility of using COLT in the process of studying certain sections of mathematics and statistics;

1.3. the indicator "graphical presentation of results" characterizes the ability to interpret the results in the form of graphs, histograms or a three-dimensional model;

1.4. the indicator "teamwork on the project" characterizes the ability to work with multiple users at the same time;

1.5. the indicator "ability to apply programming knowledge" characterizes the ability to take individual actions to perform calculations using different programming languages.

Basic data on indicators of information-didactic criteria for each of the selected COLT contains table 3.

Table 3. The information-didactic criterion for selection of cloud-oriented learning technologies and the value of its indicators.

The indicators / COLT	information support	coverage of various sections of mathematics and statistics	graphical presentation of results	teamwork on the project;	ability to apply programming knowledge	The manifestation of the criterion	The level of manifestation
CoCalc	1.93	2.67	2.07	1.80	2.00	**100%**	highly
Scilab	2.13	2.20	0.80	0.80	2.33	**60%**	suffi-ciently
WebMathematica	1.47	2.00	1.33	1.53	2.13	**80%**	highly
Wolfram|Alpha	2.33	2.27	2.33	1.53	2.33	**100%**	highly

The functional criterion characterizes the functional component of cloud-oriented learning technologies and assumes the following indicators:

2.1 the indicator "user-friendly interface" describes the convenience and comprehensibility of the interface and the computational component of the hot-water system;

2.2 the indicator "accessibility" characterizes the provision of cloud-oriented learning technology to different categories of users;

2.3 the indicator "free of charge" characterizes the possibility of free or full use of cloud-oriented learning technologies;

2.4 The indicator "multilingualism" characterizes the support of multiple languages (localization) of the interface.

The basic data on the indicators of the functional criterion for each of the selected COLT contains table 4.

The **technological** criterion is characterized as follows:

3.1 "cross-platform" indicates the possibility of using cloud-oriented learning technologies in different operating systems;

3.2 the indicator "integration with other cloud services" implies the possibility of supporting the work with calculations in different cloud services, and the possibility of further integration with other services;

3.3 "Adaptability" indicates the possibility of full use of cloud-oriented learning technologies on different devices (desktop PC, laptop, netbook, tablet, smartphone, etc.).

Table 4. The functional criterion for the selection of cloud-oriented learning technologies and the value of its indicators.

The indicators / COLT	user-friendly interface	free of charge	accessibility	multilingualism	The manifestation of the criterion	The level of manifestation
CoCalc	1.80	2.00	2.20	1.80	**100%**	highly
Scilab	2.00	1.87	2.13	1.53	**100%**	highly
WebMathematica	1.73	1.87	1.73	1.93	**100%**	highly
Wolfram|Alpha	2.13	2.53	2.20	1.60	**100%**	highly

The basic data on the indicators of the technological criterion for each of the selected COLN contains table 5.

Table 5. The technological criterion for the selection of cloud-oriented learning technologies and the value of its indicators.

The indicators / COLT	cross-platform	integration with other cloud services	adaptability	The manifestation of the criterion	The level of manifestation
CoCalc	1.53	1.53	1.93	**100%**	insufficiently
Scilab	1.53	1.53	1.53	**100%**	highly
WebMathematica	1.73	1.73	1.93	**100%**	sufficiently
Wolfram|Alpha	2.60	2.33	2.93	**100%**	sufficiently

Let's summarize the results of the study in table 6.

Table 6. Generalized results of the selection of cloud-oriented learning technologies by the manifestation of all criteria.

Criterion / COLT	Information-didactic	Functional	Technological
CoCalc	**100%**	**100%**	**100%**
Scilab	**60%**	**100%**	**100%**
WebMathematica	**80%**	**100%**	**100%**
Wolfram|Alpha	**100%**	**100%**	**100%**

4 Conclusions

Therefore, according to the research, the most appropriate, convenient and effective cloud-oriented learning technologies for the formation of professional competencies of future bachelors of statistics by the manifestation of all criteria are cloud-oriented learning technologies CoCalc and Wolfram|Alpha. In the future, according to the results of the expert evaluation, a model of using cloud-oriented learning technologies for the formation of professional competencies of future bachelors of statistics will be developed, and a methodology for using cloud-oriented learning technologies for the formation of professional competencies of future bachelors of statistics will be described.

References

1. Ukaz Prezydenta Ukrainy "Pro Natsionalnu doktrynu rozvytku osvity" (Presidential Decree "On the National Doctrine of Educational Development") (2002), https://zakon.rada.gov.ua/laws/card/347/2002. Accessed 28 Nov 2019
2. *Natsionalna dopovid pro stan i perspektyvy rozvytku osvity v Ukraini* (National report on the state and prospects of education development in Ukraine). (Pedhohichna dumka, Kyiv, 2016).
3. V.Yu. Bykov, O.M. Spirin, L.A. Luparenko, The theory and practice of social systems management 1, 3–25 (2014)
4. O.S. Golovnia, Information technology in education **24**, 119–133 (2015)
5. V.M. Demyanenko, G.P. Lavrentyeva, M.P. Shishkina, Metodychni rekomendatsii shchodo doboru i zastosuvannia elektronnykh zasobiv ta resursiv navchalnoho pryznachennia (Guidelines for the selection and use of electronic tools and resources for educational purposes). The computer at school and family 1, 44–48 (2013)
6. K.R. Kolos, Information technology in education **17**, 109–117 (2013)
7. O.M. Spirin, Kryterii zovnishnoho otsiniuvannia yakosti informatsiino-komunikatsiinykh tekhnolohii navchannia (Criteria for external evaluation of the quality of information and communication training technologies). Scientific journal of NPU named after M.P. Drahomanov. Series 2. Computer-Oriented Learning Systems **9**(16), 80–85 (2010)
8. O.V. Korotun, Dissertation, Institute of Information Technologies and Learning Tools of the NAES of Ukraine, Kyiv, 2018
9. S.H. Lytvynova, *Proektuvannia khmaro zoriientovanoho navchalnoho seredovyshcha zahalnoosvitnoho navchalnoho zakladu* (Design of the cloud-oriented educational environment of a comprehensive educational institution). (Komprint, Kyiv, 2016)
10. O.G. Glazunova, O.G. Kuzminska, T.V. Voloshina, T.P. Sayapina, V.I. Korolchuk, Khmarni servisy Microsoft ta Google: orhanizatsiia hrupovoi proektnoi roboty studentiv VNZ (Microsoft and Google cloud services: organizing group project work for university students). Open educational e-environment of Modern University 3, 199–211 (2017), http://openedu.kubg.edu.ua/journal/index.php/openedu/article/view/84/135. Accessed 28 Nov 2019
11. Z.S. Seydametova, E.I. Ablyalimova, L.M. Medzhitova, S.N. Seitvelieva, V.A. Temnenko, *Oblachnye tekhnolohyy y obrazovanye* (Cloud technologies and education). (Diaipi, Simferopol, 2012).
12. O.M. Markova, S.O. Semerikov, A.M. Striuk, The cloud technologies of learning: origin. Information Technologies and Learning Tools **46**(2), 29–44 (2015). doi:10.33407/itlt.v46i2.1234
13. A.M. Striuk, M.V. Rassovytska. Vykorystannia khmarnykh tekhnolohii u kombinovanomu navchanni informatyky studentiv inzhenernykh spetsialnostei (The use of cloud technologies in the combined teaching of computer science students of engineering specialties). Bulletin of the Alfred Nobel Dnipropetrovsk University. Series: Pedagogy and Psychology **1**(9), 221–226 (2015)
14. V. Busel (ed.), *Velykyi tlumachnyi slovnyk suchasnoi ukrainskoi movy* (The Great Interpretative Dictionary of Modern Ukrainian). (Kyiv, 2005)
15. *Profesiina osvita: slovnyk* (Vocational education: vocabulary). (Kyiv, 2000).
16. V.N. Bagrii, Kryterii ta rivni sformovanosti profesiinykh umin maibutnikh sotsialnykh pedahohiv (Criteria and levels of future social educators' professional skills). Collection of scientific works of Khmelnytskyi Institute of Social Technologies, University of Ukraine 6 (2012)
17. R.V. Torchevsky, Dissertation, Institute of Professional Technical Education of the Ukrainian National Academy of Pedagogical Sciences, 2012
18. I.M. Dichkivska, *Innovatsiini pedahohichni tekhnolohii* (Innovative pedagogical technologies). (Academia, Kyiv, 2004)
19. O.M. Spirin, T.A. Vakaliuk, Information technologies and learning tools **60**(4), 275–287 (2017)
20. O.V. Zastelo. Analiz metodiv vyznachennia uzghodzhenosti dumky hrupy ekspertiv pid chas otsiniuvannia rivnia sformovanosti inshomovnoi komunikatyvnoi kompetentnosti slukhachiv. (Analysis of methods for determining the coherence of opinion of an expert group in assessing the level of formation of the students' foreign language communicative competence). The computer at school and family **8**, 18–22 (2015)
21. O.D. Gavriluk, Porivniannia naiavnykh khmaro oriienntovanykh tekhnolohii navchannia dlia pidhotovky bakalavriv statystyky (Comparison of available cloud-oriented learning technologies for the preparation of Bachelor of Statistics). Scientific notes. Part I. Series: Pedagogical Sciences **177**, 104–107 (2019)

Neuro-network technologies as a mean for creating individualization conditions for students learning

Dmitriy Bukreiev*

Bogdan Khmelnitsky Melitopol State Pedagogical University, Department of Computer Science and Cybernetics, Melitopol, Ukraine

Abstract. The study is devoted to the analysis of the current level of use of neural network tools to create conditions for individualization of student learning. The author presents the results of the analysis of scientific developments of scientists in the questions of the use of neural network approaches in the educational process, presents a range of basic modern directions of research and development, in particular: adaptive tests, individualization of learning, improvement of the level of distance education and others. The author emphasizes that modernity is characterized by an active increase in the number of students who choose for themselves individual and distance forms of education, which leads to the need for the development of modern technological systems of distance learning. For this purpose, the scientist conducts a comparative analysis of distance learning tools, their features and capabilities in terms of implementation of an automated intellectual system of individualization of the educational process.

1 Problem statement

Nowadays in the conditions of constant decrease of time for studying of students, in connection with the fast "aging" of knowledge and the great need of employers in the qualified personnel "today and now" the problem of intensification of education in the vector of creating opportunities of qualitative distance and individual form of learning with the possibility of assessment and adjusting the learning vector for each student, based on his or her personal skills and abilities. In connection with this, the world scientific and pedagogical space has been actively engaged in recent years in solving the problems of development of the means of accumulation and implementation of electronic means of processing large volumes of information. However, the current state of the art is insufficient, so there is a need to create means of processing, transmission and presentation of information, so it is advisable to create information systems for processing and issuing the necessary information, knowledge, information. Among these developments, issues related to intelligent information systems and means of classification and presentation of information in the educational space of the 21st century.

2 Analysis of recent research and publications

When conducting research, inevitably work with information of a non-numerical and informal nature. As a result, there is a need to build a model and tools for quantitative description of processes and subjects related to pedagogy [22, 23]. Of particular importance is the quality of education, which is understood as a cumulative indicator that reflects the result of the educational institution, as well as meeting the needs and expectations of society in the formation of different competencies of the individual.

Assessment of the quality of education depends on many factors, perhaps with unknown influence. Also in this case, there is a specificity of the "product" of education – a graduate of an educational institution, which should be considered as a complex system. According to research by V. V. Osadchyi and K. P. Osadcha the introduction of electronic education, services and tools for educators, students, leaders and parents in order to cooperate with all the above mentioned participants in the educational process, development of cloud technologies [25], which are leading in the scientific space, can be considered as a major among modern trends in informatization of education. However, before determining the quality received at the end of graduate studies, it is necessary to identify the factors that have the greatest impact on the student learning process. The following are traditionally distinguished among them [24].

1. Quality of interaction with the labor market in shaping the content of education: the level of involvement of employers in the formation of educational programs; the presence of a system of targeted training of specialists to order businesses; availability of agreements with foreign universities on joint implementation of educational programs; percentage of international students in relation to all students; level of development of the system of analysis and promotion of employment of graduates. And as noted in [27], employers' requirements for engineering and technical knowledge and skills are constantly being complicated. This is due to

* Corresponding author: dessnight@gmail.com

the accelerated evolution of technical skills, the emergence of new engineering professions and the penetration of technology in all areas of human life.

2. Quality of entrants and students: level of mastering of current educational material; comprehensive assessment of residual knowledge; the level of ability to solve applied problems; the level of final certification.

3. Quality of teaching technologies, educational and methodological and logistical support of the educational process: the level of the system of continuous control of student success during the semester and technology of student rating; the level of educational material; the level of methodological support of the educational process; the level of availability of specialized laboratories; level of organization of support of students' education according to individual plans.

4. Quality of the teaching staff: the level of conformity of the teacher's basic education with regard to the length of teaching and specialization; teacher's qualification level (academic degree, rank, position); level of social-psychological relations of the teacher with students; introduction of the practice of inviting specialists from different industries to conduct training sessions.

Based on this, we can confidently say that about 50% of the total number of factors that affect students' level of knowledge and their quality as specialists in the future, focuses on the development of quality methodological materials and their adaptation to the individual qualities of students. In addition, as noted in the work of V. V. Osadchyi and I. V. Krasheninnik, the main deterrent to improving professional training in such conditions is a shorter, compared to the standard four-year, bachelor's degree. During the two years of study it is not possible to provide the necessary time for a full-fledged practical training [26], but the creation of an individual trajectory of student's learning provides an opportunity to intensify the learning load and reduces the time for obtaining general competences. This confirms the need for a defined and detailed analysis of the current level of development of neural network technologies in the vector of their use as a means to create conditions for individualization of student learning.

During a detailed analysis of the scientific background, it was concluded that in the general sense the problem of development of neural network systems is devoted to a large amount of research literature, however, taking into account the specific topic of research limitations it is advisable to analyze only those studies in which the development systems are oriented to work in the educational space, particularly in the vector of adaptive learning. Questions of studying aspects of designing adaptive information learning environments were dealt with by V. Ya. Tsap [12], A. V. Narozhny [5]; from the point of view of the development of intellectual system of quality assessment of the educational process published their works A. A. Rybanov, V. P. Shevchuk, E. A. Prikhodko [9]; O. I. Petkovsky revealed the peculiarities of developing an intellectual system for assessing the level of development of general competences in students, which in turn confirms the view that the use of adaptive learning means qualitatively improves the process of information perception by students [8].

The works of P. I. Fedoruk focused on forecasting trends in the development of educational space in accordance with the needs of today and the development of adaptive testing tools, students, the scientist emphasizes the need to solve a number of problems of test control of knowledge in modern educational systems. In his work, the author notes that the methods and tools of classical test theory, despite the great potential of this technology do not solve many of the problems posed by the current level of educational development. He emphasizes that adaptive tests, in this case, allow to solve this problem and can be effectively used for solving any problems of optimization of educational process – evaluation of efficiency of pedagogical innovations and technologies, monitoring, etc. [11]. Summarizing the results of the scientist, you can identify a number of positive features of the use of adaptive tests, so adaptive tests allow you to: adjust to the individual abilities of the student; to increase the accuracy of the assessment of the level of knowledge of each individual student by using a larger bank of questions of different levels of complexity; reduce the duration of the test and the number of questions required to achieve sufficient accuracy in assessing the student's level of knowledge; reduce student fatigue; ensure confidentiality by providing each student with an individual set of test tasks appropriate to his or her level of knowledge; simplify the procedure for making changes to the test bank. This study is key to us, as it confirms the assumption that the use of adaptive learning creates a powerful platform for creating personalized learning, which in turn significantly improves students' perception of information at every stage of learning.

Confirmation of our assumption is the work of A. S Dovbysh, who emphasized that further informatization of the educational environment is based on computerized work tools and neural network technologies and requires a fundamentally new information environment – a set of computer communication systems, databases and knowledge bases, as well as software complexes. The author note that knowledge control systems built on the basis of learning supportive decision-making systems have become widespread [4]. Such systems are a further development of expert systems. They provide a quantitative assessment of the performance of the system on the basis of selected criteria and models of alternative situations.

A number of works of Ukrainian scientists are devoted to the issues of information support of intellectual systems of vocational guidance and their use in the educational process. Research by V. V. Osadchyi reflects that the modern education system, using modern teaching methods, creates a serious alternative to traditional forms and methods of learning. With the discovery of computer intelligent systems by humanity, humanity is increasingly moving in the vector of transferring the learning process into an automated electronic process. However, studies show that a large number of human factors (such as defining a profession, choosing a life path, etc.) cannot be fully incorporated into an automated computer model. Scientist argue that the task is to at least partially support and provide the educational process with the means of intellectual information technology, reflecting the most important aspects of education and choice of profession [6]. The author emphasize that nowadays one of the main problems of education in general is the problem of effective professional orientation of young people in the conditions of dynamic labor market and educational services. We support the opinion of the authors and consider it appropriate to use intelligent Internet decision support systems for the choice of a profession to solve this problem. the user needs recommendations on how to proceed with the decision, because the choice of career path is a starting point stage of study of any discipline and leads to quality and informed learning.

From the point of view of the software of modern intellectual information systems, a number of works of scientists describes a neural network information exchange system for organizations that, using artificial intelligence concepts such as frames, production rules, and inheritance networks, aims to help people share and filter information transmitted through computer-based exchange systems using a rich set of templates. The works of H. Fujisaki, H. Kameda, S. Ohno, K. Abe, M. Iijima, M. Suzuki and others [2] are devoted to neural network systems for finding information that is the first stage of development of modern neural network system for educational needs. An interesting study, from the point of view of the above issues, is the study of B. Florian-Gaviria, C. Glahn and R. Fabregat Gesa, regarding an experiment conducted with the support of 20 teachers. In the course of the experiment, teachers used a software suite for the effective use of the European Qualifications Framework during the students' study, the result of the experiment was the confirmation that the modern pedagogical process requires the development of software in order to improve quality of the educational process. According to the results of the experiment, the authors claim that certain software helps teachers to create contextual awareness, encourages students to deeper process of thinking and understanding, increases the efficiency of their perception of progress in the study of discipline and helps in creating a model of success or failure [1]. This research is very important in our scientific work because it confirms our view that the development of software education qualitatively improves the outcome of the entire learning process and provides an opportunity to influence the perceptions of students who are not involved or little involved in training.

3 Analysis of the current state of the software

In the framework of the study, it is advisable to analyze the current state of the currently existing software oriented to the pedagogical process. The analysis showed that in recent years, the use of "virtual classes" in which students are synchronized programmatically to create the psychological effects of competition and competition, has become very popular. In our opinion, one of the most important roles in the process of distance learning is the personification, which is carried out by regular periodic testing of the level of preparation, degree and speed of mastering the material by students.

During the analysis of the software, the dynamics with which new developments in the e-learning systems market are identified, indicate its rapid development. Each such system contains some special tools that are designed for the development of disciplines. Generally, they have the following advantages:

- significantly reduces the time to develop courses;
- reduce the overall costs of the organization to develop and maintain distance courses;
- modern level of functionality and communication capabilities of the graphical interface of distance courses is ensured;
- system bugs for beginning course developers are eliminated.

The use of specialized tools for the development of distance learning courses can significantly expand the audience of potential course developers, including teachers who do not have a thorough knowledge of information technology. Here is a description of the most popular distance learning systems.

BlackBoard offers solutions for Blackboard 9.1 educational institutions. The Blackboard 9.1 solution (Fig. 1) has three main applications:

- Blackboard Learning System;
- Black Blackboard Community Portal System (a unifying portal that provides access to resources, course administration, communications tools, charts, etc. for relevant categories of users);
- Black Blackboard Transaction System (an online system that provides student identification, access and management of tuition fees and more).

The Blackboard learning system consists of the following applications:

- Course management (a content management system that provides content management, communication tools, tests, surveys, exams; and providing various additional management tools for teachers);
- Blackboard Building Blocks architecture for interoperability and customization (a component that integrates various types of content and commercial applications with the Blackboard platform, as well as containing various utilities and applications for students and teachers);
- Advanced integration and system management (a system that integrates Blackboard solutions with various information systems).

Learning Space 5.5 (Lotus / IBM) software allows you to learn and teach in asynchronous mode (access to study materials at your convenience) and participate in online classes in real time. The user can create the content of the course in any program, and then place the created material in Learning Space 5.5 (Fig. 2).

The program has a flexible system of editing and administration of the course, allows you to choose different modes of study and keep track of the current results of students. Distance learning courses are organized as a series of lessons that can be self-contained, interactive or group-based. Self-study usually contains reading material and tests that you must complete after studying the material. Interactive lessons include attending virtual class lectures, engaging in online discussion or chatting, working with a virtual whiteboard, and a web-sharing system. Group classes include offline classes and online discussions, chats. All entries left during the course discussion are available throughout the course. The online classes are scheduled on a specific date and time and are conducted by the teacher in a virtual classroom in real time. Students' current scores (grade level, grade, time lost, number of hits, etc.) are stored in the database. This information is available to the teacher at any time in the form of various forms of reports.

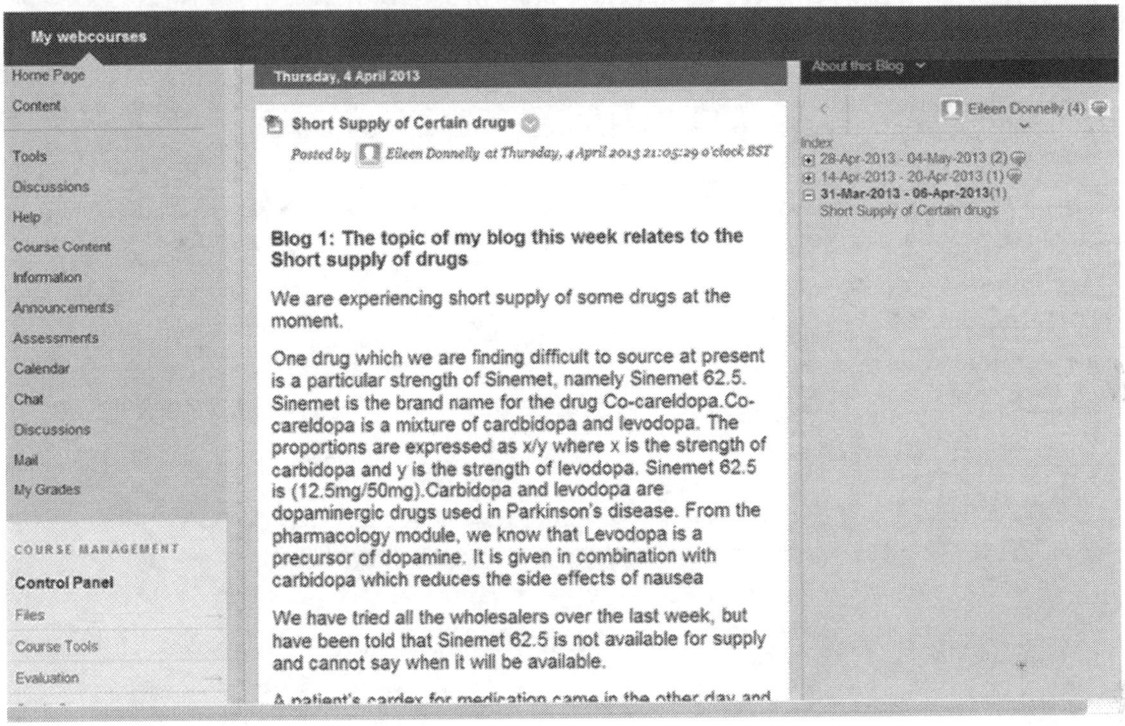

Fig. 1. View of the Blackboard Learning System.

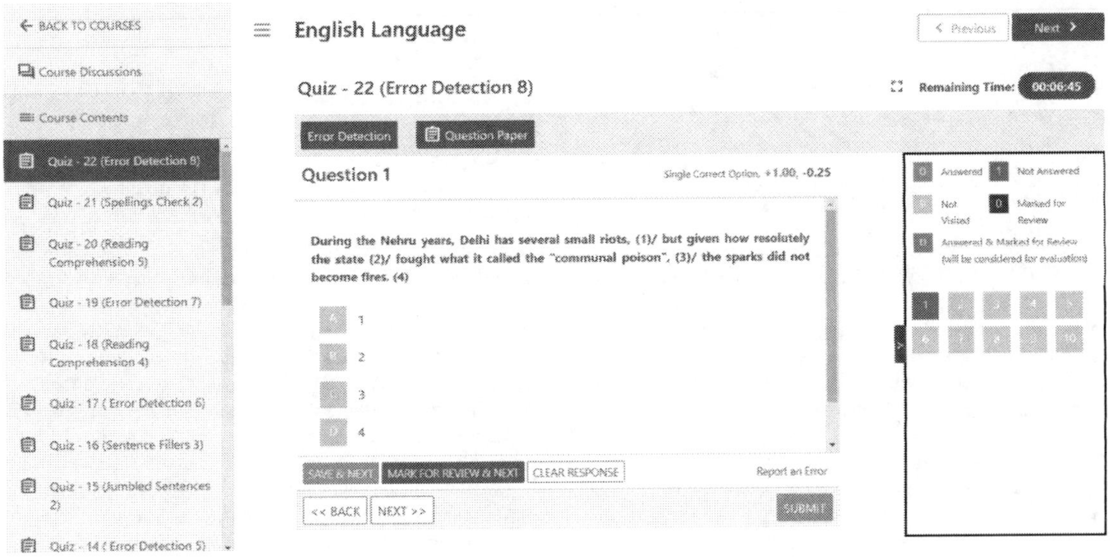

Fig. 2. View of Learning Space Learning System.

Learning Space 5.5 has two main components: the Core module and the Collaboration module. The base module consists of a Core server (on which Learning Space 5.5 software is installed and running), a database server and a web server that creates the copyright. The Learning Space 5.5 server contains the core product software and is the core of the distance learning system. It provides an interface for entering and retrieving user and course information, storing and retrieving student results. It also supports a student interface designed to attend classes and view personal enrollment and learning outcomes. The database stores user data, course information, and listener ratings that are issued automatically or at the user's request. In addition to these database queries, Learning Space 5.5 contains several predefined report formats. The database contains course structure information and the course content is retrieved from the author's web server.

Moodle (Modular Object-Oriented Dynamic Learning Environment) is a suite of open source modular software (GNU GPL) licensed to create distance learning courses and websites. This distance learning program is focused on teacher-student interaction and is also used to support full-time courses.

Moodle (Fig. 3) can be installed on any computer that supports PHP and works with MySQL, PostgreSQL, Microsoft SQL Server; the software is cross-platform [16]. The project is based on the principles of social constructivism:
- in today's learning environment, we are all potential teachers and students at the same time;
- we are successful learners, especially trying to create or explain something to people;
- observing the work of our colleagues is a major contribution to learning;
- understanding others makes it possible to study them more individually;
- the learning environment should be flexible, providing participants with an easy-to-use tool to meet their learning needs [22].

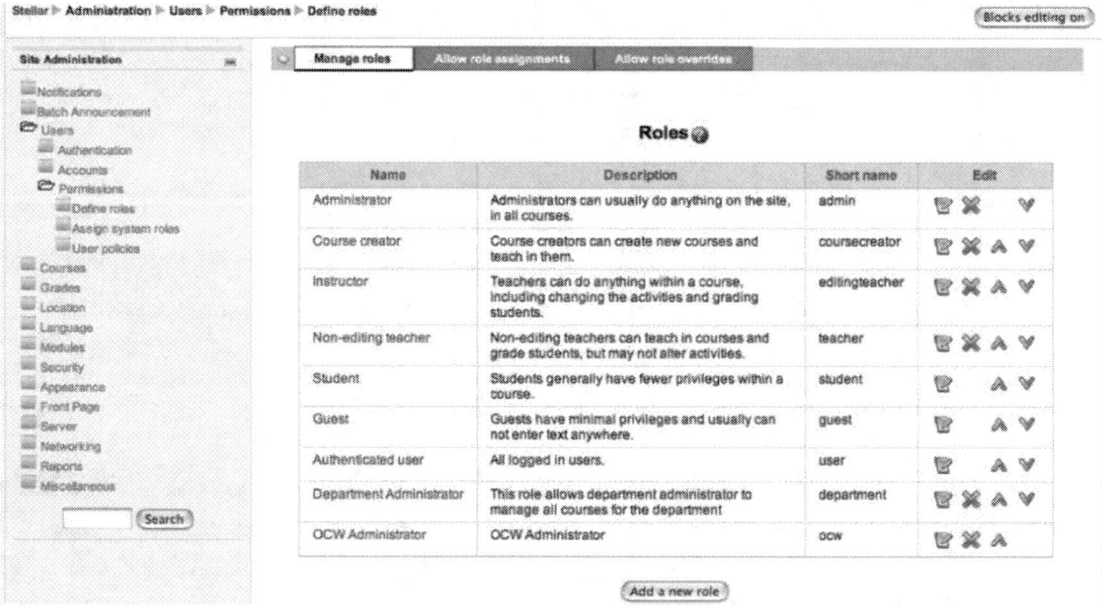

Fig. 3. View of the Moodle Learning System

SharePointLMS (Fig. 4) is a distance learning system developed on the powerful multi-functional MS Office SharePoint Server 2007. It is a comprehensive solution that integrates all users (teachers, students, administrators, etc.) into a single information and training space and provides tools cooperation. Unlike Moodle, Claroline and ATutor, the system is chargeable. It is used not only by educational institutions and training centers, but also by enterprises, organizations, and state structures [17].

However, having analyzed the existing software, we came to the conclusion that today we have many great examples of Web-based systems for distance learning, among them Blackboard, LearningSpace, Moodle and others. Such systems confidently use the benefits of information technology, but the study raises the need to determine the level of development of the aspect of the use of neural network technologies based on distance learning systems.

During further analysis, it can be concluded that e-learning should not be developed in a vacuum; rather, it should be as close as possible to the needs and desires of students and adaptable as the course progresses. Assessing open source e-learning platforms in order to find the platform that is most suitable for adaptive expansion was thus very important. The extended platform could then be used in an online learning environment. Therefore, the overall functionality of the platform is as important as adaptability, and evaluation addresses both issues.

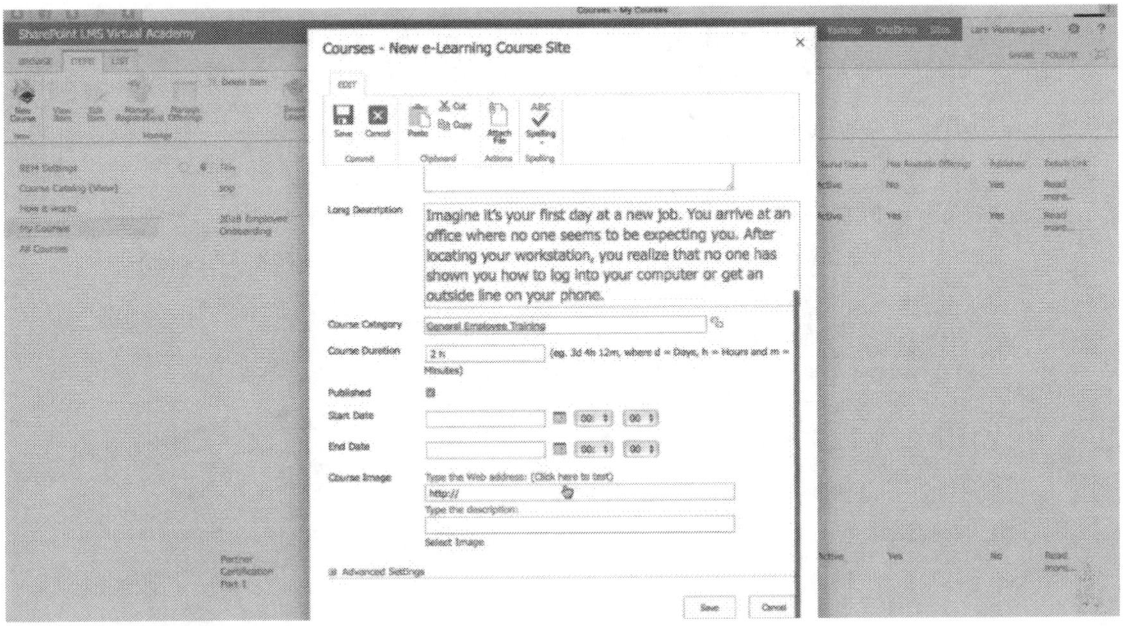

Fig. 4. View of SharePointLMS Learning System

After the preliminary assessment phase, nine open source e-learning platforms were analyzed in detail. The platforms were: Moodle, ATutor, Dokeos, dotLRN, ILIAS, LON-CAPA, OpenUSS, Sakai and Spaghettilearning. Based on the same survey, Moodle dominates the rating, achieving a top five times. The strengths of Moodle are the implementation of communication tools, their creation and administration of learning objects. An additional strength of Moodle is its comprehensive didactic concept as well as data tracking. In addition, the outstanding usability of Moodle maximizes the value in the usability category. For other platforms, ILIAS has received top honors in the technical, administration, and course management categories. Moodle achieved the best ratings. In addition, the second and third rank can be clearly assigned to ILIAS and Dokeos. According to pairwise comparisons, ATutor, LON-CAPA, Spaghettilearning and Open-USS ranked fourth, while Sakai and dotLRN came in last. The reason for the low Sakai rating is that only the main features are implemented so far. But the quality of these features is very good.

Moodle got the best overall performance as well as specific adaptation evaluation criteria.

After identifying the best platform for implementing adaptive learning, the level of technology development to create adaptive learning within individual and distance learning has been identified, and, with the technology revolution, lately, supporting adaptability in distance learning has become the interest of many researchers in the field, including technology. Agents can play a major role in empowering existing distance learning systems and providing adaptive experiences to meet students' needs. In the industry literature, agency technologies are also used in the context of e-learning systems to support adaptability and enhance the learning process. Agents are autonomous and can act reasonably in their environment. You can achieve a powerful system that is tailored to the needs of each student by using intellectual agents to develop e-learning systems.

Appropriate and up-to-date in the analysis, the Event-State-Action (ESA) model was identified. This model is a jet model that responds in real time to any changes in its environment based on predefined rules and conditions. It is used in various critical systems, such as autopilot systems and antivirus systems. It takes form if the condition takes action. She can sense the environment and respond in a timely manner according to predefined rules and conditions. Most distance learning systems are designed with a database that stores and archives all the journals and student data and activities. Database triggers are based on predefined pedagogical rules that can be updated by teachers for any new requirements without restructuring the system. In the next section, we offer our framework to support adaptability across any distance learning systems that mirror hybrid architecture using agency technology. Thus, the system consists of three components: the ESA module, the multiagent module and the e-learning environment, as shown in Fig. 5, which qualitatively influences the process of individualization of the wind of learning and can be integrated with any system of distance learning.

This environment is an important part of adopting distance and blended learning. It also provides students and teachers with the tools and tools they need to improve the learning process. Moodle and Blackboard are some of the leading e-learning platforms. However, Moodle is probably the most popular and popular platform with over 53,346 sites reaching over 70 million users in 222 countries.

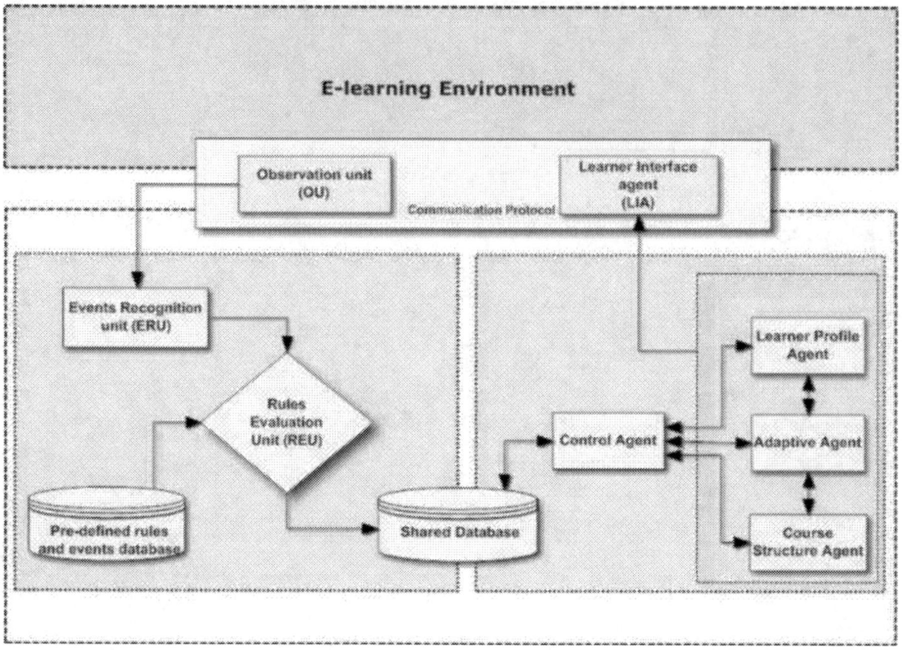

Fig. 5. Learning Automation System.

The ESA module is designed and used as a sensitive component. It is based on events that can occur in an e-learning environment and respond accordingly. However, this process is a rule-based mechanism of database triggers such that events (training activities) such as registration and course access are monitored by an observation team linked to an electronic learning environment using a communication protocol. The event is then given to the Event Recognition Department to be recognized and classified before the evaluation process takes place. The Rule Evaluation Unit is responsible for determining whether or not the current event satisfies predefined pedagogical rules to provide the multi-agent module with the necessary and necessary data through a common database to provide adaptive student learning.

4 Conclusions

During a detailed analysis of scientific developments in the vector of this problem, it was concluded that in the general understanding of the problem of the development of neural network systems, a large amount of research literature is devoted, however, given the specific topic of limitation research, it is advisable to analyze only those studies in which the development of neural networks is oriented to work in the educational space. These studies have shown that high interest in scientists raises questions about the development of educational automation tools, including adaptive testing and individualization of the curriculum to improve the quality of perception of information by students, which confirms the assumption that the development of the neural network of student individualization, which improves the quality of education, is an urgent and urgent issue that needs to be addressed more quickly. For further development, existing software for creating conditions for distance and blended education was analyzed, but in the course of the study we concluded that none of the systems for introducing distance education met the criteria of full-fledged learning. Therefore, you need to combine high-level programming languages with logical or knowledge languages, web technologies, and database management systems. In the following, we intend to apply and evaluate our approach in Moodle as a case study to provide adaptive content based on student learning styles using the Felder-Silverman learning style model (FSLSM).

References

1. B. Florian-Gaviria, C. Glahn, R. Fabregat Gesa, A Software Suite for Efficient Use of the European Qualifications Framework in Online and Blended Courses. IEEE Transactions on Learning Technologies **6**(3), 283–296 (2013). doi:10.1109/TLT.2013.18
2. H. Fujisaki et al., Principles and design of an intelligent information retrieval system over the Internet with a multimodal dialogue interface. Paper presented at the Sixth European Conference on Speech Communication and Technology (EUROSPEECH, Budapest, 1999)
3. A.A. Andreev, *Didactic Foundations of Distance Learning* (Methods and means of distance learning). (Moscow, 2019).
4. A.S. Dovbysh, *Fundamentals of designing intelligent systems* (SSU Publishing House, Sumy, 2009)
5. A.V. Narozhny. *Designing an Intelligent Programming Shell for Learning Based on the Specificity of the Credit Module System* (2006)

6. V.V. Osadchyi, Dissertation, Vinnytsia Mykhailo Kotsiubynskyi State Pedagogical University, 2005
7. E.G. Petrov, M.V. Novozhilova, I.V. Grebennik, *Methods and means of decision-making in socio-economic systems* (2004)
8. O.I. Petkovsky, *Development of a hybrid intellectual system with fuzzy neural network components for solving student's competence assessment problem* (Altai State Technical University, Barnaul, 2012)
9. A.A. Rybanov, V.P. Shevchuk, E.A. Prikhodko, in *Systematic problems of quality, mathematical modeling, information and electronic technologies*, part 4, pp. 6–7 (2003)
10. V.F. Sytnik, *Decision Support Systems* (KNEU, Kiev, 2004)
11. P.I. Fedoruk. Adaptive tests: statistical methods for analyzing the results of knowledge test. Mathematical Machines and Systems 3, 122–138 (2007)
12. V.Y. Tsap, *Designing automated learning environments as an intellectual system* (Institute of Information Technologies and Learning Tools of the NAES of Ukraine, Kiev, 2011)
13. P.S. Yakushev, *Analysis of Technologies and Systems for Controlling Electronic Learning: Report,* Implementation of educational programs of innovative type in the field of IT (2007)
14. A.V. Yakushin, The Use of LMS in the Educational Process of the Teachers College, in Modern Information Technologies and IT Education, in *Collection of Reports of the Scientific and Practical Conference*, ed. by V.A. Sukhomlin, E.N. Nikelin, pp. 189–193 (2006)
15. eFront Teacher's Guide (2019)
16. A.V. Belozubov, D.G. Nikolaev, *Moodle Distance Learning System: A Toolkit* (St. Petersburg, 2007)
17. *SharePointLMS Distance Learning System* (Belitsoft, 2019)
18. ATutor User Documentation (2019)
19. ATutor Instructor Documentation (2019)
20. Claroline – Documentation (2019)
21. A.V. Andreev, S.V. Andreeva, I.B. Dotsenko, *E-learning practice using Moodle* (TTIUFU, Taganrog, 2008)
22. E.I. Selivanov, E.V. Panyushkina, The cornerstones of the qualimetry of competencies for compliance with the requirements of federal state educational standards of secondary vocational education. Fundamental and applied research in the modern world **3**(5), 87–91 (2014)
23. E.V. Vorsina, T.A. Snigireva, Qualimetry of student's academic competence based on the analysis of educational texts. Bulletin of the N. A. Nekrasov Kostroma State University **18**(3), 124–127 (2012)
24. I.G. Bartasevich, A system of criteria for the quality of students' education in higher education. Bulletin of the Astrakhan state tech. university, 217–220 (2008)
25. V. V. Osadchyi, K. P. Osadcha, Modern realities and trends of information and communication technologies development in education. Information Technologies and Learning Tools **48**(4), 47–57 (2015). doi:10.33407/itlt.v58i2.1645
26. V. V. Osadchyi, I. V. Krasheninnik. Formation of short-cycle curricula content for future software engineers training on the basis of the labour market analysis. Information Technologies and Learning Tools **58**(2), 11–25 (2017). doi:10.33407/itlt.v58i2.1637
27. V. V. Osadchyi, S. V. Symonenko. Foreign language as a means of communicative competence development of future software engineers. Information Technologies and Learning Tools **58**(2), 38–48 (2017). doi:10.33407/itlt.v58i2.1645

Studies of impact of specialized STEM training on choice further education

Viacheslav Osadchyi[1], *Nataliia* Valko[2,*], *Liudmyla* Kuzmich[2], and *Nataliya* Abdullaeva[3]

[1]Bogdan Khmelnitsky Melitopol State Pedagogical University, 20 Hetmanska Str., Melitopol, 72300 Ukraine
[2]Kherson State University, 27 Universytetska Str., Kherson, 73000, Ukraine
[3]Secondary school No 24, 32 Karbysheva Str., Kherson, 73039, Ukraine

Abstract. The specialized training influence to the choice of further direction of study is considered in the work. The assumption that early involvement of students in the study of natural and mathematical disciplines, in particular mathematics is given. It will stimulate young people to get STEM education in future. The essential element in future professional choice, development and formation is the issue of motivation for learning. It is important that modern students are gradually losing their incentive to study. Almost a third of those who choose the appropriate study profile have contradictions between professional self-determination and the availability of the necessary knowledge for the profession, between choosing a higher education institution and being able to enter in it. The contradiction requires purposeful formation of a conscious choice of future activities. The conducted research has shown that the basis of the motivational component of the choice of STEM-learning is studying of natural and mathematical disciplines by modern teaching technologies and organizing of additional lessons system based on the projected teaching methods. It satisfies the growing needs for intelligence, knowledge, motivational beliefs to understand the specifics of the future profession.

1 Introduction

At every stage of its development, society presents to the education system certain requirements that meet the scientific and technological achievements of the time. In the reform of general and higher education, the problems of scientific substantiation, content development and teaching methods, as well as pedagogical diagnostics of learning outcomes are very relevant in improving of the teaching and educational process. The main purpose of the reform is to move from a school, knowledge is only given to a school of competencies necessary for life [1, p. 9]. The research of the relationship between the competencies of students, their intellectual abilities and their vision of further education, the use of active learning strategies for profession choice has shown the usefulness of motivational beliefs in understanding the choices of this future. Supporting such beliefs depends on a system of motivational influences in formal, non-formal and informal education.

In Ukraine there has developed the system of work with intellectually talented children in the natural-mathematical and technological disciplines: intellectual competitions; a network of specialized lyceums, the graduates demonstrate high results in STEM disciplines (Science, Technology, Engineering and Mathematics); system of extracurricular education. These are the measures on the basis of which, it is necessary to build a system of science-oriented education, to introduce the principles of scientific and engineering methods in the education. Today, STEM approaches are being implemented in many Ukrainian schools and other educational and extracurricular institutions. Non-formal STEM education in the country is Olympiads, activities of the Junior Academy of Sciences (JAS), other extracurricular institutions, various competitions and events: Intel Techno Ukraine; Intel Eco Ukraine; Sikorsky Challenge Science Festival; STEM competitions; scientific picnics, hackathons etc. [2]. At the same time, there is a shortage of specialists in engineering and sciences. Today in Ukraine these subjects do not occupy the top places in the popularity rating among young people [3]. According to the survey, the basic sciences occupy only the 10th place. STEM education, which has been actively developing in Ukraine in recent years and has been identified as a priority area for the development of education for 2020, is intended to promote the exact sciences and increase the interest of young people in scientific activities.

Education should meet the needs of society in the fields of natural and mathematical and technological fields, capable of developing innovations in bio-, nano-technologies, technologies of artificial intelligence, robotics, etc. Therefore, an important task of training should be to develop student's motivation to study natural sciences. It is important to train teachers capable of supporting such motivation and making the conditions for further interest in the sciences.

* Corresponding author: valko@ksu.ks.ua

2 Related works

Researches present the achievements in the mathematical sciences are unique, they serve as a source of the basic knowledge required by every specialist in the STEM [4]. There is the problem of learning the exact and technical disciplines, as students find them "difficult to study" [5]. According to PISA (Programme for International Student Assessment) in 2015, there was almost half of students from 12 million researches from different countries were unable to complete the simplest tasks in reading, math, science, and a number of innovative industries during world testing [1, p.15]. In the future, it may lead to the loss of interest to obtain STEM specialties.

Numerous studies of connections between competencies and academic achievement in higher education students [6, 7, 8] and others) have found the students in junior, secondary or senior school age have additionally acquired knowledge of the natural sciences (e.g. chemistry), physics, maths) have achieved higher educational attainment than the other students [9]. In addition, studies were conducted in different age groups on the ability of early intelligence, the reasons for its formation, not only because of the influence of parents, but also due to the goal, self-regulation, students' efforts. But there are few researches on the connection between motivation and achievement in engineering, technical, technological contexts, including STEM activities.

In [10] it is noted that «a shift in the focus of the classroom activities from teaching to learning» is fundamental value shift as a result of collective work on the concept of "science for all". The basis for this was:

- changing interactions between teachers, academic institutions, research centers;
- professional development of teachers of all levels, including the primary professional education of graduate students, postdoctoral fellows and entering faculty, and reeducation for midcareer faculty;
- change of teaching and learning resources from "passive" learning to "active" classroom strategies that complemented the inquiry-centered approach of modules.

Among the factors that contribute to the attraction of studying STEM-subjects occupies a special place:

- creating an appropriate support environment and an enhanced sense of community / feeling of belonging;
- enhanced accessibility for youth and their caregivers to STEM education through science clubs;
- creating long-term STEM training programs

A relatively small amount of research is devoted to examining how students studying science, technology, engineering and mathematics, developing these skills and aptitudes and what learning tools are used by teachers to contribute to this development in the STEM learning context.

Psychologists point out that young people's beliefs about their knowledge, competences often change with reference to their views and efforts to study, have such basic psychological properties as new communication opportunities, beliefs about their successes and/or failures, contribute to the development of cognitive processes etc [11]. It will help strengthen cognitive, emotional, and social skills. Important knowledge and knowledge of specialized skills in a particular field are important [12]. Do not forget about the impact of awareness of the students of their intellect on the learning purpose, efforts, strategies, academic success. Therefore, it is important to study students' assessments of their cognitive, social and other motivations, based on their own actions in the environment.

3 Research background

In the scientific field there is no single approach to the interpretation of the concept of "STEM technology". This definition is interpreted as a type of learning, a method of teaching, a technology of learning. In our point of view, it is a didactic technology allows changing by intensification of educational process in its procedural and productive aspects. In the process of realization in the conditions of general / higher vocational education, the named technology acquires pedagogical content, combining methods, techniques, teaching aids, as well as various forms of subject-subject interactions. They make the great opportunities for developing student's independence, critical thinking, mastering of cognitive and social skills in the complex [13]. The main features of STEM learning technology are:

- positive relationship: the success of each person depends on the integrity of the other members of the group; mutual responsibility and teamwork;
- direct support: young people exchange opinions, resources and materials, evaluate the execution of each team member in order to get an overall result;
- personal responsibility: everyone is responsible for the results of group activities and for their share of work, since the success of the team depends on the individual work of each of its members;
- social competence: participants learn mutual trust and respect for each other, they develop skills to manage the actions of others (leadership), to make decisions, to communicate and to overcome conflicts;
- self-assessment: students learn how to evaluate their contribution to the success of group work, as well as evaluate the group's collaborative work from the point of view of the appropriateness of the methods chosen and identify the causes of the failures;
- one of the main functions of specialized training in STEM education is to unite: uniting of disciplines, people, and forces.

The teacher acts as a conductor, facilitator, assistant, equal member of the group, who ensures necessary skills formation for the cooperation of the applicants, such as: coordinate their activities with the activities of partners; to stand in the position of others and change your own; to assist and benefit from their partners; reflect their actions and those of other group members; respect everyone's opinion; to give high priority to the achievement of the collective goal; to prevent conflicts. The teacher's main task is to instill faith in the possibility of successful achievement of the goal and to encourage students to independently search. The key competences of a teacher are in world standards. To meet the demands of the digital society, in 2018

International Society for Technology in Education (ISTE) has approved standards for ISTE teachers to improve the learning impact for all students using technologies [14].

The analysis of their work presents the insufficient theoretical and methodical level, weak subject base of pupils. Sometimes, clubs are turned into extra classes to repeat the classrooms topics. So, a lot of pupils refuse to attend clubs' activities because of their lack of interest and pattern, similarity to the ordinary lesson. However, the analysis of STEM education clubs showed that all STEM technologies lessons are interesting for the participants. They are aimed at developing logical thinking and imagination, teaching teamwork, responsibility for the results not only of group activity, but also for personal part of work, etc.

STEM education introduction in the educational process will allow students to form the most important characteristics that determine a competent specialist:

• the ability to see the problem and identify in it as many parties and connections as possible, the ability to formulate a research problem and determine the ways to solve it;

• flexibility as the ability to apply knowledge in different situations, to understand the possibility of other points of view on solving problems and stability in defending one's position;

• originality in solving problems, the ability to abstract and to concretize, to analyze and synthesize, to feel the harmony in organizing an idea.

In many countries (Canada, Finland, Singapore, and others), a variety of strategies are being used to identify, at school age, the students' ability to learn and, where appropriate, provide them with comprehensive support.

In the middle and senior classes of many secondary schools in Ukraine there is a division of classes in the following areas: mathematical classes, science classes, computer science, philological classes and others - the so-called specialized training. There are different points of view: positive and negative points. One of the motives is the idea that it leads to the formation of homogeneous classes, teaching becomes more directed.

Specialized training differs from general education with more specific professionally oriented characteristics of motives, goals, means and results of educational, productive, creative activity, which act in relation to the student in the form of certain requirements. Students should move from a general to more specific activity training. It involves a specific specialization, specifying training activities around a particular group of professions. Over time, this orientation will be narrowed down and specified in vocational training in secondary specialized or higher education institutions.

Some schools have their own focus, specialization. This specialization increases the number of lessons spent studying the subject and organizes additional activities. Based on the students' cognitive interests in the relevant profession, it is possible to orient them to a definite study profile, and vice versa, knowing the abilities and interests of students in some subjects of the educational cycle, you can offer the appropriate profile, and later the profession itself.

4 Experimental results of research

In order to study the attitude of teachers of natural sciences to innovative teaching methods use, in particular, STEM technologies, non-traditional, integrated lessons and its introduction into the educational process and cross-curricular connections, a questionnaire of teachers was conducted:

1. Is it necessary to formulate basic concepts about STEM education in students?
2. Is it possible to form an understanding of independent search in solving tasks by students in the traditional system of education?
3. Does the method of teaching in the natural sciences affect students' formation of imagination, attention, memory?
4. Does the use of STEM technologies develop students' personal responsibility, social competence?
5. Do you need a methodology for the formation and development of vocational training opportunities in basic / secondary or higher education?
6. What slows down the process of formation and development of STEM-competence of students?

The results of the answers of the five questions (Fig. 1) indicated that the interviewed teachers considered such lessons are important and necessary, but they are not always ready to use them. Teachers point out that in the traditional educational system without proper methodological support the study of STEM-disciplines is not complete. Students should be encouraged to make scientific activities, to direct and control of learning process. Otherwise scientific competences formation, understanding of natural processes will be minimal.

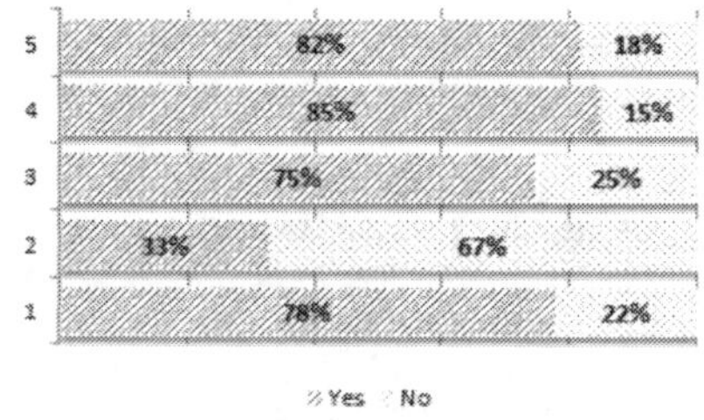

Fig. 1. Questionnaire distribution of answers.

Therefore, use of STEM technologies and innovative methods in the learning process is possible, provided the appropriate methodological support and teachers' training. It will help to develop understanding and perception of achievements in natural, technical, technological fields and contribute additional intrinsic motivation.

The sixth question has the answers:

• Lack of interest, conscious attitude, students' motivation to study.

• Low level of basic school training of students.

• Insufficient amount of methodological support.

• The topic of the subject is studied at the low level.

Researches confirmed [15, 16] that the shortcomings highlighted by teachers can be decided as follows:

• Ensuring the study of natural disciplines by systems of practical problems with STEM technologies.
• Modern methodology development of the educational process organizing in natural sciences study.
• Material base strengthening with instruments for conducting / demonstrating experiments, researches, and ICT tools.
• Improving the activities of various math, science, engineering, and modeling clubs is the foundation of STEM competencies.

As mentioned above, additional intrinsic motivation contributes to the choice of the direction of further study.

One of the experiment's stages was the study of the motivation of professional self-determination of high school students. Choosing a profession is one of the major life choices a person makes. It is important to the individual and society. Choosing a profession is a prerequisite for future successful professional activity, essentially choosing a life path, one's place in life.

The data of three schools with advanced study in mathematical, philological and non-core subjects were analyzed in the study (Fig. 2). The data of graduates' entrance from these schools in the specialty in universities / colleges for the last five years have been analyzed. We were interested in further training of these graduates. The list of branch of knowledge and specialties was taken in accordance with [17].

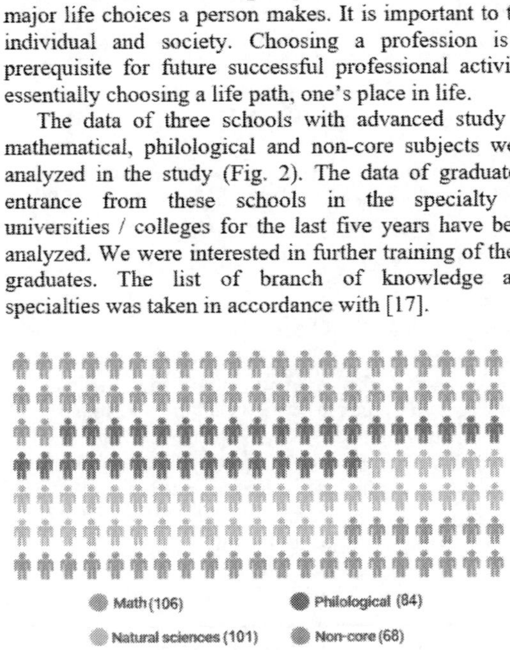

Fig. 2. Students' ratio in the study.

Some branches of knowledge have similar subject areas. For example, the branch "01 Education / Pedagogy" has a specialty "014 Secondary education (by subject specialties)". The specialty, in turn, has 16 specializations, among them "014.04 Secondary education (mathematics)", "014.05 Secondary education (biology)", "014.08 Secondary education (physics)" and others. There are also some branches such as "09 Biology", "11 Mathematics and Statistics", "14 Electrical Engineering" and others. We group the number of graduates. The results are presented in table 1.

Based on empirical research, it has been determined the graduates studied in the specialized class with advanced study of mathematics, the most popular are the specialties of branches of Tech knowledge: Information Technology, Transport.

Among the classes with a natural specialized training, more of graduates chose the branch of knowledge of Social and Behavioral sciences, Natural sciences in a second position. Among the graduates of the Philology training, the most selected are: the Humanities, Management and Administration. The branches of knowledge of Management and Administration, Health care, Law are in demand approximately equally for the three groups, both in absolute and relative terms (in the range of 8,5-13,1%), as well as "their" specialized for the appropriate group (from 15,8 to 20,8%).

Table 1. Graduates' distribution table by directions.

Branch of knowledge	Math, %	Philology, %	Sciences, %	Non-core, %	All Num
Transport	19,8	8,3	3	22,1	46
Information technology	20,8	4,8	7,9	1,5	35
Electronics and communications	0	0	4	7,4	9
Natural sciences, biotechnology	6,6	3,6	15,8	2,9	28
Production and technology	2,8	0	10,9	0	14
Agrarian sciences and food	2,8	1,2	5	16,2	20
Mechanical engineering	2,8	0	0	5,9	7
Automation and instrumentation	1,9	0	0	0	2
Architecture and construction	2,8	0	2	0	5
Management and administration	11,3	13,1	4	7,4	32
Health care	9,4	13,1	9,9	4,4	34
Law	8,5	11,9	12,9	1,5	33
Humanities	0,9	29,8	3	2,9	31
Culture and art	1,9	7,1	3	7,4	16
Service sector	1,9	2,4	0	4,4	7
International relations	1,9	1,2	1	0	4
Social and behavioral sciences	2,8	1,2	17,8	10,3	29
Education / pedagogy	0,9	2,4	0	0	3
Military activity, security	0	0	0	5,9	4
The overall result					359

The summary table 2 presented the highest number of people (%) chose specialized training closely connected to the future profession. Graduates who did not have specialized training, there is the choice for the considered directions averages 26,9% (almost a third) of considered categories of destinations.

Table 2. Table of distribution by training.

%	Tech	Science	Philology	Other
Math	51,0	21,7	2,8	24,5
Sciences	27,7	48,5	4,0	19,8
Philology	13,1	19,0	31,0	36,9
Non-core	36,8	33,8	2,9	26,5

Studies of motivation in learning have shown that "although there is not a perfect relationship between motivation and learning, there is a tendency to align attitudes, which are predispositions to behaviour, with behaviour itself. This means that initial or pre-training motivation promotes a predisposition to learn and to learning itself. At later stages, it also promotes a predisposition to transfer and make use of the learning outcomes in other contexts" [18].

The survey "Motives for choosing a profession" was conducted [19]. This technique allows determining the leading type of motivation at choosing a profession. The questionnaire consists of twenty test questions that characterize any profession. The study used the "Motives for choosing a profession" the method of K. Zamfir in the modification of A. O. Rean [20]. This test allows to determine the relevance of the following types of motivation: 1) material remuneration; 2) the pursuit of career advancement; 3) the desire not to be criticized by the manager and colleagues; 4) the desire to avoid possible punishment or trouble; 5) orientation to prestige and respect from others; 6) satisfaction with a job well done; 7) public utility of labor.

On the basis of empirical data, motivational complexes of the following types are obtained: at ISM > EPM > ENM or ISM = EPM > ENM is the optimal one in which the ISM (internal significant motives) are high; and it is sufficiently high in the case of EPM (external positive motivation) equal to or lower than ISM; it is low – if ENM is very low and close to unity.

We conducted a diagnostic survey (assessment of the level of formation of the motivational level) of choosing a specialized study of students of the Faculty (123 respondents), we obtained the following results (table 3).

Table 3. Survey's results.

	Number of respondents
Internal individually significant motives (ISM)	25
Internal socially significant motives (ISSM)	34
External positive motives (EPM)	44
External negative motives (ENM)	20

Students were asked to complete questionnaires. The results of the answers were evaluated on a 5-point scale: 1 point – "very small", 2 points – "to a very small degree", 3 points – "not to a large extent, but not to a small extent", 4 points – "in large enough, 5 points – "very large".

Students needed to evaluate the extent to which each of them influenced their choice of profession. The technique can be used to identify the predominant type of motivation (internal individually significant motives, internal socially significant motives, external positive motives and external negative motives).

Research's results presented EPM > ISSM > ISM > ENM, it is considered that the studied motivational complex is optimal. The high weight of (ISM + EPM) and the low weight of ENM (16%) directly correlate the students' satisfaction by the chosen profession.

The high level of ISM indicates the educational activities of students for the sake of knowledge, education, culture, and not for receiving rewards, praise, etc. Students with a sufficiently high motivation complex are interested in the chosen profession, but there is also a desire to receive external rewards, it is partly a means to personal gain. The presence of low indicators of the motivational complex (16% of ENM) implies indifferently, and even negative attitude to the learning process and choosing the profession as a whole.

Thus, descriptive statistics have shown that the findings of this study are consistent with our assumption that motivational beliefs are important and necessary for shaping the professional orientation of students of different degrees.

5 Conclusions and future work

The research presented the early attracting of students in advanced study of disciplines and STEM activities influence the choice of future profession. But it does not solve all the problems of professional self-determination formation in high school students. There are manifestations of self-selection of professional, life platform and goals. The research also presented the choice of the specialized training of students is influenced by the motivation for specialized-based learning. On the basis of differentiation of interests of students due to the specialized training, as well as the integration of academic subjects, students develop an individual style of mental activity, focus on deepening knowledge in the chosen field of activity, and improve mental and special abilities in the orientation to future profession. Thus, we found that the next choice of future profession is influenced by early attracting in advanced study of disciplines. Students who study mathematics, physics, and computer science more often choose a profession related to engineering, technology, or science, and have a career in STEM.

Thus, motivational beliefs are important and necessary for the formation of the professional orientation of applicants for education at various levels. Creating an academic climate emphasizes the importance of learning, mastering and improving mathematical and scientific competencies. This contributes to both productivity and perseverance in these subjects through a positive impact on students' beliefs.

So, we found the next choice of future profession is influenced by early involvement in advanced study of disciplines. Pupils who study mathematics, physics and computer science more often choose a profession related to engineering, technology or the exact sciences.

The research presented if the special attention is given to conduct STEM activities of school / student youth, to improve abilities, learning strategies, motivation of the chosen activity I educational environment. It leads to the understanding that the efforts will lead to success. The purposeful development of STEM education should be the basis for the further formation of all special competencies and students' professional training.

References

1. A. Schleicher, *World Class: How to build a 21st-century school system, Strong Performers and Successful* (OECD Publishing, Paris, 2018). doi:10.1787/4789264300002-en
2. STEAM-osvita: innovatsiina naukovo-tekhnichna systema navchannia (STEAM education: an innovative science and technology education system) (2016),

http://ippo.kubg.edu.ua/content/11373. Accessed 14 Jan 2020

3. Introductory Campaign 2019, https://vstup.edbo.gov.ua/statistics/konkurs. Accessed 14 Jan 2020
4. X. Chen, Students *Who study science, technology, engineering, and mathematics (STEM) in postsecondary education*. Stats in Brief. NCES 2009-161. National Center for Education Statistics (2009)
5. J.E. Jon, H.I. Chung, *STEM Report: Republic Of Korea. Report For The Australian Council Of Learned Academies (ACOLA)*. (Australian Council of Learned Academies, Melbourne, 2013), p. 55.
6. C.M. Phelps-Gregory, M. Frank, S.M. Spitzer, Prospective elementary teachers' beliefs about mathematical myths: a historical and qualitative examination. The Teacher Educator **55**, 1 (2019). doi:10.1080/08878730.2019.1618423
7. C.S. Dweck, Motivational processes affecting learning. American psychologist **41(**10), 1040–1048 (1986). doi:10.1037/0003-066X.41.10.1040
8. L.S. Blackwell, K.H. Trzesniewski, C.S. Dweck, Implicit theories of intelligence predict achievement across an adolescent transition: a longitudinal study and an intervention. Child development **78(**1), 246–263 (2007). doi:10.1111/j.1467-8624.2007.00995.x
9. D. Bressoud, M. Carlson, V. Mesa, C. Rasmussen, The calculus student: Insights from the MAA National Study. International Journal of Mathematical Education in Science and Technology **44**(5) (2013)
10. E. Seymour, Tracking the processes of change in US undergraduate education in science, mathematics, engineering, and technology. Science Education 86(1), 79–10 (2001). doi:10.1002/sce.1044
11. D.H. Schunk, F. Pajares, *Competence perceptions and academic functioning*. Handbook of competence and motivation **85** (2005)
12. J. Wai, D. Lubinski, C.P. Benbow, Spatial ability for STEM domains: Aligning over 50 years of cumulative psychological knowledge solidifies its importance. Journal of Educational Psychology **101**(4) (2009). doi:10.1037/a0016127
13. M. Honey, G. Pearson, H.A. Schweingruber (eds.) *STEM integration in K-12 education: Status, prospects, and an agenda for research*. (Washington, National Academies Press, 2014)
14. International Society for Technology in Education, https://www.iste.org/standards. Accessed 14 Jan 2020
15. V.V. Osadchyi, N.V. Valko, N.A. Kushnir, Determining the Level of Readiness of Teachers to Implementation of STEM-Education in Ukraine. CEUR Workshop Proceedings **2393**, 144–155 (2019), http://ceur-ws.org/Vol-2393/paper_369.pdf. Accessed 14 Jan 2020
16. A. Spivakovskiy, N. Kushnir, N. Valko, M. Vinnyk, ICT advanced training of university teachers. CEUR Workshop Proceedings **1844**, 176–190 (2017), http://ceur-ws.org/Vol-1844/10000176.pdf. Accessed 14 Jan 2020
17. Decree of the Cabinet of Ministers of Ukraine No. 266 of April 29, 2015 "On approving the list of branches of knowledge and specialties by which higher education applicants are trained" (2015), https://zakon2.rada.gov.ua/laws/show/266-2015-%D0%BF. Accessed 14 Jan 2020
18. R. Cação. Motivational Gaps and Perceptual Bias of Initial Motivation Additional Indicators of Quality for e-Learning Courses. Electronic Journal of e-Learning **15**(1), 3–16 (2017)
19. M.M. Vrublevskaya, O.V. Zykova, *Proforiyenta–tsionnaya rabota v shkole: Metodicheskiye rekomendatsii* (Magnitogorsk, Moscow State University, 2004), p. 80
20. A.A. Rean, A.R. Kudashev, A.A. Baranov *Psikhologiya adaptatsii lichnosti* (Prime-Eurosign, SPb, 2006), p. 479

The course “Development of virtual and AR software” for STEM teachers

Mykhailo Mintii*

Kryvyi Rih State Pedagogical University, 54, Gagarina Ave., Kryvyi Rih, 50086, Ukraine

Abstract. The article looks into the experience of development and using of learning resources for design of virtual reality (VR) and augmented reality (AR) systems for future STEM teachers. Unity environment for visual design, Visual Studio as a programming environment, platform of VR (Google VR) and AR (Vuforia) were chosen for development of teaching materials. The article describes the content of the created course. The research analyzes the impressions of the course participants. The research data were collected using interviews as the method of qualitative data collection. To conclude, the created course promotes the development of digital competence in the design and use of innovative learning tools.

1 Introduction

The technology of AR is well-known for most of people. It was under the close attention of Gartner analysts for many years. On July 2019, the Gartner Hype Cycle for Customer Service and Support Technologies reflected the increasing attention for such technologies as Blockchain, Field Service Drones, Mediated Interaction Matching, Things as Customers. At the peak of Hype Cycle are AR, Customer Psychographics, Workforce Engagement Management, IoT, Customer Journey Analytics, Digital Experience Platforms, Engagement Hub, 360-Degree View, Chatbots, Conversational User Interfaces, Consumer Messaging Applications, Emotion Detection/Recognition, NLP, some disappointment is observed in VideoCalling, In-App Mobile Customer Service, Robotic Process Automation Software, Voice of the Customer. According to analysts at Gartner, it will take 5-10 years for AR to reach the performance plateau (when a new technology becomes established) (Fig. 1) [1].

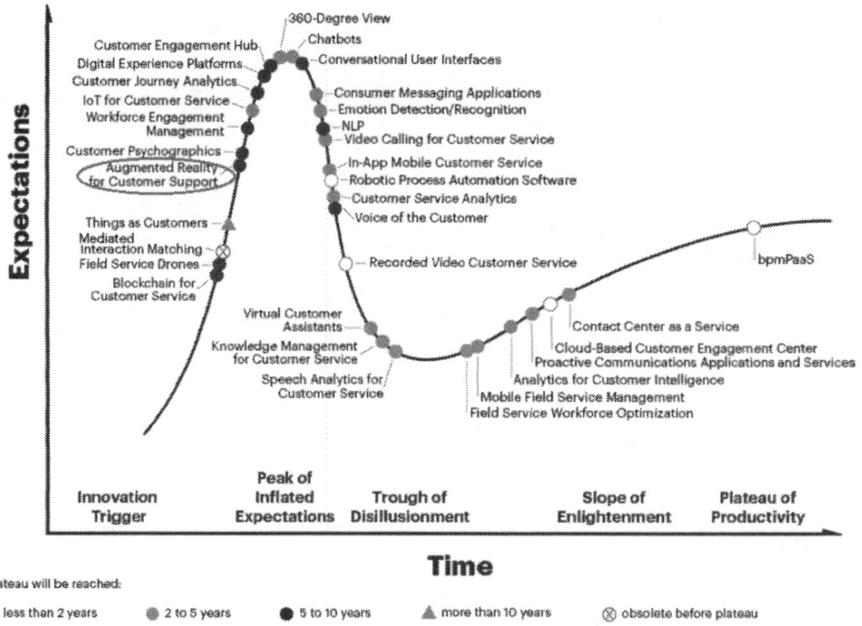

Fig. 1. The Gartner Hype Cycle for Customer Service and Support Technologies.

* Corresponding author: mykhailo.mintii@gmail.com

According to [2], among the Top 10 Strategic Technology Trends for 2020 are:

– multi-experience – in the future, this trend will become what's called an ambient experience, but currently, multi-experience focuses on immersive experiences that use AR, VR, mixed reality (MR), multichannel human-machine interfaces and sensing technologies. The combination of these technologies can be used for a simple AR overlay or a fully immersive VR experience;

– human augmentation is the use of technology to enhance a person's cognitive and physical experiences.

There are high expectations for AR and VR in education. Unfortunately, in Ukraine the technology of AR in the educational environment for training future teachers is almost absent. Despite the significant potential of AR for laboratory studies in Science and Math in secondary and high school in secondary and high school [4, 5, 6, 7, 8, 9, 10], for the training of future engineers [11, 12, 13], for extracurricular educational and research activities of students [14, 15], for foreign students' adaption [16] and even for teaching global reading of preschool children with autism spectrum disorders [17].

These and other issues were discussed at the 1st [18] and 2nd [19] International workshops on AR in Education which took place in Kryvyi Rih.

Based on the current experience of using VR and AR technologies in education and the prospects for their release into the masses in next 5-10 years, it is necessary to think about the problem of preparing for the use of these innovative technologies in the professional activities of future STEM teachers. After all, in just a few years, today's students will have to manage this process: both as software engineers and as teachers. Therefore, the main purpose of our research is to develop a training course on designing VR and AR systems for future STEM teachers, adapted to Ukrainian users and to the current level of technology development.

2 The research tasks

The object of the research is the professional training of future STEM teachers for the design of VR and AR technologies.

The subject of the research is the learning resources for the design of VR and AR systems for future STEM teachers.

The purpose of the research is to develop the learning resources for the design of VR and AR systems adapted for different types of learners.

To achieve the purpose of the research such tasks were solved:

1) an analysis of the experience of using AR tools for the development of educational materials was done;

2) the software for the design of AR tools for educational purposes were identified and the technological requirements for the course "Development of VR and AR software tools" were characterized;

3) individual components of the training complex for the design of VR and AR systems for future teachers majoring in STEM disciplines were developed.

3 Results and discussion

In the context of the study, AR is understood primarily as a synthetic environment, a kind of virtual environment in which real objects are complemented by their computer models. Therefore, among the professional requirements for developers of AR tools there are computer modeling, computer programming, computer graphics and computer ergonomics.

While solving the first problem, it was found out that at the present stage of development of information technologies, the leading means of implementing AR are mobile Internet devices – multimedia mobile devices that provide wireless access to information and communication Internet services for the collection, systematization, storage, processing, transmission, presenting all kinds of messages and data.

The use of AR technology in a mobile-oriented learning environment of higher education institutions:

– expands the capabilities of laboratory facilities used to prepare students for work with real systems;

– makes complex and expansive systems available;

– contributes to the improvement of vocational training by providing laboratory simulators with AR;

– motivates students for experimental and educational research work.

The creating of interactive training materials with the use of AR systems can be done in two main directions:

a) the use of utilities or linking markers with user-developed models;

b) the development of VR and AR software for educational purposes.

In the first direction, the developer does not require good programming skills, however, the functionality of the created tools is significantly limited by the relation to proprietary software. In the second direction, the developer needs develop tools himself. However, the functionality and adaptability of the developed tools are significantly increased.

Characterizing the technology of AR, one should point out the simplicity of reflection of virtual objects in it comparison to VR. Stages of developing an object for AR system are: 1) creating a visual model of the AR component in a 3D environment; 2) creating a simple marker in a 2D environment; 3) linking the marker with a 3D model in a software tool to maintain AR.

When the marker is recognized by software for maintaining AR, the corresponding 3D model is shown. This process is implemented according to the scheme shown in Fig. 2 [20].

To solve the second purpose of the study, we reviewed the tools for developing VR and AR and chose the most suitable for achieving the goal of the study. Among the reviewed, we note both the "old" Wikitude SDK [21] (since 2008) and the relatively new (since 2017) Apple tool – SDK ARKit [22]. The current version of ARKit allows you to develop multiplayer games with AR. ARCore [23] is a relatively new (March 2018) tool from Google, a kind of response to ARKit. Supported platforms: Android 7.0 and above, iOS 11 and above.

* Corresponding author: mykhailo.mintii@gmail.com

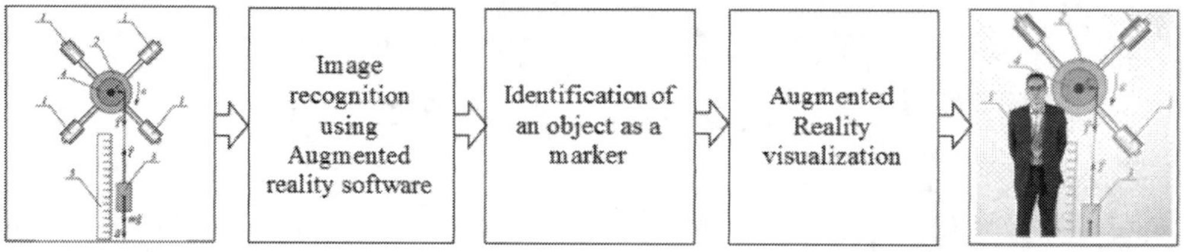

Fig. 2. AR schema.

ARCore comes with three main possibilities of combining VR and real worlds: 1) tracking the position of the phone in the environment; 2) "recognizing the environment" provides the ability of the phone to determine the size and location of horizontal surfaces; 3) lighting assessment allows the phone to evaluate the actual lighting conditions.

ARtoolKit [24] is the oldest (since 1999) SDK for the development of AR tools. It is available on Android, iOS, Linux, Windows, Mac OS, "smart glasses".

Maxst [25] is the South Korean SDK. It offers advanced tools for recognizing images and environments. Maxst is freely distributed for non-commercial use, and the free version differs from the paid version only with a watermark.

Vuforia [26] is one of the most popular platforms for developing AR. SDK implements the following functionalities: recognition of various types of visual objects (box, cylinder, plane), recognition of text and surroundings, VuMark (combination of image and QR code). Using the Vuforia Object Scanner, you can scan and create marker objects. The recognition process can be implemented using a database (local or cloud storage). Unlike other SDKs, Vuforia supports both 2D and 3D markers, including Image Target markerless, three-dimensional Multi-Target, as well as benchmark markers that select objects in the scene for recognition.

In our opinion, in the process of preparing future STEM teachers for the use of AR systems for developing interactive teaching materials it is advisable to use an integrated approach. The design with use of standard objects can be performed in a visual design environment. Providing standard objects with new properties and creating new ones can be performed in an object-oriented programming environment. At the present stage of ICT development, it is advisable to use the Unity environment [27] for visual design, Visual Studio [28] or a similar programming environment, as well as virtual platforms (Google VR or the like) and AR (Vuforia or the like).

The result of solving the third research problem is the course "Development of software tools for VR and AR". The substantive basis this course is the open course from the University of San Diego on the EdX platform [29] and the book by Jesse Glover [30].

The course consists of two substantive modules.

Content module 1. Development of VR tools

Topic 1.1. VR and Game Engines

Topic 1.2. Physical interactions and camera

Topic 1.3. 3D interface and positioning

Topic 1.4. 3D user interaction

Topic 1.5. VR navigation and introduction

Content module 2. Development of AR tools

Topic 2.1. Set up AR tools in Unity 3D

Topic 2.2. Development of a project for a photograph

Topic 2.3. Development of training materials with Vuforia

Topic 2.4. Development for promising devices

80 students were involved in the experiment, which lasted from February 20, 2019 to January 20, 2020: POKT-18m and PO-16 groups of Faculty of Information Technology of the Kryvyi Rih National University and MIM-14 and FIM-14 groups of Faculty of Physics and Mathematics of the Kryvyi Rih State Pedagogical University.

Examples of completed tasks are shown in Fig. 3-5.

So, among the participants in the experiment, a survey was conducted on the formation of competence in the design and use of innovative learning tools. Survey results are shown in Fig. 6-7.

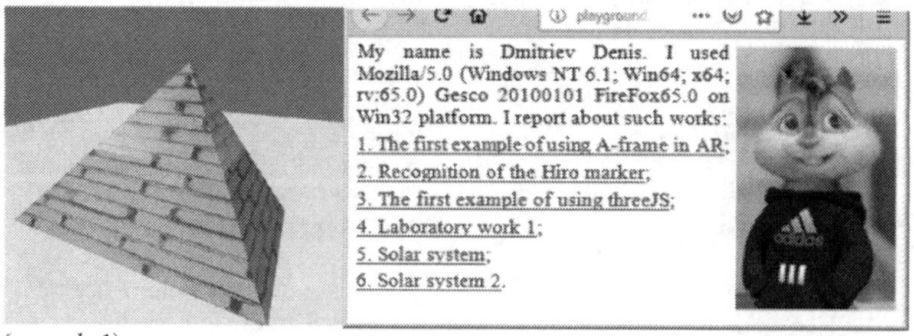

Fig. 3. The sandbox (example 1).

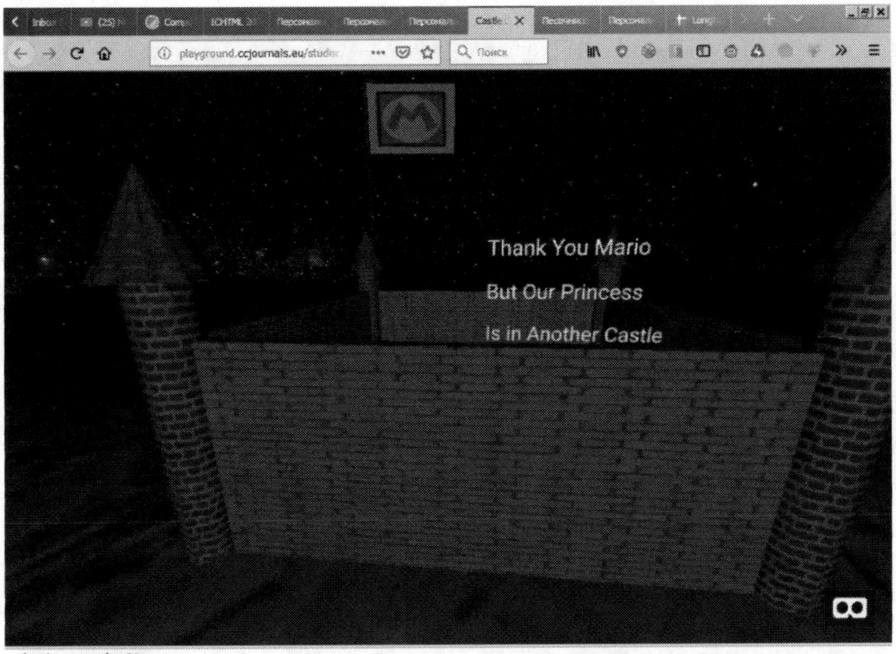

Fig. 4. The castle (example 2).

1. A-Frame first start
2. Hiro marker
3. ThreeJS
4. Shader
5. Light
6. OrbitControls
7. Object
8. **Awe marker**
   8.1. Awe marker v. 1.0
   8.2. Awe marker v. 2.0
   8.3. Awe marker v. 3.0 modification
9. Solar system
10. **Castle**
   10.1. A-Frame
      10.1.1. Castle A-Frame
      10.1.2. Castle A-Frame Marker
      10.1.3. Castle A-Frame Marker Hiro
         10.1.3.1. Task Marker for A-Frame
         10.1.3.2. Marker Hiro
   10.2. ThreeJS
      10.2.1. Castle ThreeJS
   10.3. 8th Wall
      10.3.1. Castle 8th Wall by ThreeJS
   10.4. Awe.js
      10.4.1. Castle Awe.js
         10.4.1.1. Task marker for Awe.js
11. Portal
12. ...

Fig. 5. Structure of completed tasks (example 3).

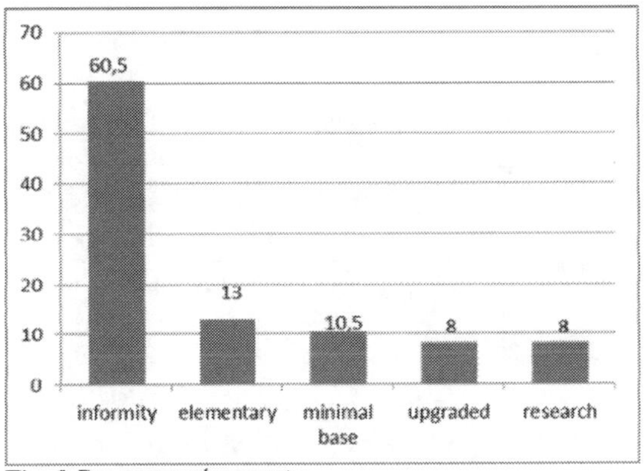

Fig. 6. Pre-course placement.

A survey was conducted after the course to obtain feedback on the impressions of the participants. The research data were collected using interview techniques in qualitative data collection method. The survey was attended by 23 participants. It should be noted that at the time of the interview all of them were no longer students, but STEM teachers, which allows us to conclude that the responses received are independent. Gender distributions of the interviewees were three men and twenty women.

Interview questions:

1. Have you had any experience with AR before studying the course?

2. What was the most interesting thing to know?

3. Would you like to improve your knowledge of AR?

4. Do you use augmented reality in your professional activities?

5. What would you suggest to change to make the course more effective?

 https://doi.org/10.1051/shsconf/20207504015

The content analysis method was used to analyze the interview data. Data analysis includes the editing, structuring, and interpretation of collected data.

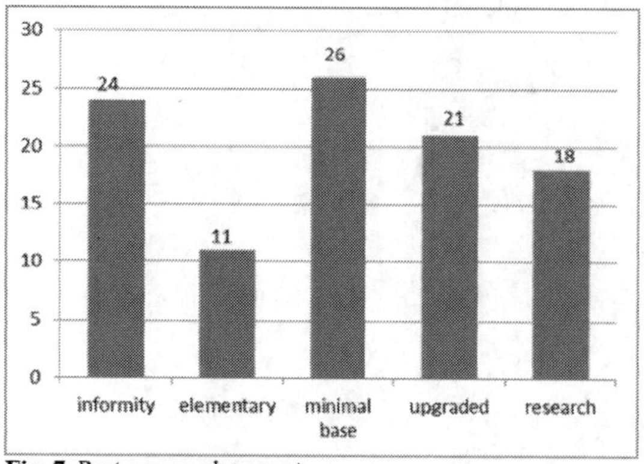

Fig. 7. Post-course placement.

1. Have you had any experience with augmented reality before studying the course?

Initially, the following answers to this question were supposed:

– Yes, I have developed applications myself.

– Yes, I used training applications.

– Yes, I used in everyday life (advertising, entertainment, etc.).

– No, I haven't.

As a result, only 4 of those respondents used AR earlier, and only in everyday life, the remaining 19 before the course had no idea about AR (Fig. 8).

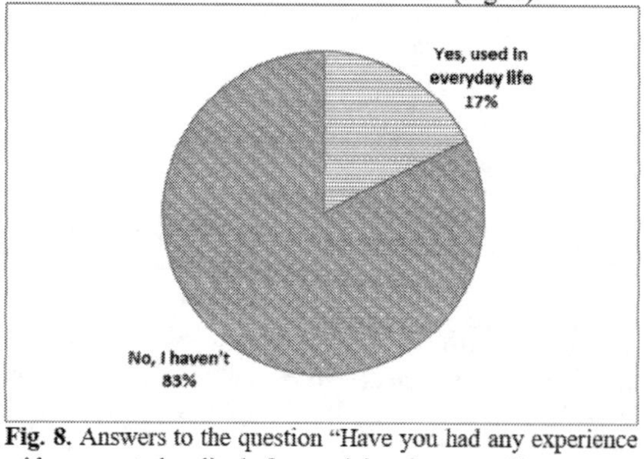

Fig. 8. Answers to the question "Have you had any experience with augmented reality before studying the course?".

2. What was the most interesting thing to know?

The meaning of 100% of respondents' answers was either to the process of development or the result of application development or the practical application of these applications.

All received answers were the application development; process of reviving pictures; convert 2D images to 3D; 3D modeling; practical application; visualization.

The most impressive answer was: "Results exceed all expectations".

3. Would you like to improve your knowledge of AR?

To this question 2 respondents gave a negative answer, 1 was difficult to answer, 15 answered in the affirmative, and 5 said that they have already improved their knowledge. Fig. 9 shows the distribution of responses as a percentage.

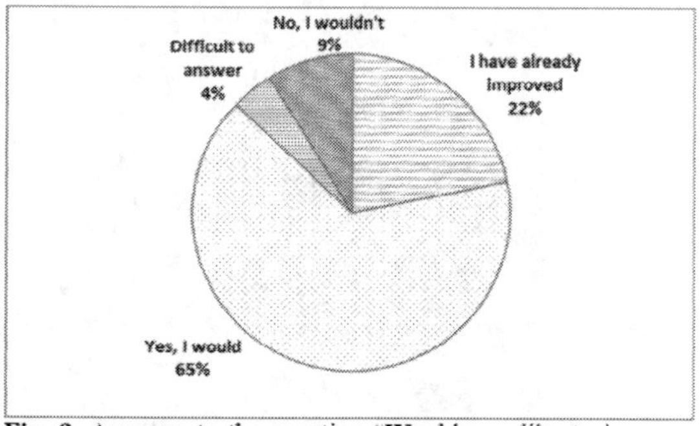

Fig. 9. Answers to the question "Would you like to improve your knowledge?".

4. Do you use AR in your professional activities?

All the received answers are:

– I'm already using it.

– I'm going to use.

– I think I will use it.

– No, I don't.

Respondents identified areas for using AR, such as a master class on the use of augmented reality for school teachers, when learning to program with high school students, when learning mathematics in specialized classes.

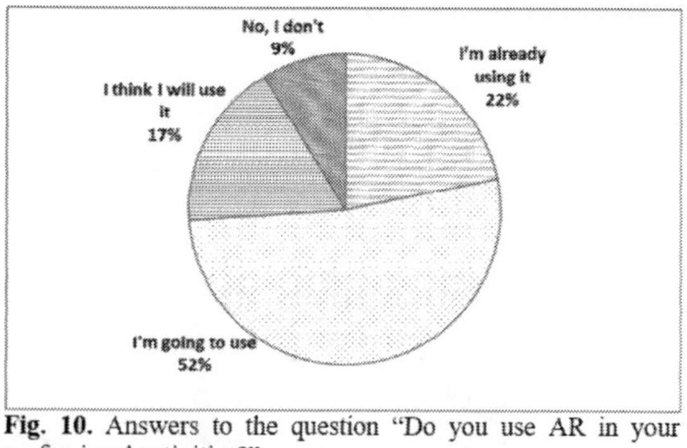

Fig. 10. Answers to the question "Do you use AR in your professional activities?".

One respondent immediately after the course during preparation for the state exam. One of the questions on the exam was "Demonstration of a fragment of a non-standard lesson". The theme of the lesson was "Creating logos. Brandguide. Brand book". The student used self-developed markers in the BlippAR environment to link the logo of the Google search (Fig. 11) with a URL-link to the Google form.

Fig. 11. The logo is an AR marker [31].

Further pupils should have united in groups, having chosen from the offered logos. Group named Historians, Lawyers, Designers, and Psychologists. Access codes for AR markers were 1150544, 1150549, 1150567 and 1150546 respectively (Fig. 12-13).

Fig. 12. Historians' and Lawyers' AR markers [32, 33].

Fig. 13. Designers' and Psychologists' AR markers [34, 35].

5. What would you suggest to change to make the course more effective?

The answers we've received:

– Reduce independent work.

– Increase classroom activities (lectures, labs, consultations).

– Extend the course for 2 semesters.

– Detail the methodological guidelines.

– Increase the number of practical tasks connected with STEM courses.

– The idea of conducting a survey using AR markers was interesting.

Conclusion

To get a complete picture of students' impressions of the course, it is necessary to reproduce exactly the answers of some students.

"Before the course, I had no idea what AR was. We enjoyed both the process and the result. And the result exceeded all expectations. The organization of the course was excellent. The presentation of the material in the lectures was available and dosed, the tasks in the laboratory classes were clear and had practical meaning."

"I use and plan to use received knowledge in the future because the AR is not only popular, but it also increases the level of understanding of the material, and what the most important is it helps to interest the student!"

Thus, the course "Development of VR and AR software" promotes the development of competence in the design and using innovative learning tools. The research is not completed, the implementation of the developed course and experimental verification of its effectiveness has been continuing.

References

1. Highlights From Gartner Hype Cycle for Customer Service and Support Technologies (2019), https://www.gartner.com/smarterwithgartner/highlights-from-gartner-hype-cycle-for-customer-service-and-support-technologies-2019/. Accessed 27 Jan 2020
2. Gartner Top 10 Strategic Technology Trends for 2020 (2019), https://www.gartner.com/smarterwithgartner/gartner-top-10-strategic-technology-trends-for-2020. Accessed 27 Jan 2020
3. 5 Trends Emerge in the Gartner Hype Cycle for Emerging Technologies (2018), https://www.gartner.com/smarterwithgartner/5-trends-emerge-in-gartner-hype-cycle-for-emerging-technologies-2018/. Accessed 4 Feb 2020
4. P.P. Nechypurenko, V.N. Soloviev, Using ICT as the Tools of Forming the Senior Pupils' Research Competencies in the Profile Chemistry Learning of Elective Course "Basics of Quantitative Chemical Analysis". CEUR Workshop Proceedings **2257**, 1–14 (2018)
5. P.P. Nechypurenko, T.V. Starova, T.V. Selivanova, A.O. Tomilina, A.D. Uchitel, Use of Augmented Reality in Chemistry Education. CEUR Workshop Proceedings **2257**, 15–23 (2018)
6. O.V. Komarova, A.A. Azaryan, Computer Simulation of Biological Processes at the High School. CEUR Workshop Proceedings **2257**, 24–32 (2018)
7. T.V. Hruntova, Yu.V. Yechkalo, A.M. Striuk, A.V. Pikilnyak, Augmented Reality Tools in Physics Training at Higher Technical Educational Institutions. CEUR Workshop Proceedings **2257**, 33–40 (2018)
8. O.V. Merzlykin, I.Yu. Topolova, V.V. Tron, Developing of Key Competencies by Means of Augmented Reality at CLIL Lessons. CEUR Workshop Proceedings **2257**, 41–52 (2018)
9. V.L. Buzko, A.V. Bonk, V.V. Tron, Implementation of Gamification and Elements of Augmented Reality During the Binary Lessons in a Secondary School. CEUR Workshop Proceedings **2257**, 53–60 (2018)
10. V.B. Shapovalov, A.I. Atamas, Zh.I. Bilyk, Ye.B. Shapovalov, A.D. Uchitel, Structuring Augmented Reality Information on the stemua.science. CEUR Workshop Proceedings **2257**, 75–86 (2018)
11. N.V. Rashevska, V.N. Soloviev, Augmented Reality and the Prospects for Applying Its in the Training of Future Engineers. CEUR Workshop Proceedings **2257**, 192–197 (2018)
12. V.S. Morkun, S.O. Semerikov, N.V. Morkun, S.M. Hryshchenko, A.E. Kiv, Defining the Structure of Environmental Competence of Future Mining Engineers: ICT Approach. CEUR Workshop Proceedings **2257**, 198–203 (2018)

13. S.O. Zelinska, A.A. Azaryan, V.A. Azaryan, Investigation of Opportunities of the Practical Application of the Augmented Reality Technologies in the Information and Educative Environment for Mining Engineers Training in the Higher Education Establishment. CEUR Workshop Proceedings **2257**, 204–214 (2018)
14. I.S. Mintii, V.N. Soloviev, Augmented Reality: Ukrainian Present Business and Future Education. CEUR Workshop Proceedings **2257**, 227–231 (2018)
15. M.V. Popel, M.P. Shyshkina, The Cloud Technologies and Augmented Reality: the Prospects of Use. CEUR Workshop Proceedings **2257**, 232–236 (2018)
16. N.O. Zinonos, E.V. Vihrova, A.V. Pikilnyak, Prospects of Using the Augmented Reality for Training Foreign Students at the Preparatory Departments of Universities in Ukraine. CEUR Workshop Proceedings **2257**, 87–92 (2018)
17. T.H. Kolomoiets, D.A. Kassim, Using the Augmented Reality to Teach of Global Reading of Preschoolers with Autism Spectrum Disorders. CEUR Workshop Proceedings **2257**, 237–246 (2018)
18. A.E. Kiv, V.N. Soloviev (ed.), CEUR Workshop Proceedings **2257** (2018)
19. A.E. Kiv, M.P. Shyshkina (ed.), CEUR Workshop Proceedings **2547** (2019)
20. Ye.O. Modlo, Yu.V. Yechkalo, S.O. Semerikov, V.V. Tkachuk, *Vykorystannia tekhnolohii dopovnenoi realnosti u mobilno oriientovanomu seredovyshchi navchannia VNZ* (Using technology of AR in a mobile-based learning environment of the higher educational institution), Naukovi zapysky, Seriia: Problemy metodyky fizyko-matematychnoi i tekhnolohichnoi osvity **11**, 93–100 (2017)
21. Wikitude Cross-Platform Augmented Reality SDK – Boost your app (2020), https://www.wikitude.com/products/wikitude-sdk/. Accessed 27 Jan 2020
22. ARKit 3 – Augmented Reality | Apple Developer (2020), https://developer.apple.com/augmented-reality/arkit/. Accessed 27 Jan 2020
23. ARCore – Google Developers | Google Developers (2020), https://developers.google.com/ar. Accessed 27 Jan 2020
24. ARToolKit Home Page (2020), http://www.hitl.washington.edu/artoolkit/. Accessed 27 Jan 2020
25. MAXST | Technology company specialized in AR (2020), http://maxst.com/#/. Accessed 27 Jan 2020
26. Vuforia: Market-Leading Enterprise AR | PTC (2020), https://www.ptc.com/en/products/augmented-reality/vuforia. Accessed 27 Jan 2020
27. Unity Real-Time Development Platform | 3D, 2D VR & AR Visualizations (2020), https://unity.com/. Accessed 27 Jan 2020
28. Visual Studio 2019 | Visual Studio (2020), https://visualstudio.microsoft.com. Accessed 27 Jan 2020
29. Virtual Reality (VR) App Development (UC San Diego, 2019), https://www.edx.org/professional-certificate/ucsandiegox-virtual-reality-app-development. Accessed 27 Jan 2020
30. J. Glover, *Unity 2018 AR Projects: Build four immersive and fun AR applications using ARKit, ARCore, and Vuforia* (Packt Publishing, Birmingham, 2018)
31. Google (2020), https://www.google.com/. Accessed 27 Jan 2020
32. BMW.com | The international BMW Website (2020), https://www.bmw.com/. Accessed 27 Jan 2020
33. Alfa Romeo – Wikipedia (2020), https://en.wikipedia.org/wiki/Alfa_Romeo. Accessed 27 Jan 2020
34. Fanta – Wikipedia (2020), https://en.wikipedia.org/wiki/Fanta. Accessed 27 Jan 2020
35. Sprite (drink) – Wikipedia (2020), https://en.wikipedia.org/wiki/Sprite_(drink). Accessed 27 Jan 2020

Methods building and printing 3D models historical architectural objects

Ihor Hevko[1*], *Olha* Potapchuk[1], *Iryna* Lutsyk[1], *Viktorya* Yavorska[2], and *Viktoriia* Tkachuk[3]

[1]Ternopil Volodymyr Hnatiuk National Pedagogical University, Department of Computer Technologies, Ternopil, 46027, Ukraine
[2]Odessa I.I Mechnikov National University, Department of Economic and Social Geography and Tourism, Odessa, 65000, Ukraine
[3]Kryvyi Rih National University, Engineering Pedagogy and Language Training Department, Kryvyi Rih, 50027, Ukraine

Abstract. The authors present methods building and printing three-dimensional models for graphical reconstruction of historical architectural objects. Procedure sequence of the methods is exemplified through building the model of the Parochial Cathedral of St. Mary of the Perpetual Assistance of the 1950s. After analyzing and assessing the most popular specialized software means, the 3DS Max environment is chosen to build a three-dimensional model. Suggested software tools enable increased accuracy, speed and granularity of fixation of complex systems and expanded databases, providing efficient instruments to deal with bulk data and being relevant to new IT achievements. Sequence and content of operations for analytical and modeling cycles are substantiated. The cathedral model is built on the basis of archive photographs and drafts. The authors describe methods and the algorithm of procedures, principles of architectural and spacious modeling to recreate the architectural object. The three-dimensional model is built by applying a stereogram miniature of the destroyed Cathedral. Reconstruction of spacious configuration of the objects is based on parallax assessment of images. Stages of project implementation are determined. There are described methods of implementing modeling by 3DS Max tools and preparing the model for 3D printing in Cura.

1 Problem statement

It is impossible to replace three-dimensional models in all spheres of human activity including industry, medicine, architecture, construction, design, education, cinema, etc. 3D model design enables assessing technical and physical properties of a modeled object before creating a real sample. The methods of studying a model allow analyzing its size, material and package contents.

The concept of an object or a project is mainly exemplified by videos or pictures based on 3-D graphics. This sets constraints on viewing, as static pictures cannot enable plot change or detailed examination.

Modern potential of 3D graphics and computer hardware capacity enable processing complex scenes on-line without reducing rendering speed and quality. This has evoked professionals' interest to 3D visualization in various activity spheres.

In architecture and bridge engineering, wider application is given to virtual buildings with inside walks and virtual cities. Photorealistic reconstruction of objects makes it possible to work with object models in museum, reconstruction and commercial projects and while studying [1].

Maintaining and promulgation of cultural heritage are essential for modern society. Development of computers and 3D graphic tools enables preserving cultural achievements not only as pictures or photographs but also as models in their original form or as electronic replicas of real-life objects [2].

A great number of architectural monuments have disappeared without any sizes, drafts or photographs left. For such historical objects, graphic reconstruction as a scientific study is the only means of identifying the lost or destroyed architectural object of a certain time period. Graphical reconstruction of architectural historical heritage reflects the whole bulk of knowledge concerning it available to date [3].

In recent years, there have been numerous museums including the virtual ones with their exhibits being computerized objects. Museums of this kind enable obtaining detailed information on historical achievements, getting to know their origin and facilitating cultural development of society. It also facilitates tourism, thus enhancing regions' economic growth [4].

2 Research aim

Nowadays, innovative technologies of 3D graphics, modeling and design enable restoring lost historical

*Corresponding author: gevko.i@gmail.com

objects. Analysis of the degree of investigation reveals only certain aspects of 3D modeling covered in modern scientific literature. Technologies for selecting software for three-dimensional modeling and methods of working with them are described in the works of D. Banach, T. Boardman, J. Jones, M. Jambruno, K. Osadcha, H. Chemerys [5]. The issues of three-dimensional modeling in architectural design are revealed in the studies of L. Borodkin, M. Rumyantsev, A. Smolin, R. Baryshev, V. Rozhko, I. Popov and others [1-3].

General problems concerning history are highlighted in works by such foreign theoreticians of design as O. Henisaretskyi, V. Hlazychev, N. Mosorova, N. Valkova and O. Lavrentieva. 3D modeling as a design and architecture tool is indirectly touched upon in publications by O. Bodnar, O. Boichuk, and V. Danylenko [6].

Thus, analysis of the researches enables concluding that there are no scientific works reflecting specific features of 3D modeling application in architecture aimed at maintaining the cultural heritage of society that makes the above-mentioned issues under-investigated.

The paper *is aimed* at disclosing peculiarities of creating and printing 3D models of historical architectural objects in the 3DS-Max environment, this being conditioned by contradictions between the problem topicality and necessity to use modern technologies of 3D modeling.

The research object involves building and printing a 3D model of the Parochial Cathedral of St. Mary of the Perpetual Assistance of the 1950s.

3 Presentation of main materials

3.1 Sequence and content of operations of the analytical and modeling cycles

3D modeling is a separate type of computer graphics, which incorporates necessary tools and techniques applied to building a model of an object in the three-dimensional space. 3D modeling techniques of a graphic object include the following main cycles: the analytical cycle (collection of input materials; calculation of object sizes and parameters) and the modeling one (building a draft of an object form; accumulation, carving, stamping, etc.).

Nowadays, 3D modeling is used in almost all fields of human activity including advertising, marketing, industry, computer games, cinema, architecture, design and animation. 3D models of buildings and facilities are an integral part of modern design providing the basis for making object prototypes with maximum granularity.

Stages of building 3D models of monuments and landscapes are specific in their character depending on set tasks and software chosen. However, the most essential components of the methods are general for different modeling objects. While setting a task for modeling, it is necessary to determine the rate of granularity and realism of the end product [7]. Realism of a model depends on selected materials for overlaying textures onto an object. Virtual 3D modeling for architectural buildings is based on solving the task of the efficient layout widespread in the theory of pattern recognition.

Nowadays, there are many software means of various parameters and applications in computer graphics. Choice of software primarily depends on the task set. After selecting functions and means required for solving the task, it is necessary to choose efficient software to build 3D models.

Architects and designers make good use of 3D graphics technologies because they are efficient and easy to use for project implementation. To select the required software environment, a survey was conducted among experts in this field and students who study the basics of 3D modeling. Based on the survey, the following software products are identified as the most popular: Blender, 3D Max, SweetHome 3D, SketchUpMake, Pro 100, FloorPlan 3D, ARCON 3D Architect, ARCHICAD, Maya, Artlantis, LUMION, Cinema 4d. It should be noted that the most appropriate is the use of environments SweetHome 3D, 3DSMax, FloorPlan 3D, ARCON 3D Architect, ARCHICAD in the architectural direction [5].

As our task is to build a model of an object, we should analyze the above-mentioned programmes to choose appropriate software. Parameters of evaluation quality are chosen according to ISO 9126:2001 Standard in which each characteristic is described by its several attributes [8].

In this case, they include functionality, user-friendliness, efficiency, the programme interface and render quality (the final image after processing) as the most important parameter. As these criteria are not equivalent, importance factors are determined for each of them relevant to the set task (Table 1).

Table 1. Assessment parameters.

Parameter	Importance factor
Functionality	3
User-friendliness	2
Efficiency	2.5
Programme interface	1.5
Render quality	4

Evaluation is performed in the system from 1 to 10 points for each parameter on the basis of working with similar programmes. Total rating evaluation equals a total of points for each parameter multiplied by a corresponding factor. Results of comparison analysis are in Fig. 1.

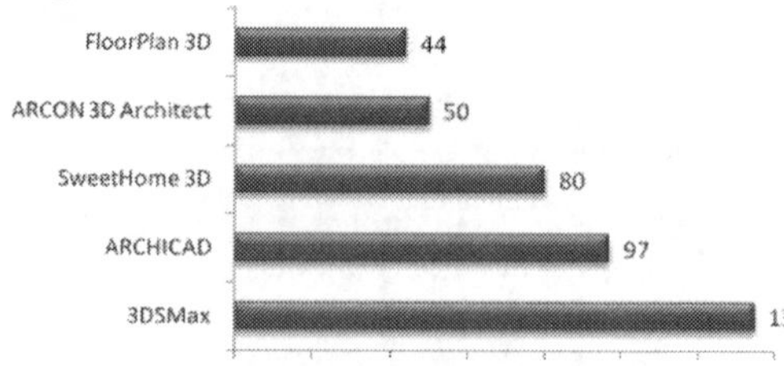

Fig. 1. Total rating evaluation of software

After analyzing and evaluating the mentioned software according to the five parameters, the first two – 3DSMax and ArchiCAD – appear the most convenient and efficient to model the Parochial Cathedral, as they are easy to use and effective in terms of time. For this reason, to build a model of the Parochial Cathedral, 3DS Max is chosen as covering all necessary tools to perform high-quality realistic rendering.

Graphic reconstruction of lost or destroyed architectural objects is a specific type of activity aimed at studying these objects in order to restore their appearance as of the time of their existence by 3D graphics means being guided by the preserved documents, drafts or photographs [1; 3].

Graphic reconstruction provides for absence of precise data on an object from a single data source. It is applied to restoring a lost appearance by means of graphic and document data through collecting and combining it from various sources. Graphic reconstruction being an activity is thought of as a set of operations including data collection, object investigation and fixation prior to modelling options of a destroyed architectural monument.

The Parochial Cathedral of St. Mary of the Perpetual Assistance of the 1950s (hereafter – the Parochial Cathedral) is one of the lost historical objects of Ternopil that decorated the city centre at the corner of Ruska and Mitskevich Streets (modern Shevchenko Boulevard). Photographs and drafts are basic data sources concerning the Cathedral.

The historical and architectural key plan of Ternopil indicates that "the majestic and delicate building in the neo-Gothic style was striking in its beauty and perfection. The slim tower-spire of 62m high was hovering over the city as if striving upward into the sky. It was even used as a fire tower built upon the project of the famous Lviv architect Professor Theodor Marian Tal′ovskyi" (here and after the translation is ours) [9].

The work written by L. Boitsun Ternopil in the course of years says that "in 1954, there were some explosions heard during several days when the Catholic Church was blasted. In 1959, a supermarket was opened there to celebrate the anniversary of the October Revolution. Many elements of the Church ornamentation were taken to Poland. Part of high reliefs of the sacred procession and the sculpture of Madonna were preserved in the Medium Church (the Church of the Nativity of Christ)" [10].

That is why, we consider it of great importance to restore this architectural monument to preserve Ternopil's cultural heritage. The 3D model is built on the basis of a stereogram miniature of the Parochial Cathedral photograph (Fig. 2).

Computer reconstruction implies restoration of a building in accordance with the data from the sources (descriptions, photographs, pictures) as well as some set parameters after comparing descriptions and data on construction technologies for cathedrals of that epoch.

Restoration of spacious configuration of objects is based on paraplex evaluation of images. It implies finding a relevant element in the right image for each element of the left image after processing a pair of images. The difference of horizontal coordinates of relevant pixels (paralex) reflects the distance to the image point [11].

Fig. 2. The photograph of the Parochial Cathedral of St. Mary of the Perpetual Assistance (1950s)

Considering relevance of the set tasks, it is expedient to distinguish such basic stages of project implementation as analysis, design engineering (application of engineering skills), design (application of design competences), and printing (Fig. 3).

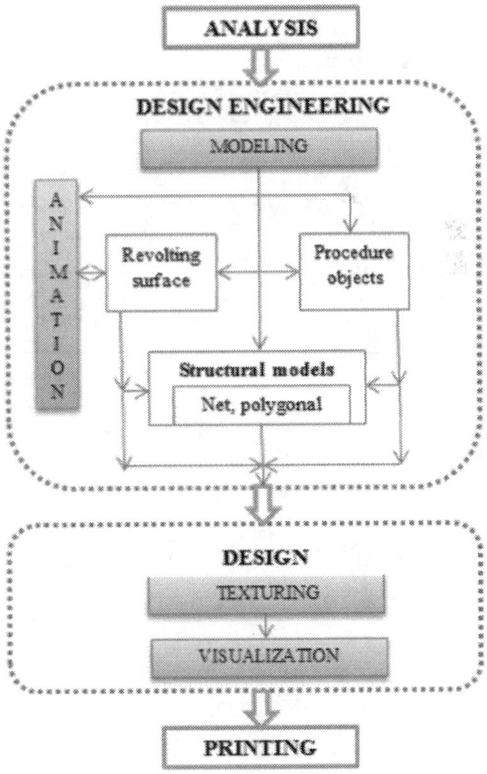

Fig. 3. Stages of project development

While analyzing, we collect necessary data on the object and required operations to build a 3D model. Design engineering involves modeling – building a 3D model and animation – providing models with movement or supplying auxiliary cameras and moving them along certain trajectories.

The design stage includes texturing (selection of materials and textures) and visualization (selection of light sources, change of perspectives, adjusting cameras, etc.). The finalizing stage of project development is creating a miniature of the Parochial Cathedral of St. Mary of the Perpetual Assistance applying 3D printing.

3.2 Building a 3D model

While building a 3D model for printing, we distinguish the following stages:

• Searching for information to create a precise model;
• Building a model in the 3DSMax environment;
• Selecting relevant sizes and building layouts of fine details;
• Preparing the model for printing.

Collection of data involves searching for cartographic materials as well as images and texts to facilitate accomplishment of the set task. Digital data are preferable followed by vector and raster images. While searching for information, we use a photograph of the Parochial Cathedral with sharp images of elements of the architectural object to create its precise model.

In applying 3D modelling methods, special attention is paid to geometrical modeling considering the type of the modeled object (engineering, design, architectural, etc.) and the technology applied [12].

Guided by detailed analysis of over 20 photographs of the Cathedral and its layout, we build a 3D model of the object. Thus, the above-described procedures result in a primary platform of the model (Fig. 4).

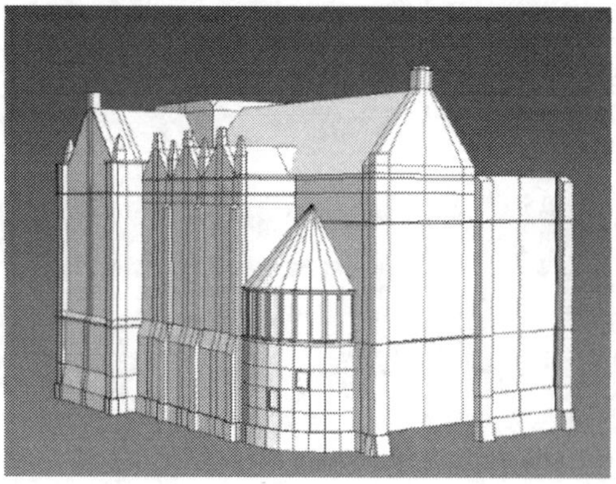

Fig. 4. The miniature of the 3D model in 3DS Max

The next actions are aimed at editing forms of the basis according to the photographs available. After completing detailed analysis of sizes and architectural features, we make amendments by means of relevant 3DS Max tools [13]. After that, the building acquires a more realistic appearance. The complex character of building the model involves numerous fine details, their asymmetry and location in different planes.

Next, we perform detailed processing of walls and domes. To reduce labour-consuming procedures of model building, repeated details like windows can be copied and dragged to the required location. If you need to resize the element, its plane or angle, then it is possible to do it using the functions of the software environment (Fig. 5).

Fig. 5. The 3D model of the Parochial Cathedral of St. Mary of the Perpetual Assistance

For the sake of convenience, we apply appropriate functions to revolving and moving the model. Thus, after completing a series of actions and operations, we obtain a three-dimensional model of the Parochial Cathedral. To make the image of the model more realistic, we perform its rendering (Fig. 6).

Fig. 6. The render image of the 3D model

Before making a printed miniature of the 3D model, we should analyze and adjust it properly. As the target result of modeling is a printed miniature, the built model should be exported into the STL-format. It should be noted that due to the intensive development of 3D printing, most specialized programs support this feature. This type supports three-dimensional objects by preserving them as a bulk of triangular data describing a surface.

3.3 Preparing the model for printing

The first stage of preparing the model for printing provides for analysis of 3D model geometry, which involves its testing for available open spaces in the polygonal net, some displacements of polygons and defects in geometry.

The next stage includes analysis of all parameters, sizes and their test for conformity with printing materials. As the built 3D model has sizes of a real-life

 https://doi.org/10.1051/shsconf/20207504016

building, it requires scaling to create its printable miniature (Fig. 7).

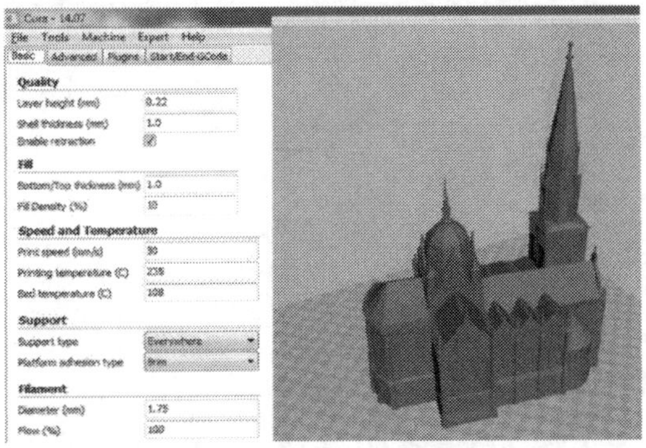

Fig. 7. Adjusting the model sizes to printing

Nowadays, there is a great variety of software for 3D printing, among which one should mention Cura, CraftWare, Slic3r, 3DTin and Repetier-Host. These software products are quite widespread due to their advanced features and relative complexity.

Yet, being guided by convenience and a relatively user-friendly interface, we apply Cura in which except for standard editing tools, printing quality adjustment and material parameters, there are functions of calculating weight of the end item, its print time, etc [14].

Basic settings of technological parameters include printing quality, filling, printing speed and temperature, parameters of printing support and plastic threads. While setting the parameters of printing quality, the most essential one is the layer height (mm) determined by the nozzle diameter and it should not exceed its half.

Shell thickness (mm) determines thickness of printing walls of thin-wall objects or objects with the reduced in-fill ratio. Shell thickness is determined by corresponding geometrical parameters of an object. For small models, the thickness of 10mm-30mm is optimal.

Economic factors of plastic consumption are determined by fill density (%). In most cases, the in-fill ratio makes 10%, yet, for inflexible models and considering structural features of a model, the in-fill ratio can reach 100%. However, printing time increases greatly.

Settings of print speed and temperatures enhance qualitative and technological parameters of printing. The most significant parameter is print speed that determines nozzle movements. As our model has many fine details, the set speed is 30mm/sec. to make printing accurate. It is caused by the fact that high print speed affects its quality because of vibration efforts on the supporting frame of a printer and accelerated wear of drive elements.

The technology also provides for printing auxiliary model elements (not specified in geometry) considering lack of possibility to form plastic mass in the air. This support is possible for both individual model elements (support type) and its platform (platform adhesion type). In this case, we select the function Brim to provide high-quality print of model elements, which are hanging (the roof, domes). The programme creates additional supports for these elements.

After setting the required parameters to make a miniature, the file is sent to the printer with automatically formed G-code and approximate print time and the amount of the required material are determined.

Fig. 8 presents a printed model of the Parochial Cathedral based on the suggested 3D modelling technology, the advantages of which are availability and low costs of produced models.

Fig. 8. The printed miniature of the Parochial Cathedral of St. Mary of the Perpetual Assistance

The methodology for creating the 3D model and printing the layout of the Parish Cathedral has been carried out by specialists of the Innovative Center for 3D-Technologies of Design and Production, which operates on the basis of the Chair of Computer Technologies of the Ternopil Volodymyr Hnatyuk National Pedagogical University.

Some specific features of the developed model indicate possible further application of the methods to reconstruction activity in order to preserve the city and the state cultural heritage.

4 Conclusions and prospects for further research

Graphic reconstruction of historical architectural objects is possible due to new technologies of 3D graphics, modeling and design in specialized computer environments. The developed methods of graphic reconstruction are exemplified by the model of the Parochial Cathedral of St. Mary of the Perpetual Assistance of the 1950s.

Stages of building 3D models of architectural monuments are based on general methods considering individual features as following from set tasks, selected software and required granularity and realism.

Stages of design implementation of the miniature of the Parochial Cathedral include the analytical stage (collection of input data; calculation of sizes and parameters of the object) and the modeling one (building a draft of an object form; accumulation, carving, stamping, etc.).

Computer reconstruction provides for restoration of a building according to the data from archive sources and parameters determined after comparing descriptions and data on construction of cathedrals of that epoch. Reconstruction of spacious configuration of objects is based on parallax evaluation of stereogram images of the destroyed Cathedral.

3DS Max is selected to build a 3D model of the object to enhance high accuracy, speed and granularity of fixing complex sets providing efficient tools of working with bulk data that incorporate new achievements of informational technologies.

Detailed analysis of images and determined sizes provides the basis for the 3DS Max model, which is then edited by relevant tools to make it more realistic. The complex character of building the model implies its numerous fine details, their asymmetry and location in different planes.

Creating a printed model of a 3D model requires its analysis and adaptation to 3D printing based on testing the model for the presence of open spaces in the polygonal mesh, defects in the geometry and checking for compliance with the print materials. To build a printed model of the Cathedral, guided by criteria of convenience and the user-friendly interface, the Cura software environment is applied.

Presented methods of building and printing three-dimensional models can be successfully applied not only to graphic reconstruction of historical architectural objects, but also to teaching and learning.

Prospects for further research include restoration of the Cathedral interior that would enable creation of an object of the virtual historical museum of the architectural monument. Yet, this problem requires auxiliary data on the Parochial Cathedral of St. Mary of the Perpetual Assistance and remains unsolved to date.

References

1. L.I. Borodkin, Komp'yuternoye 3d-modelirovaniye v issledovaniyakh po istoricheskoy urbanistike: novyye istochnikovedcheskiye podkhody (Computer 3d-modeling in studies on historical urban studies: new source-based approaches). **21**(1), 57-62 (Kostroma, 2015)
2. M.V. Rumyantsev, A.A. Smolin, R.A. Baryshev, I.N. Rudov, N.O. Pikov, Virtual'naya rekonstruktsiya ob"yektov istoriko-kul'turnogo naslediya (Virtual reconstruction of objects of historical and cultural heritage). Prikladnaya informatika **6**(36), 62-77 (2011)
3. V.M. Rozhko, *Metodyka hrafichnoi rekonstruktsii derevianoi naskelnoi arkhitektury (na prykladi pamiatok IX–XVI stolit Zakhidnoho rehionu Ukrainy)* (The technique of graphic reconstruction of wooden rock architecture (as an example of the monuments of IX-XVI centuries of the Western region of Ukraine). (Lviv Polytechnic National University, Lviv, 2013)
4. V.V. Yavorska, I.V. Hevko, V.A. Sych, O.I. Potapchuk, K.V. Kolomiyets, Features of application of information technologies in modern tourism. J. Geol. Geog. and Geoecol. **28**(3), 591–599 (2019). doi:10.15421/111956
5. K. Osadcha, H. Chemerys, Dobir zasobiv tryvymirnoho modelyuvannya dlya formuvannya hrafichnoyi kompetentnosti maybutnikh bakalavriv komp'yuternykh nauk (Selection of 3D modeling tools for forming the graphic competency of future bachelors of computer science) Information Technologies and Learning Tools **62**(6), 70–85 (2017)
6. V. Danylenko, *Dyzain Ukrainy u svitovomu konteksti khudozhno-proektnoi kultury* (Ukraine Design in the world context of art and design culture). (Kharkiv, KhDADM, 2005)
7. Yu.A. Krejdun, Kontseptsiya sozdaniya galerei virtual'nykh 3D-obrazov utrachennykh pamyatnikov istorii i arkhitektury (The concept of creating a gallery of virtual 3D images of lost monuments of history and architecture). Vestnik Altayskogo gosudarstvennogo pedagogicheskogo universiteta **18**, 71–76 (2014)
8. Ye.V. Vavilov, Seriya standartiv SQuaRE yak osnova zabezpechennya vymoh do yakosti ta otsinky prohramnykh zasobiv (SQuaRE Standards Series as a Basis for Quality Assurance and Software Evaluation). Zbirnyk naukovykh prats Odeskoyi derzhavnoyi akademiyi tekhnichnoho rehulyuvannya ta yakosti **1**, 129–139 (2015)
9. *Istoryko-arkhitekturnyi opornyi plan m. Ternopolia* (Historic-architectural Ternopil city reference plan). (Ukrzakhidproektrestavratsiia, Lviv, 2012)
10. L.S. Boitsun, *Ternopil u plyni lit. Istoryko-kraieznavchi zamalovky* (Ternopil over the years. Historical sketches). (Jura, Ternopil, 2003)
11. D. Riabokon, *Stvorennia tryvymirnykh modelei obiektiv za stereoparamy zobrazhen dlia virtualnykh muzeiv* (Creating 3D object models based on stereo image imagery for virtual museums). EVA, 61-68 (2002)
12. V.A. Lytvyn, Metodyka formuvannya informatsiynoyi kultury maybutnikh arkhitektoriv (Methods of formation of information culture of future architects). Molodyy vchenyy **2**(4), 159 (2015)
13. B.L. Smith, *Foundation 3ds Max 8 Architectural Visualization* (Apress, New York, 2006). doi:10.1007/978-1-4302-0158-8
14. V.Yu. Mitin, Obzor oborudovaniya, programmnogo obespecheniya, vozmozhnostey i etapov trekhmernoy pechati (Overview of hardware, software, features and stages of three-dimensional printing). Vestnik Permskogo Universiteta **41**(2), 67–74 (2018)

Methods and tools for teaching parallel and distributed computing in universities: a systematic review of the literature

Yuriy Sitsylitsyn*

Bogdan Khmelnitsky Melitopol State Pedagogical University, Department of Computer Science and Cybernetics, Melitopol, Ukraine

Abstract. As computer hardware becomes more and more parallel, there is a need for software engineers who are experienced in developing parallel programs, not only by "parallelizing" sequential designs. Teach students a parallelism in elementary courses in computer science this is a very important step towards building the competencies of future software engineers. We have conducted research on "teaching parallel and distributed computing" and "parallel programming" publications in the Scopus database, published in English between 2008 and 2019. After quality assessment, 26 articles were included in the analysis. As a result, the main tool for teaching parallel and distributed computing is a lab course with a C++ programming language and MPI library.

1 Introduction

Traditionally, undergraduate computer science students are taught sequential programming through a one-way programming model in higher education majors. It is typical to wait for a student to develop knowledge of serial programming before teaching parallel programming. As computer hardware becomes more and more parallel, there is a need for software engineers who are experienced in developing parallel programs, not only by "parallelizing" sequential designs. Teach students a parallelism in elementary courses in computer science this is a very important step towards building the competencies of future software engineers.

In April 2005, AMD released the AMD64 architecture Opteron 2-core processor for servers. In May 2005, Intel released the Pentium D processor x86-64 architecture, which was the first 2-core processor designed for personal computers. This was Intel's quick response to AMD's challenge. So the era of multi-core processors began. The growth of multi-core and multi-core processing has certainly added new relevance to the teaching of parallel programming. Since 2000, in the world scientific literature many articles on the topic of parallel and sequential programming and learning parallel and sequential computing.

According to Brown et al. [5], given the importance of these architectures, parallel programming becomes indispensable for undergraduate students in such specialties as computer science, computer engineering, and software engineering. Students in these specialties must be able to program in a satisfactory manner, both sequentially and in parallel. Acquaintance of students with parallelism should begin from the early periods [14], then students will consider it as a natural and general part of programming (and not as advanced and rarely used content) [4]. In addition, switching from serial to parallel programming is a difficult task [14]. There is a general opinion that the topics of parallelism should be distributed throughout the undergraduate program [3]. However, in most universities the concepts of parallelism are studied only in the last courses [18].

A tool that can be used to facilitate access from students to content is the use of distance learning, allowing the student to study the course content according to his daily routine, adapting the study to your time and allowing the teacher to monitor [24]. To conduct new research in the field of teaching parallel programming in undergraduate programs, it is important to take into account existing work and related experience, to consider the difficulties and problems that teachers face. A systematic review of the literature is a method that allows you to determine all relevant searches for a given research question. Given the above, it is necessary to organize a systematic review of the literature with the aim of analyzing scientific works related to teaching parallel programming in undergraduate programs in computer science. To do this, a study was conducted in articles indexed in the Scopus, Web of Science and Google Scholar.

1.1 Theoretical bases

The purpose of study is to examine the available literature on the development of "parallel thinking" among students in the specialty of computer science for the development of parallel programs.

All articles in the review mention such terms as parallel and distributed programming. Let us dwell briefly on these concepts.

Parallel programming is a method that aims to reduce the dependency between sections of the same code, which allows these sections to be executed out of order when executing processes and / or threads. However, this

* Corresponding author: yuriy@mdpu.org.ua

execution is not necessarily parallel, since the actual parallelism of execution depends on the parallel architecture of the computer, processors with support for multithreading or speculation at the thread level, and / or using parallelism libraries (i.e. parallel programming).

Parallel programming explores the potential parallelism that exists in code using parallelism libraries that allow explicit execution of processes and / or threads to simultaneously execute program code. In this case, it is important that a parallel architecture exists (for example, a multi-core processor), so that parallel execution is actually supported. Sometimes the terms "parallel computing" and "parallel programming" are used as synonyms, but it is important to note that parallel computing does not imply simultaneity, but the potential for simultaneity. Thus, parallel programming involves identifying and reducing dependencies between sections of the same code, and parallel computing involves using parallelism libraries to examine these sections in parallel architectures.

Distributed programming is a method used to increase the scalability of parallel code execution using a communications network. In this context, for parallel codes, it is necessary to use message passing for collective communication, that is, the use of primitives, such as sending and receiving. Distributed programming is not necessarily used for concurrency, because processing can be distributed and sequential, but in the context of this article, the distributed potential of a cluster computer system offers the conditions for studying concurrency on a larger scale. For this reason, the concept of parallel and distributed programming is also used for systems in which several nodes with parallel architectures (for example, several cores) are connected by a high-performance communication network. In a cluster architecture of this type, parallel programming is performed by shared (operational) memory (intra-node communication) and transmission of network messages (external communication).

This research is trying to understand the methods and tools of teaching parallel programming in higher education institutions. To study existing literature, the following questions were used:

1. What methods and tools are used in teaching parallel and distribution programming?
2. What do the authors define as problems in teaching parallel and distribution programming?

1.2 Materials and methods

This systematic literature review utilized the PRISMA guidelines and flow chart. PRISMA guidelines include a 27-item checklist and a four-phase flow diagram outlining the items essential for transparency in conducting literature reviews [16].

In order to be included in this review, it is necessary to carry out research, to be peer-reviewed and to be published in English in a scientific journal article between 2000 and 2019. Research also needs to solve at least one of the research questions.

The following bases were used for the search: Scopus (https://www.scopus.com), Web of Science (https://apps.webofknowledge.com/) and Google Scholar (https://scholar.google.com). The last search was performed on January 29, 2020.

The search phrases used were: "teaching parallel and distributed programming", "teaching parallel and distributed computing".

The following criteria were used sequentially against article abstracts to select studies for inclusion:

Criteria 1: Study published between 2000 and 2020 in English

Criteria 2: Study published in scholarly journal

Criteria 3: Research was conducted for undergraduate computer science students

Criteria 4: Extracted data aligns with current study focus and research questions.

The results of the database search we entered in the table.

Table 1. The results of the database search.

Search terms	Database	Hits
"teaching parallel and distributed programming"	Scopus	41
	Web of Science	82
	Google Scholar	5
"teaching parallel and distributed computing"	Scopus	52
	Web of Science	78
	Google Scholar	51

Next, we remove articles from the results that do not mention teaching parallel programming. We bring all found articles into a single list. Then we delete articles that do not describe studies at the undergraduate or graduate level. Next, in a single list, we delete duplicate articles obtained from different databases. As a result, we get 26 articles.

2 Methods and tools for teaching parallel and distributed computing

After the screening, we received 26 research articles.

If we look at the articles in terms of practical experience in teaching parallel and distributed computing, it can be said that 85% of articles are written on the basis of their own experience and 15% of review articles in which the authors provide the results of theoretical studies on the teaching of parallel and distributed computing.

When considering methods of teaching parallel and distributed computing, the authors of the article pay attention to the following methods: a course of laboratory work that addresses selected topics or paradigms of parallel and distributed programming [1], [26]; course of lectures and laboratory work – here the authors give a complete theme of the course, which they teach: topics of lectures and laboratory work [3], [15]; some authors [2]; [8] introduce in the course of laboratory work with visualization of parallel computing, which according to the authors should promote the development of parallel thinking in students; some authors consider it appropriate to use the project method for laboratory work on real practical tasks [21] or team [31]. For a better understanding of the stages of concurrent programming, one of the authors [23] proposes to introduce a course on parallel programming in the form of a game. According to the author, this will improve the formation of students in "parallel thinking". We have summarized the methods of teaching parallel and distributed computing into a graph (Figure 1).

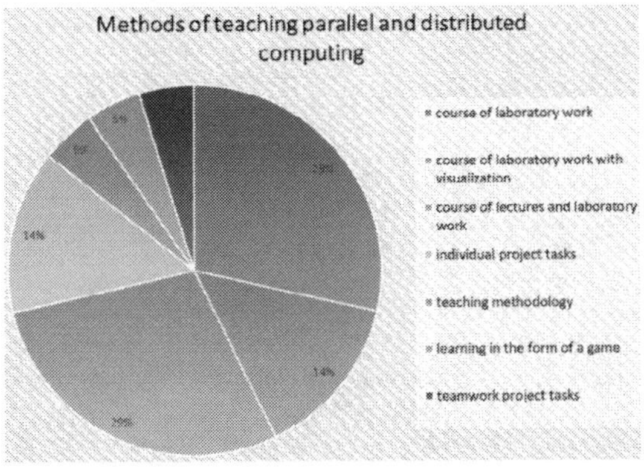

Fig. 1. Methods of teaching parallel and distributed computing.

The graph shows that most authors – 71%, use the method of laboratory work for practical training in parallel and distributed computing: one lab work reveals one topic or paradigm of parallel programming. Only 19% of authors consider it advisable to use the project method of teaching parallel programming and only one article about the use of the game method in teaching parallel and distributed computing.

Let us now consider the tools of parallel and distributed computing. First, consider the use of hardware. Based on the tools used by the authors of the articles, we divided them as follows: use of personal computer [6], cluster personal computers [17], personal computer with a set of video cards [26], single board computer [20], cluster of single board computers [10]. We reduce the use of learning hardware by parallel and distributed computation to the diagram (Figure 2).

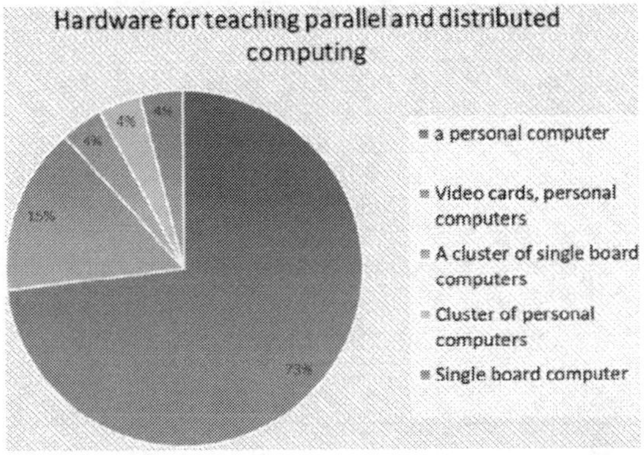

Fig. 2. Methods of teaching parallel and distributed computing.

The analysis of the chart shows that 88% of authors-teachers of higher education institutions use a personal computer to teach students parallel and distributed computing – that is, each student uses a personal computer (in computer class or own) to develop parallel and distributed programs. writing, debugging and running parallel and distributed programs. For 15% of authors, one or more discrete graphics cards must be present on a personal computer. Only 4% of authors use the latest technologies in the form of single-board computers for teaching parallel and distributed computing in their lab work. Laboratory work using clusters of single-board computers or personal computers is the same percentage.

Now let's consider the software for teaching students learning parallel and distributed computing. When designing lab work to teach parallel and distributed computing, the authors use most of the programming languages or additional libraries in most. Therefore, instead of a pie chart, we used a column in which the B axis contains the number of references in programming language articles (Figure 3).

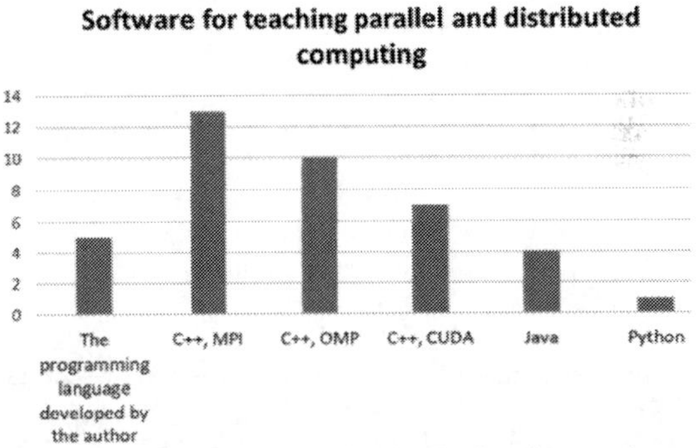

Fig. 3. Tools for teaching parallel and distributed computing.

Analyzing the diagram we can conclude that for laboratory work in teaching parallel and distributed computing, most authors use the C++ language together with the MPI library. The Messaging Interface (MPI) is a standardized messaging library. The messaging paradigm was developed for shared memory architecture. It offers various messaging technologies that provide shared memory. In messaging, processes are connected through a set of point-to-point primitives and collective communications. MPI is the actual standard for developing high-performance science programs [19]. MPI offers APIs for developing scalable C++ applications that are portable, efficient and flexible. The disadvantages of learning parallel and distributed computing using the MPI library are the difficulty of mastering the basic paradigms of MPI by students who previously worked only with sequential programming.

In second place by the number of mentions, the authors of the article are using the C++ programming language together with the OMP library. One of the most popular interfaces that supports multi-platform multi-core shared memory is OpenMP [10]. Using the simple semantics of this interface, a programmer can enable concurrent multicore computing in their applications. The OpenMP library uses runtime library routines, compiler directives, and environment variables in C++. The authors note that it is easier for students to learn about using this library than MPI libraries, but program development using the OMP library is limited to use only in concurrent programming whereas the MPI library can be used in concurrent, distributed, and hybrid programming.

Recently, new features in hardware for parallel and distributed computing, such as the use of video cards with a library of parallel and distributed computing CUDA. The structure of the CUDA program reflects the coexistence of the host (CPU) and one or more devices (GPUs) in the computer. Each CUDA output file may contain a mixture of both host and device code. By default, any traditional C program is a CUDA program that contains only the host code. You can add device features and data declarations to any source file C. Device features or data declarations are clearly identified by CUDA special keywords. These are usually features that exhibit a great deal of data concurrency [2]. The CUDA library is a good tool for teaching students how to develop parallel programs based on data concurrency – complex mathematical calculations for matrices. The disadvantages include the availability of expensive equipment – graphics cards and the need to combine with other technologies to develop a program for distributed computing.

Some teachers [27] use the Java programming language to simplify learning the basic principles and paradigms of parallel programming. As the author points out, Java's programming language advantages over C or Fortran include higher-level programming concepts, improved compile-time and runtime checking, resulting in faster problem detection and debugging. In addition, Java's automatic garbage collection during operation saves the programmer from many lower-level language pitfalls. Built-in thread support provides a way to insert concurrency into Java applications. The Java Development Kit (JDK) includes a large set of libraries that developers can use to quickly develop applications. Another interesting argument in favor of Java is its large pool of developers – the main reason is that Java is taught as one of the main languages in many universities around the world [27]. The disadvantages of learning parallel and distributed computing in the Java programming language can be attributed to the much slower speed of running parallel programs than when used to develop the C++ programming language.

One author explores the use of Python programming language to teach students parallel programming [19]. The author points out that Python is gaining popularity in academia as the best language for teaching beginners to consistent programming. Python syntax is clean, easy and easy to understand. At the same time, it is a high-level programming language that supports the paradigms of many programs, such as imperative, functional and object-oriented. Therefore, by default, it's almost obvious to think that Python is also the right language for teaching parallel programming paradigms. In conclusion, A. Marowka's study concludes: Python is still not mature enough to teach concurrent programming. It does not support adequate multithreaded programming, which is the main paradigm of parallel programming today. However, it has relevant modules that support the messaging paradigm and heterogeneous programming. Moreover, Python has drawbacks that make it an inappropriate language for training inexperienced programmers, such as the lack of visual debuggers and profiles [19].

It should be noted separately several experimental studies on the implementation in the parallel programming of integrated programming environments or programming languages developed by the authors of the studies themselves [12, 23, 28]. Y. Wepathana states that students should consider different concurrent architectures and programming models in their learning process. To achieve this, it is advisable to use an integrated system that has different concurrent architectures and supporting programming languages. The process of developing such environments or programming languages is still ongoing, and working with them, despite the student's learning success, has many limitations, as the authors of the research note.

Consider the authors' vision of the process of organizing learning by parallel and distributed computing. For this purpose, we will put the following questions in the articles: the number of subjects of the curriculum on which parallel and distributed calculations are taught or required to be taught; what course you need to study concurrent programming; what are the directions of parallel programming. Let us summarize these questions into a table.

Table 2. Font styles for a reference to a journal article.

	Author, year of publication	Number of items	What course to study	Computer science only
1	Adams, 2013	one	not specified	Yes
2	Anderson, 2010	one	magistracy	Yes
3	Arroyo, 2013	few	at all	Yes
4	Breuer, 2012	one	first	Yes
5	Cesar, 2015	one	magistracy	No
6	Delistavrou, 2011	one	not specified	No
7	Eijkhout, 2018	one	not specified	Yes
8	Franczak,	one	not specified	Yes
9	Gardner, 2017	one	not specified	Yes
10	Gregg, 2012	one	school	
11	Grossman,2017	one	second	Yes
12	Kuhail, 2018	one	not specified	Yes
13	Lin, 2013	few	at all	Yes
14	Marowka, 2008	one	third	Yes
15	Marowka, 2017	one	not specified	Yes
16	Matthews, 2018	one	not specified	Yes
17	Mosin, 2018	one	not specified	Yes
18	Muresano, 2010	one	not specified	Yes
19	Ontañón, 2017	one	not specified	Yes
20	Paprzycki, 2006	one	first	Yes
21	Prasad, 2018	one	not specified	Yes
22	Shafi, 2014	one	not specified	Yes
23	Wepathana, 2015	one	not specified	Yes
24	Wilkinson, 2013	one	not specified	Yes
25	Wilkinson, 2016	few	first and second	Yes
26	Younis, 2019	one	third	Yes

Concerning the question, it is necessary to consider parallel and distributed calculations of more authors (26 out of 29), which should be enclosed in one subject. It is only the authors that need to develop "parallel thinking" in all streams of all undergraduate courses [3].

More real authors do not change in explicit age for any course, it is necessary to hang up the objects that exist with parallel and different exchanges. There are no others who share the same opinion: from the beginning of the course to graduate school. One of the authors provided an example of an experiment in basic parallel high school programming [11].

In the third question, we try to execute more authors (26 out of 29) by conducting experiments in their own fields of work.

3 Conclusions

At present, the main method of teaching parallel and distributed computing is a laboratory course, where each work studies a separate topic or paradigm of parallel programming, but to improve teaching, teachers try to implement a design method in the laboratory.

The main tool for teaching parallel and distributed computing is a lab course with a C programming language and MPI library. But many teachers find that the use of such tools leads to the loss of interest in parallel programming by students, so many authors publish the results of experiments to improve learning by means of developing their own environments or the organization of cluster computing systems.

References

1. J. Adams, R. Brown, E. Shoop, Patterns and Exemplars: Compelling Strategies for Teaching Parallel and Distributed Computing to CS Undergraduates, in *2013 IEEE International Symposium on Parallel & Distributed Processing, Workshops and Phd Forum* (2013). doi:10.1109/IPDPSW.2013.275
2. N. Anderson, J. Mache, W. Watson, Learning CUDA: Lab Exercises and Experiences, in *OOPSLA '11* (2010), pp. 201–202. doi:10.1145/2048147.2048206
3. M. Arroyo, Teaching Parallel and Distributed Computing to Undergraduate Computer Science Students, in *2013 IEEE International Symposium on Parallel & Distributed Processing, Workshops and Phd Forum* (2013). doi:10.1109/IPDPSW.2013.276
4. S. A. Bogaerts, One step at a time: Parallelism in an introductory programming course. Journal of Parallel and Distributed Computing **105**, 4–17 (2017). doi:10.1016/j.jpdc.2016.12.024
5. R. Brown, Strategies for preparing computer science students for the multicore world, in *ITiCSE-WGR '10*. June 2010, pp. 97–115. doi:10.1145/1971681.1971689
6. A. Breuer, M. Bader, Teaching Parallel Programming Models on a Shallow-Water Code, in *2012 11th International Symposium on Parallel and Distributed Computing*. doi:10.1109/ISPDC.2012.48
7. E. Cesar, A. Cortés, A. Espinosa, T. Margalef, J. C. Moure, A. Sikora, R. Suppi, Teaching Parallel Programming in Interdisciplinary Studies, in *Euro-Par 2015: Parallel Processing Workshops*. LNCS 9523 (2015), pp. 66–77. doi:10.1007/978-3-319-27308-26
8. C. T. Delistavrou, K. G. Margaritis, Towards an Integrated Teaching Environment for Parallel Programming, in *2011 15th Panhellenic Conference on Informatics*. doi: 10.1109/PCI.2011.16
9. V. Eijkhout, Teaching distributed memory programming from mental models, in *2011 15th Panhellenic Conference on Informatics*, p. 107. doi:10.1109/PCI.2011.16
10. T. Franczak, A. Nkansah, T. Marrinan, M. E. Papka, A Path from Serial Execution to Hybrid Parallelization for Learning HPC, in *Proceedings of the 2017 Workshop on Education for High-Performance Computing*. doi:10.1145/1734263.1734339
11. W. B. Gardner, Should We Be Teaching Parallel Programming?, in *WCCCE '17*. doi:10.1145/3085585.3085588
12. C. Gregg, L. Tychonievich, J. Cohoon, K. Hazelwood, J. EcoSim: A language and experience teaching parallel programming in elementary school, in *SIGCSE '12*, February 2012, pp. 51–56. doi:10.1145/2157136.2157155
13. M. Grossman, M. Aziz, H. Chi, A. Tibrewal, S. Imam, V. Sarkar, Pedagogy and tools for teaching parallel computing at the sophomore undergraduate level. J. Parallel Distrib. Comput. **105**, 18–30 (2017). doi:10.1016/j.jpdc.2016.12.026
14. Y. Ko, B. Burgstaller, B. Scholz, Parallel from the beginning: The case for multicore programming in the computer science undergraduate curriculum, in *SIGCSE '13*, pp. 415-420. doi:10.1145/2445196.2445320
15. M. A. Kuhail, S. Cook, J. W. Neustrom, P. Rao, Teaching Parallel Programming with Active Learning. IPDPSW. (2018). doi:10.1109/IPDPSW.2018.00069
16. A. Liberati, D. G. Altman, J. Tetzlaff, C. Mulrow, The PRISMA statement for reporting systematic reviews and metanalyses of studies that evaluate health care interventions: explanation and elaboration. Journal of Clinical Epidemiology **62**, 1–34 (2009). doi:10.1016/j.jclinepi.2009.06.006
17. H. Lin, Teaching Parallel and Distributed Computing Using a Cluster Computing Portal, in *2013 IEEE International Symposium on Parallel & Distributed Processing, Workshops and Phd Forum*. doi:10.1109/IPDPSW.2013.35
18. A. Marowka, Think Parallel: Teaching Parallel Programming Today. IEEE Distributed Systems Online **9**(8) (2008). doi:10.1109/MDSO.2008.24
19. A. Marowka, On parallel software engineering education using Python. Education and Information Technologies. (2018). doi:10.1007/s10639-017-9607-0
20. S. J. Matthews, J. C. Adams, R. A. Brown, E. Shoop, Portable Parallel Computing with the

Raspberry Pi, in *SIGCSE '18*, February 2018, pp. 92–97. doi:10.1145/3159450.3159558

21. S. Mosin, N. Pleshchinskii, I. Pleshchinskii, D. Tumakov, Technique for Teaching Parallel Programming via Solving a Computational Electrodynamics Problems. RuSCDays 2018: Supercomputing (2018). doi:10.1016/j.jpdc.2018.02.023
22. R. Muresano, D. Rexachs, E. Luque, Learning parallel programming: a challenge for university students. Procedia Computer Science **1**, 875–883 (2010). doi:10.1016/j.procs.2010.04.096
23. S. Ontañón, B. Char, J. Zhu, E. Freed, Designing Visual Metaphors for an Educational Game for Parallel Programming, in *CHI EA '17*. doi:10.1145/3027063.3053253
24. K. Osadcha, O. Sysoieva, Condition, technologies and prospects of distance learning in the higher education of Ukraine. Information technologies and learning tools **70**, 271–284 (2019)
25. M. Paprzycki, Education: Integrating Parallel and Distributed Computing in Computer Science Curricula. IEEE distributed systems online **7**, 6 (2006). doi:10.1109/MDSO.2006.9
26. S. K. Prasad, A. Gupta, A. L. Rosenberg, A. Sussman, C. Weems, Topics in Parallel and Distributed Computing: Introducing Algorithms, Programming, and Performance within Undergraduate Curricula (2018). doi:10.1007/978-3-319-93109-8
27. A. Shafi, B. Carpenter, Teaching Parallel Programming Using Java, in *EduHPC '14*, November 2014, pp. 56–63. doi:10.1109/EduHPC.2014.7
28. Y.M.R.D. Wepathana, G. Anthonys, L.S.K. Udugama, Compiler for a simplified programming language aiming on Multi Core Students' Experimental Processor, in *ICIIS* (2015). doi:10.1109/ICIINFS.2015.7399025
29. B. Wilkinson, J. Villalobos, C. Ferner, Pattern Programming Approach for Teaching Parallel and Distributed Computing, in *SIGCSE '13*, March 2013, pp. 409–414. doi:10.1145/2445196.2445319
30. B. Wilkinson, C. Ferner, The Suzaku Pattern Programming Framework, in *IPDPS Workshops 2016*, pp. 978–986
31. A. A. Younis, R. Sunderraman, M. Metzler, A. G. Bourgeois, Case Study: Using Project Based Learning to Develop Parallel Programing and Soft Skills, in *IPDPSW* (2019). doi:10.1109/IPDPSW.2019.00059

Using spreadsheets as learning tools for computer simulation of neural networks

Serhiy Semerikov[1,*], *Illia* Teplytskyi[1], *Yuliia* Yechkalo[2], *Oksana* Markova[2], *Vladimir* Soloviev[1], and *Arnold* Kiv[3]

[1]Kryvyi Rih State Pedagogical University, 54, Gagarina Ave., Kryvyi Rih, 50086, Ukraine
[2]Kryvyi Rih National University, 11, Vitali Matusevich St., Kryvyi Rih, 50027, Ukraine
[3]Ben-Gurion University of the Negev, Beer Sheva, Israel

Abstract. The article substantiates the necessity to develop training methods of computer simulation of neural networks in the spreadsheet environment. The systematic review of their application to simulating artificial neural networks is performed. The authors distinguish basic approaches to solving the problem of network computer simulation training in the spreadsheet environment, joint application of spreadsheets and tools of neural network simulation, application of third-party add-ins to spreadsheets, development of macros using the embedded languages of spreadsheets; use of standard spreadsheet add-ins for non-linear optimization, creation of neural networks in the spreadsheet environment with-out add-ins and macros. The article considers ways of building neural network models in cloud-based spreadsheets, Google Sheets. The model is based on the problem of classifying multi-dimensional data provided in "The Use of Multiple Measurements in Taxonomic Problems" by R. A. Fisher. Edgar Anderson's role in collecting and preparing the data in the 1920s-1930s is discussed as well as some peculiarities of data selection. There are presented data on the method of multi-dimensional data presentation in the form of an ideograph developed by Anderson and considered one of the first efficient ways of data visualization.

1 Introduction

The Fourth Industrial Revolution (Industry 4.0) has become a system-related challenge for the scientific community [46]. Industry 4.0 is primarily characterized by evolution and convergence of nano-, bio-, information and cognitive technologies to enhance high quality transformations in economic, social, cultural and humanitarian spheres. Professionals dealing with development and introduction of the sixth techno-logical paradigm technologies determine to a great extent whether our country is able to ride the wave of Industry 4.0 innovations. Therefore, extensive implementation of ICT is a top priority of Ukraine's higher education updating in order to form a professionally competent specialist able to ensure the country's innovative development.

According to the Decree of the Cabinet of Ministers of Ukraine "Certain issues of specifying medium-term priorities of the national-level innovative activity for 2017-2021" (2016), developing modern ICT and robotics, particularly cloud technologies, computer training systems and technologies of mathematical informatics (intellectual simulation, informational security, long-term data storage and "big data" management, artificial intelligence systems) are nationally and socially important directions of the innovative activity [28; 29]. The Decree of the Cabinet of Ministers of Ukraine "Certain issues of specifying medium-term priorities of the sectoral-level innovative activity for 2017-2021" (2017) specifies that these directions accompanied by smart web-technologies and cloud computing make the basis for creating and defining themes for scientific researches and technical (experimental) developments as well as for forming the state order of training ICT specialists.

For the past 25 years, the authors have been developing the concept of systematic computer simulation training at schools and teachers' training universities. The concept ideas have been generalized and presented in the textbook [53]. Spreadsheets are chosen to be the leading environment for computer simulation training, their application discussed in articles [47]. Using spreadsheet processors (autonomous, integrated and cloud-based) as examples, the authors demonstrate components of teaching technology of computer simulation of determined and stochastic objects and processes of various nature. The systematic training of simulation provides for changing and integrating simulation environments ranging from general (spreadsheets) to specialized subject-based ones. While teaching computer simulation of intellectual systems specialized languages and programming environments are traditionally used. They can be easily mastered by first-year students [1]. One of the most wide-spread languages, Scheme, is offered to be applied to teaching computer simulation of classical mechanics at universities [51]. Extensive application of artificial intelligence in everyday life calls for students' early acquaintance with its models

* Corresponding author: semerikov@gmail.com

and methods including neural network-based while teaching informatics at secondary schools [48]. It conditions the need for developing training methods of computer simulation of neural networks in the general-purpose simulation environment, i.e. spreadsheets.

2 Literature review and problem statement

The first description of spreadsheet application to teaching neural network simulation of visual phenomena dates back to 1985 and belongs to Thomas T. Hewett, Professor of the Department of Psychology of Drexel University [17]. In [16] there are de-scribed simple models of microelectrode recording of two neuron types of neural activity – receptors and transmitters localized in two brain-hemispheres. Thomas T. Hewett offered psychology students to independently choose coefficients of intensifying or reducing input impulses to achieve the desired output: "... the simulations can be designed in such a way that the student is able to "experiment" with a simulation-experiment both in the sense of discovering the characteristics of an unknown model and in the sense of modifying various components of a known model to see how the simulation is affected" [16, p. 343]. This approach implies simultaneous studying a neural network and understanding its functioning as psychology students conclude the laws of the neural impulse spread by applying the trial-and-error method.

In his article [8], James J. Buergermeister, Professor of Hospitality and Tourism Management of University Wisconsin-Stout, associates electronic spreadsheet application with basic principles of training technology and methods of data processing (Fig. 1). The author does not work out the methods of applying electronic spread-sheets to neural network simulation in detail, yet, the presented scheme reveals such basic steps as data obtainment, semantic coding, matching with an etalon, etc.

Since 1988, Murray A. Ruggiero, one of the pioneers of autotrading, has been developing Braincel, an application for Microsoft Excel 2.1C, which is a set of twenty macros to solve tasks of image recognition by artificial neural network tools [23]. At the beginning of 1991, Murray A. Ruggiero received a patent "Embedding neural networks into spreadsheet applications" [45], which describes an artificial neural net-work with a plurality of processing elements called neurons arranged in layers. They further include interconnections between the units of successive layers. A network has an input layer, an output layer, and one or more "hidden" layers in between, necessary to allow solutions of non-linear problems. Each unit (in some ways analogous to a biological neuron: dendrites – input layer, axon – output layer, synapses – weights [43], soma – summation function) is capable of generating an output signal which is determined by the weighted sum of input signals it receives and an activation function specific to that unit. A unit is provided with inputs, either from outside the network or from other units, and uses these to compute a linear or non-linear output. The unit's output goes either to other units in subsequent layers or to outside the network. The input signals to each unit are weighted by factors derived in a learning process.

When the weight and activation function factors have been set to correct levels, a complex stimulus pattern at the input layer successively propagates between the hid-den layers, to result in a simpler output pattern. The network is "taught" by feeding it a succession of input patterns and corresponding expected output patterns. The net-work "learns" by measuring the difference at each output unit between the expected output pattern and the pattern that it just produced. Having done this, the internal weights and activation functions are modified by a learning algorithm to provide an output pattern which most closely approximates the expected output pattern, while minimizing the error over the spectrum of input patterns. Neural network learning is an iterative process involving multiple lessons. Neural networks have the ability to process information in the presence of noisy or incomplete data and yet still generalize to the correct solution.

In his patent, Murray A. Ruggiero details a network structure (multi-level), an activation function (sigmoidal), a coding method (polar), etc. He presents a mathematical apparatus for network training and determines a method of data exchange between a spreadsheet processor nucleus and an add-in to it. The patent author suggests storing input data in columns, maximum and minimum values for each column of input data, the number of learning patterns. Data can be normalized or reduced to the polar range [0; 1] both in spreadsheets and add-ins.

In his article of 1989, Paul J. Werbos, the pioneer of the backpropagation method for artificial neural network training [55] demonstrates how to make the corresponding mathematical apparatus simpler to use it directly in the spreadsheet processor. The cycling character of training is supported by a macro that exchanges data between lines to avoid restrictions on the number of iterations because of the limited number of lines on a sheet of a separate spreadsheet. Some other authors suggest applying a similar approach of macros application [14; 57].

The authors of [24] in Chapter 2 "Neural Nets in Excel" give an example of applying the non-linear optimization tool, Microsoft Excel Solver, to forecasting stock prices using the "grey-box" concept, in which the model is evident, yet, the details of its realization are hidden.

In their article of 1998 [15], Tarek Hegazy and Amr Ayed from the Department of Civil Engineering at University of Waterloo distinguish the corresponding steps. Unlike [44], the authors suggest using bipolar data normalization (over the range of [–1; 1]) and a hyperbolic tangent as an activation function. Three addins for Microsoft Excel are used to determine weighting factors – the standard Solver and third-party add-ins (NeuroShell2 and GeneHunter by Ward Systems Group). Experiment results reveal that the best result is provided by the optimizing general-purpose tool (Solver) and not by specialized ones. In spite of the fact that "Journal of Construction Engineering and Management" does not refer to educational editions, the article [15] and the paper [7] by their structure and focus on details can be considered the first description of methodic of using spreadsheets for neural network simulation.

In their article of 2012 [43], Thomas F. Rienzo and Kuriakose K. Athappilly from Haworth College of Business at Western Michigan University consider model illustrating the process of machine learning as networks examine training data would provide another. Authors incorporate the stepwise learning processes of artificial neural network in a spreadsheet containing (1) a list or table of training data for binary input combinations, (2) rules for target outputs, (3) initial weight factors, (4) threshold values, (5) differences between target outputs and neural network transformation values, (6) learning rate factors, and (7) weight adjustment calculations. Unlike the previous ones, this model is invariant to the spreadsheet and does not call for applying any third-party addins.

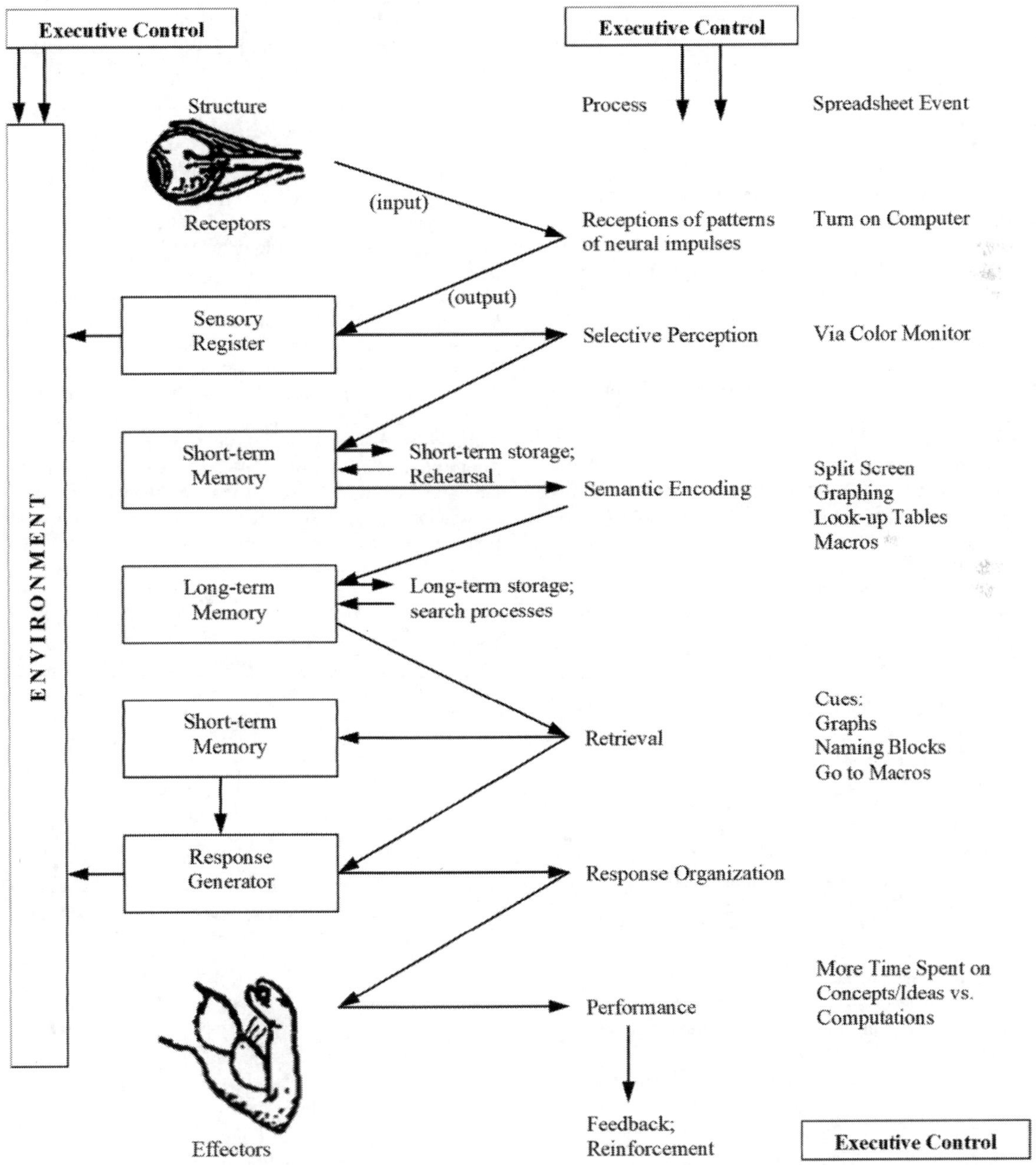

Fig. 1. The information-processing model using spreadsheet events (according to [8])

In [28] the role of neural network simulation in the training content of the special course "Foundations of Mathematical Informatics" is discussed. The course is developed for students of technical universities (future IT-specialists) and aimed at breaing a gap between theoretic computer science and its practical application to software, system and computing engineering. CoCalc is justified as a training tool for mathematical informatics in general and neural network modelling in particular. The elements of CoCalc techniques for studying the topic "Neural network and pattern recognition" within the special course "Foundations of Mathematical Informatics" are shown.

The authors of [47] distinguish basic approaches to solving the problem of network computer simulation training in the spreadsheet environment, joint application of spreadsheets and tools of neural network simulation. In [48], there are opportunities for applying spreadsheets to introducing essentials of machine learning [31] at secondary and higher school as well as some elements of their application to solving problems of pattern classification. Thus, using spreadsheets as a tool for teaching basics of machine learning creates conditions for early and simultaneously deeper mastering of corresponding models and methods of mathematical informatics [2].

The conducted review makes it possible to find the following solutions of the problem of computer simulation teaching to neural networks in the spreadsheet environment:

- joint application of spreadsheets and neural network tools [32], in which data is exported to the unit calculating weighting factors imported to spreadsheets and used in calculations;
- application of third-party add-ins for spreadsheets ([15; 23; 45]), according to which structured spreadsheet data is processed in the add-in, calculation results are arranged in spreadsheet cells;
- macros development ([7; 14; 55; 57]) enables direct software control over neural network training and creation of a user's specialized interface;
- application of standard add-ins for optimization ([15; 24]) calls for transparent network realization and determination of an optimization criterion (minimization of a squared deviation total of the calculated and etalon outputs of the network);
- creation of neural networks in the spreadsheet environment without add-ins and macros [43] requires transparent realization of a neural network with evident de-termination of each step of adjustment of its weighting factors.

The advantage of the first solution is its flexibility as one can choose any relevant combinations of the simulation environments, yet, their integration level is usually insufficient. The closed character of the second solution and its binding to a certain software platform make it relevant to be applied to solving various practical tasks and irrelevant for neural network simulation training as a network becomes a black box for a user. The fourth solution is partially platform-dependent as a neural network becomes a grey box for a user. The final solution is totally mobile and offers an opportunity to regard the model as a white box, thus making it the most relevant for initial mastering of neural network simulation methods.

3 The aim and objectives of the study

The research is aimed at considering mathematical models of neural networks realized in spreadsheet environment. To accomplish the set goal, the following tasks are to be solved: (1) to study historical models of neural networks; (2) to distinguish learning tools of computer simulation of neural networks in the spreadsheet environment; (3) to substantiate the chosen dataset to develop a model; (4) to develop a demonstration model of an artificial neural network using cloud-based spreadsheets.

4 Early neural networks models: from William James to Walter Pitts

Russell C. Eberhart and Roy W. Dobbins [12] suggest dividing the history of artificial network development into four stages. The first stage, the Age of Camelot, starts with "The Principles of Psychology" (1890) by the American psychologist, William James, who formulates the elementary law of association: "When two elementary brain processes have been active together or in immediate succession, one of them, on reoccurring, tends to propagate its excitement into the other" [22, p. 566]. The elementary law of association (the elementary principle) is closely related to the concepts of associative memory and correlational learning. In the authors' opinion [12], James seemed to foretell the notion of a neuron's activity being a function of the sum of its inputs, with past correlation history contributing to the weight of interconnections: "The amount of activity at any given point in the brain-cortex is the sum of the tendencies of all other points to discharge into it, such tendencies being proportionate (1) to the number of times the excitement of each other point may have accompanied that of the point in question; (2) to the intensity of such excitements; and (3) to the absence of any rival point functionally disconnected with the first point, into which the discharges might be diverted" [22, p. 567].

In "Psychology" (1892), an abridged re-edition of "The Principles of Psychology", James formulates basic principles of the image recognition theory: "We know, in short, a lot about it, whilst as yet we have no acquaintance with it. Our perception that one of the objects which turn up is, at last, our qucesitum, is due to our recognition that its relations are identical with those we had in mind, and this may be a rather slow act of judgment. Every one knows that an object may be for some time present to his mind before its relations to other matters are perceived. Just so the relations may be there before the object is." [21, p. 275].

"The Bulletin of Mathematical Biophysics" has been an advanced platform for approbating network models and methods since the moment of its foundation by Nicolas Rashevsky [11]. It should be no surprise as Rashevsky invented one of the first models of the neuron [40] and started the idea of artificial neural networks. The basic idea was to use a pair of linear differential equations and a nonlinear threshold operator:

$$\text{Input} = I(t)$$
$$\begin{cases} \dfrac{\partial e}{\partial t} = AI(t) - ae \\ \dfrac{\partial j}{\partial t} = BI(t) - bj \end{cases}$$
$$\text{Output} = H(e - j - \theta)$$

where θ is the threshold, e and j could represent excitation and inhibition or the amount or concentration of two substances within a neuron, $H(x)$ is the Heaviside operator (takes positive values to 1, and non-positive values to 0). This gives an easy way to model the all-or-none firing of a neuron – Rashevsky showed that this simple model was able to model many of the known experimental results for the behavior of single neurons. He also made the point that networks of these model neurons could be connected to give quite complicated behavior and even serve as a model for a brain [11].

In his article of 1941 [56], Gale J. Young shows that the Rashevsky's two-factor model of nerve excitation can account for sustained inhibition or enhancement by a sequence of stimulus pulses, and for the decrease in the reinforcement period with each successive pulse of the train.

Developing Rashevsky's ideas, his student Alston Scott Householder, who gave his name to the known linear transformation describing a reflection about a plane or hyperplane containing the origin, and a class of root-finding algorithms used for functions of one real variable with continuous derivatives up to some order, in his article of 1940 [19], suggests a parameter measuring the "strength" of the inhibitory neurons acting among the terminal synapses. In [20], he describes the activity parameter as a characteristic of the fiber which is assumed to be different from zero, but it may be either positive (when the fiber is excitatory in character) or negative (when the fiber is inhibitory in character).

Thus, at the beginning of 1942, the theory of biological neural networks based on Rashevsky's continuous two-factor model was created and intensively developed. As remembered by J. A. Anderson and E. Rosenfeld, at the boundary of two decades, Walter Pitts was introduced to Nicolas Rashevsky by Rudolf Carnap, and accepted in to his mathematical biology group [10]. In his early publication, Pitts suggests "a new point of view in the theory of neuron networks is here adumbrated in its relation to the simple circuit: it is shown how these methods enable us to extend considerably and unify previous results for this case in a much simpler way" [36, p. 121]. With due consideration of Householder's articles, Pitts determines the total conduction time of a fiber as the sum of its conduction time and the synaptic delay at the postliminary synapse. Pitts was the first to use spreadsheet abstraction and discrete description of neural network functioning by determining a corresponding algorithm: Pitts was the first to use spreadsheet abstraction and discrete description of neural network functioning by determining a corresponding algorithm: "The excitation-pattern of [neural circuit] C may be described in a *matrix* E, of n rows and an infinite number of columns, each of whose elements e_{rs} represents the excitation at the synapse s_r during the interval $(s, s+1)$. The successive entries in the excitation *matrix* E may be computed recursively from those in its first column – these are the quantities λ_r – by the following rule, whose validity is evident: Given the elements of the p-th column, compute those of the p+l-st thus: if the element e_{ip} is negative or zero, place σ_{i+1} in the i+l-st row and p+1-st column, or in the first row of the p+l-st column if $i=n$. Otherwise put $\sigma_{i+1}+a_i e_{ip}$, in this place. We shall say that C is in a *steady-state* during a series of n intervals $(s, s+1)$, ..., $(s+n-1, s+n)$ if, for every p between s and $s+n$, the p-th and $p+n$-th columns of E are identical. If s is the smallest integer for which this is the case, we shall say that the steady state begins at the interval $(s, s+1)$" [36, pp. 121–122]. The suggested algorithm describes a parallel neural network [36, p. 122]. Rather than analyzing the steady-state activity of networks, Pitts was more concerned with initial nonequilibrium cases, and how a steady state could be achieved [2, p. 18].

The results provided by Pitts in his articles on the linear theory of neuron networks (the static problem [38] and the dynamic problem [37]), enabled him to draw two essential conclusions: (1) it is possible to find a set of independent networks each of which consists of n simple circuits with one common synapse (*rosettes*), such that network arises by running chains from the centers of the rosettes to various designated points outside: but none back, so that the state of the whole network is determined by the states of the separate rosettes independently – Pitts calls networks of this kind *canonical networks* [37, p. 29]; (2) given any finite network, it is possible to find a set of independent rosettes such that the excitation function of network for every region is a linear combination of those of the rosettes – i. e., we can reduce any network to a canonical network having the same excitation function [37, p. 31]. Thus, in his article of 1943, Pitts solves the inverse network problem, "which is, given a preassigned pattern of activity over time, to construct when possible a neuron-network having this pattern" [37, p. 23] by allowing creating problem-oriented neural networks. Tara H. Abraham indicates that adopting Householder's model of neural excitation, Pitts develops a simpler procedure for the mathematical analysis of excitatory and inhibitory activity in a simple neuron circuit, and aimed to develop a model applicable to the most general neural network possible [2].

"Psychometrika", the official journal of the Psychometric Society (both founded in 1935 by Louis Leon Thurstone, Edward Lee Thorndike and Joy Paul Guilford), is devoted to the development of psychology as a quantitative rational science. It has become another mouthpiece of Nicolas Rashevsky and his students, whose articles examine statistical methods, discuss mathematical techniques, and advance theory for evaluating behavioral data in psychology, education, and the social and behavioral sciences generally. Pitts's article "A general theory of learning and conditioning" has been published in this journal. Part I [34] deals only with the case where the stimuli and responses are wholly independent, so that transfer and generalization do not occur, and proposes a law of variation for the reaction-tendency, which takes into ac-count all of classical conditioning and the various sorts of inhibition affecting it. Part II [35] extends a mathematical theory of non-symbolic learning and conditioning, still under the hypothesis of complete independence, to cases where reward and punishment are involved as motivating factors. The preceding results are generalized to the case where stimuli and responses are related psychophysically, thus constituting a theory of transfer, generalization, and discrimination.

Another article of 1943, "A logical calculus of the ideas immanent in nervous activity" [30], published again in "Bulletin of Mathematical Biophysics", has resulted from cooperation of Warren Sturgis McCulloch and Walter Pitts and is considered one of the most famous papers on artificial neural networks. They stated five physical assumptions for nets without circles [30, p. 118]:

1. The activity of the neuron is an "all-or-none" process [any nerve has a finite threshold and the intensity of excitation must exceed this for production of excitation – once produced, the excitation proceeds independently of the intensity of the stimulus].
2. A certain fixed number of synapses must be excited within the period of latent addition [time during which the neuron is able to detect the values present on its in-puts, the synapses – typically less than 0.25 msec] in order to excite a neuron at any time, and this number is independent of previous activity and position on the neuron.
3. The only significant delay within the nervous system is synaptic delay [time delay between sensing inputs and acting on them by transmitting an outgoing pulse, – typically less than 0.5 msec].
4. The activity of any inhibitory synapse absolutely prevents excitation of the neuron at that time.
5. The structure of the net does not change with time.

The neuron described by these five assumptions is known as the McCulloch-Pitts neuron [12, p. 17]. In the same way as propositions in propositional logic can be "true" or "false," neurons can be "on" or "off" – they either fire or they do not: this formal equivalence allowed them to argue that the relations among propositions can correspond to the relations among neurons, and that neuronal activity can be represented as a proposition [29, p. 19].

In [30], there is a set of theorems that "does in fact provide a very convenient and workable procedure for constructing nervous nets to order, for those cases where there is no reference to events indefinitely far in the past in the specification of the conditions" [30, pp. 121–122]. McCulloch and Pitts appear to be the first authors since William James to describe a massively parallel neural model. The theories they developed were important for a number of reasons, including the fact that any finite logical expression can be realized by networks of their neurons.

Combining simple "logical" neurons in chains and cycles, the authors show that the brain is able to perform any logical operation and arbitrary logical calculations. The paper is essential for developing computing machines as it allows creating a universal computer operating with logical expressions (in the hands of John von Neumann, the McCulloch-Pitts model becomes the basis for the logical design of digital computers [11, p. 180]): "It is easily shown: first, that every net, if furnished with a tape, scanners connected to afferents, and suitable efferents to perform the necessary motor-operations, can compute only such numbers as can a Turing machine; second, that each of the latter numbers can be computed by such a net; and that nets with circles can be computed by such a net; and that nets with circles can compute, without scanners and a tape, some of the numbers the machine can, but no others, and not all of them. This is of interest as affording a psychological justification of the Turing definition of computability and its equivalents, Church's λ-definability and Kleene's primitive recursiveness: If any number can be computed by an organism, it is computable by these definitions, and conversely." [30, pp. 121–122].

In the same issue of "Bulletin of Mathematical Biophysics", in which [30] was published, Herbert Daniel Landahl (the first doctoral student in Rashevsky's mathematical biology program at the University of Chicago, who became the second President of the Society for Mathematical Biology in 1981), Warren Sturgis McCulloch and Walter Pitts published a short (about 3 pages), yet essential addition [25], suggesting a method for converting logical relations among the actions of neurons in a net into statistical relations among the frequencies of their impulses. In the presented theorem, they detailed transition from Boolean calculations (in "true" and "false") to probabilistic ones (numbers within [0; 1]): the conjunction sign ∨ is replaced by +, the disjunction sign (single dot) is replaced by ×, negation ~ is replaced by «1 –», etc. The correspondence expressed by this theorem connects the logical calculus of the [30] with previous treatments of the activity of nervous nets in mathematical biophysics and with quantitatively measurable psychological phenomena.

The monograph by Householder and Landahl "Mathematical Biophysics of the Central Nervous System" has become a kind of conclusion of the discussed period [18]. In Paul Cull's opinion, there is no unambiguous answer to the question which model is better, the Rashevsky continuous model or the McCulloch-Pitts discrete model: "For some purposes, one model is better, but for other purposes, the other model is better. Rashevsky and Landahl were quick to notice, that in physics, one often averaged over a large set of discrete events to obtain a continuous model which represented the large scale behavior of a system, and so they posited that the continuous neuron model might be suitable for modeling whole masses of neurons even if each individual neuron obeyed a discrete model. In the hands of Householder and Landahl, this observation led to the idea of modeling psychological phenomena by neural nets with a small number of continuous model neurons. In particular, they found that the cross-couple connection [Fig. 2] was extremely useful. For such problems as reaction time, enhancement effects, flicker phenomena, apparent motion, discrimination and recognition, they were able to fit these models to experimental data and to use their models to predict phenomena that could be measured and verified" [11, p. 180].

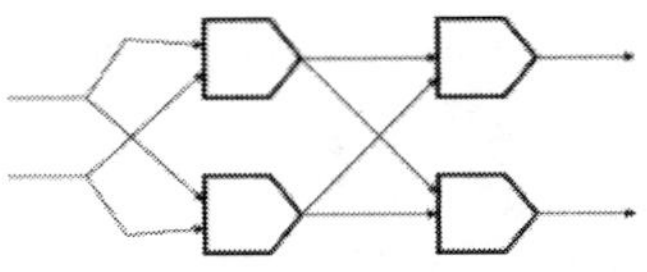

Fig. 2. This cross-couple connection of four neurons is capable of modeling a large number of phenomena (according to [46])

In 1945, Rashevsky wrote about [30] and [25]: "authors show that by applying logical calculus, it is possible to construct any complicated network having given properties. One could attempt to construct by the method of McCulloch and Pitts a network that would represent all modes of logical reasoning, and then apply the usual methods of mathematical biophysics to derive some quantitative relations between different manifestations of the processes of logical thinking" [39, p. 146]. "It seems somewhat awkward to have to construct by means of Boolean algebra first a "microscopic circuit" and then obtain a simpler one by a transition to the "macroscopic" picture. We should expect that a generalization of the application of Boolean algebra should be possible so as to permit its use for the construction of networks in which time relations are of a continuous rather than of a quantized, nature" [41, p. 211].

Rashevsky intensively develops the apparatus created by McCulloch and Pitts in his further papers. In [42] a theory of such neural circuits is developed which provide for formal logical thinking. Predicate apparatus application enables Rashevsky synthesizing huge neural networks from single-type fundamental elements of McCulloch-Pitts.

Telson Wei develops another approach to matrix representation of a neural net-work [54]. The structure of a complete or incomplete neural net is represented here by several matrices: the intensity matrix *E*, the connection matrix *D*, the structural matrix *T*, the diagonal inverse threshold-matrix *H*, and activity vector ***a*** from [26; 27].

In their paper [49], Alfonso Shimbel and Anatol Rapoport (pioneered in the modeling of parasitism and symbiosis, researching cybernetic theory) develop a probabilistic approach to the theory of neural nets: neural nets are characterized by certain parameters which give the probability distributions of different kinds of synaptic connections throughout the net. In their further papers, they consider steady states in random nets and contribution to the probabilistic theory of neural nets: randomization of refractory periods and of stimulus intervals, facilitation and threshold phenomena, specific inhibition and various models for inhibition.

The last joint article by Pitts and McCulloch, "How we know universals the perception of auditory and visual forms", in "Bulletin of Mathematical Biophysics" came out in 1947. "Numerous nets, embodied in special nervous structures, serve to classify information according to useful common characters. In vision they detect the equivalence of apparitions related by similarity and congruence, like those of a single physical thing seen from various places. In audition, they recognize timbre and chord, regardless of pitch. The equivalent apparitions in all cases share a common figure and define a group of transformations that take the equivalents into one another but pre-serve the figure invariant. So, for example, the group of translations removes a square appearing at one place to other places; but the figure of a square it leaves invariant. ... We seek general methods for designing nervous nets which recognize figures in such a way as to produce the same output for every input belonging to the figure. We endeavour particularly to find those which fit the histology and physiology of the actual structure." [33, pp. 127–128]

Thus, the models and methods developed by Pitts and McCulloch have created a foundation for designing a new type of computers – neurocomputers based on human brain principles and able to solve tasks of recognizing distorted (noisy) images.

5 Edgar Anderson and his Iris data set

Edgar Shannon Anderson (November 9, 1897 – June 18, 1969) was born in Forestville, New York. According to George Ledyard Stebbins, from an early age he exhibited both superior intelligence and a great interest in plants, particularly in cultivating them and watching them grow [50, p. 4].

He went to Michigan Agricultural College at the age of sixteen, just before his seventeenth birthday, knowing already that he wanted to be a botanist. After completing his degree, he accepted a graduate position at the Bussey Institution of Harvard University. After leaving Harvard with his doctor's degree in 1922, Anderson spent nine years at the Missouri Botanical Garden, where he was a geneticist and Director of the Henry Shaw School of Gardening; at the same time he was Assistant Professor, later Associate Professor, of Botany at Washington University in St. Louis. During this period, he developed the beginnings of his highly original and effective methods for looking at and recording variation in plant populations, as well as his keen interest in the needs and progress, both scientific and personal, of students in botany. His training in genetics had given him habits of precision and mathematical accuracy in ob-serving and recording variation in natural populations that were entirely foreign to the taxonomists of that period [50, p. 5].

Through contacts with Jesse Greenman, Curator of the Garden Herbarium, he became aware of the enormous complexity and extent of the variation present in any large plant genus and of the need for understanding the origin of species as a major step in evolution. On extensive field trips he began to realize that a great amount of genetic variation exists within most natural populations of plants. This realization led him to the conclusion that "if we are to learn anything about the ultimate nature of species we must reduce the problem to the simplest terms and study a few easily recognized, well differentiated species" [6, p. 243].

He first selected *Iris versicolor*, the common blue flag, because he believed it to be clearly defined, and it was common and easily observed. Initially, this appeared to be a mistaken choice, since he soon found that *Iris versicolor* of the taxonomic manuals was actually two species, which, after preliminary analysis, he could easily tell apart. He then set himself the task of finding out, by a careful analysis of populations throughout their geographic areas, how one of these species could have evolved from the other. He recorded several morphological characters in more than 2,000 individuals belonging to 100 populations, data far more extensive than those that any botanist had yet obtained on a single species.

In order to enable these data to be easily visualized and compared, he constructed the first of his highly original and extremely useful series of simplified diagrams or ideographs (Fig. 3). By examining them, he reached the conclusion that the variation within each of his two species was of another order from the differences between them; no population of one species could be imagined as the beginning of a course of evolution toward the other. He therefore concluded that speciation in this example was not a continuation of the variation that gave rise to differences between populations of one species, and started to look for other ways in which it could have taken place. The current literature offered a possible explanation: hybridization followed by chromosome doubling to produce a fertile, stable, true-breeding amphidiploid. To apply this concept to *Iris*, he had to find a third species that would provide an alter-nate parent for one of those studied. Going to the herbarium, he found it: an undescribed variety of *Iris setosa*, native to Alaska.

All of his data, including counts of chromosome numbers, agreed with the hypothesis that *Iris versicolor* of northeastern North America had arisen as an amphiploid, one parent being *Iris virginica* of the Mississippi Valley and the Southeast Coast and the other being *Iris setosa* var. interior of the Yukon Valley, Alaska. This was one of the earliest demonstrations that a plant species can evolve by hybridization accompanied or followed by chromosome doubling. Moreover, it was the first one to show that amphiploid or allopolyploid species could be used to support hypotheses about previous distribution of species.

Anderson's research into Iris resulted in all the techniques in his later successful work, namely:
1. careful examination of individual characteristics of plants growing in nature and progeny raised in the garden;
2. reduction of this variation to easily visualized, simple terms by means of scatter diagrams and ideographs;
3. extrapolation from a putative parental species and supposed hybrids to reconstruct the alternative parent;
4. development of testable hypotheses by synthesizing data from every possible source.

The *Iris* research was Anderson's chief accomplishment during his first period at the Missouri Botanical Garden. Toward the end of this period, in 1929-1930, he received a National Research Fellowship to study in England. There he was guided chiefly by geneticist J. B. S. Haldane, but he also studied cytology under C. D. Darlington and statistics with R. A. Fisher. Haldane introduced him to the mutants of *Primula sinensis*, which he analyzed in collaboration with Dorothea De Winton. Their joint research was the first effort in plant material to relate pleiotropic gene action to growth processes.

In 1931 Anderson went to Harvard, where he stayed until 1935, as an arborist at the Arnold Arboretum. He returned to the Missouri Botanical Garden in 1935 and remained there for the rest of his life. Returning to his study of the genus Iris, he and several students analyzed a complex variation pattern of populations found in the Mississippi delta region [5].

Anderson integrated his new experience with past memories, popular accounts of his methods of research, and his general philosophy of life in the book "Plants, Man and Life" [4] published in 1952. It is a combination of scientific knowledge, folklore of Latin American and other countries, and Anderson's comments on early herbalists and the habits of taxonomists and botany professors, plus a bit of philosophy. One of his chief contributions to plant science, the pictorialized scatter diagram, is presented for the first time in its final form in a chapter entitled, characteristically, "How to Measure an Avocado" (Fig. 3).

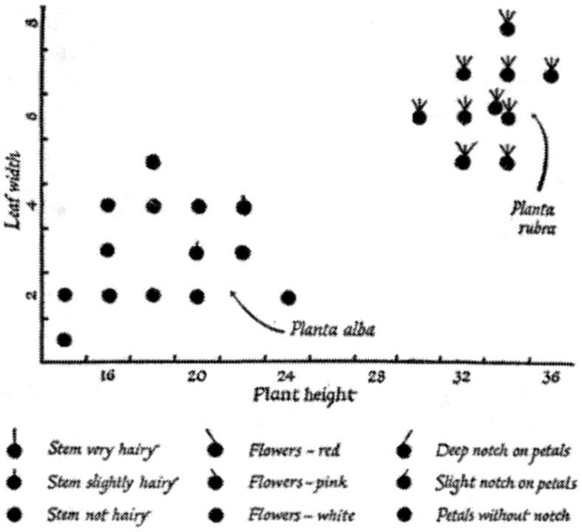

Fig. 3. Anderson's pictorialized scatter diagram [4, p. 97]

Anderson's article of 1936 [3] was his last work dedicated to the problem of *Iris* origin and classification. In his introduction to the article, Anderson not only expressed his gratitude to his English teachers, but also directly indicated that "Dr. Wright, Prof. J. B. S. Haldane, and Dr. R. A. Fisher have greatly furthered the final analysis of the data, though they are in no way responsible for the imperfections of the work or of its presentation." [3, p. 458].

In 1936, Sir Ronald Aylmer Fisher published the article "The Use of Multiple Measurements in Taxonomic Problems" indicating that "Table I shows measurements of the flowers of fifty plants each of the two species *Iris setosa* and *I. versicolor*, found growing together in the same colony and measured by Dr E. Anderson, to whom I am indebted for the use of the data" [13, p. 179-180]. Fisher's article contained only three references two of which to Anderson's works – that of 1935 [5] and that of [3] marked with "(in the Press)". In 1936, Fisher was not the member of the editorial board of "Annals of the Missouri Botanical Garden". The only way of his being aware of Anderson's article [3], was their personal correspondence.

The set of data used by Fisher and collected by Anderson was introduced as "Iris flower data set" (or "Iris data set" and "Iris data"). The phrase "Fisher's Iris data set" traditionally expresses Fisher's role as the founder of linear discriminant analysis, but not the authorship of the data set.

Although Anderson never published these data, he described [5] how he collected information on irises: "For some years I have been studying variation in irises but never before have I had the good fortune to meet such quantities of material for observation. On the simple assumption that if current theories are true, one should be able to find evidence of continuing evolution in any group of plants, I have been going around the world looking as sharply as possible at variation in irises. On any theory of evolution the differences between individuals get somehow built up, in time, into the differences between species. That is to say that by one process or another the differences which exist between one plant of *Iris versicolor* and its neighbor are com-pounded into the greater difference which distinguishes *Iris versicolor* from *Iris setosa canadensis*. It is a convenient theory and if it is true, we should be able to find the beginnings of such a compounding going on in our present day species. For that reason I have studied such irises as I could get to see, in as great detail as possible, measuring iris standard after iris standard and iris fall after iris fall, sitting squat-legged with record book and ruler in mountain meadows, in cypress swamps, on lake beaches, and in English parks. The result is still merely a ten year's harvest of dry statistics, only partially winnowed and just beginning to shape itself into generalizations which permit of summarization and the building of a few new theories to test by other means.

I have found no other opportunity quite like the field from De Verte to Trois Pistoles. There for mile after mile one could gather irises at will and assemble for comparison one hundred full-blown flowers of *Iris versicolor* and of *Iris setosa canadensis*, each from a different plant, but all from the same pasture, and picked on the same day and measured at the same time by the same person with the same apparatus. The result is, to ordinary eyes, a few pages of singularly dry statistics, but to the bio-mathematician a juicy morsel quite worth looking ten years to find.

After which rhapsody on the beauty of variation it must immediately be emphasized that *Iris setosa canadensis* varies but little in comparison with our other native blue flags. *Iris versicolor* in any New England pastures may produce ground colors all the way from mauve to blue and with hafts white or greenish or even sometimes quite a bright yellow at the juncture with the blade. *Iris setosa canadensis* by contrast is prevailingly uniform, its customary blue grey occasionally becoming a little lighter or a little darker or even a little more towards the purple, and its tiny petals producing odd variants in form and pattern, but presenting on the whole only a fraction of the variability of *Iris versicolor* from the same pasture.

The reasons for this uniformity are not far to seek. Its lower chromosome number is one, but a discussion of that and its bearings on the whole problem would be a treatise in itself. More important probably is the fact that by geological and biological evidence, *Iris setosa canadensis* is most certainly a remnant, a relict *[sic]* of what was before the glacial period a species widely spread in northern North America.

If we take a map and plot thereon all known occurrences of *Iris setosa* and *Iris setosa canadensis*, we shall find the former growing over a large area at the northwest corner of the continent, and the latter clustering in a fairly restricted circle about the Gulf of St. Lawrence, while in the great intervening stretch of territory, none of these irises has been collected. This is a characteristic distribution for plants which were almost exterminated from eastern North America by the continental ice sheet, but while *[sic]* managed to persist in the unglaciated areas about the Gulf of St. Lawrence from which center they have later spread. In Alaska the species itself, *Iris setosa*, is apparently quite as variable as our other American irises."

So, we should pay tribute to Edgar Anderson by naming this data set after him – Anderson's Iris data set.

6 Model development

As indicated in [28], the special course "Foundations of Mathematical Informatics" final control of knowledge is a credit by the presentation of individual education and research projects on the artificial neural networks built by using CoCalc. Students can be offered to use cloud-based spreadsheets, Google Sheets, with the Solver additional cloud-based component (add-in) which is similar to "Solver" in Excel Online.

Let us consider the corresponding application method by taking a Anderson's Iris data set to solve the pattern classification problem. Anderson's Iris is composed of data on 150 measurements of three Iris species (Fig. 4) – *Iris setosa*, *Iris virginica* and *Iris versicolor*) – including 50 measurements for each species.

Fig. 4. Anderson's Irises

There were measured four features (Fig. 5): sepal length (SL), sepal width (SW), petal length (PL), and petal width (PW).

To draw a grounded conclusion on the *Iris* type, we build a three-layered neural network with the following architecture (Fig. 6):
- the input layer is a four-dimensional arithmetical vector (x_1, x_2, x_3, x_4) the components of which are corresponding measured features of Anderson's Irises (SL, SW, PL, PW) normalized according to the network activation function;
- the hidden layer has dimension 9 (the minimal required number according to Kolmogorov–Arnold representation theorem) and is described by the vector $(h_1, h_2, h_3, h_4, h_5, h_6, h_7, h_8, h_9)$;

- the output layer is a three-dimensional arithmetical vector (y_1, y_2, y_3) the components of which are probabilities indicating the correspondence of the data set to one of the three *Iris* types.

Fig. 5. Measurement features of Anderson's Irises

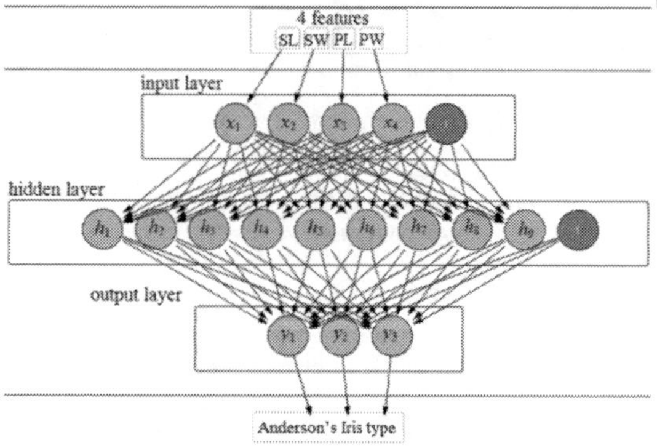

Fig. 6. Architecture of the neural network to solve the problem of Anderson's Iris classification

The bias neuron equal to 1 (marked red in Fig. 6) is added to the neurons of the input and hidden layers. The bias neurons are noted for not having synapses so they cannot be located in the output layer.

Let us first introduce Anderson's Irises into spreadsheets with the following values of cells: A1 is `Iris Data`, A2 is `SL`, B2 is `SW`, C2 is `PL`, D2 is `PW`, E2 is `Species`.

The table cells A3:E152 include Anderson's Irises (Fig. 7).

	A	B	C	D	E
1	*Iris Data*				
2	SL	SW	PL	PW	Species
3	5.10	3.50	1.40	0.20	*setosa*
4	4.90	3.00	1.40	0.20	*setosa*
5	4.70	3.20	1.30	0.20	*setosa*
6	4.60	3.10	1.50	0.20	*setosa*
7	5.00	3.60	1.40	0.20	*setosa*
8	5.40	3.90	1.70	0.40	*setosa*

Fig. 7. The fragment of the spreadsheet of Anderson's Irises

We cannot input the data of the given set into the input layer as the value of the four characteristics is beyond the range limits [0; 1]. The next step is normalization of columns A, B, C and D to meet the given range and coding of *Iris* types from column E.

Each *Iris* type is coded by the three-dimensional arithmetical vector: for *i*-Iris (*Iris setosa* is 1, *Iris versicolor* is 2, *Iris virginica* is 3) we set the *i*-th component in 1, and the other ones – in 0. To do this, we introduce the following values into the cells: G1 is `encoding`, G2 is `setosa`, H2 is `versicolor`, I2 is `virginica`, G3 is `=if($E3=G$2,1,0)`.

Next, we copy the formula from the cell G3 to the range G3:I152 and obtain the following model codes for the three *Iris* types: for *Iris setosa* – (1, 0, 0), for *Iris virginica* – (0, 0, 1) and for *Iris versicolor* – (0, 1, 0).

Each column is normalized separately. To perform this, we find minimum and maximum values by introducing the following values: E154 is `min`, E155 is `max`, A154 is `=min(A3:A152)`, A155 is `=max(A3:A152)`.

We apply the cells A154:A155 to the range B154:D155 and introduce the following values into the cells: K1 is `normalization`, K2 is x_1, L2 is x_2, M2 is x_3, N2 is x_4, K3 is `=(A3-A$154)/(A$155-A$154)`.

The latter formula is applied to the range K3:N152. Its essence is explained by:

$$\text{normalization} = \frac{\text{value - min}}{\text{max - min}}.$$

This approach results in the minimum value normalized to 0, while the maximum one – to 1.

According to the chosen architecture, we add the bias neuron to the four neurons of the input layer by introducing its name (x_5) into the cell O2 and its value (1) into the range O3:O152. On this stage, the input layer is formed as x_1, x_2, x_3, x_4, x_5.

The next step includes transmission of a signal from the input layer to the hidden one of the neural network. We denote the weight coefficient of the synapse connecting the neuron x_i ($i = 1, 2, 3, 4, 5$) of the input layer with the neuron h_j ($j = 1, 2, ..., 9$) of the hidden layer by w^{xh}_{ij}, while the weight coefficient connecting the neuron h_j of the hidden layer with the neuron y_k ($k = 1, 2, 3$) of the input layer is denoted by w^{hy}_{jk}. In this case, the force of the signal coming to the neuron h_j of the hidden layer is determined as a scalar product of signal values on the input signals and corresponding weight coefficients. To determine a signal going further to the output layer, we apply the logistic function of activation $f(S) = 1/(1+e^{-S})$, where S is a scalar product. The formulae for determining the signals on the hidden and output layers will look like:

$$h_j = f\left(\sum_{i=1}^{4+1} x_i w^{xh}_{ij}\right), y_k = f\left(\sum_{j=1}^{9+1} x_i w^{hy}_{jk}\right).$$

Accordingly, two matrices should be created. The matrix w^{xh} of 5×9 contains weight coefficients connecting five neurons of the input layer (the first four contain normalized characteristics of Anderson's Irises, while the fifth one is the bias neuron) with the neurons of the hidden layer. The matrix w^{hy} of 10×3 contains weight coefficients connecting ten neurons of the hidden layer (nine of which are calculated and the tenth one is the bias neuron) with the neurons of the output layer. For the "untaught" neural network, initial values of the weight coefficients can be set either randomly or left undetermined or equal to zero. To realize the latter, we fill the cells with the following values: R1 is w^{xh}, Q2 is `input/hidden`, R2 is `1`, S2 is `=R2+1`, Q3 is `1`, Q4 is `=Q3+1`, R3 is `0`, R9 is w^{hy}, Q10 is `hidden/output`, R10 is `1`, S10 is `=R10+1`, Q11 is `1`, Q12 is `=Q11+1`, R11 is `0`.

To create the matrices, we should copy the cells R3 into the range R3:Z7, R11 – into R11:T20, S2 – into T2:Z2, Q4 – into Q5:Q7, S10 – into T10, Q12 – into Q13:Q20 (Fig. 8).

To calculate the scalar product of the vector row of the input layer values by the matrix vector-column of the weight coefficients w^{hy}, we should apply the matrix multiplication function: AB1 is `calculate the hidden layer`, AB2 is h_1, AC2 is h_2, AD2 is h_3, AE2 is h_4, AF2 is h_5, AG2 is h_6, AH2 is h_7, AI2 is h_8, AJ2 is h_9, AK2 is h_{10}, AB3 is `=1/(1+exp(-mmult($K3:$O3,R$3:R$7)))`, AK3 is `1`.

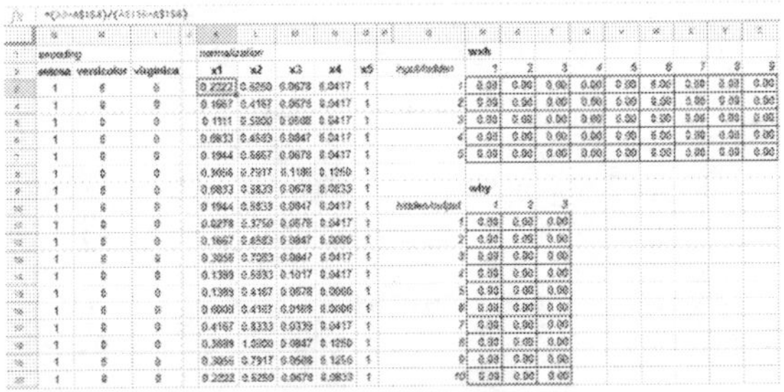

Fig. 8. The fragment of the spreadsheet after coding and normalization of the output data and creation of the matrices of the weight coefficients

Next, we copy the cell AK3 into the range AK4:AK152, while AB3 – into AB3:AJ152.

Considering the fact that all the matrix elements of the weight coefficients w^{xh} equal to zero, after duplicating the formulae, the calculated elements of the hidden layer will be equal to 0.5.

In the same way, we calculate the output layer elements: AM1 is `calculate the output layer`, AM2 is y_1, AN2 is y_2, AO2 is y_3, AM3 is `=1/(1+exp(-mmult($AB3:$AK3,R$11:R$20)))`.

Next, we copy the cell AM3 to the range AM3:AO152 (Fig. 9).

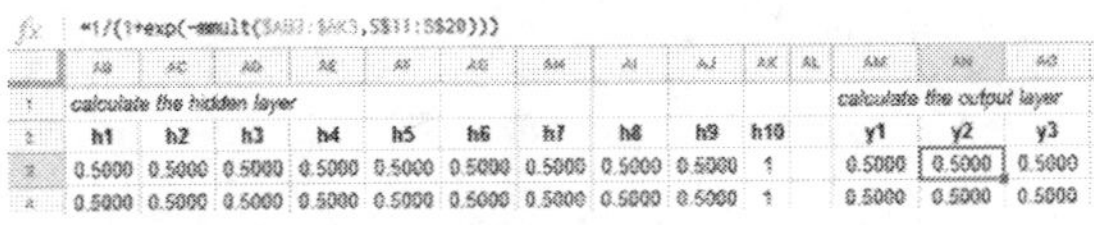

Fig. 9. The fragment of the spreadsheet of calculating the hidden and output layers with initial values of the weight coefficients

Neural network training is performed by varying weight coefficients so that with each training step the difference between the calculated values of the output layer and the desired (reference ones) reduces. To solve the problem, the three-dimensional vectors resulted from coding of the three *Iris* types are reference.

To find the difference between the calculated and the reference output vectors we apply the Euclidean distance: AQ2 is `distance`, AR2 is `sum of distances`, AQ3 is `=sqrt((AM3-G3)^2+(AN3-H3)^2+(AO3-I3)^2)`, AR3 is `=sum(AQ3:AQ152)`.

Next, we copy the cell AQ3 to the range AQ4:AQ152. The cell AR3 contains general deviation of the calculated output vectors from the reference ones.

Under this approach, the neural network training can be treated as an optimization problem in which the target function (the sum of distances in the cell AR3) will be minimized by varying the matrix weight coefficients w^{xh} (the range R3:Z7) and w^{hy} (the range R11:T20). To solve this problem, application of cloud-based spreadsheets (Google Sheets) is not enough and it is necessary to install an additional cloud-based component (add-in) Solver.

Adjustment of the add-in Solver to solve the set goal: the target function (Set Objective) is minimized (To: Min) by changing the values (By Changing) of the matrix weight coefficients in the range (Subject To) from –10 to +10 by one of the optimization methods (Solving Method).

To reduce the total distances, the actions with Solver can be done repeatedly as it is expedient to experiment with combination of various optimization methods by changing the variation limits of the weight coefficients. It is not necessary to try to reduce the value of the total distances to zero as this can be a greater (quite smaller) value (Fig. 10).

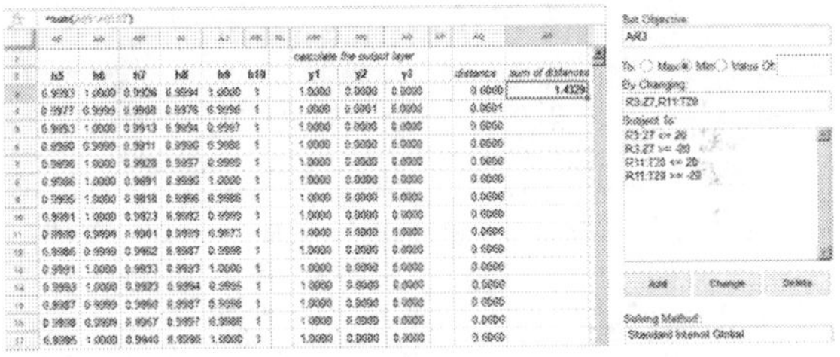

Fig. 10. Optimization results

On the assumption of the chosen coding method, the output vector actually contains three probabilities: y_i denotes the probability of the given sample being the *i*-type Iris, where $i = 1$ for *Iris setosa*, 2 for *Iris versicolor* and 3 for *Iris virginica*. Then, to find out which *Iris* type describes the input vector (SL, SW, PL, PW), the most probable component should be determined.

To do this, we fill the cells in the following way: AT2 is `Calculated Iris species`, AT3 is `=if(max(AM3:AO3)=AM3,$G$2,if(max(AM3:AO3)=AN3,$H$2, $I$2))`, AU3 is `=if(AT3=E3, "right!", "wrong")`.

Next, the range AT3:AU3 is copied to the range AT4:AU152.

The obtained result enables us to visualize pattern recognition simulated in spread-sheets. The built model will be considered relevant in all 150 cases, the column AU contains the value "right!".

To check the limits of the built model application, we try to input the vector not coinciding with any reference input vector. For this, we copy the table row 152 to 158 and delete the content of the cells E158:I158, AQ158, AU158. We introduce averaged values borrowed from the description of *Iris versicolor* in the article by Anderson [33, p. 463]: 5.50, 2.75, 3.50 and 1.25. The reference values $x_1 = 0.3333$, $x_2 = 0.3125$, $x_3 = 0.4237$, $x_4 = 0.4792$ are conveyed to the input layer, while on the hidden layer there are calculated $h_1 - h_9$ and the values of the output layer $y_1 = 0.0000$, $y_2 = 1.0000$, $y_3 = 0.0000$. As the maximum value of the output layer 1.0000 corresponds to the other *Iris* type, we can conclude that *Iris versicolor* is identified.

7 Conclusions

1. Extensive application of artificial intelligence in everyday life calls for students' early acquaintance with its models and methods including neural network-based while teaching informatics at secondary schools. It conditions the need for developing training methods of computer simulation of neural networks in the general-purpose simulation environment, i.e. spreadsheets.
2. Basic solutions of the problem of computer simulation training of neural networks in the spreadsheet environment include: 1) joint application of spreadsheets and network simulation tools; 2) application of third-party add-ins to spreadsheet processors; 3) macros development using embedded languages of spreadsheet processors; 4) application of standard spreadsheet add-ins for non-linear optimization; 5) creation of neural networks in the spreadsheet environment without add-ins and macros.
3. Neural network simulation competences should be formed through mastering models based on the historical and genetic approach. The review of papers on computational neuroscience of its early period allows determining three groups of models, which are helpful for developing corresponding methods: the continuous two-factor model of Rashevsky, the discrete model of McCulloch and Pitts, and the discrete-continuous models of Householder and Landahl.
4. Edgar Anderson appeared to be not a simple botanist whose data were the basis for Fisher's known method. Anderson's Irises resulted from his long experience of working out relevant models to describe changes in specific populations by means of a limited number of characteristics. Yet, Anderson had also coped with the opposite problem of building simple multi-dimensional data interpretation 40 years before Chernoff faces appeared [9].
5. The described methods of applying cloud-based spreadsheets as a tools for training mathematical informatics can enable solution of all basic problems of neural net-work simulation. The only limitation is not so much the volume of a spreadsheet as the memory space and the speed of the device processing it. In the special course projects if the limitation is overcome, this becomes a stimulus for replacing the simulation environment by a more relevant one [52].

References

1. H. Abelson, G.J. Sussman, J. Sussman, *Structure and Interpretation of Computer Programs*, 2nd edn. (MIT Press, Cambridge, 1996)
2. T.H. Abraham, (Physio)logical circuits: The intellectual origins of the McCulloch-Pitts neural networks. Journal of the History of the Behavioral Sciences. **38**(1), 3–25 (2002). doi:10.1002/jhbs.1094
3. E. Anderson, The Species Problem in Iris. Annals of the Missouri Botanical Garden. **23**(3), 457–469+471–483+485–501+503–509 (1936). doi:10.2307/2394164.
4. E. Anderson, *Plants, Man and Life* (University of California Press, Boston, 1952)
5. E. Anderson, Bulletin of the American Iris Society. **59**, 2–5 (1935)
6. E. Anderson, The Problem of Species in the Northern Blue Flags, Iris versicolor L. and Iris virginica L. Annals of the Missouri Botanical Garden. **15**(3), 241–332 (1928). doi:10.2307/2394087
7. A.S. Ayed, Master thesis, Memorial University, 1997
8. J.J. Buergermeister, in Restructuring Training and Education through Technology, ed. by D.W. Dalton. 32nd Annual Conference of the Association for the Development of Computer-Based Instructional Systems, San Diego, California, October 29-November 1, 1990. (ADCIS International, Columbus, 1990), pp. 214–220
9. H. Chernoff, Journal of the American Statistical Association. **68**(342), 361–368 (1973)
10. J.D. Cowan in *Talking nets: An oral history of neural networks*, ed. by J.A. Anderson, E. Rosenfeld (MIT Press, Cambridge, 1998), pp. 97–124
11. P. Cull, The mathematical biophysics of Nicolas Rashevsky. BioSystems. **88**(3), 178–184 (2007). doi:10.1016/j.biosystems.2006.11.003
12. R.C. Eberhart, R.W. Dobbins, in *Neural Network PC Tools: A Practical Guide*, ed. by R.C. Eberhart, R.W. Dobbins (Academic Press, San Diego, 1990), pp. 9–34
13. R.A. Fisher, The Use of Multiple Measurements in Taxonomic Problems. Annals of Eugenics. **7**(2), 179–188 (1936). doi:10.1111/j.1469-1809.1936.tb02137.x
14. R.S. Freedman, R.P. Frail, F.T. Schneider, B. Schnitta, in *Proceedings First International Conference on Artificial Intelligence Applications on Wall Street*, Institute of Electrical and Electronics Engineers, New York, 9–11 Oct. 1991
15. T. Hegazy, A. Ayed, Neural Network Model for Parametric Cost Estimation of Highway Projects. Journal of Construction Engineering and Management. **124**(3), 210–218 (1998). doi:10.1061/(ASCE)0733-9364(1998)124:3(210)
16. T.T. Hewett, Teaching Students to Model Neural Circuits and Neural Networks Using an Electronic Spreadsheet Simulator. Behavior Research Methods,

Instruments, & Computers. **17**(2), 339–344 (1985). doi:10.3758/BF03214406

17. T.T. Hewett, *Using an Electronic Spreadsheet Simulator to Teach Neural Modeling of Visual Phenomena*. (Drexel University, Philadelphia, 1985)
18. A.S. Householder, H.D. Landahl, *Mathematical Biophysics of the Central Nervous System* (Principia Press, Bloomington, 1945)
19. A.S. Householder, A neural mechanism for discrimination: II. Discrimination of weights. Bulletin of Mathematical Biophysics. **2**(1), 1–13 (1940). doi:10.1007/BF02478027
20. A.S. Householder, A theory of steady-state activity in nerve-fiber networks I: Definitions and Preliminary Lemmas. Bulletin of Mathematical Biophysics. **3**(2), 63–69 (1941). doi:10.1007/BF02478220
21. W. James, *Psychology* (Henry Holt and Company, New York, 1892)
22. W. James, *The Principles of Psychology* (Henry Holt and Company, New York, 1890)
23. S.J. Johnston, InfoWorld. **13**(7), 14 (1991)
24. D.A. Kendrick, P.R. Mercado, H.M. Amman, *Computational Economics* (Princeton University Press, Princeton, 2006)
25. H.D. Landahl, W.S. McCulloch, W. Pitts, A statistical consequence of the logical calculus of nervous nets. Bulletin of Mathematical Biophysics. **5**(4), 135–137 (1943). doi:10.1007/BF02478260
26. H.D. Landahl, R. Runge, Outline of a matrix calculus for neural nets. Bulletin of Mathematical Biophysics. **8**(2), 75–81 (1946). doi:10.1007/BF02478464
27. H.D. Landahl, A matrix calculus for neural nets: II. Bulletin of Mathematical Biophysics. **9**(2), 99–108 (1947). doi:10.1007/BF02478296
28. O. Markova, S. Semerikov, M. Popel, CoCalc as a Learning Tool for Neural Network Simulation in the Special Course "Foundations of Mathematic Informatics". (CEUR Workshop Proceedings, 2018), http://ceur-ws.org/Vol-2104/paper_204.pdf. Accessed 30 Nov 2018
29. O.M. Markova, S.O. Semerikov, A.M. Striuk, H.M. Shalatska, P.P. Nechypurenko, V.V. Tron, Implementation of cloud service models in training of future information technology specialists. (CEUR Workshop Proceedings, 2019), http://ceur-ws.org/Vol-2433/paper34.pdf. Accessed 10 Sep 2019
30. W.C. McCulloch, W. Pitts, A logical calculus of the ideas immanent in nervous activity. Bulletin of Mathematical Biophysics. **5**(4), 115–133 (1943). doi:10.1007/BF02478259
31. T.M. Mitchell, Key Ideas in Machine Learning. http://www.cs.cmu.edu/%7Etom/mlbook/keyIdeas.pdf. Accessed 28 Jan 2019
32. O.S. Permiakova, S.O. Semerikov, *Zastosuvannia neironnykh merezh u zadachakh prohnozuvannia* (The use of neural networks in forecasting problems), in *Materials of the International Scientific and Practical Conference "Young scientist of the XXI century"*, KTU, Kryviy Rih, 17–18 November 2008
33. W. Pitts, W.S. McCulloch, How we know universals the perception of auditory and visual forms. Bulletin of Mathematical Biophysics. **9**(3), 127–147 (1947). doi:10.1007/BF02478291
34. W. Pitts, A general theory of learning and conditioning: Part I. Psychometrika. **8**(1), 1–18 (1943). doi:10.1007/BF02288680
35. W. Pitts, A general theory of learning and conditioning: Part II. Psychometrika. **8**(2), 131–140 (1943). doi:10.1007/BF02288697
36. W. Pitts, Some observations on the simple neuron circuit. Bulletin of Mathematical Biophysics. **4**(3), 121–129 (1942). doi:10.1007/BF02477942
37. W. Pitts, The linear theory of neuron networks: The dynamic problem. Bulletin of Mathematical Biophysics. **5**(1), 23–31 (1943). doi:10.1007/BF02478116
38. W. Pitts, The linear theory of neuron networks: The static problem. Bulletin of Mathematical Biophysics. **4**(4), 169–175 (1942). doi:10.1007/BF02478112
39. N. Rashevsky, Mathematical biophysics of abstraction and logical thinking. Bulletin of Mathematical Biophysics. **7**(3), 133–148 (1945). doi:10.1007/BF02478314
40. N. Rashevsky, Outline of a physico-mathematical theory of excitation and inhibition. Protoplasma. **20**(1), 42–56 (1933). doi:10.1007/BF02674811
41. N. Rashevsky, Some remarks on the boolean algebra of nervous nets in mathematical biophysics. Bulletin of Mathematical Biophysics. **7**(4), 203–211 (1945). doi:10.1007/BF02478425
42. N. Rashevsky, The neural mechanism of logical thinking. Bulletin of Mathematical Biophysics. **8**(1), 29–40 (1946). doi:10.1007/BF02478425
43. T.F. Rienzo, K.K. Athappilly, Introducing Artificial Neural Networks through a Spread-sheet Model. Decision Sciences Journal of Innovative Education. **10**(4), 515–520 (2012). doi:10.1111/j.1540-4609.2012.00363.x
44. M.A. Ruggiero, *Cybernetic Trading Strategies: Developing a Profitable Trading System with State-of-the-Art Technologies* (John Wiley & Sons, New York, 1997)
45. M. Ruggiero, US Patent 5,241,620, 31 Aug 1993
46. K. Schwab, N. Davis, *Shaping the Fourth Industrial Revolution* (Portfolio Penguin, London, 2018)
47. S.O. Semerikov, I.O. Teplytskyi, Yu.V. Yechkalo, A.E. Kiv, Computer Simulation of Neural Networks Using Spreadsheets: The Dawn of the Age of Camelot. (CEUR Work-shop Proceedings, 2018), http://ceur-ws.org/Vol-2257/paper14.pdf. Accessed 21 Mar 2019
48. S.O. Semerikov, I.O. Teplytskyi, Metodyka uvedennia osnov Machine learning u shkilnomu kursi informatyky (Methods of introducing the basics of

Machine learning in the school course of informatics), in *Problems of informatization of the educational process in institutions of general secondary and higher education*. Ukrainian scientific and practical conference, Kyiv, October 09, 2018. (Vyd-vo NPU imeni M. P. Drahomanova, Kyiv, 2018), pp. 18–20

49. A. Shimbel, A. Rapoport, A statistical approach to the theory of the central nervous system. Bulletin of Mathematical Biophysics. **10**(2), 41–55 (1948). doi:10.1007/BF02478329
50. G.L. Stebbins, *Edgar Anderson 1897-1969*. (National Academy of Sciences, Washington, 1978)
51. G.J. Sussman, J. Wisdom, *Structure and interpretation of classical mechanics*, 2nd edn. (MIT Press, Cambridge, 2015)
52. I.O. Teplytskyi, O.I. Teplytskyi, A.P. Humeniuk, New computer technology. **6**, 67–68 (2008)
53. I.O. Teplytskyi, *Elementy kompiuternoho modeliuvannia* (Elements of computer simulation), 2nd edn. (KSPU, Kryvyi Rih, 2010)
54. T. Wei, On matrices of neural nets. Bulletin of Mathematical Biophysics. **10**(2), 63–67 (1948). doi:10.1007/BF02477433
55. P.J. Werbos, Maximizing long-term gas industry profits in two minutes in Lotus using neural network methods. Transactions on Systems Man and Cybernetics. **19**(2), 315–333 (1989). doi:10.1109/21.31036
56. G. Young, On reinforcement and interference between stimuli. Bulletin of Mathematical Biophysics. **3**(1), 5–12 (1941). doi:10.1007/BF02478102
57. T. Zaremba, in *Neural Network PC Tools: A Practical Guide*, ed. by R.C. Eberhart, R.W. Dobbins (Academic Press, San Diego, 1990), pp. 251–283

Reduction of programs execution time for tasks related to sequences or matrices

Oleksandr Mitsa[1], *Yurii* Horoshko[2,*], and *Serhii* Vapnichnyi[1]

[1]Uzhhorod National University, Uzhhorod, Ukraine
[2]T.H. Shevchenko National University "Chernihiv Colehium", Chernihiv, Ukraine

Abstract. The article discusses three approaches to reducing runtime of the programs, which are solutions of Olympiad tasks on computer science, related to sequences or matrices. The first approach is based on the representation of some sequences in matrix form and then the program of calculating the members of the sequence will have asymptotics equal to the time complexity of the exponentiation algorithm and will be O(log (n)). The second approach is to upgrade the known code to obtain significant reduction of the program runtime. This approach is very important to know for scientists who write code for scientific researches and are faced with matrix multiplication operations. The third approach is based on reducing time complexity by search for regularities; the author's task is presented and this approach is used to solve it.

Introduction

Sports programming has now become a promising intellectual sport. Every year, the number of pupils and students interested in Olympiads in computer science, as perhaps the most common type of sports programming, is growing. There are many Olympiads and other competitions held by the largest IT companies. Relevant HR specialists from these companies have been monitoring the results of various competitions and specific participants for many years. The most promising and successful participants are offered internships, combined with university study and the opportunity to gain full-time employment at the company after training. Also, many former Olympians are organizing successful projects related not only to programming and IT. Due to their participation in the Olympics, they were able to develop resistance to complex psychological stress. After spending so much time training, they have learned how to evaluate the likelihood of victory and defeat, have mastered existing and developed their own methods of dealing with stressful situations, the doubts and anxieties experienced by Olympic athletes in varying degrees.

Participation in Olympiads, tournaments and other competitions help students to improve their skills [1]. At first glance, it seems that to achieve solid results at the Olympics, it is enough to study a certain number of existing algorithms and theoretical material, and then only to successfully use them in competitions, leaving others no chance of winning. But it is not. Tasks in competitions are usually formulated in such a way that it is not enough to guess which algorithm to use to solve it. Almost always, in order to obtain a complete solution, it is necessary to upgrade the known algorithm, to supplement it, to combine several algorithms in one program, and to take some steps to reduce the time complexity of the solution [2].

This paper proposes three ways to reduce the runtime for computer science tasks that require the use of sequences and / or arrays:
- performing calculations using a matrix representation of sequences;
- reducing program execution time by using the features of the programming language;
- reducing time complexity by looking for regularities.

The first two techniques need to be learned to show the best results in standard situations. The third approach already needs a creative approach, has no general recommendations and is often used with the first two approaches.

1 Performing calculations using a matrix representation of sequences

Matrix data representation allows to use such algorithm as rapid exponentiation, that will significantly accelerate the program's work to find the desired element. One of these sequences that can be written in matrix form is the second-order linear recurrent sequences named after Edward Luke. These are pairs of sequences $\{U_n(P,Q)\}$ and $\{V_n(P,Q)\}$, whose recurrence relationship is written as follows:

$$U_0(P,Q) = 0, \quad U_1(P,Q) = 1,$$
$$U_{n+2}(P,Q) = P * U_{n+1}(P,Q) - Q * U_n(P,Q), n \geq 0$$
$$V_0(P,Q) = 2, \quad V_1(P,Q) = P,$$
$$V_{n+2}(P,Q) = P * V_{n+1}(P,Q) - Q * V_n(P,Q), n \geq 0 \quad (1)$$

* Corresponding author: horoshko_y@ukr.net

Partial variants of Luke's sequences are well studied and have their own names. In particular, the sequence $\{U_n(1,-1)\}$ is better known as the Fibonacci sequence, and the sequence $\{U_n(2,-1)\}$ – as the Pell sequence. The Pellet sequence is used to quickly find $\sqrt{2}$, Pythagorean triples, etc. The Pellet sequence numbers themselves in the ratio approach the silver intersection, similar to the Fibonacci sequence numbers in the ratio approach the gold intersection. Another known sequence is the sequence $\{U_n(3,2)\}$, which is called the Mersenne sequence. It is the numbers of this sequence that are the largest known prime numbers. The numbers of this sequence can be easily verified using the Luke-Lemmer test. They are also used to effectively construct long-period pseudorandom number generators called the Mersenne vortex.

A slightly less well-known practical application, compared to the sequences discussed above, is the sequence $\{U_n(1,-1)\}$, which is called the Jacobsthal sequence. Elements of this sequence are easy to find by different schemes. The most known is the recurrence ratio:

$$J_n = \begin{cases} 0, & n = 0; \\ 1, & n = 1; \\ J_{n-1} + 2J_{n-2}, & n > 1. \end{cases} \quad (2)$$

One can also use the following recursive records

- $J_{n+1} = 2J_n + (-1)^n;$ (3)
- $J_{n+1} = 2^n - J_n.$ (4)

There is a known relation of the Jacobsthal sequence with the Pascal triangle [3]. It consists in the rule of choosing in the line of the Pascal triangle certain numbers, the sum of which will be the number of the Jacobsthal sequence (Fig. 1).

1
1 1
1 2 1
1 3 3 1
1 4 6 4 1
1 5 10 10 5 1
1 6 15 20 15 6 1
1 7 21 35 35 21 7 1
1 8 28 56 70 56 28 8 1

-
- 1
1 - -
- - 3 -
- 4 - - 1
1 - - 10 - -
- - 15 - - 6 -
- 7 - - 35 - - 1
1 - - 56 - - 28 - -

Fig. 1. Relation of the Jacobsthal sequence with the Pascal triangle.

This relation can be represented as a formula as follows:

$$J(n) = \sum_{(n+k) \bmod 3=1} C(n,k) = \sum_{(n+k) \bmod 3=2} C(n,k) \quad (5)$$

The Jacobsthal sequence is also used in the problem of convergence of certain centers of a triangle on the Eulerian line of an arbitrary triangle [3]. The various relations that arise between Jacobsthal numbers are well explored in [4]. Our work below discusses a problem which has an effective solution that is based on the use of elements of the Jacobsthal sequence.

Any sequence from the Luke family of sequences is easily represented in matrix form. For example, the Fibonacci sequence has a known matrix representation [5]:

$$\begin{pmatrix} F_{n+1} & F_n \\ F_n & F_{n-1} \end{pmatrix} = \begin{pmatrix} 1 & 1 \\ 1 & 0 \end{pmatrix}^n, \quad (6)$$

which can be overwritten as

$$\begin{pmatrix} F_n \\ F_{n+1} \end{pmatrix} = \begin{pmatrix} 1 & 1 \\ 1 & 2 \end{pmatrix} \begin{pmatrix} F_{n-2} \\ F_{n-1} \end{pmatrix}, \quad (7)$$

or

$$\begin{pmatrix} F_{2n} \\ F_{2n+1} \end{pmatrix} = \begin{pmatrix} 1 & 1 \\ 1 & 2 \end{pmatrix}^n \begin{pmatrix} 1 \\ 1 \end{pmatrix}. \quad (8)$$

Other sequences under consideration are specified similarly, and the time complexity of the program of finding members of the sequence will be equal to the time complexity of the exponentiation algorithm and will be O(log(n)).

The ability to find directly the value of an element of the Fibonacci sequence by the formula

$$F_n = \left[\frac{\varphi^n}{\sqrt{5}}\right] \text{ or } F_n = \left\lfloor \frac{\varphi^n}{\sqrt{5}} + \frac{1}{2} \right\rfloor, \text{ where } \varphi = \frac{1+\sqrt{5}}{2}, \quad (9)$$

faces the problem of cumulative computational error and is of little use. On the other hand, some other sequences, such as the Jacobsthal sequence, have a convenient formula

$$J_n = \frac{2^n - (-1)^n}{3}. \quad (10)$$

Also is known the formula writing to find elements of a Fibonacci sequence across a continuum of size n x n:

$$F_{n+1} = det \begin{vmatrix} 1 & 1 & 0 & \cdots & 0 \\ -1 & 1 & 1 & \cdots & 0 \\ 0 & -1 & 1 & \cdots & 0 \\ \vdots & \vdots & \vdots & \cdots & \vdots \\ \vdots & \vdots & \vdots & \cdots & 1 \\ 0 & 0 & 0 & \cdots & 1 \end{vmatrix}. \quad (11)$$

If the n-th element of the sequence equals to the sum of k previous elements

$$A_n = A_{n-1} + A_{n-2} + \cdots + A_{n-k} \qquad (12),$$

then such a sequence is written in the following matrix form

$$\begin{pmatrix} A_n \\ A_{n-1} \\ A_{n-2} \\ \vdots \\ A_{n-k+1} \end{pmatrix} = \begin{pmatrix} 1 & 1 & 1 & \cdots & 1 & 1 \\ 1 & 0 & 0 & \cdots & 0 & 0 \\ 0 & 1 & 0 & \ddots & 0 & 0 \\ \vdots & \vdots & \vdots & \vdots & \vdots & \vdots \\ 0 & 0 & 0 & \cdots & 1 & 0 \end{pmatrix}^{n-k} \cdot \begin{pmatrix} A_k \\ A_{k-1} \\ A_{k-2} \\ \vdots \\ A_1 \end{pmatrix}. \quad (13)$$

The matrix with the help of which the calculations will be made will be of dimension k x k.

Therefore, performing calculations to find members of sequences using the matrix form of their representation significantly reduces the time complexity of the corresponding algorithms.

2 Reducing program execution time by using the features of the programming language

When solving a problem, it is very important to use the features of the programming languages in which the solution is implemented. In particular, let us take the well-known problem of multiplying two matrices. Consider two implementations of this operation.

Table 1. Two implementations of multiplication of two matrices.

Well-known variant

```
for (int i = 0; i < n; i++)
  for (int j = 0; j < n; j++)
    for (int k = 0; k < n; k++)
      c[i][j] += a[i][k] * b[k][j];
```

Accelerated variant

```
for (int i = 0; i < n; i++)
  for (int k = 0; k < n; k++) {
    long long x = a[i][k];
    for (int j = 0; j < n; j++)
      c[i][j] += x * b[k][j];
  }
```

Table 1 describes the finding of the product of matrices **C** = **A** · **B**, all matrices of dimensions n x n. If in the well-known variant the second and third cycles are swapped and the element of the first matrix is fixed in the usual variable, then the multiplication operation rate for matrices of dimension 1000 x 1000 will increase more than 15 (!) times. As the dimension increases, the advantage of the accelerated version will increase further. Further improvement steps are possible [6], but they will not give such a tangible advantage.

Scientific problems [7] also require the use of a matrix recording form for a particular model. Another recommendation is to keep the values that are constantly repeated in the calculation in a regular array. In particular, the values of trigonometric functions, if there is enough memory to store them, should be stored in the array, not re-calculated every time. This will significantly reduce the running time of the program.

3 Reducing time complexity by looking for regularities.

Solving regularity search tasks often leads to the identification of known sequences. Consider the problem proposed by Oleksandr Mitsa at the 13th Open Student International Programming Olympiad named after S. O. Lebedev and V. M. Glushkov "KPI-OPEN 2018" [8]. The title of this task is "Counter Racing"" This task combines, under a completely new perspective, two tasks that are well known to the general public – the Joseph Flavius problem [9] and the task No 2808 taken from the well-known E-Olymp site [10], which is described as the Choriv counter.

3.1 The task

The legendary Shchek and Choriv decided to arrange a competition for their counters.

Shchek counter was created on the base of the story of Josephus Flavius, when N people are in a circle and every second person is taken out of the circle. The remaining person number will be the result of the counter. For example, when there are 5 people in a circle, people will be taken out in the order of their numbers – 2, 4, 1, 5 and the result will be number 3.

Choriv's counter was based on a completely different principle. He took the number N and wrote out in a row all the numbers from 1 to N. Then he cross out the numbers that are in odd positions. Further, he lined them up anew, but then crossed out those that are in even positions. These actions were repeated until one number remained, which would be the result. For example, for N=5, the numbers with odd numbers – 1, 3, 5 are first crossed out, then from remaining numbers – 2, 4 – the number, which is in the even position, is crossed out, that is, 4. Therefore, the result will be 2.

For the full objectivity of determining the winner, it was decided to compete counts for each natural value from 1 to N. If, as a result, for some value the result of the Shchek counter is greater than the result of the Choriv counter, the Shchek will receive one point, if less, one point will receive Choriv, in case of a draw – the current account will not change.

It is need to determine the game score for a given number N.

Input format

Enter the number N ($1<N<10^{18}$).

Output *format*

Display the score of the competition.

Table 2. Example to the task.

Standard input	*Standard output*
10	3 6
100	48 51

Note. In the first example, the Shchek counter will win only at values 3, 5 and 7, at value 1 it will be a draw and in other cases the Choriv counter will win.

3.2 Solution of the problem

We first examine the regularities in the first counter. To do this, we use the scheme proposed in [9] and refine it. First, let's consider how one can reduce the dimension of the problem twice with an even value of *N*.

Table 3. Simulation of Joseph Flavius problem with an even value of *N*.

1	~~2~~	3	4	5	~~6~~	7	~~8~~	9	~~10~~
1		2		3		4		5	

Table 3 shows that the dimension of the problem has decreased by 2 times and the formula for the transition from old to new values will look like

$$T(N) = 2 * T\left(\frac{N}{2}\right) - 1. \quad (14)$$

For an odd value of N we use the same scheme and note that the value 1 in this case will never be a solution (Table 4).

Table 4. Simulation of Joseph Flavius problem with an odd value of *N*.

~~1~~	~~2~~	3	4	5	~~6~~	7	~~8~~	9	~~10~~	11
		1		2		3		4		5

Again we see that the dimension of the problem has decreased by 2 times and it is easy to deduce the formula of transition from old to new values, which will look like

$$T(N) = 2 * T\left(\frac{N}{2}\right) + 1. \quad (15)$$

To summarize, we give a complete scheme of recalculation

$$T(N) = \begin{cases} 1, & if\ N = 1; \\ 2 * T\left(\frac{N}{2}\right) - 1, & if\ N - even; \\ 2 * T\left(\frac{N}{2}\right) + 1, & if\ N - odd. \end{cases} \quad (16)$$

But even this formula is not enough to solve the problem as a whole. Therefore, we write the solutions for both counters for values *N* from 1 to 30 (Table 5).

Table 5. Tables for preliminary research.

N	1	2	3	4
Counter 1	1	1	3	1
Counter 2	1	2	2	2
Winner	Draw	Choriv	Shchek	Choriv
N	5	6	7	8
Counter 1	3	5	7	1
Counter 2	2	6	6	6
Winner	Shchek	Choriv	Shchek	Choriv
N	9	10	11	12
Counter 1	3	5	7	9
Counter 2	6	6	6	6
Winner	Choriv	Choriv	Shchek	Shchek
N	13	14	15	16
Counter 1	11	13	15	1
Counter 2	6	6	6	6
Winner	Shchek	Shchek	Shchek	Choriv
N	17	18	19	20
Counter 1	3	5	7	9
Counter 2	6	6	6	6
Winner	Choriv	Choriv	Shchek	Shchek
N	21	22	23	24
Counter 1	11	13	15	17
Counter 2	6	22	22	22
Winner	Shchek	Choriv	Choriv	Choriv
N	25	26	27	28
Counter 1	19	21	23	25
Counter 2	22	22	22	22
Winner	Choriv	Choriv	Shchek	Shchek
N	29	30	31	32
Counter 1	27	29	31	1
Counter 2	22	22	22	22
Winner	Shchek	Shchek	Shchek	Choriv

From Table 5, one can make the following observation: in the first counter, for N, which is a degree of two, always the answer is 1, and for the following *N* the answer is incremented by 2. That is, if the closest to *N* degree of two is equal 2^k, then the answer is easily determined by formula

$$T(N) = 1 + 2(N - 2^k) \quad (17)$$

Also note that the number of values of the degree of two for the input values from 1 to 10^{18} is only 60.

Let us proceed to the analysis of the second counter. Table 5 shows that the number of answers is negligible. Moreover, it can be seen that for 1 the answer will be 1, then from 2 to 5 the answer will be 2, and from 22 and to the next value to be investigated - the answer will be 22.

Let's simulate this task and write down the values of the answers that occur in it. These will be the following values

1, 2, 6, 22, 86, 342, 1366, 5462, 21846, 87382, 349526, 1398102, 5592406, 22369622,...

With these values in front of you, it is easy to determine the scheme of their calculation

$$P(k) = 4*P(k\text{-}1) - 2, \text{ where } P(1) = 1. \quad (18)$$

It is also understood that in the interval $[P(k), P(k+1)\text{-}1]$ the answer will be $P(k)$. Moreover, there will be very few such values. So, in the interval from 1 to 1018 there will be only 31 of them.

Thus, one of the schemes of the solution could be the following. At each interval from the number $N = 2^k+1$ to $N = 2^{k+1}$ we see how many numbers of the sequence 2 it contains, and accordingly we consider this when forming the account together with the sequence 1. Of course, the last interval will go only to the number *N*.

The described option will have the following solution in C++ programming language:

```
#include <iostream>

using namespace std;
```

```
int main() {
  long long N;
  cin >> N;
  long long p = 1, q = 2, Choriv = 0;
  while (2 * p + 1 <= N) {
    p = 2 * p + 1;
    if (p > 4 * q - 2) {
      long long pp = p;
      while (pp > 4 * q - 3)
        pp = (pp - 1) / 2;
      pp = 2 * (4 * q - 3 - pp) - 1;
      if (pp > q)
        Choriv += (pp - q) / 2 + 1;
      q = 4 * q - 2;
    }
    Choriv += (p - q) / 2 + 1;
  }
  if (N >= 4 * q - 2) {
    long long pp = p;
    while (pp > 4 * q - 3)
      pp = (pp - 1) / 2;
    pp = 2 * (4 * q - 3 - pp) - 1;
    if (pp > q)
      Choriv += (pp - q) / 2 + 1;
    q = 4 * q - 2;
  }
  p = 2 * (N - p) - 1;
  if (p > q) Choriv += (p - q) / 2 + 1;
  cout << Choriv << " " <<
          N - Choriv - 1 << endl;
  return 0;
}
```

But if to continue the research, one can get a simpler way of solving the problem under consideration.

Note that in this game only when $N = 1$, both players will draw. With all other values, N wins either the first or the second. So let's translate the game results to 0-1 form. Write the sequence in which the i-th element is 1 if the second player wins and 0 - otherwise, starting with the game for $N = 2$:

1, 0, 1, 0, 1, 0, 1, 1, 1, 0, 0, 0, 0, 0, 1, 1, 1, 0, 0, 0, 1, 1, 1, 1, 1, 0, 0, 0, 0, 0, 1, 1, 1, 1, 1, 1, 1, 1, 1, 1, 1, 0, 0, 0, 0, 0, 0, 0, 0, 0, 0, 0, 0, 0, 0, 0, 0, 0, 0, 0, 0, 0, 1, 1, 1, 1, 1, 1, 1, 1, 1, 1, 1, 0, 0, 0, 0, 0, 0, 0, 0, 0, 0, 0, 1, 1, 1, 1, 1, 1, 1, 1, 1, 1, 1, 1, 1, 1, 1, ...

Then we write down the quantities of consecutive numbers:

1, 1, 1, 1, 1, 1, 3, 5, 3, 3, 5, 5, 11, 21, 11, 11, 21, 21, ...

The next step is to divide this sequence into blocks of 6 elements and notice that each block consists of two sequential Jacobshtal numbers

1, 1, 3, 5, 11, 21, 43, 86, 171, ...

That is, the new scheme for solving this problem will be as follows. Each time, by choosing two values of the Jacobshtal sequence, we will form a current account. Of course, in forming the final account we will use only those values in the last six, which are limited by the number of lots N.

For example, we calculate the score for the number of lots $N = 40$. Note that this will require the use of the first two sixths and part of the first value of the third sixth. The score after 31 games will be described by summing the elements of the first two sixes. To count the number of games won by the first player, we count the items in even positions $1 + 1 + 1 + 5 + 3 + 5 = 16$, and to count the number of games won by the second player, we count the items in odd positions $1 + 1 + 1 + 3 + 3 + 5 = 14$. Then we consider that the third six starts with 11 second player victories, of which we need to count 9. That is, the final score will be 16:23 in favor of the second player.

The described solution in C++ programming language will be quite simple and compact:

```
#include <iostream>

using namespace std;

long long J[61], N, Choriv, Shchek, rem;

void Score (long long &X, long long Y) {
  if (rem > Y) {
    X += Y; rem -= Y;
  }
  else {
    X += rem; rem = 0;
  }
}

int main() {
  cin >> N;
  int i = 1;
  J[1] = 1; J[2] = 1;
  while(Shchek+Choriv+3*(J[i]+J[i+1])<N) {
    Shchek += 2 * J[i] + J[i + 1];
    Choriv += J[i] + 2 * J[i + 1];
    i += 2;
    J[i] = J[i - 1] + 2 * J[i - 2];
    J[i + 1] = J[i] + 2 * J[i - 1];
  }
  rem = N - 1 - Shchek - Choriv;
  Score(Shchek, J[i]);
  Score(Choriv, J[i + 1]);
  Score(Shchek, J[i]);
  Score(Choriv, J[i]);
  Score(Shchek, J[i + 1]);
  Score(Choriv, J[i + 1]);
  cout << Choriv << " " << Shchek;
  return 0;
}
```

It should be noted that in this task the number of involved Jakobshtal numbers does not exceed 60. The Score procedure allows to realize an ending of the task when the last block is not fully used.

Thus, the considered problem (of increased complexity) used in the international competition is a combination of two well-known tasks under a completely new perspective that has not been used before. Two variants of its solution are presented. In a more flexible and efficient way, it was enough to find the regularity given by the elements of the Jacobsthal sequence.

Conclusions

This paper discusses three approaches to reducing solution execution times for computer science tasks that require some knowledge of sequences and / or arrays to solve them. The first approach is to write the sequence in the matrix form and then use the rapid matrix exponentiation. This allows to quickly identify a particular element of a sequence. The second approach is essentially to improve the code of the program, which is considered traditional, and can significantly speed up the program, significantly (for matrices 1000 by 1000 more than 15 times) reducing the time of finding matrices using a fairly simple method. This approach is tested for C++, the most popular sports programming language. It is effective in solving sports programming tasks, because in this area, rapid methods of matrix multiplication are rarely used due to the excessive size of their code. It is also very important to know for scientists who write code for scientific researches and are faced with matrix multiplication operations. To demonstrate the third approach, we present a rather complex authorial task and show that its solution can be based on finding members of the well-known Jacobsthal sequence. The approaches presented in the paper can be used both individually and in combination.

The work will be interesting to pupils, students and teachers interested in programming, especially sports, and for scientists who write code for scientific researches.

References

1. S.S. Zhukovsky, Methodical Aspects of Preparing Gifted Schoolchildren for Informatics Olympiads. Problems of modern pedagogical education **47**(V), 62–69 (2015)
2. Y.V. Horoshko, O.V. Mitsa, V.I. Melnyk, Information Technologies and Learning Tools **71**(3), 40–52 (2019)
3. P. Barry, Irish Mathematical Society Bulletin **51**, 45–57 (2003)
4. Z. Čerin, Journal of Integer Sequences **10**(07.2), 5 (2003)
5. D.E. Knuth, *The Art of Computer Programming, Volumes 1-4A Boxed Set*, 3rd edn (Addison-Wesley, Reading, 2011)
6. I. Ermolaev, Umnozhenie matritc: effektivnaia realizatciia shag za shagom (Matrix multiplication: effective implementation step by step) (2019), https://habr.com/ru/post/359272/. Accessed 21 Mar 2019
7. P.I. Stetsyuk, Ellipsoid methods and r-algorithms (Evrika, Chisinau, 2014).
8. 13th Open International Student Programming Olympiad "KPI-OPEN 2018" named after S.O. Lebediev and V.M. Glushkov "KPI-OPEN 2018" (2018), http://kpi-open.org/. Accessed 28 Nov 2019
9. R.L. Graham, D.E. Knuth, O. Patashnik, *Concrete Mathematics*, 2nd edn (Addison-Wesley, Reading, 1994)
10. Internet portal of organizational and methodological support of distance Olympiads on programming for gifted youth of educational institutions of Ukraine E-Olimp (2019), https://www.e-olymp.com/uk/problems/2808. Accessed 28 Nov 2019

AUTHOR INDEX